CRC
Handbook
of
Laser Science
and
Technology

Volume I
Lasers and Masers

Editor

Marvin J. Weber, Ph.D.

Laser Program
Lawrence Livermore National Laboratory
University of California
Livermore, California

CRC Press, Inc.
Boca Raton, Florida

Library of Congress Cataloging in Publication Data

Main entry under title:

Lasers and masers.

  (Handbook of laser science and technology;
v. 1)
  Bibliography: p.
Includes index.
1. Lasers. 2. Masers. I. Weber, Marvin J.,
1932-   . II. Series.
TA1675.L38   621.36'6   AACR2
ISBN 0-8493-3501-9

This book represents information obtained from authentic and highly regarded sources. Reprinted material is quoted with permission, and sources are indicated. A wide variety of references are listed. Every reasonable effort has been made to give reliable data and information, but the author and the publisher cannot assume responsibility for the validity of all materials or for the consequences of their use.

Direct all inquiries to CRC Press, Inc., 2000 Corporate Blvd., N.W., Boca Raton, Florida, 33431.

© 1982 by CRC Press, Inc.
Second Printing, 1985
Third Printing, 1986
Fourth Printing, 1987

International Standard Book Number 0-8493-3501-9

Library of Congress Card Number 81-17086
Printed in the United States

FOREWORD

This handbook series updates and considerably extends the earlier compilation of laser data published in 1971, and is a testament to the continued vigorous growth of quantum electronics and its applications.

When lasers were first envisaged in the late 1950s, their broad applicability was already easily imagined in principle. As man's first amplifier of light, giving electronic-like control over ultraviolet, visible, and infrared wavelengths, the laser clearly had the possibility of application to many fields where electronics and optics were already effective, in addition to some new ones. Nevertheless, rapid development was not really certain and many specific applications which are now important were not even envisaged. To the thoughtful as well as to the unsophisticated observer, what has actually happened in this field over the last two decades must hence in any case be startling, or perhaps border on the miraculous. The many talented physicists, engineers, chemists, and entrepreneurs who grasped the significance of quantum electronics and applied their ingenuity to it have now constructed a science, a versatile collection of laboratory instruments, numerous applications in almost every conceivable industry, and a mass of information which this series presents both through review of ideas and in tabulated form. The summaries and tabulations of lasers in these first two volumes, furthermore, are given us by individuals who have been important, first-hand participants in these two decades of remarkable development.

While the first three lasers made to oscillate were solid-state systems, I have always regarded gas lasers as prototypes because isolated atoms and molecules somehow seem more elementary. The He-Ne laser, the earliest gaseous type is furthermore the most ubiquitous, showing up nowadays in many scientific laboratories and in common applications such as surveying or information transfer. But all families of gas lasers have also become prolific; they are complex and so numerous that a separate volume is devoted to their coverage. Although gas systems are not frequently very tunable, anyone now wanting specific wavelengths for spectroscopy or other purposes has many from which to choose, and a substantial variety of operating characteristics. The array of molecular frequencies in the far infrared is especialy large and a systematic tabulation hence particularly helpful.

Solid-state lasers have proliferated from Maiman's first ruby system to a wide variety of crystalline species, including the relatively new and versatile color center lasers, glass lasers, and the unusual Raman fiber lasers. Semiconductor lasers represent a quite different family of solid state systems, the small current-excited ones perhaps destined to be most abundant and cheapest of all.

It was clear from the beginning that masers and lasers could be tuned — by a few cm^{-1}. However, the development of liquid dye lasers and of some other systems with very high gain-band width product per unit length soon made broad tunability a reality. Liquid lasers, especially organic dye lasers, are now almost indispensible tools of the trade, and the present codification of their many types and charcteristics will be useful.

In the section on other lasers are chapters on the intriguing and powerful free electron lasers and on the prospects for X-ray lasers. The latter have seemed almost impossible in the past, but now appear imminent in some form, and are likely to provide some of the excitement of future years.

Masers are treated here along with miscellaneous special types of lasers, for masers now seem primarily to fulfill rather specialized functions such as providing our most sensitive microwave receivers and, at least by some measures, our best clocks. As man builds them, they are all low-power devices. However, in this realm he has been far outdone by nature, for many varieties of molecular masers have now been discovered in interstellar and circumstellar gases. They were naturally created long before man ever knew that molecules exist and are more powerful than any he will ever build, for some emit amplified microwaves almost as powerful as the total output of the sun. An excellent summary of such masers is included.

The first volume ends appropriately with the inclusion of a careful section on laser safety.

The function of a handbook of laser science and technology is to help provide a basis both for the practical minded user and for those who will be pushing the field into future stages of development, whose nature has constantly invited speculation. How many more surprises are in store for us as the field develops? Perhaps the present state of quantum electronics corresponds to middle adolescence. The period of most rapid and startling growth is possibly over, but there is still growth ahead of types which will be critical to development of its full potential. And after that, there will be a long period of additional maturation and refinement, undoubtedly the longest period and probably one when lasers will be still much more widely used. This volume can, and I hope will, be of assistance both to the profitable employment of what has so far developed and to progress towards such later stages.

C. H. Townes

PREFACE

In the ten years since the publication of the *CRC Handbook of Lasers with Selected Data on Optical Technology,* the growth in the number and diversity of lasers and their applications has continued at a rapid pace. These developments have prompted an update and expansion of the original volume into the *CRC HANDBOOK SERIES OF LASER SCIENCE AND TECHNOLOGY.* The first two volumes of the series are devoted to lasers in all media. Later volumes are planned on optical materials and on laser instrumentation and operating techniques.

The object of this series is to provide a readily accessible and concise source of data in tabular and graphical form for workers in the areas of laser research and development. Each volume has an advisory board of eminent scientists and engineers with broad experience in the area of quantum electronics. Individual sections in the volumes are contributed by active researchers who have critically evaluated the data.

Volumes I and II contain extensive tables of experimental data on lasers and complete references to the original work. This is the primary emphasis. Many books and review articles describing the physics and operation of various lasers already exist. Therefore textual material is in general only that required to explain the data, to summarize the general operation and characteristics of each type of laser, and to describe important specific lasers or properties not covered elsewhere.

I originally planned a single volume covering lasers in all media, however it became so large that it has been divided into two volumes. Of the various lasing media, gases contribute the largest number of laser transitions. Volume II is devoted exclusively to gas lasers. This volume covers all remaining types of lasers and masers. Only primary lasers are considered (with the exception of fiber Raman lasers); frequency converted lasers are not included.

A separate section is devoted to special lasers which are either in an early stage of development, as in the case of the free electron lasers, or not demonstrated but widely discussed, as in the case of X-ray and gamma-ray lasers. These are included for the sake of completeness. The tabulated data is necessarily very limited, however comprehensive references are provided.

To broaden the coverage of stimulated emission, microwave masers — both in the laboratory and in space — are also covered. This is also a reminder that in the early 1960s lasers, because of their heritage, were called optical masers. Popular alternative meanings attributed to the acronym MASER in those early days included *M*oney *Ac*quisition *S*cheme for *E*xpensive *R*esearch or *M*ore *A*merican *S*cientists *E*at *R*egularly.

As the uses of lasers in diverse fields continue to proliferate, anyone considering the use of various lasers should be cognizant of associated hazards — ocular, electrical, and toxic. These hazards are reviewed in the concluding section.

It is a pleasure to acknowledge the dedicated efforts and cooperation of the many contributors to this volume. The numerous suggestions and guidance of the Advisory Board throughout the preparation of this volume are also greatly appreciated. Finally, I wish to thank Marsha Baker and Pamela Woodcock of the CRC Press for their help and technical expertise in the editing of the book and my wife, Pauline, for her assistance and constant encouragement.

Marvin J. Weber
Danville, California
February, 1981

THE EDITOR

Marvin J. Weber is an Assistant Associate Program Leader, Basic Research, of the Laser Program at the Lawrence Livermore National Laboratory, University of California, Livermore, California.

Dr. Weber received the A.B., M.A., and Ph.D. degrees in physics from the University of California, Berkeley, in 1954, 1956 and 1959, respectively. After graduation, Dr. Weber joined the Research Division of Raytheon Company where he was a Principal Scientist and became Manager of Solid-State Lasers. In 1966, Dr. Weber was a Visiting Research Associate in the Department of Physics, Stanford University.

In 1973, Dr. Weber joined the Laser Program of the Lawrence Livermore National Laboratory. His activities have included studies of the physics, characterization, and development of optical materials for high power lasers Dr. Weber has published numerous research papers and review articles in the areas of lasers, luminescence, optical spectroscopy, and magnetic resonance in solids and holds several patents on solid-state laser materials.

Dr. Weber is a Fellow of the American Physical Society and a member of the Optical Society of America and the American Ceramics Society. He has served as a consultant for the Division of Materials Research, National Science Foundation, and is a member of the Advisory Editorial Board of the *Journal of Non-Crystalline Solids*.

ADVISORY BOARD

CONTRIBUTORS

Stephen R. Chinn, Ph.D.
Senior Scientist
Optical Information Systems, Inc.
Elmsford, New York

Robin DeVore
Health and Safety Division
Los Alamos Scientific Laboratory
Los Alamos, New Mexico

Raymond C. Elton, Ph.D.
Head, Optical Interactions Group
Optical Sciences Division
Naval Research Laboratory
Washington, D.C.

Michael Ettenberg, Ph.D.
Head, Opto-Electronic Devices and
  Systems Research Group
RCA David Sarnoff Research Center
Princeton, New Jersey

James K. Franks
Physicist
U.S. Army Environmental Hygiene
  Agency
Aberdeen Proving Ground, Maryland

Victor L. Granatstein, Ph.D.
Head, High Power Electromagnetic
  Radiation Branch
Plasma Physics Division
Naval Research Laboratory
Washington, D.C.

Henry Kressel, Ph.D.
Staff Vice President
Solid State Research
RCA Corporation
RCA David Sarnoff Research Center
Princeton, New Jersey

William F. Krupke, Ph.D.
Chief Scientist for Lasers
Laser Program
Lawrence Livermore Laboratory
Livermore, California

Chinlon Lin, Ph.D.
Member of Technical Staff
Physical Optics and Electronics
  Research Department
Bell Laboratories
Holmdel, New Jersey

Linn F. Mollenauer, Ph.D.
Research Scientist
Bell Laboratories
Holmdel, New Jersey

James M. Moran, Ph.D.
Radioastronomer
Smithsonian Astrophysical Observatory
Professor of the Practice of Astronomy
Harvard University
Cambridge, Massachusetts

Peter F. Moulton, Ph.D.
Staff Member
Quantum Electronics Group
MIT Lincoln Laboratory
Lexington, Massachusetts

Robert K. Parker, Ph.D.
Head, Microwave and Millimeter Wave
  Tube Technology Branch
Electronics Technology Division
Naval Research Laboratory
Washington, D.C.

Adrian E. Popa
Manager, Optical Circuits Department
Hughes Research Laboratories
Malibu, California

Donald Prosnitz, Ph.D.
Physicist
Lawrence Livermore National
  Laboratory
Livermore, California

Harold Samelson, Ph.D.
Allied Corporation
Electro-Optical Products
Warren, New Jersey

David H. Sliney
Chief, Laser Branch
Laser Microwave Division
U.S. Army Environmental Hygiene
  Agency
Aberdeen Proving Ground, Maryland

Phillip A. Sprangle, Ph.D.
Head of Plasma Theory Branch
Plasma Physics Division
Naval Research Laboratory
Washington, D.C.

Richard Steppel, Ph.D.
President
Exciton Chemical Company, Inc.
Dayton, Ohio

Stanley E. Stokowski, Ph.D.
Physicist
Lawrence Livermore Laboratory
Livermore, California

Rogers H. Stolen, Ph.D.
Member Technical Staff
Bell Telephone Laboratories
Holmdel, New Jersey

Charles H. Townes, Ph.D.
University Professor of Physics
Department of Physics
University of California
Berkeley, California

HANDBOOK OF LASER SCIENCE AND TECHNOLOGY

VOLUME II: GAS LASERS

SECTION 1: NEUTRAL GAS LASERS—Christopher C. Davis

SECTION 2: IONIZED GAS LASERS—William B. Bridges

SECTION 3: MOLECULAR GAS LASERS

3.1 Electronic Transition Lasers—Charles K. Rhodes and Robert S. Davis
3.2 Vibrational Transition Lasers—Tao-Yuan Chang
3.3 Far Infrared Lasers—Paul D. Coleman; David J. E. Knight

SECTION 4: TABLE OF LASER WAVELENGTHS—Marvin J. Weber

HANDBOOK OF LASER SCIENCE AND TECHNOLOGY

VOLUME I: LASERS AND MASERS

TABLE OF CONTENTS

Section 1
Introduction

1.1 Types and Comparisons of Laser Sources

1.1 TYPES AND COMPARISONS OF LASER SOURCES

William F. Krupke

INTRODUCTION

Light Amplification by Stimulated Emission of Radiation was first demonstrated by Maiman in 1960, the result of a population inversion produced between energy levels of chromium ions in a ruby crystal when irradiated with a xenon flashlamp.[1] In the ensuing two decades, population inversions and coherent emission have been generated in literally thousands of substances (neutral and ionized gases, liquids, and solids) using a variety of incoherent excitation techniques (optical pumping, electrical discharge, gasdynamic-flow, electron-beam, chemical reaction, nuclear decay). The number and types of laser sources has been further expanded many-fold by utilizing one laser source (primary) to generate coherent radiation in a second medium, either by optically producing a population inversion in the second medium or as the result of nonlinear scattering in the second substance. Recently, laser action has even been achieved by passing a dilute electron beam through a periodic magnetic field (free-electron laser, or FEL[2]). By properly choosing the kinetic energy of the electron beam and the periodicity of the magnetic (wiggler) field, the output wavelength of the FEL can be varied, in principle, from the ultraviolet to the far infrared spectral region.

The extrema of laser output parameters which have been demonstrated to date, and the laser media used, are summarized in Table 1.1.1. Note that the extreme power and energy parameters listed in this table were attained with laser systems (such as a master-oscillator-power-amplifier, or MOPA system) rather than with simple laser oscillators.

To be sure, no single laser source can simultaneously provide this spectacular set of characteristics. Each laser gain medium possesses a unique ensemble of energy levels (electronic/vibrational/rotational) which are dynamically coupled to each other through various radiative and nonradiative processes. These structural and kinetic features determine a laser's nominal operating wavelength(s), its spectral tuning range, its possible output waveforms, and its energy and power scalability. Laser efficiency is determined by the degree to which appropriate pump excitation energy can be generated, fed selectively into the upper laser level(s), and subsequently extracted coherently before deleterious decay processes otherwise remove this excitation energy. It is the very richness of energy level schemes and transition probabilities provided in nature that results in such a large number of lasers with such a wide variety of output characteristics.

Given this considerable diversity in laser properties, it is the purpose of this introductory section to order laser sources into basic classes and to describe the principal characteristics that define the classes and their subdivisions.

CLASSES OF LASER SOURCES

Laser sources are commonly classified in terms of the state-of-matter of the active medium: gas, liquid, and solid. Each of these classes is further subdivided into one or more types as shown in Table 1.1.2. A well-known representative example of each type of laser is also given in Table 1.1.2 together with its nominal operation wavelength and the method(s) by which it is pumped.

Gas Lasers

Gas lasers are conveniently described in terms of six basic types, two involving electronic transitions in atomic active species (neutral and ionic), three based on neutral

Table 1.1.1
EXTREMA OF OUTPUT PARAMETERS OF LASER DEVICES OR SYSTEMS

Parameter	Value	Laser medium
Peak power	2×10^{13} W (collimated)	Nd:glass
Peak power density	10^{18} W/cm^2 (focused)	Nd:glass
Pulse energy	$>10^4$ J	CO_2, Nd:glass
Average power	10^5 W	CO_2
Pulse duration	3×10^{-13} sec	Rh6G dye;
	continuous wave (cw)	various gases, liquids, solids
Wavelength	60 nm $\leftrightarrow$ 385 μm	Many required
Efficiency (nonlaser pumped)	70%	CO
Beam quality	Diffraction limited	Various gases, liquids, solids
Spectral linewidth	20 Hz (for 10^{-1} sec)	Neon-helium
Spatial coherence	10 m	Ruby

molecular active species (differentiated by laser action occurring on electronic, vibrational, and rotational transitions), and one based on molecular-ion active species. Gas lasers are pumped using a wide variety of excitation methods, including several types of electrical discharges (cw, pulsed, dc or rf, glow or arc), electron beam excitation, gasdynamic expansion, electrically or spontaneously induced chemical reactions, and optical pumping using primary lasers.

Liquid Lasers

Liquid lasers are commonly described in terms of three distinct types: organic dye lasers which are most well-known for their spectral tunability, rare-earth chelate lasers which utilize organic molecules, and lasers utilizing inorganic solvents and trivalent rare earth ion active centers. As indicated in Table 1.1.2, liquid lasers are optically pumped using three basic methods: flashlamps, pulsed primary lasers, or cw primary lasers.

Solid State Lasers

Solid state lasers are subdivided, first by the type of solid used — a dielectric insulator or a semiconductor. Dielectric insulators may take the form of an impurity doped crystal or an impurity doped amorphous material such as glass. More recently, solid state lasers have been developed using insulating crystals in which the active specie has been fully substituted into the lattice (stoichiometric materials) and using insulator crystals in which color centers (specific types of lattice defects) serve as the active centers. Lasers utilizing dielectric insulators are almost exclusively pumped optically, either with flashlamps, cw arc-lamps, or with other laser sources.

Semiconductor lasers are usually differentiated in terms of the means by which the hole-electron pair population inversion is produced. Semiconductor lasers can be pumped optically (usually with other laser sources), by electron-beams, or more commonly by injection of electrons in a p-n junction. A more comprehensive representation of laser sources is given in Figures 1.1.1a and 1.1.1b. The wavelengths of important molecular electronic lasers (excimers, dihalogens, etc.), molecular vibration-rotation lasers, solid state lasers, neutral and ionized atomic lasers, and simple binary semiconductor lasers operating between 100 and 12000 nm are shown. Other important lasers which are known particularly for their ability to be tuned in wavelength (dye lasers, excimers, tertiary semiconductor lasers, color-center lasers, and phonon-terminated lasers), will be discussed later.

Table 1.1.2
CLASSES, TYPES, AND RERESENTATIVE EXAMPLES OF LASER SOURCES

Class	Type (characteristic)	Representative example	Nominal operating wavelength (nm)	Method(s) of excitation
Gas	Atomic, neutral (electronic transition)	Neon-Helium (Ne-He)	633	Glow discharge
	Atomic, ionic (electronic transition)	Argon (Ar^+)	488	Arc discharge
	Molecule, neutral (electronic transition)	Krypton fluoride (KrF)	248	Glow discharge; e-beam
	Molecule, neutral (vibrational transition)	Carbon dioxide (CO_2)	10600	Glow discharge; gasdynamic flow
	Molecule, neutral (rotational transition)	Methyl fluoride (CH_3F)	496000	Laser pumping
	Molecule, ionic (electronic transition)	Nitrogen ion (N^+_2)	420	E-beam
Liquid	Organic solvent (dye-chromophor)	Rhodamine dye (Rh6G)	580—610	Flashlamp; laser pumping
	Organic solvent (rare earth chelate)	Europium:TTF	612	Flashlamp
	Inorganic solvent (trivalent rare earth ion)	Neodymium: $POCl_4$	1060	Flashlamp
Solid	Insulator, crystal (impurity)	Neodymium:YAG	1064	Flashlamp, arc lamp
	Insulator, crystal (stoichiometric)	Neodymium:UP(NdP_5O_{14})	1052	Flashlamp
	Insulator, amorphous (impurity)	Neodymium:glass	1061	Flashlamp
	Semiconductor (p-n junction)	GaAs	820	Injection current
	Semiconductor (electron-hole plasma)	GaAs	890	E-beam, laser pumping

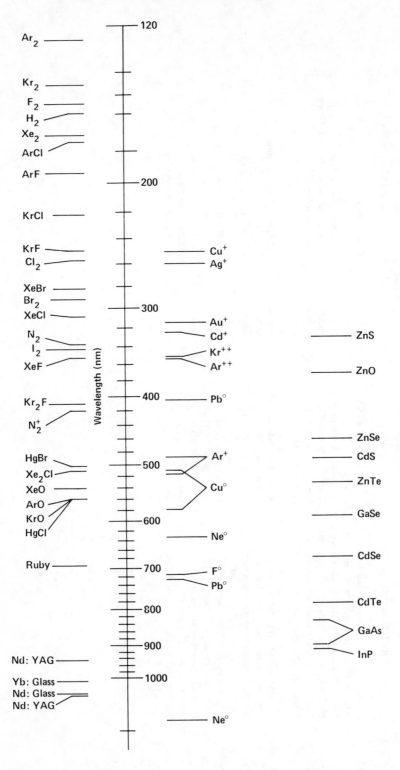

FIGURE 1.1.1a.   Wavelengths of lasers operating in the 120 to 1200 nm spectral region.

TEMPORAL MODES OF LASER OPERATION

By suitably designing the excitation source and/or by controlling the Q of the laser

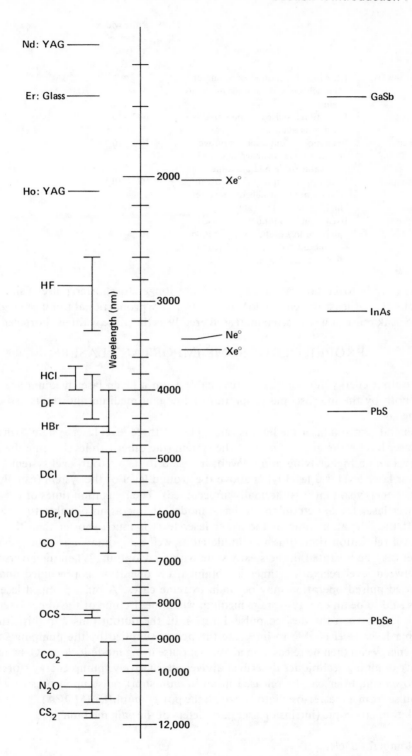

FIGURE 1.1.1b. Wavelength of lasers operating in the 1300 to 12,000 nm spectral region.

resonator structure, a given laser gain medium can be induced to provide output radiation in a variety of temporal forms:

Form	Technique	Pulse width range (sec)
Continuous wave	Excitation is continuous; resonator Q is held constant at some moderate value	∞
Pulsed	Excitation is pulsed; resonator-Q is held constant at some moderate value	10^{-8}—10^{-3}
Q-Switched	Excitation is continuous or pulsed; resonator Q is switched from a very low value to a moderate value	10^{-8}—10^{-6}
Cavity dumped	Excitation is continuous or pulsed; resonator Q is switched from a very high value to a low value	10^{-7}—10^{-5}
Mode locked	Excitation is continuous or pulsed; phase or loss of the resonator modes are modulated at a rate related to the resonator transit time	10^{-12}—10^{-9}

Energy tends to be conserved in going from longer to shorter pulses using these techniques (or at least the energy loss is not inversely proportional to pulse length) so that the peak pulse power is often increased greatly as the pulse width is shortened.

PROPERTIES OF SOME IMPORTANT LASERS

The output energy and/or power obtainable from a given laser medium are determined both by the microscopic properties of the gain medium and by its associated "scaling laws".

In general terms, a laser medium is said to be a "three-level laser system" when the lower laser level is the ground-state of the system (the other two levels being the upper laser level and a higher-lying pump level); it is said to be a "four-level system" when the lower laser level is a level lying above the ground level of the system (usually with sufficient energy so that it is thermally unoccupied). The relaxation times of the upper and lower laser levels determine the basic modes of operation possible for the laser itself. If the relaxation time of the lower laser level is much shorter than the upper laser level relaxation time (due to stimulated as well as spontaneous processes) then the laser may be operated in the steady state with a cw output. When the inverse relation between level relaxation times is obtained, cw operation is precluded and self-terminated pulsed operation may occur in extreme cases. A pulse-pumped laser medium is said to be an energy-storage medium when the lifetime of the upper laser level is much longer than the desired pulse-duration of the output pulse. In this situation the upper laser level is able to integrate the power supplied by the pumping source. Stored energy can then be released in an output pulse using mode-locking, Q-switching, or cavity dumping techniques described above; alternatively, pump energy stored in a laser power-amplifier can be released in an intense short pulse upon passing a weak short pulse from a master-oscillator through the power amplifier (MOPA).

The key microscopic (intrinsic) laser parameters of the gain medium are

- Nominal wavelength
- Stimulated emission cross-section
- Spectral gain-bandwidth and type of saturation (homogeneous/inhomogeneous)
- Saturation fluence or flux
- Radiative and kinetic lifetimes of upper and lower laser levels

and the characteristic specific excitation parameters are

- Population inversion density
- Small signal gain coefficient
- Input and output power (energy) densities

Values for these parameters are given in Tables 1.1.3 and 1.1.4, respectively, for important cw and pulsed lasers of each class. References 3 to 15 are cited in the tables as key literature sources dealing with the lasers used to illustrate the various classes and types of lasers. These tables also provide nominal performance data (size, input, output, and efficiency parameters) for commercial or laboratory laser devices. Note that larger laser outputs than those given in Tables 3 and 4 may be obtained according to scaling laws discussed more fully in appropriate sections of this handbook.

AVERAGE POWER SCALING

The increase in the average power output of a laser as it is made bigger is determined primarily by the rate at which waste heat generated in the laser process can be removed from the laser medium and/or the active volume enclosed by the optical resonator. In average power producing lasers of practical interest, removal of waste heat is accomplished by either convection or conduction, the choice depending on the class of laser medium involved. For both gaseous and liquid laser media, scaling to high average power is achieved using convective flow of the waste heat (and spent laser medium) out of the active volume defined by the laser resonator. In the case of gas lasers, the flow may be supersonic (as in the CO_2 gasdynamic laser[16] which has resulted in the highest average output power yet achieved) or it may be subsonic (as in CO_2 molecular,[17] neutral atomic,[18] and rare gas halide excimer lasers[19]). In the case of liquid dye lasers, significant average power has been obtained using a confined transverse flow[20] of the organic dye laser medium through the optically pumped laser volume, as well as by using a free-flowing transverse jet stream.[21]

In the case of solid state lasers, the laser medium itself cannot be rapidly and continuously moved through the volume of space defined by the laser resonator and cooling of the laser medium must be accomplished by conduction of waste heat to an exterior surface. This surface can then be cooled using a gaseous or liquid cooling fluid flowing across it. Crystalline materials generally exhibit relatively high thermal conductivities (0.1 to 2.0 W/cm - K) which are strongly temperature dependent compared to those of amorphous-glasses (0.005 to 0.01 W/cm - K) which are essentially temperature independent. Nd:YAG solid state lasers have been successfully scaled in average power to the kilowatt range[14] using cylindrical laser rods cooled by water flowing along their lateral surfaces. Scaling the average power of dielectric solid state lasers (glass and crystalline) has also been accomplished using the "axial gradient laser" configuration[22] in which flat disks of laser material are pumped through their broad faces and cooled through these same surfaces by flowing a liquid or gas across them. Other "face-pump" geometries[23] leading to higher average output powers from solid state glass lasers have also been developed. Solid state semiconductor laser materials exhibit both high heat capacities and thermal conductivities (similar to crystalline dielectric materials); however, the maximum average power achievable with a single diode is severely constrained by large optical loss coefficients, limited junction thicknesses (spatial extent of population inversion) and by electrical power transfer considerations.[8] Single diode CW powers in excess of 100 mW have been obtained at room temperature and in excess of 500 mW have been obtained at a temperature of 77 K. To extend the

Table 1.1.3
PROPERTIES AND PERFORMANCE OF SOME CONTINUOUS WAVE (CW) LASERS

Parameter	Unit	Gas			Liquid	Solid	
		Neon helium	Argon ion	Carbon dioxide	Rhodamine 6G dye	Nd:YAG	GaAs
Excitation method		DC discharge	DC discharge	DC discharge	Ar$^+$ laser pump	Krypton arc lamp	DC injection
Gain medium composition		Neon: helium	Argon	CO_2:N_2:He	Rh 6G:H_2O	Nd:YAG	p:n:GaAs
Gain medium density	Torr	0.1:1.0	0.4	0.4:0.8:5.0	—	—	—
	ions/cc	—	—	—	2(18):2(22)	1.5(20):2(22)	2(19):3(18):3(22)
Wavelength	nm	633	488	10600	590	1064	810
Laser cross-section	cm^{-2}	3(−13)	1.6(−12)	1.5(−16)	1.8(−16)	7(−19)	6(−15)
Radiative lifetime (upper level)	sec	~1(−7)	7.5(−9)	4(−3)	6.5(−9)	2.6(−4)	~1(−9)
Decay lifetime (upper level)	sec	~1(−7)	~5.0(−9)	~4(−3)	6.0(−9)	2.3(−4)	~1(−9)
Gain bandwidth	nm	2(−3)	5(−3)	1.6(−2)	80	0.5	10
Type, gain saturation		Inhomogeneous	Inhomogeneous	Homogeneous	Homogeneous	Homogeneous	Homogeneous
Homogeneous saturation flux	W cm^{-2}	—	—	~20	3(5)	2.3(3)	~2(4)
Decay lifetime (lower level)	sec	~1(−8)	~4(−10)	~5(−6)[b]	<1(−12)	<1(−7)	<1(−12)
Inversion density	cm^{-3}	~1(9)	2(10)	2(15)	2(16)	6(16)	1(16)
Small signal gain coefficient	cm^{-1}	~1(−3)	~3(−2)	1(−2)	4	5(−2)	40
Pump power density	W cm^{-3}	3	900	0.15	1(6)	150	7(7)
Output power density	W cm^{-3}	2.6(−3)	~1	2(−2)	3(5)	95	5(6)
Laser size (diameter:length)	cm:cm	0.5:100	0.3:100	5.0:600	1(−3):0.3	0.6:10	5(−4):7(−3):2(−2)[a]
Excitation current/voltage	A/V	3(−2):2(3)	30:300	0.1:1.5(4)	—	90:125	1.0/1.7
Excitation current density	A cm^{-2}	0.15	600	6(−3)	—	140	4.5(3)
Excitation power	W	60	9(3)	1.5(3)	4	1.1(4)	1.7
Output power	W	0.06	10	240	0.3	300	0.12
Efficiency	%	0.1	0.1	13	7	2.6	7
Reference	[—]	[3]	[4]	[5]	[6]	[7]	[8]

[a] Junction thickness:width:length.
[b] Pressure dependent.

average power output substantially beyond these levels appeal is made to laser-diode arrays.[24]

PEAK POWER SCALING

Increasing the peak-power output of a laser is constrained ultimately by the optical damage properties of the laser medium itself or of the optical materials required to make the laser operate. The subjects of optically induced damage of "transparent" dielectric materials[25] and optical breakdown in transparent gases[26] are extensive and complex. Readers are referred to References 25 and 26 (and additional references contained therein) for a summary of important aspects of optical damage and breakdown. In broad terms it is found that optical damage thresholds of dielectric materials depend on the wavelength of the incident radiation and the duration over which it is applied. Damage thresholds of surfaces are generally lower than those of the bulk material; damage thresholds of antireflection (AR) coatings are generally lower than those of high-reflectivity coatings. For nanosecond-long laser pulses at both one and ten microns, the best highly reflecting optical elements exhibit damage thresholds of a few (5 to 10) J/cm^2; for shorter pulses, thresholds tend to be reduced as the square root of the pulse-length; for longer pulses, energy damage fluences tend to increase less rapidly than the square root of the pulse duration. For pulse durations of about a microsecond and longer, linear absorption and thermal diffusion processes dominate damage thresholds. In this pulse regime, power loading becomes a more relevant parameter characterizing optical damage.

For cw lasers, water-cooled metal mirrors have been developed for use at 10 μm which can accept incident power fluxes of many kW/cm^2. In the visible spectral region, uncoated metals are not sufficiently reflective for use on high average power mirrors and multilayer dielectric coatings on metal or on transparent substrates are employed. Incident power fluxes of many kW/cm^2 can also be sustained by such mirrors, primarily because of the low residual absorption of the materials used. In the ultraviolet spectral region ($\lambda < 400$ nm) relatively little is known at present about optical damage.[27] However, the discovery and rapid development of ultraviolet excimer lasers is now stimulating considerable research and development of damage-resistant UV optical components.

Although scaling of power and energy are constrained by intrinsic laser properties and by optical damage mechanisms, laser output power and/or energy will ultimately be limited by external factors associated with the use of the laser (e.g., system size, weight, available prime power, and cost). Figure 1.1.2 plots the peak-powers and pulse-durations of a number of large Nd:Glass,[28,29,30] CO_2,[31,32] and Nd:YAG[33] laser systems designed and built for special purposes.

LASER COSTS

The costs of lasers and laser systems vary widely and cannot be readily generalized. However, sufficient development of CO_2 lasers for use in commercial applications has now occurred to give some perspective on how the cost per watt ($/W) of a given type of laser may decrease as the total power output of the laser is increased (economy of scale). Using the 1975 Laser Focus Buyers Guide as a reference source, Jensen et al.[34] found that averaging over for all types of CO_2 lasers, the cost in dollars/watt varied inversely with the total laser power, P, according to the expression $80 + (10^4/P)$ with P in watts. This expression indicates that for laser powers above 1 kW, CO_2 laser radiation costs a constant $80/watt. Other types of lasers may show similar unit cost behavior as the power or energy is scaled, but quantitative data is sparse. Two studies

Table 1.1.4
PROPERTIES AND PERFORMANCE OF SOME PULSED LASERS

Parameter	Unit	Gas				Liquid	Solid	
		Carbon dioxide		Krypton fluoride		Rhodamine 6G	Nd:YAG	Nd:glass
Excitation method		TEA-discharge	E-beam/sust.	Glow discharge	E-beam	Xenon flashlamp	Xenon flashlamp	Xenon flashlamp
Gain medium composition		CO_2:N_2:He	CO_2:N_2:He	He:Kr:F_2	Ar:Kr:F_2	Rh6G:alcohol	Nd:YAG	Nd:Glass
	torr	100:50:600	240:240:320	1070:70:3	1235:52:3			
Gain medium density	ions/cc	—	—	—	—	1(18):1.5(22)	1.5(20):1(22)	3(20):2(22)
Wavelength	nm	10600	10600	249	249	590	1064	1061
Laser cross-section	cm^{-2}	2(−18)	2(−18)	2(−16)	2(−16)	1.8(−16)	7(−19)	2.8(−20)
Radiative lifetime (upper level)	sec	4(−3)	4(−3)	7(−9)	7(−9)	6.5(−9)	2.6(−4)	4.1(−4)
Decay lifetime (upper level)	sec	~1(−4)	5(−5)	2(−9)	3(−9)	6.0(−9)	2.3(−4)	3.7(−4)
Gain bandwidth	nm	1	1	2	2	80	0.5	26
Homogeneous saturation fluence	J/cm^{-2}	0.2	0.2	4(−3)	4(−3)	2(−3)	0.6	~5
Decay lifetime (lower level)	sec	5(−8)^a	1(−8)^a	<1(−12)	<1(−12)	<1(−12)	<1(−7)	<1(−8)
Inversion density	cm^{-3}	3(17)	6(17)	4(14)	2(14)	2(16)	4(17)	3(18)
Small signal gain coefficient	cm^{-1}	2(−2)	4(−2)	8(−2)	4(−2)	4	0.3	8(−2)
Medium excitation energy density	J/cm^{-3}	0.1	0.36	0.15	0.13	2.8	0.15	0.6
Output energy density	J/cm^{-3}	2(−2)	1.8(−2)	1.5(−3)	1.2(−2)	0.85	5(−2)	2(−2)
Laser dimensions	cm:cm:cm	4.5:4.5:87	10:10:100	1.5:4.5:100	8.5:10:100	1.2φ25	0.6φ7.5	0.6φ8.3
Excitation current/voltage	A/V	6(4)/3.3(3)	2.2(4)/4(4)	2.54(4)/1.5(5)	1.2(4)/2.5(5)	2(5)/2.5(4)		
Excitation current density	A cm^{-2}	8.5	22	170	11.5	2.6(3)		
Excitation peak power	W	2(8)	9(8)	4(9)	3(9)	5.4(9)	4(4)	9(4)
Output pulse energy	J	35	180	1	102	32	0.1	1.0
Output pulse length	sec	1(−6)	4(−6)	2.5(−8)	6(−7)	3.2(−6)	2(−8)	1(−4)
Output pulse power	W	3.5(7)	4(7)	4(7)	2(8)	1(7)	5(6)	1(4)

Efficiency	%	17	5	1	10[b]	0.2	1.5	3.7
Reference	[ ]	[9]	[10]	[11]	[12]	[13]	[14]	[15]

[a] Pressure dependent.
[b] Intrinsic efficiency ≡ energy output/energy deposited in gas.

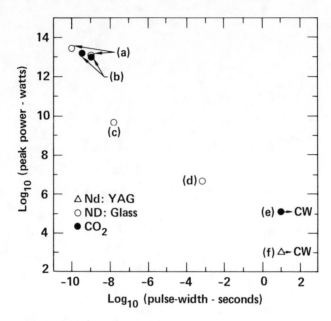

FIGURE 1.1.2. Output peak power and pulse-durations achieved by several large MOPA laser systems: (a) Nd:glass, Reference 28; (b) CO_2 e-beam sustainer, Reference 31; (c) Nd:glass, Reference 29; (d) Nd:glass, Reference 30; continuous-wave output power of two large laser oscillators: (e) CO_2 gasdynamic, Reference 32; (f) Nd:YAG, Reference 33.

of the cost of lasers have recently been conducted in connection with the use of lasers for the commercial separation of uranium isotopes. The study by Mail et al.[35] examined the costs of copper vapor and copper halide lasers, dye, carbon dioxide, rare-gas-halide, and neodymium-YAG lasers. The study by Schofner and Hoglund[36] examined the cost of pulsed and CW CO_2 lasers, conventional and frequency-doubled neodymium-YAG lasers, and to a very limited extent KrF excimer lasers. As large-scale commercial applications of lasers become more numerous and mature, additional cost scaling models and data bases will surely become available in the field.

TUNABLE LASERS

Spectral tunability is a particularly useful property of many laser sources. Semiconductor diode lasers, organic dye lasers, and color center lasers are particularly known for this property. The nominal spectral regions in which these types of lasers operate are shown in Figure 1.1.3. Using seven different dye types (coumarins, rhodamines, oxazines, etc.) and various solvents, the spectral region from 350 to 1000 nm can be spanned with tunable dye lasers. A single dye/solvent combination typically can be tuned several hundred wavenumbers (cm⁻¹) away from the spectral peak of the gain curve. Best dye laser performance is currently achieved with the yellow-orange rhodamine dye. Power and energy availability tend to roll-off to the biue and to the red, although useful amounts of energy and power can be achieved in these spectral regions.

As shown in Figure 1.1.3, semiconductor lasers of various types collectively span the spectral region from 330 nm to beyond 15 μm. A semiconductor diode of specified composition will exhibit a spectral gain peak at a specific wavelength and can be tuned several tens of nanometers around this wavelength. By varying the composition of a semiconductor diode it is possible to adjust the wavelength of its spectral gain peak.

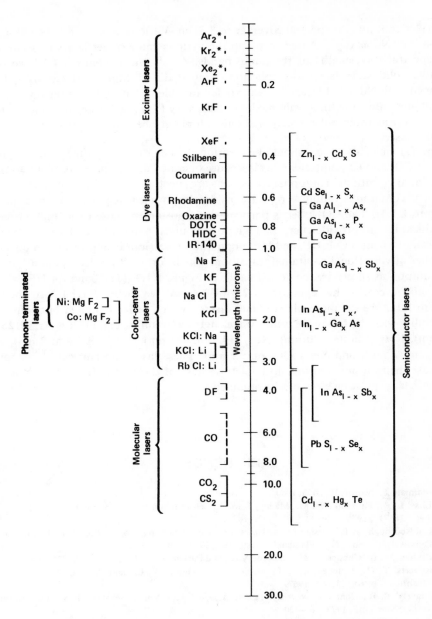

FIGURE 1.1.3.   Spectral tuning ranges of various types of tunable lasers.

Zinc and cadmium sulfides and selenides are used to span the UV-visible spectral region; diodes based on Ga, Al, As, P, and Sb span the region from 650 to 1500 nm; indium based diodes cover the region from 1 to 5 μm, and diodes of the type $PbS_{1-x}Se$, $Cd_{1-x}Hg_xTe$, and PbSnTe work out into the middle infrared spectral region (5 to 20 μm). Depending on the type of diode, tuning can be accomplished using an applied magnetic field, by changing the current passing through the diode, or by applying pressure to the diode.[37] Spectral linewidths as small as 10^{-6} cm^{-1} have been observed with tunable diode lasers. Output powers are typically several tens of microwatts, but hundreds of microwatts can be obtained in certain cases.

Recently discovered and developed alkali-halide color center lasers[38] have become important and practical tunable lasers in the near infrared spectral region (0.9 to 3.3 μm). Li, Na, and K fluorides cover the region from 0.9 to 1.5 μm and Na, K, and Rb chlorides continue the spectral coverage to 3.3 μm. Various laser sources (krypton ion, argon ion, Nd:YAG) are used to pump the various color center lasers depending on the spectral wavelengths of the pump bands. As shown in Figure 1.1.3, two other tunable solid state lasers — "phonon-terminated" Ni:MgF$_2$ and Co:MgF$_2$ lasers pumped with Nd:YAG lasers — can provide tunable radiation in the region from 1.6 to 2.0 μm.[39] Because these gain media possess very long energy storage times (5 to 10 msec), considerable pulse energies should ultimately be obtained from this type of tunable laser.

Several molecular lasers should be mentioned when discussing tunable lasers. When pulsed CO$_2$ and CS$_2$ molecular gas lasers are operated at high pressure ($\geqslant$ 15 atm), the individual vibrational-rotational transitions are spectrally broadened sufficiently to allow for continuous tuning across a considerable spectral region in the mid-infrared (Figure 1.1.3). The CO$_2$ laser is particularly important because of its high efficiency and its ability to provide energetic high power pulses.[40] In the case of several diatomic molecules, such as CO, DF, and HF, population inversions can be produced on a large number of vibrational-rotational transitions which form a "quasicontinuum" of spectral output, as shown by the dotted bands in Figure 1.1.3. (The band for HF, which is not shown, covers the region from 2.6 to 3.4 μm; see Figure 1.1.1.) Also shown in Figure 1.1.3 are the molecular excimer lasers[41] which provide limited but continuous tunability with the following center wavelengths and tuning bandwidths Ar*$_2$ (126 nm; 20 nm), Kr*$_2$ (146 nm; 20 nm), Xe*$_2$ (172 nm; 20 nm), ArF* (193 nm; 2 nm), KrF* (249 nm; 2 nm), and XeCℓ (308 nm; 2 nm). These lasers are particularly important because their output is in the ultraviolet, because of their high efficiency ($\geqslant$ 1%), and because large pulse energies can be obtained.

REFERENCES

1. **Maiman, T.,** *Nature,* 187, 493, 1960.
2. **Elias, L. R., Fairbank, W. M., Madey, J. M. J., Schwettman, H. A., and Smith, T. L.,** *Phys. Rev. Lett.,* 36, 717, 1976.
3. **Goldsborough, J. P.,** Design of gas lasers, in *Laser Handbook,* Vol. 1, Arechi, F. T. and Schulz-Dubois, E. D., Eds., North Holland, Amsterdam, 1972, 598—629.
4. **Bridges, W. B., Chester, A. N., Halsted, A. S., and Parker, J. V.,** *Proc. IEEE,* 59, 724, 1971.
5. **Roberts, T. G., Hutcheson, G. J., Ehrlich, J. J., Hales, W. L., and Barr, T. A., Jr.,** *IEEE J. Quantum Electron.,* 3, 605, 1967.
6. **Snavely, B. B.,** Continous-wave dye lasers, in *Dye Lasers,* Vol. I, Schaefer, F. P., Ed., Springer-Verlag, New York, 1973, 86—120.
7. **Foster, J. D. and Osterink, L. M.,** *J. Appl. Phys.,* 41, 3656, 1970.
8. **Kressel, H.,** Semiconductor lasers: devices, in *Laser Handbook,* Vol. I, Arechi, F. T. and Schulz-Dubois, E. D., Eds., North Holland, Amsterdam, 1972, 441—495.
9. **Dumanchin, R., Michon, M., Farcy, J. C., Boudinet, G., and Rocca-Serra, J.,** *IEEE J. Quantum Electron.,* 8, 163, 1972.
10. **Hoffman, J. M., Bingham, F. W., and Moreno, J. B.,** *J. Appl. Phys.,* 45, 1798, 1974.
11. **Watanabe, S., Shiratori, S., Sato, T., and Kashiwagi, H.,** *Appl. Phys. Lett.,* 33, 141, 1978.
12. **Rokni, M., Mangano, J. A., Jacob, J. H., and Hsia, J. C.,** *IEEE J. Quantum Electron.,* 14, 464, 1978.
13. **Schaefer, F. P.,** Principles of dye laser operation, in *Dye Lasers,* Vol. 1, Schaefer, F. P., Ed., Springer-Verlag, New York, 1973, 1—85.
14. **Koechner, W.,** *Solid State Laser Engineering,* Springer-Verlag, New York, 1976.
15. **Snitzer, E. and Young, C. G.,** Glass lasers, in *Lasers,* Vol. 2, Levine, A. K. and DeMaria, A. J., Eds., Marcel Dekker, New York, 1973.
16. **Gerry, E. T.,** Gasdynamic lasers, *SPIE J.,* 9, 61, 1971.
17. **Brown, C. O. and Davis, J. W.,** *Appl. Phys. Lett.,* 21, 480, 1972.
18. **Targ, R. and Sasnett, M. W.,** *Appl. Phys. Lett.,* 19, 537, 1971.

19. Wang, C. P., *Appl. Phys. Lett.,* 32, 360, 1978.
20. Peterson, O. G., Tuccio, S. A., and Snavely, B. B., *Appl. Phys. Lett.,* 17, 245, 1970.
21. Runge, P. K., *Opt. Commun.,* 5, 311, 1972.
22. Matovich, E., The axial gradient laser, *Proc. DOD Conf. Laser Technol.,* San Diego, 1970, 311—362.
23. Hulme, G. J. and Jones, W. B., *SPIE,* 69, 38, 1975.
24. Crow, J. D., Comerford, L. D., Laff, R. A., Bradey, M. J., and Harper, J. S., *Optics Lett.,* 1, 40, 1977.
25. Smith, W. L., *Opt. Eng.,* 17, 489, 1978.
26. Smith, D. C. and Meyerand, R. G., Laser radiation induced gas breakdown, in *Principles of Laser Plasmas,* Bekefi, G., Ed., John Wiley & Sons, New York, 1976, 457—507.
27. Newnam, B. E. and Gill, D. H., Proc. tenth symposium on laser induced damage materials, *NBS Spec. Publ.,* No. 541, 190, 1978.
28. Speck, D. R., Bliss, E. S., Glaze, J. A., Johnson, B. C., Manes, K. R., Ozarski, R. G., Rupert, P. R., Simmons, W. W., Swift, C. D., Thompson, C. E., *IEEE J. Quantum Electron.,* 15, 9, 1979.
29. Hagen, W. P., *J. Appl. Phys.,* 40, 511, 1969.
30. Batanov, V. A., Bufetov, I. A., Gusev, S. B., Ershov, B. V., Kolisnichenko, P. I., Malkov, A. N., Pinemov, Yu. P., and Federov, B. V., *Sov. J. Quant. Electron.,* 4, 853, 1975.
31. Schappert, G. T., Singer, S., Ladish, J., and Montgomery, M. D., Digest of Technical Papers, Tenth Int. Quantum Electron. Conf., IEEE, Piscataway, N.J., 1978, 668.
32. *Laser Focus,* 1977, 14.
33. Koechner, W., *Solid State Laser Engineering,* Springer-Verlag, New York, 1976, 548—549.
34. Jensen, R. J., Marinuzzi, J. G., Robinson, C. P., and Rockwood, S. D., *Laser Focus,* 1976, 51.
35. Mail, R. A., Markovich, F. J., and Carr, R. H., Cost Analysis of Lasers for a Laser Isotope Separation System, Final Report CR-2-718, General Research Corp., 1977; also UCRL 13744-2, University of California, Lawrence Livermore Laboratory, Livermore, Calif., 1977.
36. Schofner, F. M. and Hoglund, R. L., Laser Cost Experience and Estimation, Report K/OA-4038, June 28, 1977, Oak Ridge Gaseous Diffusion Plant, Union Carbide, Oak Ridge National Laboratory, Oak Ridge, Tenn.
37. Hinkley, E. D., Nill, K. W., and Blum, F. A., Infrared spectroscopy with tunable lasers, in *Laser Spectroscopy of Atoms and Molecules,* Walther, H., Ed., Springer-Verlag, New York, 1974, 127—190.
38. Mollenauer, L. F. and Olsen, D. H., *J. Appl. Phys.,* 46, 3109, 1975.
39. Moulton, P. F. and Mooradian, A., *Appl. Phys. Lett.,* 35, 838, 1979.
40. Mazurenko, Yu. T., Rubinov, Yu. A., and Shakhverdov, P. A., *Sov. J. Opt. Technol.,* 46, 341, 1979.
41. Ewing, J. J., Excimer Lasers, in *Laser Handbook,* Vol. 3, Stitch, M. L., Ed., North-Holland, New York, 1979.

Section 2
Solid State Lasers

2.1.1 PARAMAGNETIC ION LASERS

P. F. Moulton

INTRODUCTION

The active media of the paramagnetic ion crystalline lasers described in this section are host crystals doped with a relatively small percentage of ions from the iron, rare-earth (lanthanide) or actinide group of the periodic table. Stoichiometric lasers in which the paramagnetic ions are a primary constituent of the crystal are treated in Section 2.1.2.

The following discussion is intended to provide an introduction to paramagnetic ion lasers. This subsection has not been heavily referenced; details on the specific lasers discussed can be found in references to Tables 2.1.1.2 - 2.1.1.8. For general information a number of sources are available including reviews by DiBartolo[1,2] which explain the properties of paramagnetic ions in solids and summarize current research in that area. Two recent publications by Weber[3,4] review the field of rare-earth and actinide lasers. Laser crystal growth is the subject of a somewhat earlier article by Nassau.[5] Properties of the Nd:YAG laser, perhaps the most popular paramagnetic ion system, are discussed thoroughly by Danielmeyer.[6] An excellent reference on the engineering aspects of paramagnetic ion lasers, covered only briefly in this chapter, has been written by Koechner.[7] Finally, a recent publication by Kaminskii[28] provides much information on basic research into paramagnetic ion lasers, particularly that performed in the Soviet Union.

CHARACTERISTICS OF DIFFERENT CLASSES OF PARAMAGNETIC-ION LASERS

Paramagnetic ions in general have unfilled inner electronic shells. At present only those ions from the iron, rare-earth or actinide groups have been used as dopants for paramagnetic ion crystalline lasers. It is convenient to classify the wide variety of such laser systems into categories determined not only by the periodic-table group but also the valence of the ion and the type of electronic transition involved. The categories include:

Trivalent Rare Earths, 4f - 4f Transitions

Ions from the rare-earth (RE) group differ in electronic structure only by the number of electrons in the 4f inner shell. For a given RE ion the energetically lowest-lying excited electronic states involve configurations in which all electrons remain in that shell. Figure 2.1.1.1 indicates that for trivalent ions such $4f^n$ configurations (n is the number of electrons in the 4f shell for the ionic ground state) are maintained for energies up to the UV region.

In practically all cases lasers using trivalent RE ions operate on transitions between 4f energy levels. The Russell-Saunders notation used for a given 4f free-ion level is

$$^{2S+1}(L)_J$$

where S, L, and J are the net spin, orbital, and total angular-momentum quantum numbers for all the 4f electrons. Letters are used to denote the value of L by the following convention:

L	0	1	2	3	4	5	6	7	8	....
Letter	S	P	D	F	G	H	I	K	L	....

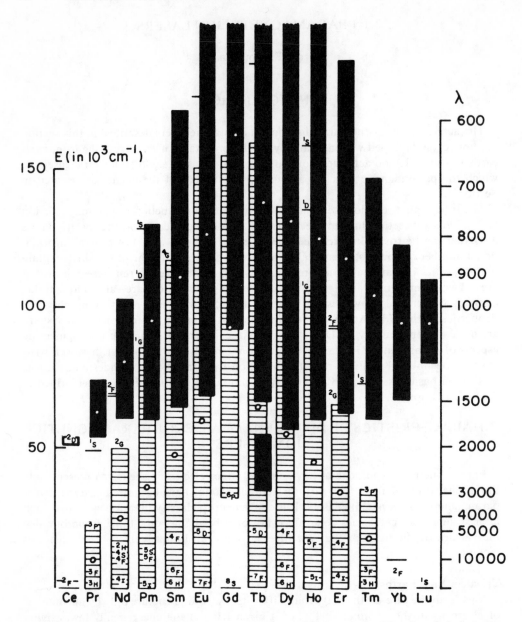

FIGURE 2.1.1.1.   Schematic diagram of the 4f^n (white) and 4f^n-^{1}5d (black) configuration of trivalent rare-earth ions. ^{a}(From Dieke, G. H. and Crosswhite, H. M., *Appl. Optics*, 2, 675, 1963. With permission.)

Thus the level $^4F_{3/2}$ has S = 3/2, L = 3, and J = 3/2. Because the wavefunctions for 4f states are relatively small in spatial extent and shielded by the filled 5s and 5p outer shells the 4f^n-configuration energies are only slightly perturbed when RE ions are placed in a crystalline environment. The perturbation is small compared to spin-orbit and electrostatic interactions among the 4f electrons. The primary change in level structure is a splitting (caused by the Stark effect of the crystal field) into many closely spaced levels of each of the (2J + 1)-degenerate free-ion levels. In crystals the free-ion levels are then referred to as manifolds. The energy levels shown in Figure 2.1.1.2A (such a figure is often called a Dieke^9 diagram) though observed in one crystal are appropriate, within reasonable accuracy, to all other host crystals. The approximate extent of splitting in each of the manifolds is indicated by the width of the lines in the

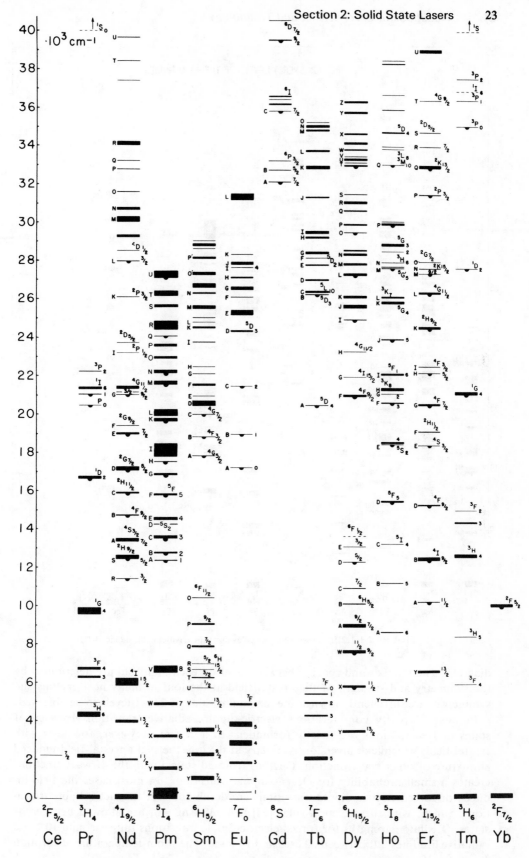

FIGURE 2.1.1.2A.   Observed energy levels of the trivalent rare-earth ions.[16]

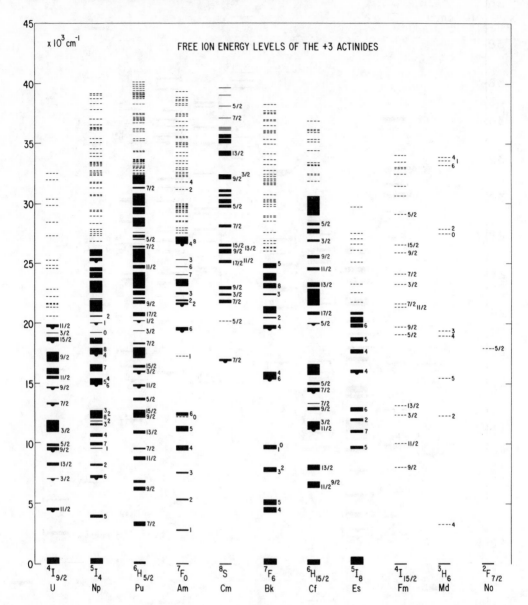

FIGURE 2.1.1.2B.   Observed energy levels of the trivalent actinide ions.[16]

diagram. The number and spacing between levels in each manifold are determined by the symmetry and strength of the crystal field at the ion; in many host crystals site symmetries are sufficiently low for the maximum value of $2J + 1$ levels to be observed.

Because of parity considerations electric-dipole radiative transitions between 4f states of free RE ions are strictly forbidden. In a crystal, however, for hosts with crystal fields which lack inversion symmetry at the ion, the slight crystal-field-induced admixture of other wavefunctions (such as from 5d states) with the 4f wavefunctions creates a small probability for electric-dipole transitions. In most cases the field-induced or forced-electric-dipole transitions are dominant over magnetic-dipole transitions, which, with the selection rules $|\Delta J| \leqslant 1$, $\Delta S = \Delta L = 0$, are allowed between 4f states. Oscillator strengths for forced-electric-dipole transitions range around 10^{-6} and radiative lifetimes fall between 50 μsec and 10 msec. The room-temperature absorption spectra of trivalent rare earths doped in $Y_3Al_5O_{12}$(YAG), a common host crystal, appear in Figures 2.1.1.3A to N, and show the absorption from numerous 4f-transitions between the ground and excited states. Also shown are the absorption spectra of several multiply-doped crystals in which energy transfer among ions occurs (discussed in the section entitled "Laser Operation"). Included are crystals containing Cr^{3+} ions which exhibit the broad absorption features of 3d-3d transitions discussed below. The broad absorption from 4f-5d transitions of Ce^{3+}, also described below, are evident.

The 10 to 100 cm^{-1}-linewidths typically observed at 300 K from 4f-4f transitions between distinct levels result from two types of effects. First, the crystal field may vary somewhat from one ion site to another, creating a distribution of energy levels which leads to inhomogeneous broadening. The crystal-field variance can be caused by strains, defects, or other impurity ions, all of which are randomly distributed in the crystal, or for mixed or disordered crystals (e.g., CaF_2-BaF_2), by the fundamental nature of the lattice. Also, some crystals may have more than one type of site for the RE ions.

The second line-broadening mechanism, a homogeneous one, is related to the coupling between phonons (lattice vibrations) and ions which occurs because phonons perturb the crystal field at the ion. One result of the coupling is the existence of one-phonon processes, which involve a photon-induced change in ion electronic state accompanied by the emission or absorption of a phonon. Higher-order interactions can cause several phonons to be absorbed or emitted. Such phonon-assisted processes are often referred to as vibronic and lead to asymmetrical lineshapes which differ depending on whether a photon is absorbed or emitted. When the ion-phonon interaction is relatively strong, absorption and emission occur in clearly disjoint wavelength regions (Stokes shift) provided the crystal temperature is sufficiently low such that processes involving emission of a phonon are most probable. Phonon-scattering, Raman-like processes also occur which result in phonons gaining or losing energy when a photon is absorbed or emitted by an ion. These processes create symmetrical, essentially Lorentzian lineshapes. Finally, electronic transitions can take place entirely as a result of phonon emission or absorption. Since the rates for such nonradiative processes can be large (see below) lineshapes involving phonon-relaxed levels can be significantly lifetime broadened. Generally, for 4f-4f transitions of RE ions homogeneous linewidths are primarily determined by phonon-scattering and nonradiative processes.

Whether or not a line is inhomogeneous or homogeneous depends on the nature of the host crystal and the crystal temperature. In cases where the host is not a mixed or disordered crystal and there is only one type of site for the active ion, transitions are homogeneously broadened for unstrained crystals at room temperature.

The nonradiative relaxation of 4f energy levels by phonons is of crucial importance to laser operation, providing, in 4-level systems, rapid decay of the lower laser level and, in general, allowing significant population of the upper laser state by decay from excited higher-energy states. Measurements of the nonradiative decay rate between two levels have shown that for processes involving emission of many phonons the rate falls exponentially as the energy difference between the upper and lower level increases, independent of the particular RE ion. However, the rate does depend strongly on the particular host crystal, being greater for host crystals with higher maximum phonon frequencies. Figure 2.1.1.4 shows the nonradiative relaxation rate as a function of energy difference for a variety of crystals. For small energy differences, comparable to phonon energies, rates may rise to 10^{11} to 10^{12} sec^{-1}.

Fluorescence is readily observed from ionic levels where the radiative decay rate is large compared to rates for phonon relaxation and other nonradiative processes. Such levels are indicated by a semicircle in Figure 2.1.1.2 and are generally characterized by a large energy separation from the next-lower level. Fluorescence spectra from the

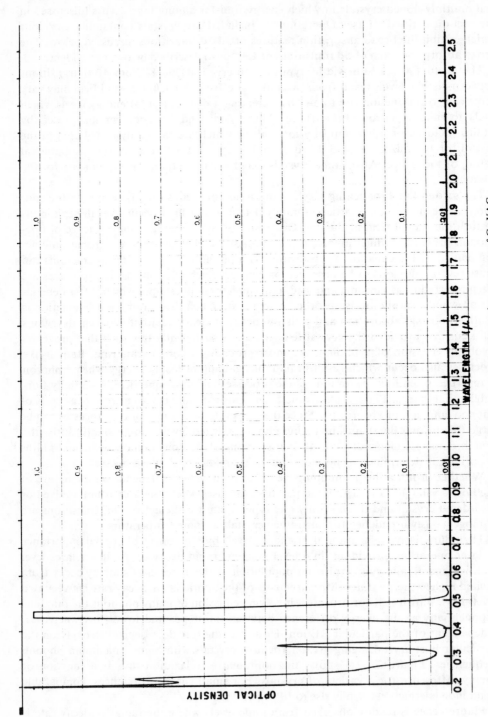

FIGURE 2.1.1.3A.   Room-temperature absorption spectrum of Ce:YAG.

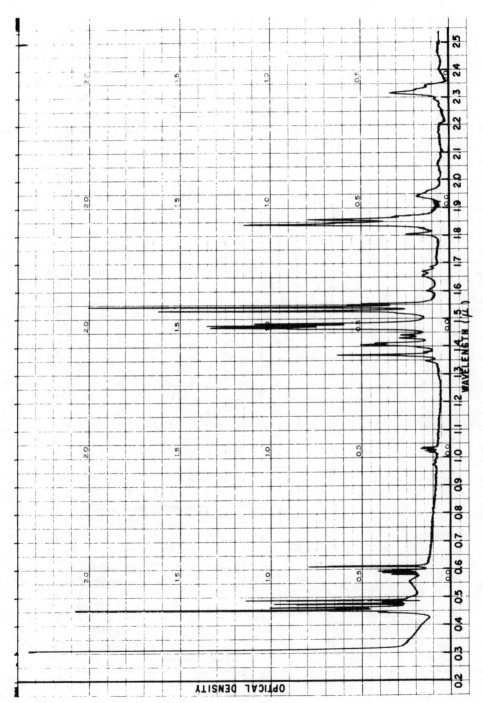

FIGURE 2.1.1.3B.   Room-temperature absorption spectrum of Pr:YAG.

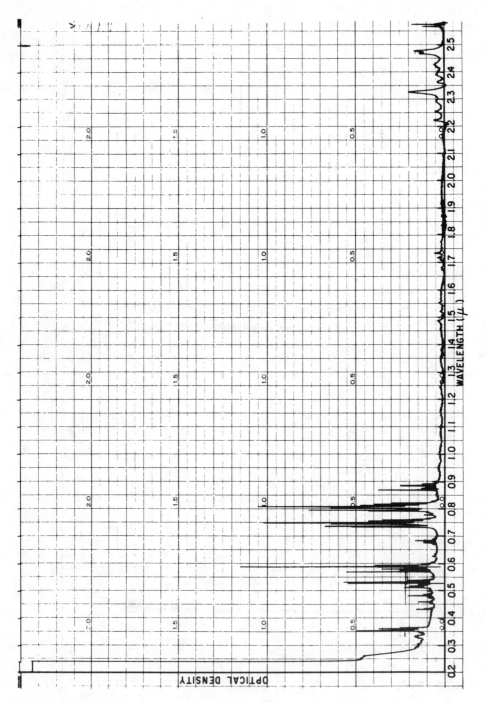

FIGURE 2.1.1.3C.   Room-temperature absorption spectrum of Nd:YAG.

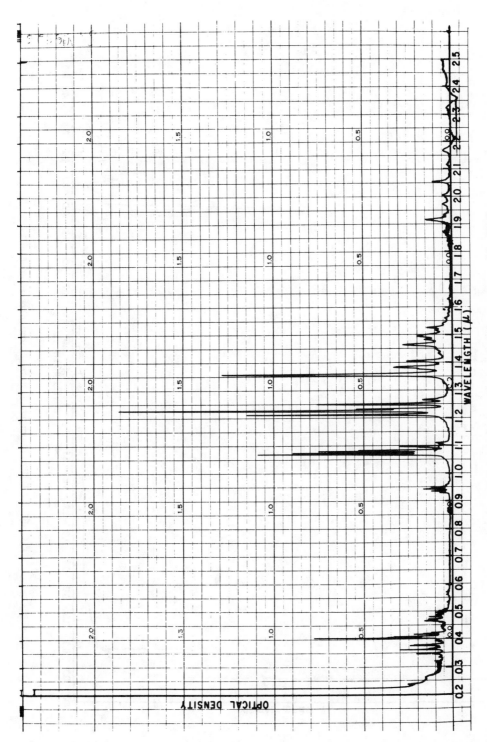

FIGURE 2.1.1.3D.   Room-temperature absorption spectrum of Sm:YAG.

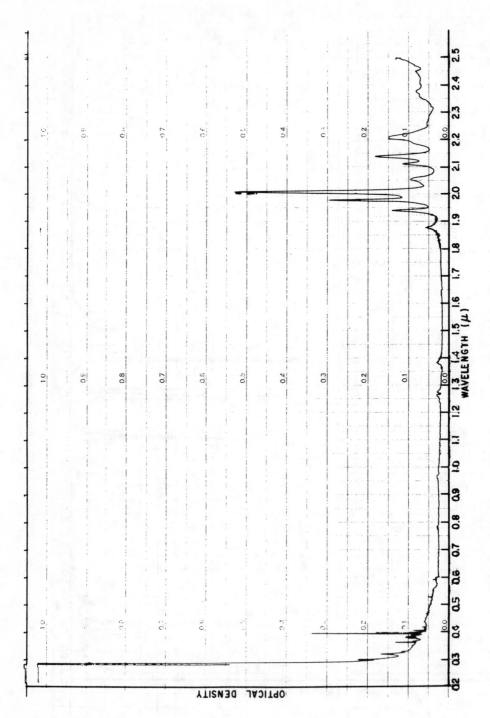

FIGURE 2.1.1.3E.   Room-temperature absorption spectrum of Eu:YAG.

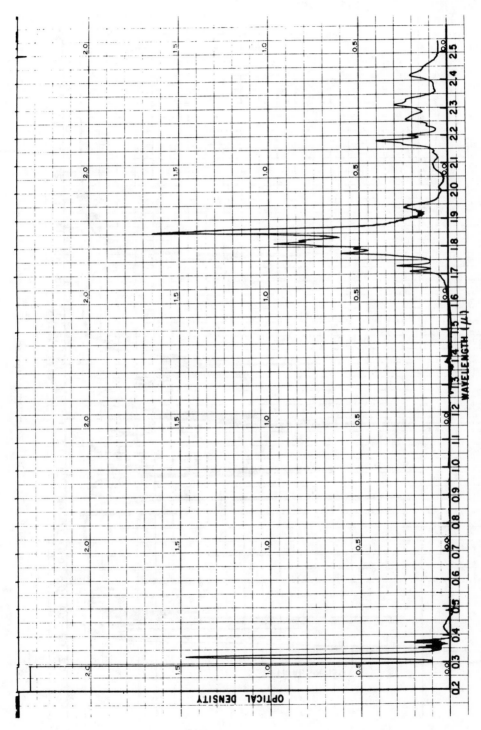

FIGURE 2.1.3F.   Room-temperature absorption spectrum of Tb:AG.

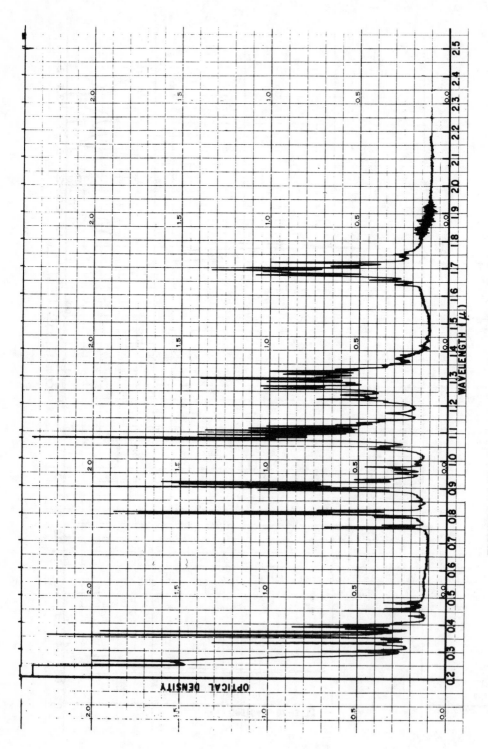

FIGURE 2.1.1.3G.   Room-temperature absorption spectrum of Dy:YAG.

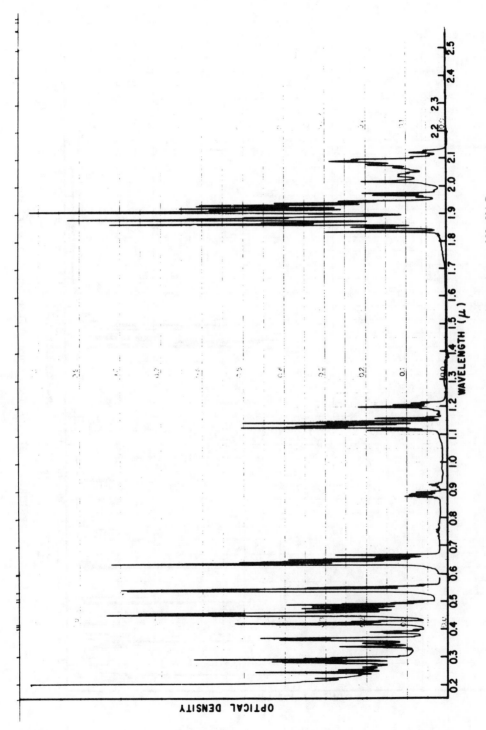

FIGURE 2.1.1.3H.   Room-temperature absorption spectrum of Ho:YAG.

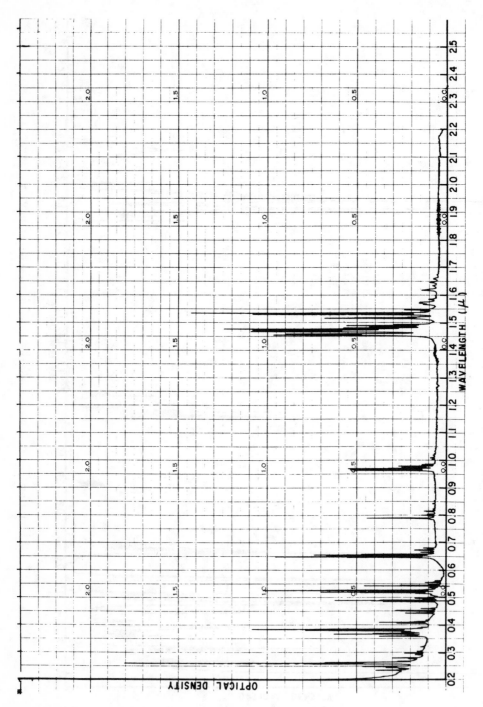

FIGURE 2.1.1.3I.   Room-temperature absorption spectrum of Er:YAG.

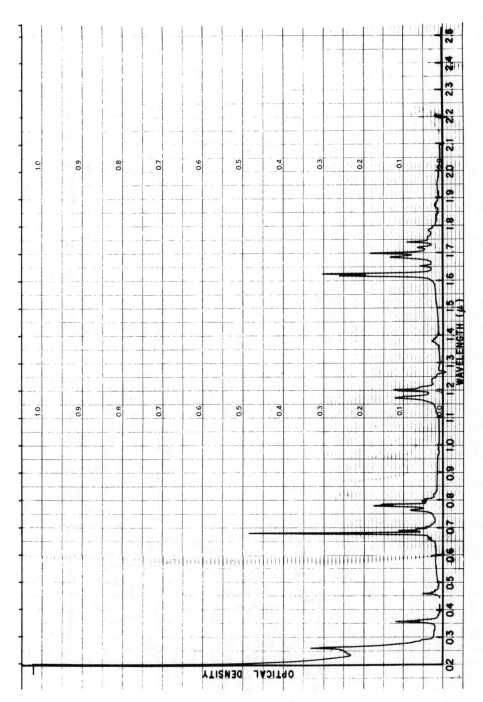

FIGURE 2.1.1.3J.   Room-temperature absorption spectrum of Tm:YAG.

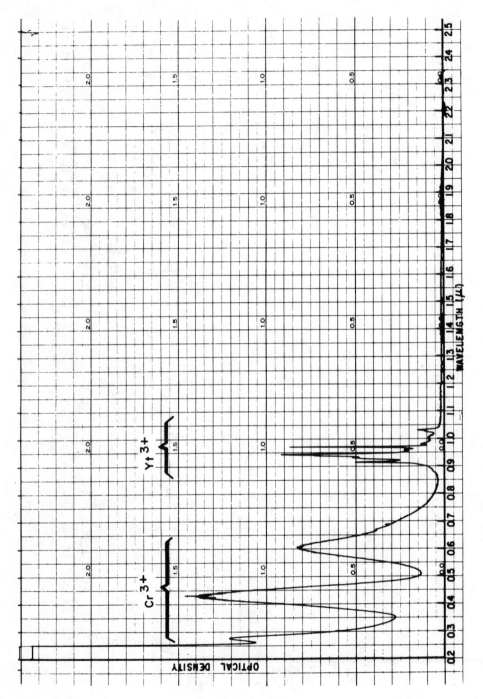

FIGURE 2.1.1.3K.   Room-temperature absorption spectrum of Cr,Yb:YAG.

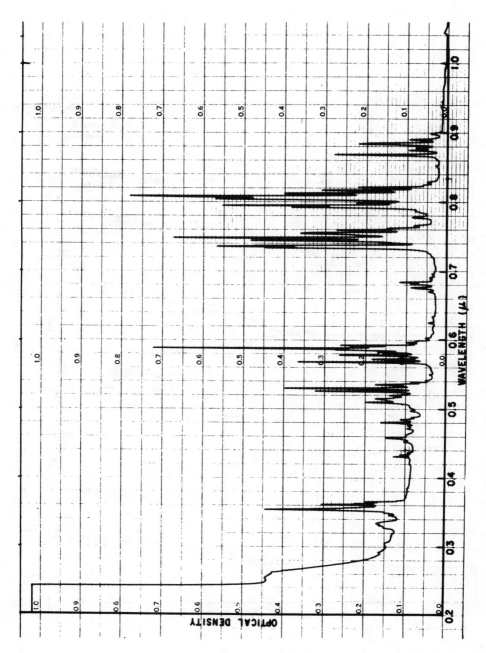

FIGURE 2.1.1.3L.   Room-temperature absorption spectrum of Nd:YAG (expanded).

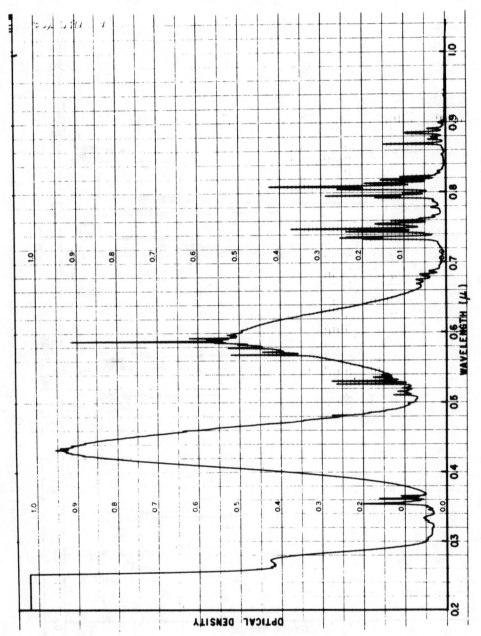

FIGURE 2.1.1.3M.   Room-temperature absorption spectrum of Nd,Cr:YAG.

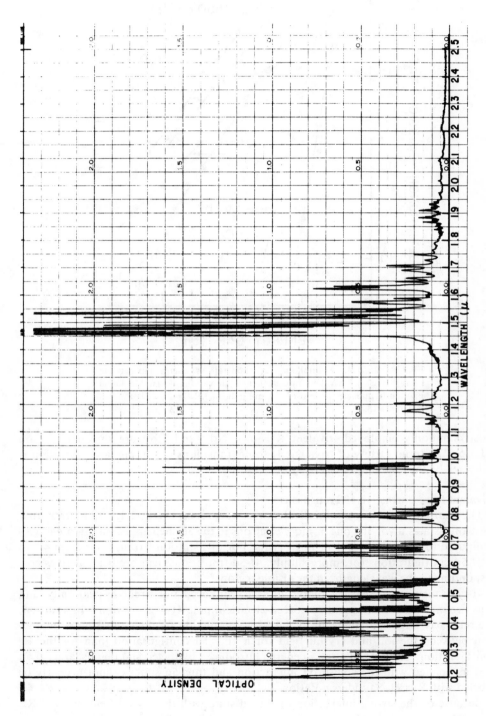

FIGURE 2.1.1.3N.   Room-temperature absorption spectrum of Er,Tm,Ho:YAG.

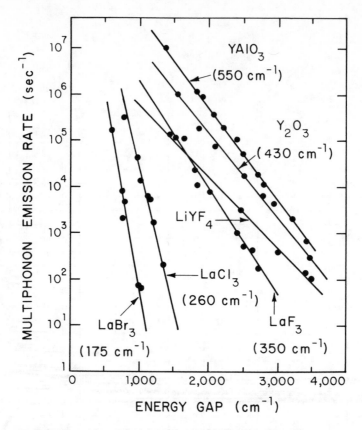

FIGURE 2.1.1.4. Multiphonon emission rate vs. energy gap between energy levels of trivalent rare-earth ions in various hosts. Maximum phonon frequencies for most hosts are indicated.[17] Data on $LiYF_4$ from Reference 18.

most commonly employed trivalent RE ions, Nd^{3+} and Ho^{3+} are shown in Figure 2.1.1.5. The large variety of trivalent 4f to 4f transitions exhibiting laser operation is indicated by the level diagrams of Figure 2.1.1.6.

Divalent Rare Earths, 4f-4f Transitions

The configuration diagram for divalent rare-earth ions, Figure 2.1.1.7, indicates that electronic states involving one or more electrons in the 5d shell appear at lower energies compared to those for trivalent rare earths. For divalent ions, transitions from $4f^n$ to $4f^{n-1}$ 5d configurations (or, more simply, 4f-5d transitions) can occur at energies corresponding to the wavelengths in the visible region.

4f-5d transitions are electric-dipole allowed, with oscillator strengths near unity, and thus are much more intense than forced-electric-dipole 4f-4f transitions. In addition, the relatively large spatial extent of the 5d wavefunction results in a stronger interaction between the ion and crystal field for a $4f^{n-1}$ 5d configuration than for the $4f^n$ configuration. The perturbation in energy introduced by the crystal field is larger than that associated with the spin-orbit interaction and thus 5d states cannot properly be labeled with Russell-Saunders notation.

Because of the strong interaction of the 5d state with the crystal field, when a RE ion makes a 4f-5d transition the positions of neighboring host-crystal ions generally change slightly after the transition occurs, in order to establish a lower energy state for the coupled system of ion and host crystal. Such a change greatly increases the

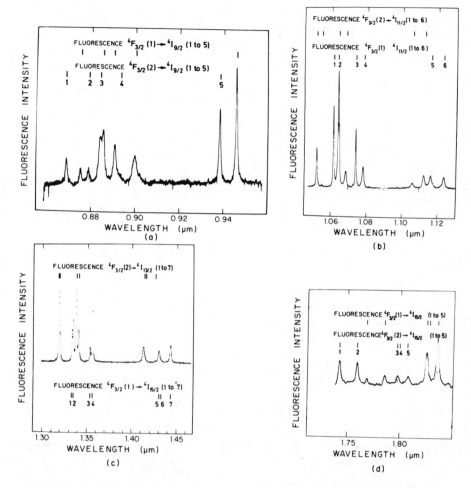

FIGURE 2.1.1.5A.   Fluorescence spectra of Nd:YAG from the $^4F_{3/2}$ level.[19] (From Daniel-meyer, H., Blatte, M., and Bolmer, P., *Appl. Phys.,* 1, 269, 1973. With permission.)

probability for phonon-assisted processes and leads to large (~ 1000 cm^{-1}) linewidths for 4f-5d transitions.

Divalent ions excited into 5d states can decay by phonon emission into 4f states. Optical excitation into the broad, intense absorption bands associated with transitions to 5d levels followed by decay to lower-lying 4f levels thus provides an excellent means for populating the upper laser level for the systems, shown in Figure 2.1.1.8, operating on 4f-4f transitions. Such transitions in host crystals such as CaF_2 are magnetic-dipole-allowed, not forced-electric-dipole, because the crystal field at divalent sites does have inversion symmetry, and fluorescence lifetimes are typically on the order of 10 msec.

It should be noted that one major problem associated with divalent RE ions is chemical in nature. RE ions are most stable in the trivalent state, which can make growth of crystals doped with appreciable amounts of divalent RE ions difficult. For example, Dy^{2+} ions in CaF_2 are typically obtained by subjecting a Dy^{3+}-doped crystal to ionizing radiation, such as γ-rays.

Iron Group, 3d-3d Transitions

Ions within the iron group differ in electronic structure solely by the number of electrons in the 3d shell. Elements within the iron group belong to the more general class of transition metals. The relatively large spatial extent of the 3d wavefunctions

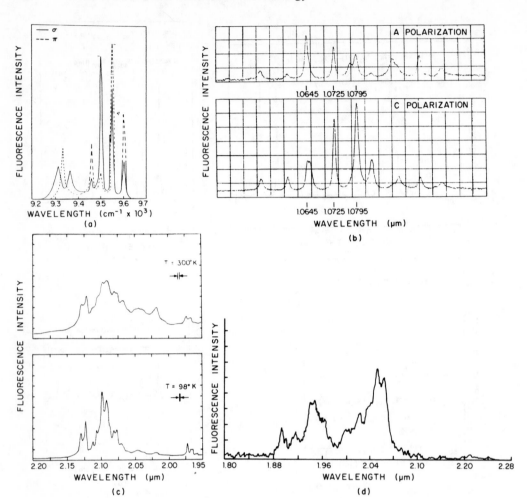

FIGURE 2.1.1.5B.   Fluorescence from (a) $^4F_{3/2} \rightarrow {}^4I_{11/2}$ transitions in Nd:LiYF$_4$,[20] and (b) Nd:YAlO$_3$[21] and fluorescence from $^5I_7 \rightarrow {}^5I_8$ transitions of Ho^{3+} in (c) YAG:[22] and (d) LiYF$_4$.[23] (From (a) Warmer, A. L., Linz, A., and Gabbe, D. R., *J. Phys. Chem. Sol.*, 30, 1483, 1969; (b) Massey, G. A., *IEEE J. Quantum Electron.*, QE-8, 669, 1972. With permission.

and lack of any shielding electrons in higher shells leads to a significant interaction between the 3d^n configuration and the crystal environment. High-energy configurations involving states other than 3d play no direct role in laser operation of iron group ions and thus will not be considered. As for the 5d states of rare earths, Russell-Saunders notation is not appropriate for the various 3d^n (or simply, 3d) energy levels of ions in a crystal. Since the substitutional site symmetry and the strength of the ion-host interaction vary among differing host crystals the position and number of 3d levels are host dependent. However, for a large class of host crystals a good first approximation is that the ion is surrounded by an octahedral-symmetry Coulombic field created by six equal charges located on the face centers of a cube. (Another common class of crystals has ion sites with tetrahedral symmetry, but includes no laser hosts at this writing.) It should be noted that more accurate calculations of energy levels involve molecular-orbital theory to describe the effect of the host environment on the ion. The various energy levels in the octahedral field approximation are denoted by

$$^{2S+1}(R)$$

where R denotes the different group-theoretical representations of the octahedral group, namely (in the notation conventionally used for 3d levels) A_1, A_2, E, T_1, and T_2. S is the total electron spin. An additional subscript is sometimes used (e.g., A_{2g}) to indicate the parity of the wavefunction associated with the level (g for even parity, u for odd parity) but for all configurations formed solely with 3d electrons the parity is even. Formally, levels with differing representations have associated wavefunctions which behave differently when subjected to the various symmetry operations of the octahedral group. This leads to varying interactions with the octahedral crystal field and thus differing energy levels. In most host crystals it is possible to treat any deviation from a strict octahedral field as a small perturbation on the calculation of configuration energies. Such a perturbation will generally split up levels degenerate in the octahedral field. The designation of levels which result from any deviation from octahedral symmetry is properly done using group-theoretical representations for the actual symmetry group appropriate at the ion site. Additional structure in the energy levels (10 to 1000 cm^{-1}) appears as a result of the spin-orbit interactions. In some cases, the coupling between ion and host causes orbitally degenerate or nearly degenerate ion states to split because of certain displacements from equilibrium of the surrounding ions. This (Jahn-Teller) effect can significantly reduce level splitting caused by both spin-orbit coupling and deviations from octahedral symmetry, and add new structure by removal of orbital degeneracies.

Transitions between 3d levels of ions in an octahedral field are electric-dipole forbidden (because the levels all have the same parity) but are magnetic-dipole allowed. For $\Delta S = 0$ (spin-allowed) transitions oscillator strengths are $\sim 10^{-6}$; for $\Delta S \neq 0$ transitions strengths are $\sim 10^{-8}$. Typical spin-allowed lifetimes are ~ 10 msec for infrared transitions. Generally, host crystal fields, if not strictly octahedral, still have inversion symmetry and thus will not allow forced-electric-dipole transitions. Two notable exceptions are the Cr^{3+}-hosts Al_2O_3 and $BeAl_2O_4$, both of which have low-symmetry Cr^{3+} sites which increase the oscillator strength for all transitions by mixing in odd-parity wavefunctions from other configurations.

The strong interaction between the 3d energy levels of the iron group and the crystal field has a profound effect on transition lineshapes. Certain transitions lead to narrow lines similar to those observed from 4f-4f transitions while others lead to broad (~ 1000 cm^{-1}) bands. If two energy levels have associated wavefunctions composed of basically the same combination of 3d electron orbitals then the interaction between these levels and the crystal field is similar. Thus, when a transition occurs between the levels there is no change in the positions of neighboring host-crystal ions. The narrow linewidths observed for such transitions are due to such effects as phonon scattering and inhomogeneous broadening, discussed previously in reference to 4f-4f transitions. If the transition takes place between two levels made up of different combinations of orbitals then the surrounding ions do shift position, and, as for 4f-5d transitions, linewidths are large because of the high probability for phonon-assisted transitions.

The energy-level diagrams and laser transitions for several of the reported iron-group lasers are shown in Figure 2.1.1.9. The first three diagrams serve to illustrate the important effect of the host crystal on the energy-level structure of the ions. It should be noted that Cr^{3+} and V^{2+} ions are isoelectronic, i.e., have the same number of 3d electrons, and thus have, for the same crystal field, essentially identical energy levels. The ^2E-level energy is nearly independent of the magnitude of the crystal field and the linewidth of the $^4A_2 \rightarrow {}^2E$ transition is small. The $^4T_2$-level energy increases almost linearly with the crystal field; $^4A_2 \rightarrow {}^4T_2$ transitions have large linewidths characteristic of vibronic processes.

In Cr^{3+}:Al_2O_3 (ruby) the crystal field is large enough to place the $^4T_2$ level higher in energy than the ^2E level. Optical excitation into the $^4T_2$ level (and higher levels) decays by multiphonon emission into the ^2E level and laser action occurs on the $^2E \rightarrow {}^4A_2$ transition. (The deviation of the Cr^{3+} site in Al_2O_3 from octahedral symmetry removes the

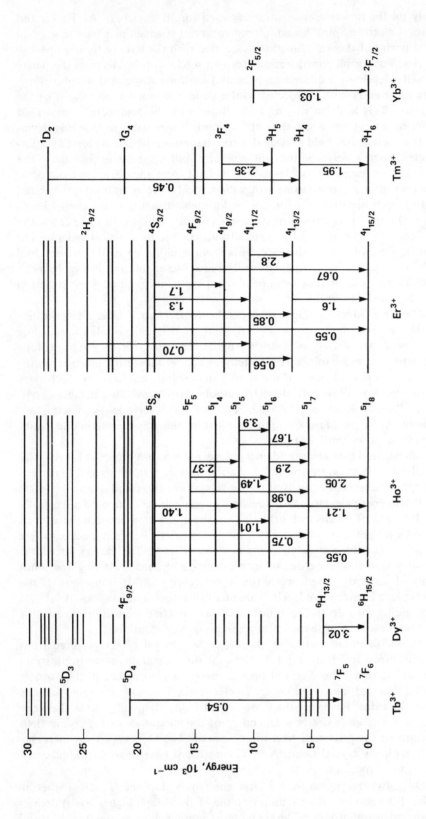

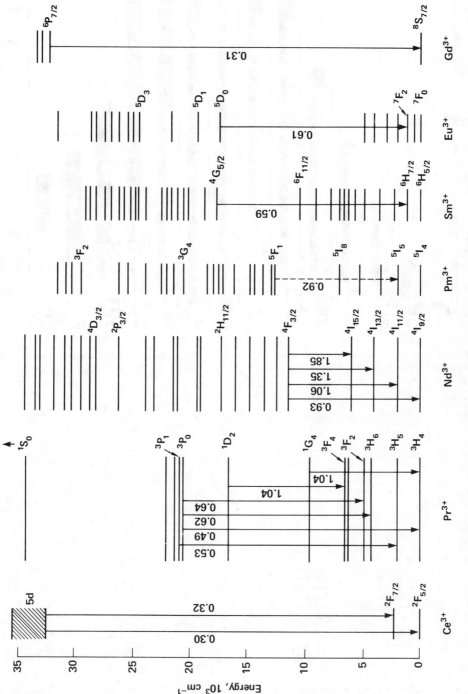

FIGURE 2.1.1.6.   Energy-level diagrams for laser transitions of trivalent rare-earth ions.[4] Approximate wavelengths of transitions are given in µm.

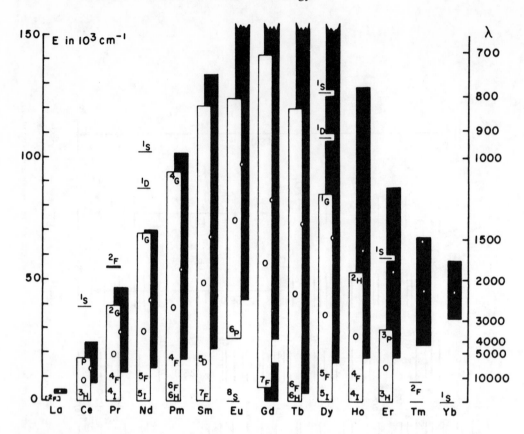

FIGURE 2.1.1.7.   Schematic diagram of the $4f^n$ (white) and $4f^{n-1}$ 5d (black) configuration of divalent rare-earth ions. (From Dieke, G. H. and Crosswhite, H. M., *Appl. Optics*, 2, 675, 1963. With permission.)

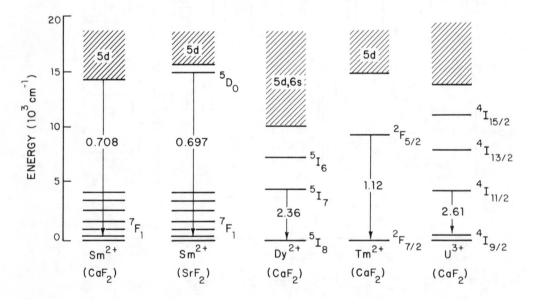

FIGURE 2.1.1.8.   Energy-level diagrams for laser transitions of divalent rare-earth and trivalent actinide ions. Approximate wavelengths of transitions are given in μm.

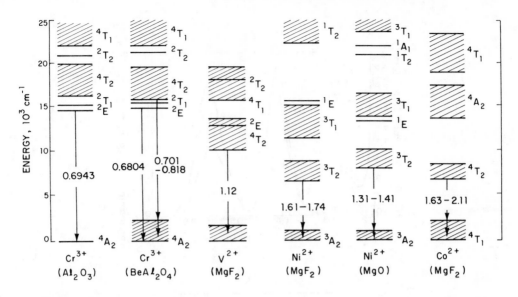

FIGURE 2.1.1.9.   Energy-level diagrams for laser transitions of iron-group ions in various hosts. Approximate wavelengths of transitions are given in μm.

two-fold orbital degeneracy of the ^2E state and produces two distinct lines from the $^2E \rightarrow ^4A_2$ transition, often referred to as R_1 and R_2.) In $Cr^{3+}:BeAl_2O_4$ (alexandrite) the crystal field is somewhat smaller than in ruby. While the $^4T_2$ level is still above the ^2E level, the two states are close enough in energy such that at 300 K and above, excitation is shared between the two levels when a quasithermal distribution is established among excited states. Laser operation can thus be obtained on either the narrow-linewidth $^2E \rightarrow ^4A_2$ transition or the vibronic $^4T_2 \rightarrow ^4A_2$ transition. Finally, in $V^{2+}:MgF_2$ the crystal field is relatively small, the $^4T_2$ level is well below the ^2E level and laser operation is obtained only on the vibronic $^4T_2 \rightarrow ^4A_2$ transition.

The two Ni^{2+}-doped crystals MgO and MgF_2, the level diagrams of which are shown in Figure 2.1.1.9, exhibit laser operation on vibronic $^3T_2 \rightarrow ^3A_2$ transitions. The stronger crystal field in MgO shifts many of the levels to higher energies compared to MgF_2. In $Co^{2+}:MgF_2$, laser operation is on the vibronic $^4T_2 \rightarrow ^4T_1$ transition.

Figure 2.1.1.10 shows absorption and fluorescence spectra from a variety of iron-group-doped crystals. For ruby and alexandrite the broad absorption bands resulting from vibronic transitions are evident in Figure 2.1.1.10a, as are the narrow line $^2E \rightarrow ^4A_2$ transitions. As mentioned above, the broad vibronic emission observed from alexandrite, Figure 2.1.1.10b, results from $^4T_2 \rightarrow ^4A_2$ transitions. For transitions which are primarily vibronic there is some probability that the transition may take place without the absorption or emission of phonons. The probability for such purely electronic or zero-phonon transitions varies depending on the specific levels involved, the host crystal and the temperature, being highest at low temperatures. Narrow zero-phonon lines are evident in the 77 K absorption-fluorescence spectra of $Ni^{2+}:MgF_2$, $Ni^{2+}:MgO$ and $Co^{2+}:MgF_2$ shown in Figure 2.1.1.10c. For these spectra it is clear that for vibronic transitions absorption and fluorescence processes are separated into differing wavelength regions, an important characteristic for laser operation. The coupling between phonons and electronic levels which gives rise to vibronic transitions also increases the probability for radiationless multiphonon decay of excited levels. Such decay tends to have a rate which increases rapidly above a characteristic temperature, determined by the particular combination of transition and host crystal. The fluorescence from $Co^{2+}:MgF_2$ for example, has a very low quantum efficiency at 300 K because of strong competition from multiphonon decay.

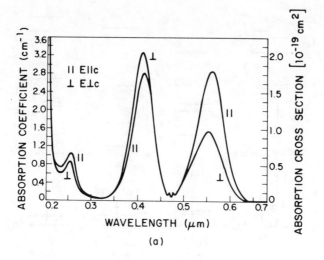

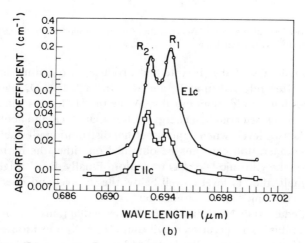

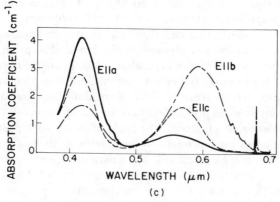

FIGURE 2.1.1.10A.   Absorption spectra for (a) Cr:Al$_2$O$_3$ (ruby) main pump bands[24] and (b) "R" lines[25] (Cr^{3+} concentration 1.58 × 10^{19} cm^{-3}). Absorption spectra for Cr:BeAl$_2$O$_4$ (alexandrite) is shown in (c).[26](From (a) Maiman, T. H., et al., *Phys. Rev.*, 123, 1151, 1961; (b) Cronemeyer, D. C., *J. Opt. Soc. Am.*, 56, 1703, 1966. With permission.)

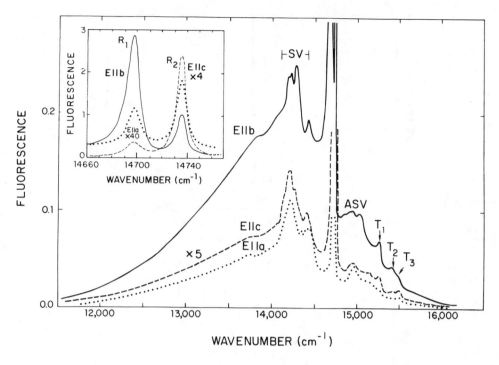

FIGURE 2.1.1.10B.   Fluorescence spectra for alexandrite.[26]

Rare Earths, 5d→4f Transitions

As mentioned in the discussion above on divalent rare earths, 4f→5d absorption, because of the strong interaction of the 5d state with the host crystal, has a broad linewidth. For most of the divalent ion-host combinations rapid nonradiative decay of the excited 5d state into lower-energy 4f states results in low quantum efficiency for broadband 5d→4f emission. One exception to this case occurs for Sm^{2+} in CaF_2 (Figure 2.1.1.8), where the lowest-energy 5d level falls below the energy of 4f levels from which fluorescence would normally be observed. Laser operation is thus obtained from a 5d→4f transition in $Sm^{2+}:CaF_2$.

5d→4f fluorescence is also readily observed from trivalent Ce-doped crystals. The Ce^{3+} ion has only one electron in the 4f shell; the next excited state above the ground-state 4f configuration (which exhibits a ~ 2000 cm^{-1} spin-orbit splitting) must be a high-energy 5d state. Thus there are no nearby 4f states to which the 5d state can decay nonradiatively and the fluorescence quantum efficiency is high. The absorption and fluorescence spectra from two Ce^{3+}-doped laser crystals are shown in Figure 2.1.1.11. The 5d state is split into distinct levels by the crystal field, producing several absorption bands, but rapid nonradiative decay to the lowest 5d level from higher levels results in efficient fluorescence from only that lowest level. As for 3d levels, the energies of 5d levels vary greatly among differing host crystals. The two peaks in fluorescence are due to the split 4f ground state. The fluorescence lifetime of the 5d→4f transition of Ce^{3+}, $\sim 10^{-8}$ sec, is the shortest of any of the various paramagnetic ion lasers.

Emission from 5d-4f transitions in many-electron trivalent RE ions has been observed in the VUV region,[10] but to date no laser action from such transitions has been obtained.

Actinides, 5f-5f Transitions

The actinide group of the periodic table is similar to the RE group, except that ions within the group differ in the number of 5f instead of 4f electrons. Much of the dis-

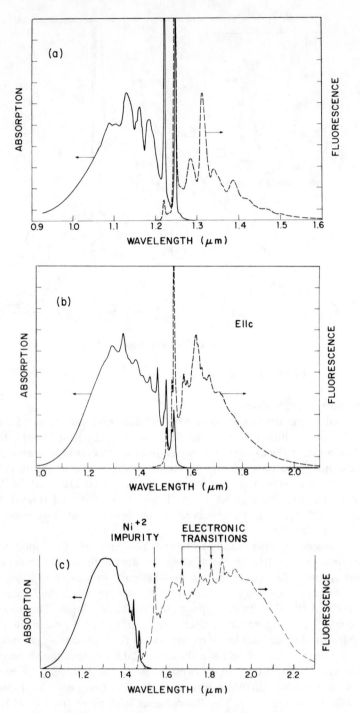

FIGURE 2.1.1.10C.   Absorption and fluorescence spectra[27] for (a) Ni:MgO, (b) Ni:MgF$_2$ and (c) Co:MgF$_2$.

cussion of RE transitions applies to actinide transitions; an energy level diagram for the various trivalent actinides is shown in Figure 2.1.1.2B. The Russell-Saunders notation for differing levels is less appropriate for the actinides because of the larger spin-orbit interaction among 5f electrons. The short nuclear lifetime of some of the actinides definitely prevents their use in lasers, and the radioactive nature of others

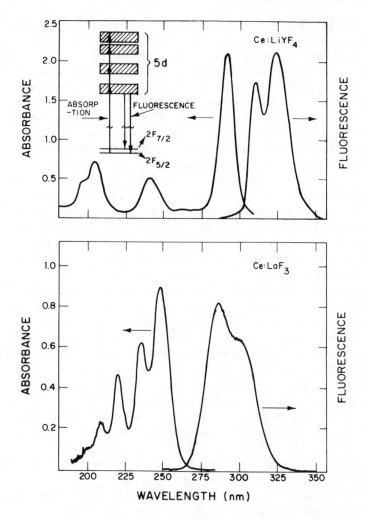

FIGURE 2.1.1.11.   Absorption and fluorescence spectra for Ce:LiYF₄[28] and Ce:LaF₃.[29] Inset shows partial energy-level diagram for Ce³⁺ ions.

with relatively long (years) half-lives presents practical problems in crystal growth and handling. The only reported actinide laser ion, U^{3+}, has the same number of f-state electrons as Nd^{3+}. The laser level diagram for U^{3+} in CaF_2 is shown in Figure 2.1.1.8.

LASER OPERATION

General

It is clear from Table 2.1.1.2 that a great many different combinations of transitions, paramagnetic ions, and host crystals have exhibited laser operation. In practice, however, only a limited variety of paramagnetic ion lasers are commonly used for either research or industrial applications. Many factors determine whether or not a given ion and host will operate effectively as a laser, and many of the lasers listed in Table 2.1.1.2 fail to satisfy all the necessary criteria for practical systems. Of major importance for any system is the need for a low pump-power or energy threshold for stimulated emission. It is desirable to have a minimal initial population of ions in the desired laser lower, or terminal state. In strict 4-level lasers this population is sufficiently small such that the effect on threshold of absorption from the lower state can be ignored. At the other extreme, the 3-level laser, the terminal level is the ground state of the ion and the threshold is almost entirely determined by the requirement (for equally degenerate states) that more ions be in the upper laser level than the ground state. Many paramagnetic-ion lasers fall in the region between the two extremes; in some cases laser operation may be 4-level in nature at low temperature, where thermal excitation of the terminal level is negligible, but tend toward 3-level operation at higher temperatures. For vibronic or phonon-assisted transitions, 4-level operation is obtained between two electronic levels when vibronic absorption is negligible in the vibronic emission region. This condition is achieved when the difference between the zero-phonon-line energy and the desired emission energy is large compared to the thermal energy.

For a given transition of an ion, thresholds will be lowest for host crystals in which the nonradiative decay rate between the upper and lower laser levels is small compared to radiative rate, and transition linewidths are narrow and homogeneously broadened, maximizing the cross section for stimulated emission. For certain combinations of ion and host crystal, excited-state absorption will occur, which effectively reduces the gain cross section for the desired laser transition. Another problem, particular to 4-level systems, occurs if the lifetime of the terminal level is long enough to allow appreciable buildup of population in that level during laser operation. In some cases the lifetime of the level may be so long that only pulsed, self-terminating operation is possible.

Optical pumping is almost exclusively used for excitation of the upper level of paramagnetic ion lasers. When incoherent, broadband emission sources such as tungsten or gas-filled lamps are used for pumping, efficient low-threshold operation is achieved if there are numerous absorbing energy states above the upper laser level which, when excited, subsequently decay by phonon emission into the upper level.

The total amount of radiation absorbed from broadband pump sources clearly increases with ion concentration in a given size host crystal. This situation is balanced by the fact that above concentration levels determined by both the specific ion and crystal, multipole or exchange interactions between ions produces new nonradiative channels for energy decay. These reduce fluorescence lifetime and thus raise the laser threshold. Such concentration quenching is minimal for certain crystals which make up an entire class of materials discussed in Section 2.1.2.

Energy-transfer interactions among ions can be used to improve the optical pumping process and make practical certain laser schemes. Such interactions occur when a host crystal is doped with two or more different ions. In the ideal case one set of ions (sensitizers) absorbs pump radiation at wavelengths where the laser-active ions (activators) have little or no absorption; the sensitizer excitation is then transferred to the activators, thereby augmenting direct excitation of the activators. If there is some energy mismatch between sensitizer and activator levels phonon emission or absorption can occur to conserve energy. Table 2.1.1.3 lists the various transfer schemes successfully employed in lasers. Many of the schemes involve transfer between 4f levels of rare earths, but transfer from 3d to 4f ($Cr^{3+} \rightarrow Nd^{3+}$), 5d to 4f ($Ce^{3+} \rightarrow Nd^{3+}$), and 3d to 3d ($Mn^{2+} \rightarrow Ni^{2+}$) levels has been demonstrated. In addition, transfer from molecular-like complexes (VO_4) and defects in the host crystal has been employed. The energy-level diagrams for several of the processes are shown in Figure 2.1.1.12. Included are successive transfer systems (Figure 2.1.1.12e) in which the sensitizer transfers energy to the activator twice to "upconvert" infrared wavelength excitation into visible-wavelength laser operation.

The major alternative to optical pumping by incoherent sources is pumping by another laser. Such an arrangement is useful because much greater excitation intensities can be realized with laser pumping, an important condition when the desired transition has a small cross section and/or the upper-state lifetime is short. Also, in some cases there may be only a few, spectrally narrow absorbing transitions which serve to populate the upper level, resulting in weak excitation by broadband pump sources. Laser

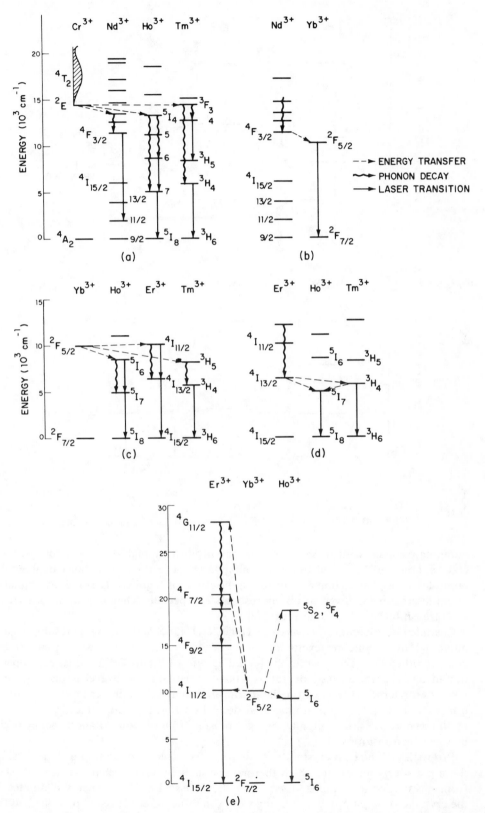

FIGURE 2.1.1.12.   Energy-level diagrams for a variety of energy-transfer processes among trivalent rare-earth ions (see Table 2.1.1.3).

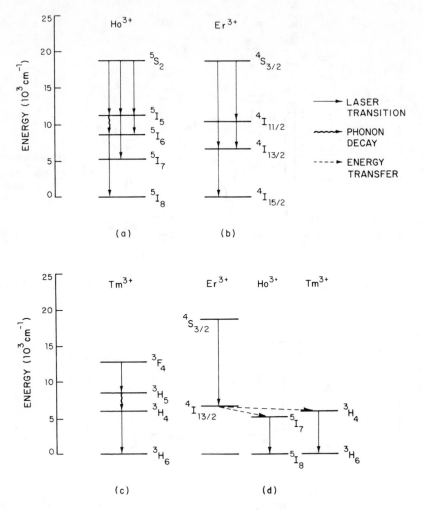

FIGURE 2.1.1.13.   Cascade-laser energy-level diagrams (see Table 2.1.1.4).

pumping resonant with these transitions can produce enough excitation for laser operation, however. For fixed-frequency pump lasers, a fortunate resonance between the pump-laser wavelength and absorption in the crystal is required. In some cases tunable pump lasers are employed to obtain resonance, especially when excitation of narrow 4f-4f transitions in RE crystals is desired.

Cascade laser operation, covered in Table 2.1.1.4, achieves the same results as laser pumping, by channeling excitation into specific levels which may not be readily pumped otherwise. As shown in the level diagrams of Figure 2.1.1.13, cascade operation involves successive laser operation from one crystal at two wavelengths. For some schemes the terminal level of one laser transition may be the upper level of the other transition. In other systems nonradiative decay between levels may precede laser action by the second transition. Finally, energy transfer from one ion to another may occur between the transitions.

Properties of the host crystal desirable for effective laser operation include, besides those previously mentioned, high thermal conductivity and mechanical strength. The former is needed to minimize thermal gradients associated with energy dissipated in the crystal during pumping and lasing cycles, and the latter is needed to withstand the stresses necessarily generated by thermal gradients. The host crystal must be capable of being grown in a size large enough for practical systems, with a low scattering loss at the laser wavelength and minimal optical distortion due to refractive-index inhomogeneities. Such properties must be maintained when the crystal is doped with the paramagnetic-ion impurities. For high-power operation a high threshold for optical damage at both the surface and in the bulk of the crystal is necessary.

Common Systems

The Nd^{3+} ion, as is evident from Table 2.1.1.2, has exhibited laser operation in the largest number of host crystals. There are several reasons for this. Low-threshold, room-temperature 4-level operation is readily obtained on the ~ 1.06 μm, $^4F_{3/2} \rightarrow ^4I_{11/2}$ transition, which invariably has the largest cross section of all the fluorescent transitions. Optical pumping of the $^4F_{3/2}$ level is efficient because of the relatively large number of closely spaced levels at higher energies, which create absorption at visible and near-infrared wavelengths. Excitation into any of these levels decays rapidly down into the $^4F_{3/2}$ state by nonradiative processes because of the relatively small energy gaps between the various levels. The energy of the $^4F_{3/2} \rightarrow ^4I_{15/2}$ transition is sufficiently large, however, so that the fluorescence from the $^4F_{3/2}$ level has a relatively high quantum efficiency. In most host crystals nonradiative relaxation of the terminal $^4I_{11/2}$ state is fast, avoiding a "bottleneck" effect.

Of the many host crystals used for Nd^{3+} lasers, one material, $Y_3Al_5O_{12}$, a garnet-structured crystal commonly referred to as YAG, has been studied most extensively and has dominated in applications. YAG has relatively high thermal conductivity (higher than other Nd^{3+} hosts), high mechanical strength, good optical quality, and can be grown in crystal sizes large enough for most applications. There is only one substitutional site for Nd^{3+} ions and thus absorption and emission lines are homogeneously broadened. The maximum useful doping level for Nd^{3+} ions in YAG, $\sim 1\%$, limited by the onset of concentration quenching, is sufficiently high for reasonably efficient absorption of radiation from common incoherent pump sources such as tungsten-filament lamps and xenon- or krypton-arc discharge lamps. Because of the nature of the phonon interaction with the Nd^{3+} transitions in YAG, room-temperature line-widths observed are among the narrowest of known host crystals, which minimizes the threshold. Other host crystals have been found which have certain properties superior to YAG, but few have the combination of favorable characteristics that YAG exhibits.

There are some RE-ion lasers other than those based on Nd^{3+} which have gone beyond the demonstration stage of development, or show promise of some utility. The $^5I_7 \rightarrow ^5I_8$ transition of Ho^{3+}, at a wavelength of approximately 2.1 μm, has been made to operate in a number of crystals. Energy transfer from Er^{3+} and Tm^{3+} ions co-doped in the crystal has been exploited to greatly enhance the efficiency of optical pumping. One major problem with the $^5I_7 \rightarrow ^5I_8$ transition is the low $^5I_8$-level energy above ground, ranging in different hosts from 200 to 600 cm^{-1}. Strict 4-level laser operation cannot be obtained at room temperature and pump thresholds for even the most efficiently multiply-doped crystals are considerably higher than those for the Nd:YAG, 1.06 μm laser. Much improved operation is attainable for multiply-doped laser crystals cooled to 77 K, where the highest cw efficiency for any lamp-pumped laser has been observed (see below). The two most widely used Ho^{3+} hosts are YAG and $LiYF_4$ (YLF). CW operation of Ho^{3+} lasers has not yet been observed at room temperature. The $^5I_6 \rightarrow ^5I_7$ laser transition of Ho^{3+} at approximately 2.9 μm has been demonstrated in a variety of hosts at 300 K but is limited to pulsed operation because of the long $^5I_7$-state lifetime.

Lasers based on the Er^{3+} ion have operated on a variety of transitions. Of most interest are the $^4S_{3/2} \rightarrow ^4I_{13/2}$, $^4S_{3/2} \rightarrow ^4I_{9/2}$ and $^4I_{11/2} \rightarrow ^4I_{13/2}$ transitions at approximate wavelengths of 0.85, 1.7, and 2.9 μm, respectively. All are 4-level systems and operate at room temperature. Like the 2.9-μm Ho^{3+} laser, all have limitations imposed by a long terminal-state lifetime.

The Cr^{3+} iron-group ion doped in Al_2O_3 is the medium in which laser operation was first demonstrated by Maiman[11] in 1960. $Cr:Al_2O_3$, or ruby, operates as a three-level system and thus, per unit volume, has a comparatively high threshold. Fortunately the thermal conductivity and mechanical strength of Al_2O_3 are both high, superior to any other existing laser host crystal, and thus successful operation of the ruby laser is possible. For all but a few specialized applications the much-lower-threshold, higher average-power-output Nd:YAG laser has replaced the ruby laser, however. Efficient frequency-doubling techniques for 1.06 μm radiation have in many cases eliminated the need for the 0.69-μm ruby laser where visible radiation is required.

The $Cr:BeAl_2O_4$ (alexandrite) laser, which is under development at this writing, has the potential for becoming an important device. Because of the level structure of Cr^{3+} in $BeAl_2O_4$, discussed in the preceding section on types of transitions, 4-level operation on the vibronic $^4T_2 \rightarrow {}^4A_2$ transition is possible. Thus, not only are thresholds considerably lower than those for ruby, but broadly tunable oscillation centered around 0.75 μm can be achieved. The vibronic absorption bands, as for ruby, allow for efficient coupling to incoherent pump sources, and the properties of the $BeAl_2O_4$ host crystal compare favorably with those of Al_2O_3.

Lasers using the divalent iron-group ions Ni^{2+}, Co^{2+}, and V^{2+} operate on vibronic transitions and, like alexandrite, can be tuned over a broad range. In particular, cw tunable operation in the 1.6 to 2.0 μm region results from cryogenically cooled $Ni:MgF_2$ and $Co:MgF_2$ crystals pumped by a Nd:YAG laser. Such lasers may become useful sources for applications where tunable infrared emission is required.

Operating Characteristics

The discussion of operating characteristics is divided into sections determined by the different laser operating modes.

Pulse-Pumped Operation

The majority of paramagnetic-ion lasers have been limited to pulse-pumped operation, primarily because of high thresholds. The pump source is almost always a xenon- (or in some cases a krypton-) gas discharge lamp, driven by an LC discharge network. Pulsed pumping without Q-switching is often referred to as normal-mode or long-pulse operation. The initial portion of the output pulse of such lasers typically exhibits relaxation oscillations, or spiking, as the amounts of energy stored in the laser medium and in the optical cavity oscillate back and forth in an attempt to reach a stable operating point. In the absence of any perturbations the approximate time scale in which the relaxations damp out is the lifetime of the upper laser level. Many lasers with particularly long lifetimes (ruby is a good example) exhibit spiking over the entire pulse because of inevitable perturbations, such as those caused by interactions between different laser cavity modes, which continually excite new oscillations.

Figure 2.1.1.14 shows the laser-pulse output energy vs. electrical-energy input to the flashlamp for a variety of different systems. Simple laser theory predicts that above threshold the output energy should be linear in flashlamp energy, i.e., the slope efficiency of a system, defined as the derivative of output energy with respect to flashlamp energy, should be constant. In actual systems optical excitation of the laser crystal may not be uniform, with typically the center of the crystal receiving higher excitation than the outside. For multitransverse-mode operation, which is common for lamp-pumped systems because the excited crystal volume is generally many times that of the TEM_{oo} cavity mode, different transverse modes thus have different thresholds. The distribution of thresholds leads to a nonlinear output vs. input curve, especially near the

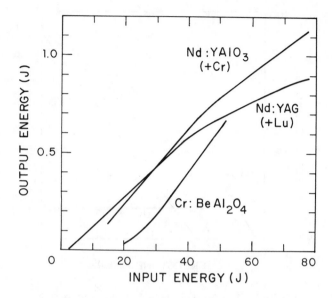

FIGURE 2.1.1.14A.   Long-pulse laser performance from 5 × 50 mm rods of Nd:YAlO₃ and Nd:YAG[30] at ~1.06 μm and from a 6.3 × 75 mm rod of Cr4:BeAl₂O₄ (alexandrite) at 0.75 μm.[31]

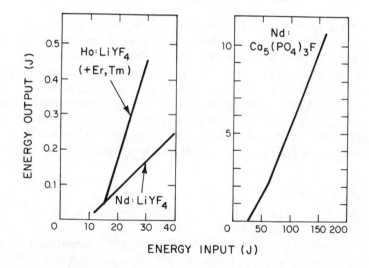

FIGURE 2.1.1.14B.   Long-pulse laser performance of a 3 × 20 mm rod of Ho:LiYF₄ (YLF) at 2.06 μm,[32] a 6.3 × 75 mm rod of Nd:YLF at 1.05 μm[33] and a 6.3 × 250 mm rod of Nd:Ca₅(PO₄)₃F (FAP) at 1.06 μm.[34]

threshold region. At high pumping levels the slope efficiency may be reduced because of crystal heating effects. Also, as the lamp input energy increases the peak in the spectral distribution of lamp emission shifts toward shorter wavelengths, which can either increase or decrease efficiency depending on the overlap between the absorption spectrum of the crystal and the lamp spectrum. All of the above effects contribute to the variations in slope efficiency with lamp energy apparent in Figure 2.1.1.14.

In Figure 2.1.1.14A it is interesting to note that, for the particular systems compared, although Nd:YAG has by far lowest threshold, the higher slope efficiencies exhibited by Nd:YAlO₃ (sensitized by Cr³⁺ ions) and by alexandrite lead to essentially the same output energies at the 50 J input level. Figure 2.1.1.14B shows the perform-

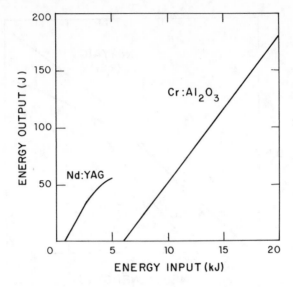

FIGURE 2.1.1.14C.   Long-pulse laser performance of a 10 × 100 mm rod of Nd:YAG at 1.06 μm[35] and a 20 × 100 mm rod of Cr:Al$_2$O$_3$ (ruby) at 0.69 μm. [36](From Koechner, W., *Solid State Laser Engineering*, Springer-Verlag, New York, 1976. With permission.)

ance of Nd^{3+} and Ho^{3+} ions in the host crystal LiYF$_4$, with the evident beneficial effect of Er^{3+} and Tm^{3+} sensitization for the Ho^{3+} laser. Also apparent in the figure is the high slope and overall efficiency of the Nd:Ca$_5$(PO$_4$)$_3$F or Nd:FAP laser, which surpasses all other published lamp-pumped crystalline lasers. The relatively poor thermal properties of FAP limit the crystal to low-average-power applications, however. Finally, in Figure 2.1.1.14C the performance of very large-crystal-size Nd:YAG and ruby lasers is shown. With present technology much larger ruby than Nd:YAG crystals can be grown; the ruby laser can ultimately generate greater output energies than Nd:YAG simply because larger laser rods capable of handling higher pump energies can be fabricated.

CW Operation

As is evident from Table 2.1.1.5, a considerably smaller number of paramagnetic-ion lasers have been operated in the cw mode. Many of the systems listed in Table 2.1.1.2 are not capable of cw operation for fundamental reasons, such as the existence of a long-lifetime terminal state. Others have thresholds which exceed the capabilities of cw excitation sources or the stress limits of the host crystal. In many cases the lack of a laser crystal of sufficient size or optical quality prevents cw operation.

Figure 2.1.1.15 shows the power output vs. power input curve for a variety of cw lasers. The discussions in the previous section on the causes of nonlinear input-output relationships also apply to this data. Figure 2.1.1.15A presents the performance of several efficient krypton-arc-lamp-pumped Nd:YAG lasers, all operating in many transverse modes. The output of ~300 W obtained from a 6.3 mm-dia. × 100 mm-long laser rod is close to the limit set by the mechanical strength of the crystal. The 1100-W output power indicated in Figure 2.1.1.15A was from eight laser rods, each pumped by a lamp, and placed in series in a common laser cavity; this level of output is the highest yet reported for any cw paramagnetic ion laser. The most efficient cw system performance reported is shown in Figure 2.1.1.15B, for a liquid-nitrogen-cooled Ho:YAG laser, sensitized by Er^{3+} and Tm^{3+} ions. A slope efficiency of 6.5% and an overall efficiency of 5% was realized with optical pumping by a tungsten-fila-

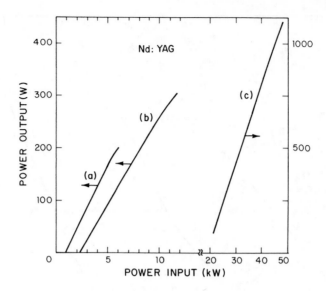

FIGURE 2.1.1.15A. CW performance of Nd:YAG lasers at 1.06 μm, pumped by krypton arc lamps. (a) 6.3 × 100 mm rod in a single-elliptical pump cavity, (b) same as (a) except a double-elliptical cavity,[37] (c) output from eight 6 × 75 mm rods in series, each pumped by a single lamp.[38]

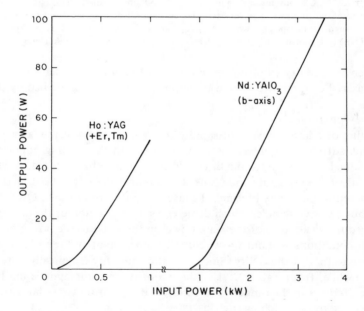

FIGURE 2.1.1.15B. CW output at 2.1 μm from a 4 × 70 mm rod of Er, Tm, Ho:YAG, cooled by liquid nitrogen and pumped by a tungsten-filament lamp,[39] and cw output at 1.08 μm from a 6.3 × 75 mm rod of Nd:YAlO₃.[40]

ment lamp. Data on a krypton-lamp-pumped Nd:YAlO₃ laser also appears in Figure 2.1.1.15B, and is of interest primarily because, unlike the Nd:YAG laser, the output is linearly polarized because of the anisotropic nature of the YAlO₃ crystal. The behavior of some lower-output cw systems is indicated by Figure 2.1.1.15C. The liquid-nitrogen-cooled Ho:Er:Tm:LiYF₄ laser was pumped by a tungsten-filament lamp, the alex-

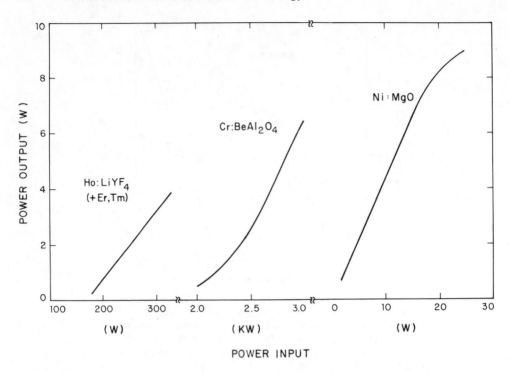

FIGURE 2.1.1.15C.   CW outputs at 2.06 μm from a 3 × 50 mm rod of Er, Tm, Ho:LiYF (YLF), cooled by liquid nitrogen and pumped by a tungsten-filament lamp,[41] at 0.75 μm from a 3 × 64 mm rod of Cr:BeAl₂O₄ (alexandrite) pumped by two mercury capillary arc lamps,[42] and at 1.32 μm from a 4 × 5 × 18.5 mm crystal of Ni:MgO, cooled by liquid nitrogen and pumped by a 1.06-μm Nd:YAG laser.[43]

andrite laser by mercury capillary arc lamps and the Ni:MgO laser by a cw, 1.06-μm Nd:YAG laser. For the latter system the input power is that absorbed by the crystal.

Q-Switched Operation

Q-switching of a laser consists of rapidly changing the loss of the laser cavity from a high value, sufficient to prevent laser operation with the existing population inversion, to a value low enough for the threshold condition to be exceeded. A straightforward analysis of the laser rate equations shows that if the upper laser level lifetime is large compared to the cavity lifetime, the laser output in response to Q-switching consists of a short, energetic pulse which depletes a major fraction of the energy stored in the laser medium. The ratio of the peak power in the pulse to the power output under equilibrium conditions without Q-switching is approximately the ratio of the upper-level lifetime to that of the cavity. Since the latter quantity is typically in the range 10 to 100 nsec while, from Table 2.1.1.2, the majority of paramagnetic ion lasers have upper-state lifetimes in the range 0.1 to 10 msec, it is clear that very large peak powers can be generated by Q-switching such lasers.

Q-switching can be accomplished under either pulsed or cw pumping conditions. For pulsed excitation the length of the pump pulse is usually tailored to maximize stored energy in the medium. The occurrence of Q-switching is set at the point of maximum inversion. Typically a single output pulse is obtained under these conditions; the pulse energy can almost equal that obtained without Q-switching. The ultimate limit on stored energy is determined by mechanical properties of the host crystal. For 4-level lasers with large gain cross sections other effects limit stored energy well below this point, however. As the inversion increases the unsaturated gain of the laser medium becomes so high that the attainable high-loss state of the cavity is incapable of

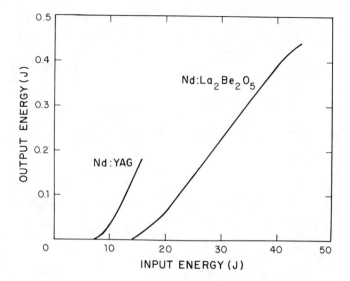

FIGURE 2.1.1.16A.   Q-switched performance of a 6.3 × 75 mm rod of Nd:YAG, operating at 1.06 μm in an unstable laser cavity,[44] and of the same size rod of Nd:La$_2$Be$_2$O$_5$ (BEL) operating at 1.07 μm.[45]

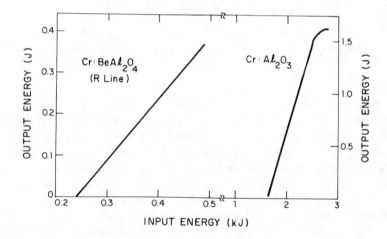

FIGURE 2.1.1.16B.   Q-switched performance of a 6.3 × 75 mm rod of Cr:BeAl$_2$O$_4$ (alexandrite) operating at 0.75 μm[46] and of a 10 × 100 mm rod of Cr:Al$_2$O$_3$ (ruby) operating at 0.69 μm.[47]

preventing oscillation. Even if infinite cavity loss could be obtained, at high levels of inversion spontaneous emission from the upper level is amplified in traveling through the medium and grows to an intensity which depletes the inversion. The effect of amplified spontaneous emission appears as an apparent saturation in the stored energy at high pumping levels. Finally, optical damage to either the laser crystal or other components in the laser cavity may determine the maximum available energy from a system.

Table 2.1.1.6 lists reported performance for a variety of Q-switched systems. Figure 2.1.1.16 shows the output energy vs. input energy curves for several Q-switched flash-pumped lasers. Pulsewidths are in the 10 to 20 nsec range. Figure 2.1.1.16A includes data on an unstable-resonator Nd:YAG laser. Such a resonator extracts almost all the available output energy in a single transverse mode. The high gain of the 1.065-μm

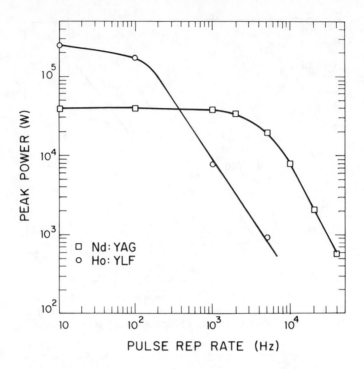

FIGURE 2.1.1.17.   Peak power vs. repetition rate for cw pumped, Q-switched Nd:YAG[48] and Ho:LiYF$_4$ (YLF) lasers.[49] The cw output of the Nd:YAG laser was 8.5 W in a TEM$_{00}$ mode while that of the Ho:YLF laser was ~4 W.

Nd:YAG laser transition limits extractable energy because of effects discussed above, and the maximum output shown of ~0.2 J is typical of Q-switched systems. Other Nd^{3+} hosts have smaller gain cross sections and thus can generate higher Q-switched energies; the capabilities of one such crystal, La$_2$Be$_2$O$_5$(BEL), are also shown in Figure 2.1.1.16A. The problems in obtaining high energy output are somewhat alleviated for 3-level lasers since there is no gain in the laser medium until about half of the active ions are excited; optical damage tends to be the limiting factor. Good energy storage is accompanied by high thresholds compared to 4-level systems, however. In addition only about half of the stored energy can be extracted, because of the self-terminating nature of 3-level operation. The performance of two Q-switched 3-level lasers is shown in Figure 2.1.1.16B.

Pulse-pumped amplifiers have been used to overcome the limitations on Q-switched output energy from Nd:YAG and ruby oscillators. 2.7 J of energy at a 10 Hz pulse rate has been obtained from a Nd:YAG system composed of an oscillator and four stages of amplification.[12] A ruby laser with an oscillator and two amplifiers has produced 14 J at a 0.33 Hz rate.[13]

CW-pumped lasers can also be Q-switched and can operate at much higher pulse rates than possible with pulsed pumping. Peak powers and output energies are considerably lower, however, because of the lower excitation intensities obtained with cw pumps. The peak power output from a cw-excited laser is essentially constant up to a rate which is approximately the inverse of the upper-state lifetime. Above that rate the power decreases because the depleted inversion does not have time to increase back to the cw limit before the next Q-switching event. The peak power vs. repetition-rate behavior of two cw-pumped lasers is shown in Figure 2.1.1.17. The difference in performance is the result of the much longer lifetime of the Ho^{3+} upper laser level compared to the Nd^{3+} level.

Mode-Locked Operation

Typically, if a laser operates on several longitudinal or transverse cavity modes at once the phase relationship between different modes is random. Mode-locking of a laser consists of providing a mechanism to establish a definite phase relationship among modes. The discussion here is limited to locking of longitudinal modes. In such a process the temporal output of the laser becomes modulated by the coherent inter-action among the modes, at the mode separation frequency, ν, which is the inverse of the cavity round-trip time. When a large number, n, of modes is locked the output is in the form of a train of short pulses, each of a width of roughly the round-trip time divided by n. The gain bandwidth of the laser limits n and the shortest pulses are clearly created in large-bandwidth systems. The pulse width can be increased by insert-ing etalons in the laser cavity to reduce the bandwidth. It is interesting to note that the mode-locking process may not simply establish a phase coherence among the modes that oscillate in the free-running laser. Rather, because of the coherent interaction between the modes and the locking mechanism, a much greater number of longitudinal modes may be present in the mode-locked laser. Mode-locking of paramagnetic-ion lasers has been observed in pulse-pumped Q-switched systems and in cw-pumped sys-tems operating both cw and repetitively Q-switched.[14] Table 2.1.1.7 presents data on a number of systems. Pulse-pumped lasers have employed passive mode-locking tech-niques, usually saturable absorbers such as dyes in solutions, which force the laser modes to phase-lock in order to generate the short pulses needed to overcome the absorber loss. Active mode-locking, the insertion of an intracavity element which mod-uates either the amplitude or the phase of the laser cavity field at frequency ν, has been used for both pulsed and cw pumping. At present, from Table 2.1.1.7, the pulse-widths achieved by common paramagnetic ion lasers are shorter than those observed from gas lasers but longer than the 1 to 10 psec pulses from mode-locked dye, color-center or semiconductor lasers. The broad gain-bandwidth of vibronic transitions could make possible much shorter pulse generation in the future, however.

Tunable Operation

Broadly tunable operation is in principle possible from lasers operating on vibronic 3d-3d and 5d-4f transitions. Table 2.1.1.8 indicates results to date of tuning experi-ments with such systems and Figure 2.1.2.18 presents tuning curves for a number of different curves. The simplest form of tuning consists of varying the crystal tempera-ture, which, primarily because of the change in absorption from the lower level, has the effect of shifting the wavelength of maximum gain and thus laser operation.[15] Because of local maxima present in the lineshape of some vibronic transitions, temper-ature tuning may not always provide continuous tunability over the wavelength range covered. A generally more useful and controllable tuning technique consists of varying the wavelength response of the laser cavity with intracavity prisms, birefringent filters, or gratings.

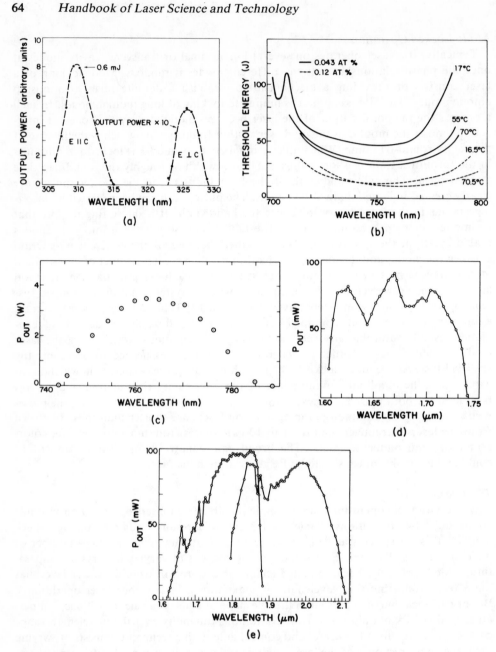

FIGURE 2.1.1.18.   Tuning curves for (a) Ce:LiYF$_4$ (YLF) laser,[28] (b) pulse-pumped Cr:BeAl$_2$O$_4$ (alexandrite) laser,[50] (c) cw pumped alexandrite laser,[42] (d) Ni:MgF$_2$ laser,[51] and (e) Co:MgF$_2$ laser.[51]

REFERENCES TO TEXT AND FIGURES

1. **DiBartolo, B.,** *Optical Interactions in Solids,* John Wiley & Sons, New York, 1968.
2. **DiBartolo, B.,** *Luminescence of Inorganic Solids,* Plenum Press, New York, 1978.
3. **Weber, M. J.,** Rare-earth lasers, in *Handbook on the Physics and Chemistry of Rare Earths,* Gschneidner, K. A., Jr. and Eyring, L., Eds., North Holland, Amsterdam, 1979, 275.
4. **Weber, M. J.,** Lanthanide and actinide lasers, in *The Chemistry of the Lanthanides and Actinides,* Edelstein, N., Ed., American Chemical Society, Washington, D.C., 1980, with additions.
5. **Nassau, K.,** The chemistry of laser crystals, in *Applied Solid State Science,* Vol. 2, Wolfe, R., Ed., Academic Press, New York, 1974, 174.
6. **Danielmeyer, H. G.,** Progress in Nd:YAG lasers, in *Lasers,* Vol. 4, Levine, A. K. and DeMaria, A. J., Eds., Marcel Dekker, New York, 1976, 1.
7. **Koechner, W.,** *Solid-State Laser Engineering,* 1976.
8. **Kaminskii, A. A.,** *Laser Crystals,* Springer-Verlag, New York, 1980.
9. **Dieke, G. H. and Crosswhite, H. M.,** The spectra of the doubly ionized and triply ionized rare earths, *Appl. Optics,* 2, 675, 1963.
10. **Yang, K. H. DeLuca, J. A.,** VUV fluorescence of Nd^{3+}-, Er^{3+}-, and Tm^{3+}-doped trifluorides and tunable coherent sources from 1650 to 2600 Å, *Appl. Phys. Lett.,* 29, 499, 1976.
11. **Maiman, T. H.,** Stimulated optical and radiation in ruby, *Nature (London),* 187, 493, 1960.
12. **Kogan, R. M. and Crow, T. G.,** A high brightness, one joule, frequency-doubled Nd:YAG laser, in *Digest of Technical Papers CLEA 1977 IEEE/OSA,* Washington, D.C., 1977.
13. **Koechner, W.** *Solid State Laser Engineering,* Springer-Verlag, New York, 1976, 563.
14. **Siegman, A. E. and Kuizenga, D. J.,** Active mode-coupling phenomena in pulsed and continuous lasers, *Opto-Electronics,* 6, 43, 1974.
15. **McCumber, D. E.,** Theory of phonon-terminated optical masers, *Phys. Rev. A,* 134, A299, 1964.
16. **Carnall, W. T.,** Argonne National Laboratory, private communication.
17. **Weber, M. J.,** Lawrence Livermore Laboratory, private communication.
18. **Jenssen, H. P.,** Phonon Assisted Laser Transitions and Energy Transfer in Rare Earth Laser Crystals, Crystal Physics Laboratory Tech. Rep. No. 16, Center for Materials Science and Engineering, MIT, Cambridge, Mass., 1971.
19. **Danielmeyer, H. G., Blatte, M., and Balmer, P.,** Fluorescence quenching in Nd:YAG, *Appl. Phys.,* 1, 269, 1973.
20. **Harmer, A. L., Linz, A., and Gabbe, D. R.,** Fluorescence of Nd^{3+} in lithium yttrium fluoride, *J. Phys. Chem. Sol.,* 30, 1483, 1969.
21. **Massey, G. A.,** Measurements of device parameters for $Nd:YAlO_3$ lasers, *IEEE J. Quantum Electron.,* QE-8, 669, 1972.
22. **Karpick, J. T. and DiBartolo, B.,** Thermal Dependence of Energy Transfer Processes Between Rare-Earth Ions in Crystals, Tech. Rep. AFAL-TR-71-114, Air Force Avionics Laboratory, 1971.
23. **Chicklis, E. P.,** unpublished data.
24. **Maiman, T. H., et al.,** Stimulated optical emission in fluorescent solids. II. Spectroscopy and stimulated emission in ruby, *Phys. Rev.,* 123, 1151, 1961.
25. **Cronemeyer, D. C.,** Optical absorption characteristics of pink ruby, *J. Opt. Soc. Am.,* 56, 1703, 1966.
26. **Walling, J. C., Jenssen, H. P., Morris, R. C., O'Dell, E. W., and Peterson, O. G.,** Tunable laser performance in $Be\,Al_2O_4$; Cr^{3+}, *Opt. Lett.,* 4, 182, 1979.
27. **Moulton, P. F.,** MIT Lincoln Laboratory, unpublished.
28. **Ehrlich, D. J., Moulton, P. F., and Osgood, R. M., Jr.,** Ultraviolet solid-state Ce-YLF laser at 325 nm, *Opt. Lett.,* 4, 185, 1978, and unpublished data.
29. **Ehrlich, D. J., Moulton, P. F., and Osgood, P. M., Jr.,** Optically pumped $Ce:LaF_3$ laser at 286 nm, *Opt. Lett.,* 5, 339, 1980.
30. **Bass, M. and Weber, M. J.,** Nd, $Cr:YAlO_3$ laser tailored for high energy Q-switched operation, *Appl. Phys. Lett.,* 17, 395, 1970.
31. **Walling, J. C. and Sam, C. L.,** unpublished data, 1980.
32. **Chicklis, E. P., et al.,** Development of Multiply Sensitized Ho:YLF as a Laser Material, Tech. Rep. ECOM-73-0066F, U. S. Army Electronics Command, 1974.
33. **Le Goff, D., BeHinger, A., and Labadens, A.,** Etude d'un oscillateur a blocage de modes utilisant uncristal de $L:YF_4$ dope au neodyme, *Opt. Commun.,* 26, 108, 1978.
34. **Ohlmann, R. C., Steinbrugge, K. B., and Mazelsky, R.,** Spectroscopic and laser characteristics or neodymium-doped calcium fluorophosphate, *Appl. Optics,* 7, 905, 1968.
35. **Koechner, W.,** *Solid-State Laser Engineering,* Springer-Verlag, New York, 1976, 119.
36. **Koechner, W.,** *Solid-State Laser Engineering,* Springer-Verlag, New York, 1976, 106.
37. **Foster, J. D. and Osterink, L. M.,** Thermal effects in a Nd:YAG laser, *J. Appl. Phys.,* 41, 3656, 1970.

38. **Koechner, W.**, *Solid State Laser Engineering,* Springer-Verlag, New York, 1976, 548.
39. **Beck, R. and Gurs, K.**, Ho laser with 50-W output and 6.5% slope effeciency, *J. Appl. Phys.,* 46, 5224, 1975.
40. **Weber, M. J. and Bass, M.**, "Research and Development of Yttrium Aluminate Lasers," Technical Report AFML-TR-32, Air Force Materials Laboratory, Wright-Patterson Air Force Base, Dayton, O., 1972.
41. Model 517 CW Laser Performance, Defensive Systems Division, Sanders Associates, Nashua, N.H.
42. **Walling, J. C., Peterson, O. G., and Morris, R. C.**, Tunable CW alexandrite laser, *IEEE J. Quantum Electron.,* QE-16, 120, 1980.
43. **Moulton, P. F., Mooradian, A., Chen, Y., and Abraham, M. M.**, unpublished.
44. **Andreou, D.**, Construction and performance of a 20-pps unstable Nd:YAG oscillator, *Rev. Sci. Instrum.,* 49, 586, 1978.
45. **Jenssen, H. P., Begley, R. F., Webb, R., and Morris, R. C.**, Spectroscopic properties and laser performance of Nd^{+3} in lanthanum beryllate, *J. Appl. Phys.,* 47, 1496, 1976.
46. **Walling, J. C. and Peterson, O. G.**, High gain laser performance in alexandrite, *IEEE J. Quantum Electron.,* QE-16, 119, 1980.
47. **Koechner, W.**, *Solid-State Laser Engineering,* Springer-Verlag, New York, 1976, 424.
48. **Koechner, W.**, *Solid-State Laser Engineering,* Springer-Verlag, New York, 1976, 408.
49. Model 517 Q-switched Laser Performance, Defensive Systems Division, Sanders Associates, Nashua, N. H.
50. **Walling, J. C.**, Allied Chemical, private communication.
51. **Moulton, P. F. and Mooradian A.**, Broadly tunable CW operation of $Ni:MgF_2$ and $Co:MgF_2$ lasers, *Appl. Phys, Lett.,* 35, 838, 1979, and unpublished data.

Table 2.1.1.1
HOST CRYSTALS

A listing of the physical properties of some of the many demonstrated laser host crystals is given in this table. Data has been obtained from a variety of sources including "Crystal Structures" by R. W. G. Wyckoff (Interscience Publishers, New York, 1963), "Handbook of Chemistry and Physics", edited by R. C. Weast (CRC Press, Inc., Boca Raton), commercial data sheets, and papers in which laser operation with the crystals is discussed. The latter references are cited in Table 2.1.1.2.

Ions which are partially replaced in the host crystal by dopant ions are listed under substitutional site symmetry. In some crystals there are several possible substitutional sites and each is indicated.

The methods of growth listed are those that have yielded crystals of laser size and quality and include Verneuil (flame fusion), Czochralski (vertical pulling), Bridgman-Stockbarger (Crucible), and zone melting (floating zone). Acronyms for materials are included in parentheses.

Table 2.1.1.1
INSULATING CRYSTAL LASER HOSTS

Material	Symmetry	Space group	Structure	Substitutional site symmetry	Lattice constant (Å)	Density (g/cm³)	Melting point (°C)	Hardness (Knoop scale)	Thermal conductivity (W/cm-C)	Thermal expansion (10⁻⁶/°C)	Transparency range (µm)	Refractive index	Method of growth
Oxides													
Al_2O_3	rhombo-hedral	D_{3d}^5 $R3c$	corundum	$C_{3v}(Al^{3+})$	$a_o = 4.759$ $c_o = 12.989$	3.987	2040	2100	$0.35 \|$c-axis $0.33 \perp$c-axis	$5.31 \|$c-axis $4.78 \perp$c-axis	0.14–6.5	$1.763(n_o)$ $1.755(n_e)$	Czochralski Verneuil
Er_2O_3	cubic	T_h^7 $Ia3$	bixbyite	$C_2(Er^{3+})$* $C_{3i}(Er^{3+})$	10.547	8.64							Verneuil
Gd_2O_3	mono-clinic	C_{2h}^3 $C2/m$		$C_5(Gd^{3+})$	$a_o = 14.061$ $b_o = 3.566$ $c_o = 8.760$	7.407	2350			9.05		2.1	Verneuil
La_2O_3	hexag-onal	D_{3d}^3 $C\bar{3}m$		$C_{3v}(La^{3+})$	$a_o = 3.937$ $c_o = 6.130$		2307						Verneuil

*There are three sites of C_2 symmetry and one of C_{3i} symmetry in the unit cell. Because the C_{3i} sites have inversion symmetry, electric-dipole transitions are forbidden.

Table 2.1.1.1 (continued)
INSULATING CRYSTAL LASER HOSTS

Material	Symmetry	Space group	Structure	Substitutional site symmetry	Lattice constant (Å)	Density (g/cm³)	Melting point (°C)	Hardness (Knoop scale)	Thermal conductivity (W/cm-°C)	Thermal expansion (10⁻⁶/°C)	Transparency range (μm)	Refractive index	Method of growth
MgO	cubic	O_h^5 $Fm3m$	rocksalt	$O_h(Mg^{2+})$	4.212	3.58	2800	692	0.6	13.8	0.25–9.5	1.75	arc melting
Y_2O_3	cubic	T_h^7 $Ia3$	bixbyite	$C_2(Y^{3+})$ $C_{3i}(Y^{3+})$	10.604	5.06	2410		0.27	7	0.23–8	1.91	Verneuil
$BeAl_2O_4$	ortho-rhombic	D_{2h}^{16} $Pnma$	olivine	$C_s(Al^{3+})$† $C_i(Al^{3+})$	$a = 9.404$ $b = 5.476$ $c = 4.427$	3.69	1870	~2000	0.23	4.4(∥a) 6.8(∥b) 6.9(∥c)	0.14–	1.746(n_a) 1.748(n_b) 1.756(n_c)	Czochralski
$La_2Be_2O_5$ (BEL)	mono-clinic	C_{2h}^6 $C2/C$		$C_1(La^{3+})$	$a = 7.536$ $b = 7.347$ $c = 7.439$	6.061	1361	~890	0.047	7.0(∥a) 7.9(∥b) 9.5(∥c)		1.9641(n_a) 1.9974(n_b) 2.0348(n_c)	Czochralski
La_2O_2S	hexag-onal	D_{3d}^3 $P\bar{3}m$		$C_{3v}(La^{3+})$	$a_o = 4.051$ $c_o = 6.943$	5.73	2070	780	0.05	6∥c-axis 3⊥c-axis	0.35–8	2.2	Bridgman-Stockbarger
$Gd_3Ga_5O_{12}$	cubic	O_h^{10} $Ia3d$	garnet	$D_2(Gd^{3+})$ $C_{3i}(Ga^{3+})$	12.376		~1825		0.09				Czochralski
$Y_3Al_5O_{12}$ (YAG)	cubic	O_h^{10} $Ia3d$	garnet	$D_2(Y^{3+})$ $C_{3i}(Al^{3+})$**	12.01	4.55	1970	1380	0.13	6.9	0.3–5.5	1.823	Czochralski
$Y_3Ga_5O_{12}$	cubic	O_h^{10} $Ia3d$	garnet	$D_2(Y^{3+})$ $C_{3i}(Gd^{3+})$	12.277				0.09			1.93	Czochralski
$YAlO_3$ (YAP, YALO)	ortho-rhombic	D_{2h}^{16} $Pbnm$	distorted perovskite	$C_{1h}(Y^{3+})$ $C_i(Al^{3+})$	$a_o = 5.179$ $b_o = 5.329$ $c_o = 7.370$	5.35	1875	1325	0.11	9.5∥a-axis 4.3∥b-axis 10.8∥c-axis	0.3–5.8	1.97(n_a) 1.96(n_b) 1.94(n_c)	Czochralski

Fluorides

BaF$_2$	cubic	O_h^5 $Fm3m$	fluorite	$O_h(Ca^{2+})$*	6.19	4.83	1280	82		18.4	0.23–12	1.473	Bridgman-Stockbarger
CaF$_2$	cubic	O_h^5 $Fm3m$	fluorite	$C_{4v}(Ca^{2+}+F^-)$* $C_{3v}(Ca^{2+}+O^{2-})$	5.464	3.179	1360	163	0.097	19.5	0.13–12	1.433	Bridgman-Stockbarger
SrF$_2$	cubic	O_h^5 $Fm3m$	fluorite		5.78	4.24	1400	4(Moh)				1.438	
MgF$_2$	tetragonal	D_{4h}^{14} $P4/mmm$	rutile	$D_{2h}(Mg^{2+})$	$a_0 = 4.621$ $c_0 = 3.053$	3.177	1255	~5(Moh)	0.2	8.8∥c-axis 13.1⊥c-axis	0.15–9.6	1.38	vertical gradient freeze
MnF$_2$	tetragonal	D_{4h}^{14} $P4/mmm$	rutile	$D_{2h}(Mn^{2+})$	$a_0 = 4.715$ $c_0 = 3.131$		930					1.476	vertical gradient freeze
ZnF$_2$	tetragonal	D_{4h}^{14} $P4/mmm$	rutile	$D_{2h}(Zn^{2+})$	$a_0 = 4.703$ $c_0 = 3.134$	4.84	947					1.497	vertical gradient freeze
CeF$_3$	hexagonal	D_{6h}^3 $C6/mcm$	tysonite	$C_2(Ce^{3+})$	$a_0 = 4.115$ $c_0 = 7.288$	6.16	1325–1425	4.5(Moh)			0.3–9.5	1.62	Bridgman-Stockbarger
HoF$_3$	orthorhombic	Pnma		$C_s(Ho^{3+})$			1100	743				1.58	Czochralski
LaF$_3$	hexagonal	D_{6h}^3 $C6/mcm$	tysonite	$C_2(La^{3+})$	$a_0 = 4.148$ $c_0 = 7.354$	5.94	1490	4.5(Moh)	0.051			1.6	Czochralski
LiYF$_4$ (YLF)	tetragonal	C_{4h}^6 $I4_1/a$	scheelite	$S_4(Y^{3+})$	$a_0 = 5.26$ $c_0 = 10.94$					9.1∥c-axis 8.5⊥c-axis		1.634 (n_o) 1.631 (n_e)	Czochralski

Tungstates

CaWO$_4$	tetragonal	C_{4h}^6 $I4_1/a$	scheelite	$S_4(Ca^{2+})$*	$a_0 = 5.242$ $c_0 = 11.372$	3.99		4–5(Moh)	0.06	13(∥a) 8(∥c)	0.12–8	1.448 (n_o) 1.470 (n_e)	top-seeded solution
SrWO$_4$	tetragonal	C_{4h}^6 $I4_1/a$	scheelite	$S_4(Sr^{2+})$	$a_0 = 5.40$ $c_0 = 11.90$	6.18	1566						Czochralski
NaGd(WO$_4$)$_2$	tetragonal	C_{4h}^6 $I4_1/a$	scheelite	$S_4(Gd^{3+})$			1400						Czochralski
NaLa(WO$_4$)$_2$	tetragonal	C_{4h}^6 $I4_1/a$	scheelite	$S_4(La^{3+})$	$a_0 = 5.357$ $c_0 = 11.743$								Czochralski

* There are 16 octahedral (C_{3i} symmetry) and 24 tetrahedral (S_4 symmetry) Ga^{3+} sites in the garnet unit cell. Cr^{3+} impurities substitute predominantly in octahedral sites.

† Laser-active site.

‡ Trivalent rare earths enter the fluorite lattice substitutionally at Ca^{2+} sites. Charge compensation occurs in many different ways resulting in different site symmetries (see, for example, the discussion by K. Nassau in "Applied Solid State Science," Vol. 2, edited by R. Wolfe, Academic Press, New York, 1971, p. 268.

* When used as a host for trivalent rare earths, Na$^+$ ions at Ca^{2+} sites or Nb^{5+} ions at W^{6+} sites are frequently added for charge compensation.

Table 2.1.1.1 (continued)
INSULATING CRYSTAL LASER HOSTS

Material	Symmetry	Space group	Structure	Substitutional site symmetry	Lattice constant (Å)	Density (g/cm³)	Melting point (°C)	Hardness (Knoop scale)	Thermal conductivity (W/cm-°C)	Thermal expansion (10⁻⁶/°C)	Transparency range (µm)	Refractive index	Method of growth
Molybdates													
$CaMoO_4$	tetragonal	C_{4h}^6 $I4_1/a$	scheelite	$S_4(Ca^{2+})$	$a_0 = 5.226$ $c_0 = 11.43$	4.256	1430	4.3 (Moh)	0.04	25.5∥c-axis 19.4∥a-axis		1.97	Czochralski
$LaNa(MoO_4)_2$	tetragonal	C_{4h}^6 $I4_1/a$	scheelite	$S_4(La^{3+})$	$a_0 = 5.328$ $c_0 = 11.699$	4.79		4.5 (Moh)	0.022				Czochralski
$PbMoO_4$	tetragonal	C_{4h}^6 $I4_1/a$	scheelite	(Pb^{2+})	$a_0 = 5.435$ $c_0 = 12.110$	6.816	1070						Czochralski
$SrMoO_4$	tetragonal	C_{4h}^6 $I4_1/a$	scheelite	$S_4(Sr^{2+})$	$a_0 = 5.38$ $c_0 = 11.97$	4.15	1490		0.040∥a-axis 0.041∥c-axis			1.94	Czochralski
$Gd_2(MoO_4)_3$	ortho-rhombic	$Pba2$		(Gd^{3+})	$a_0 = 10.3858$ $b_0 = 10.4186$ $c_0 = 10.7004$								Czochralski
$KY(MoO_4)_3$		D_{2h}^{14} $Pbna$		$S_4(Y^{3+})$	$a_0 = 5.07$ $b_0 = 18.23$ $c_0 = 7.95$						0.33–5.5		Czochralski
Niobates													
$Ca(NbO_3)_2$	ortho-rhombic	D_{2h}^{14} $Pbcn$	fersmite	$C_2(Ca^{2+})*$		4.72	1560	5.5 (Moh)			0.3–5.5	2.07(α) 2.10(β) 2.18(γ)	Czochralski
$LiNbO_3$	rhombo-hedral	C_{3v}^6 $R3c$	illmenite		$a_0 = 5.47$ $b_0 = 5.112$ $c_0 = 13.816$	4.628	1253	5 (Moh)	0.056		0.37–5	2.286 (n_0) 2.200 (n_e)	Czochralski
$LaNbO_4$	mono-clinic	C_2	ferguso-nite**	(La^{3+})	$a_0 = 5.56$ $b_0 = 11.54$ $c_0 = 5.20$		1650				0.3–6	2.187(n_g) 2.094(n_m) 2.077(n_p)	Czochralski

Vanadates

$\mathrm{Ca_3(VO_4)_2}$	monoclinic	C_{2h}^{6} $C2/c$		$(\mathrm{Ca^{2+}})$	$a_o=8.35$, $b_o=10.77$, $c_o=7.00$	3.55	1470	4.3 (Moh)	0.014	55∥c-axis 35∥a-axis		1.89	Czochralski
$\mathrm{YVO_4}$	tetragonal	D_{4h}^{19} $I4_1/amd$	zircon	D_{2d} $(\mathrm{Y^{3+}})$	$a_o=7.133$, $c_o=6.291$	4.23	1750–1900	5.5 (Moh)	0.052∥c-axis 0.051⊥c-axis	7.3⊥c-axis	0.45–4.8	1.86 (n_o) 1.88 (n_e)	Bridgman-Stockbarger

Phosphates, Silicates, Germanates

$\mathrm{Ca_5(PO_4)_3F}$ (FAP)	hexagonal	C_{6h}^{2} $P6_3/m$	apatite	$C_{1h}(\mathrm{Ca^{2+}}\text{-I})$ $C_3(\mathrm{Ca^{2+}}\text{-II})$	$a_o=9.368$, $c_o=6.884$	3.189	1705	540	0.021[0001] 0.019$[10\bar{1}0]$	10[0001] 9.4$[10\bar{1}0]$		1.63	Czochralski
$\mathrm{CaY_4(SiO_4)_3O}$ (SOAP)	hexagonal	$P6_3/m$	apatite				> 2000					~1.8	Czochralski
$\mathrm{Ba_2MgGe_2O_7}$ (BMAG)	tetragonal	D_{2d}^{3} $P\bar{4}2_1m$	akermanite	$(\mathrm{Ba^{2+}})$†	$a_o=7.84$, $c_o=5.01$								
$\mathrm{CaY_2Mg_2Ge_3O_{12}}$ (CAMGAR)	cubic	O_h^{10} Ia3d	garnet	$D_2(\mathrm{Y^{3+}})$	12.307	4.809	~1000		0.06	$\sim 7 \times 10^{-6}$	0.2–8	1.83	top-seeded solution

* When used as a host for trivalent rare earths, $\mathrm{Na^+}$ or $\mathrm{Ti^{4+}}$ ions are added for charge compensation.

** Undergoes a polymorphic transformation from a tetragonal system (Scheelite) at 500–600°C.

† When used as a host for trivalent rare earths, $\mathrm{Na^+}$ or $\mathrm{K^+}$ are used for charge compensation.

Table 2.1.1.2
LASER CRYSTALS

Table 2.1.1.2 lists all reported paramagnetic-ion lasers other than stoichiometric devices discussed in Section 2.1.2. The listing is divided into four groups determined by the type of paramagnetic-ion; the divisions are trivalent rare earths, divalent rare earths, the iron group and the actinides. Each group is then further divided into sections determined by the particular dopant ion, with the order of ions the same as that in the periodic table. An explanation of the data in each column is as follows:

Host — Crystals are listed alphabetically. In cases where sensitizer ions have been added the approximate concentration for each ion is given. Note that in some cases, conforming to the style of the original reference, the crystal formula includes sensitizer ions. Unless indicated, rare-earth sensitizers are in the trivalent state, as is the Cr ion.

Laser transition — The upper laser state is shown first. A discussion of the notation appears in the second section of this chapter. Abbreviations used are

Nd1	$^4F_{3/2} \rightarrow$	$^4I_{11/2}$
Nd2	$^4F_{3/2} \rightarrow$	$^4I_{13/2}$
Ho1	$^5I_7 \rightarrow$	$^5I_8$
Ho2	$^5I_6 \rightarrow$	$^5I_7$
Er1	$^4I_{13/2} \rightarrow$	$^4I_{15/2}$
Er2	$^4I_{11/2} \rightarrow$	$^4I_{13/2}$
Tm1	$^3H_4 \rightarrow$	$^3H_6$
Yb1	$^4F_{5/2} \rightarrow$	$^4F_{7/2}$

For noncubic host crystals the orientation of the rod axis (e.g., $\parallel$ a) or the polarization of the laser output (σ or π) may appear along with the transition identification.

Laser wavelength — The shortest-wavelength transitions are listed first and for a given transition (in most cases) room-temperature wavelengths appear first. When a range of wavelengths is shown, indicating temperature or resonant-cavity tuning, it is necessary to check the original reference to determine if tuning was truly continuous over the range. (More information on broadly tunable lasers appears in Table 2.1.1.8.)

Terminal level energy — The energy difference between the ground state of the ion and that of the lower laser level appears in this column. In many cases because of incomplete spectroscopic data this energy is known only approximately. For vibronic (phonon-assisted) lasers terminal level energy is generally not a well-defined quantity and the letter "V" is shown.

Laser ion concentration — This quantity may be weight or atomic percent and may be that present in the melt or in the crystal; the original reference should be consulted. The same statement applies to the sensitizer concentrations shown. A range of concentrations indicates that a variety of crystals with differing concentrations were studied.

Fluorescence lifetime — The lifetime is that of the upper laser state. In cases where a measurement was made as a function of temperature or laser-ion concentration the appropriate value of either quantity is shown in parentheses.

Fluorescence linewidth — The value indicated is for full width at half-maximum. A range of values appears when the crystal temperature was varied.

Mode of operation — P stands for pulsed operation and CW for continuous operation.

Temperature — Where a range appears the measurement was carried out with a variable-temperature system.

Rod size — Generally the dimensions are diameter × length. A single number refers to the rod length and three numbers give the width × height × length for a rectangular-cross-section crystal.

Optical pump — The key to pump sources is as follows:

Xe	—	xenon arc lamp
Kr	—	krypton arc lamp
Hg	—	mercury arc lamp
W	—	tungsten filament lamp
Sun	—	solar pumping
ArL	—	argon-ion laser
DNGL	—	frequency-doubled Nd-doped glass laser
DyL	—	dye laser
KrL	—	krypton-ion laser
KrFL	—	krypton-fluoride excimer laser
NdYL	—	Nd:$Y_3Al_5O_{12}$ laser
XeL	—	xenon-ion laser
XeFL	—	xenon-fluoride excimer laser.

Threshold — For lamp pumping the threshold is electrical energy or power into the lamp for the pulsed or CW mode, respectively. For solar or laser pumping threshold is typically energy or power absorbed in the laser crystal. Since the threshold is dependent on a number of factors such as the crystal quality, ion concentrations, optical resonator configuration and losses and, for pulsed operation the shape and duration of the pump pulse, the values given may not necessarily be the minimum obtainable.

References — The reference list is a representative but not necessarily exhaustive survey of papers on the various lasers. For historical interest the first reports of operation from the more important laser crystals have been included along with more recent papers.

Table 2.1.1.2
LASER CRYSTALS

Host	Laser transition	Laser wavelength (μm)	Terminal level energy (cm⁻¹)	Laser ion concentration (%)	Fluorescence lifetime (msec)	Fluorescent linewidth (cm⁻¹)	Mode of operation	Temperature (K)	Crystal size (mm)	Optical pump	Threshold	Ref.
TRIVALENT RARE-EARTH LASERS												
CERIUM (Ce^{3+}, $4f^1$)												
LaF_3	$5d \rightarrow {}^2F_{7/2}$	0.286	>	0.05	18 nsec	—	P	300	5×5×15	KrFL	0.008	1
$LiYF_4$	$5d \rightarrow {}^2F_{5/2}$	0.306	>	1	40 nsec	—	P	300	5×5×9	KrFL	0.003	1,2
		→0.315										
	$5d \rightarrow {}^2F_{7/2}$	0.323	>									
		→0.328										
PRAESODYMIUM (Pr^{3+}, $4f^2$)												
$Ca(NbO_3)_2$	${}^1G_4 \rightarrow {}^3H_4$	1.04	—	—	—	—	P	77	3×30	Xe	20-25	3
$CaWO_4$	${}^1D_2 \rightarrow {}^3F_4$	1.0468	~6700	0.5	0.023	—	P	77	3×36	Xe	20	4
$LaBr_3$	${}^3P_1 \rightarrow {}^3H_5$	0.532	~2130	1	0.007	—	P	<<300		DyL	—	5
	${}^3P_0 \rightarrow {}^3H_6$	0.621	~4200		0.01			<<300				
	${}^3P_2 \rightarrow {}^3F_3$	0.632	~6300		2×10^{-5}			<70				
	${}^3P_0 \rightarrow {}^3F_2$	0.647	~4900		0.01			<<300				
$LaCl_3$	${}^3P_0 \rightarrow {}^3H_4$	0.4892	~34	1	14.7	—	P	5.5-14	~10 μm	DyL	—	6
	${}^3P_1 \rightarrow {}^3H_5$	0.5298	2137		~0.001			35				

Material	Transition	λ (μm)	(cm⁻¹)				Mode	T (K)	Dimensions	Pump			Ref
LaF_3	$^3P_0 \rightarrow\ ^3H_6$	0.6164	4230			14.7		12			4×10^{-7}		
	$^3P_0 \rightarrow\ ^3F_2$	0.6452	4923					65			1.2×10^{-6}		
	$^3P_0 \rightarrow\ ^3H_6$	0.5985	~4200	1		–	P	8	0.6 × 32	Xe	1.0×10^{-6}	60	7
$LiYF_4$	$^3P_0 \rightarrow\ ^3H_4$	0.479	0	0.2	0.05	22	P	300	2 × 6 × 8	DyL	10^{-3}	0.002	8
$SrMoO_4$	$^1G_4 \rightarrow\ ^3H_4$	1.04	–	–	–	–	P	65	–	–	1.2×10^{-6}	–	9
NEODYMIUM (Nd^{3+}, $4f^3$)													
$Ba_{0.75}Ca_{0.25}Nb_2O_6$	Nd1	1.062	~2000	–		–	P	295	–	Xe		380	10
BaF_2	Nd1	1.060	~2000	–		–	P	77	3 × 52	Xe		1600	11
$BaF_2 - CeF_3$ (5%)	Nd1	1.0543	~2000	2	0.31	105	P	300	6 × 15	Xe		28	12
$BaF_2 - GdF_3$ (5%)	Nd1	1.0526	~2000	2	0.3	75	P	300	6 × 20	Xe		25	12
$BaF_2 - LaF_3$	Nd1	1.0534	1992	0.5–4.5	~0.43	75	P	300	(5–6) × (13–80)	Xe		2.7	13
		1.0538		(0.5%)		30		77				15	
		1.0580	2035			35						50	
		1.0534 → 1.0563	1992			>75		300 → 920					
	Nd2	1.3185	~4000	2		~55		300	6 × 40			10	14
		1.3280				~80						7	
		1.3290	3956			~30		77				30	

Table 2.1.1.2 (continued)
LASER CRYSTALS

NEODYMIUM (Nd^{3+}, $4f^3$) (Continued)

Host	Laser transition	Laser wavelength (μm)	Terminal level energy (cm^{-1})	Laser ion concentration (%)	Fluorescence lifetime (msec)	Fluorescent linewidth (cm^{-1})	Mode of operation	Temperature (K)	Crystal size (mm)	Optical pump	Threshold	Ref.
$BaF_2 - YF_3$ (5%)	Nd1	1.0521	~2000	2	0.3	~60	P	300	5 × 15	Xe	13	12
	Nd2	1.3200	~4000			~100					55	
$BaMgGe_2O_7$	Nd1	1.05436	1895	2	0.4	40	P	300	5 × 25	Xe	5.1	15
$Ba_{0.25}Mg_{0.75}Y_2Ge_3O_{12}$	Nd1	1.0615	2014	2	~0.3	30	P	300	5 × 19	Xe	22.5	16
$Ba_2NaNb_5O_{15}$	Nd1	1.0613	~2000	0.1	0.2	55	P	300	3.5 × 21	Xe	23	17
$Ba_2ZnGe_2O_7$	Nd1	1.05437	1890	2	0.34	48	P	300	5 × 27	Xe	4.5	18
$Bi_4Ge_3O_{12}$	Nd1	1.0644	1932	1	0.22	24	P	300	6 × 37	Xe	1.3	10, 19
		1.0638	~2000		(0.3%)	-		77			0.65	
		1.06425	1931			3					~4.5	
	Nd2	1.3418	3875			29		300			~4.5	
$Bi_4Si_3O_{12}$	Nd1	1.0629	1924	1	0.23	20	P	300	4 × 17	Xe	~7	20
		1.0629	1928		(0.3%)	3		77			1	
	Nd2	1.3407	3873			29		300			~7	
$Bi(Si, Ge)_3O_{12}$	Nd1	1.0635	1928	1	0.25	35	P	300	3 × 10	Xe	100	21
$CaAl_4O_7$	Nd1	1.0596	~2100	0.4	0.28	30	P	300	6 × 25	Xe	-	22

Material	Site	λ (µm)	(cm⁻¹)				Mode	Temp	Size	Lamp	Thresh.	Ref
		1.0638				11					–	
		1.0786				28					8	
		1.05895				3.5		77			70	
		1.06585				1.7					10	
		1.07655				4.8					35	
		1.0772				6.2					40	23
	Nd2	1.3420	~4100			16		300			21	
		1.3710	~4300			20					15	
		1.3400	~4100			4		77			65	
		1.3675				6					50	
CaAl$_{12}$O$_{19}$	Nd1	1.0497	~2000	1–2	0.34	20	P	300		Xe	50	24
Ca$_{0.25}$Ba$_{0.75}$(NbO$_3$)$_2$	Nd1	1.062	~2000	0.25–2	–	–	P	295	–	Xe	380	10
CaF$_2$(I)	Nd1	1.0448	2034	0.4–0.6	1.1	9	P	120	3 × 25	Xe	78	25
		1.0661	–			–					–	
		1.0448	2031			~2		77	6.35 × 75		400	26, 27
		1.0457	2032								30	
		1.0467	2028								160	
		1.0448	2031					50			240	
		1.0456	2032								70	
		1.0466	2028								160	
		1.0480	2053								1350	

Table 2.1.1.2 (continued)
LASER CRYSTALS

NEODYMIUM (Nd^{3+}, 4f^3) (Continued)

Host	Laser transition	Laser wavelength (μm)	Terminal level energy (cm^{-1})	Laser ion concentration (%)	Fluorescence lifetime (msec)	Fluorescent linewidth (cm^{-1})	Mode of operation	Temperature (K)	Crystal size (mm)	Optical pump	Threshold	Ref.
		1.0507	2061			~3					870	
		1.0648	2191			~5					1100	28
		1.0370	2030					460 → 500	6.5 × 75		—	
		1.0461 → 1.0468	2032			20 →		300 → 530			—	
		1.0628 → 1.0623	—			30 →		300 → 560			—	
CaF$_2$(II)	Nd1	1.0885 → 1.0889	~2000	0.2-0.5	1.25	25	P	300 → 420	6 × 75	Xe	170	28, 29
CaF$_2$ − CeF$_3$ (30%)	Nd1	1.0657 → 1.0640	~2000	0.5-1	0.7	~100	P	300 → 700	5 × 45	Xe	86	28
	Nd2	1.3190	~4100	0.5	0.7	~100	P	300	5.5 × 45	Xe	300	22
CaF$_2$ − CeO$_2$	Nd1	~1.0885	~2000	1	~1.2	~25	P	300	6.5 × 45	Xe	850	30
CaF$_2$ − GdF$_3$ (10%)	Nd1	1.0654	~2000	2	—	110	P	300	5 × 21	Xe	30	31
	Nd2	1.3185	~4000			100					90	
CaF$_2$ − LaF$_3$ (10%)	Nd1	1.0645	~2000	2	—	100	P	300	5 × 21	Xe	25	31
	Nd2	1.3190	~4000			120					70	
CaF$_2$ − SrF$_2$ (50%)	Nd1	1.0369	~2000	0.5	1.4	~20	P	300	5 × 20	Xe	140	32
CaF$_2$ − SrF$_2$ − BaF$_2$	Nd1	1.0535	~2000	1	0.4	~100	P	300	5 × 30	Xe	30	33

Material	Ion	λ (μm)	(cm⁻¹)					Temp (K)	Size (mm)	Pump	Output	Ref
$-YF_3 - LaF_3$		1.0623			0.48	~120					150	28
		1.0535 → 1.0547				100→		300→550				
		1.0623 → 1.0585				120→		300→700				
$CaF_2 - YF_3$	Nd1	1.0461	2032	0.2-12		30	P	300	5×40-75	Xe	180	34,35
		1.0540	-		(0.2%)	65					6	
		1.0632	2170			90			5×40-75		7	
		1.0540 → 1.0540	-			65		300→430	5×45			28
		1.0632 → 1.0603	2170			90→		300→950				
$CaF_2 - YF_3$	Nd2	1.3270	~4040	1-2		70	P	300	5×39	Xe	12	36
		1.3370	~4090			80					10	
		1.3585	~4200			120					20	
		1.3255	~4040			45		77			40	
		1.3380	~4090			55					70	
		1.3600	~4200			60					200	
$CaF_2 - YF_3 - NdF_3$	Nd1	1.0632	~2170	16	0.4	100	P	300	6×40	Xe	250	37
$Ca_3Ga_2Ge_3O_{12}$	Nd1	1.0597	~2000	0.8	0.21	-	P	300	2.5×5×9	Xe	4.5	38
		1.0639									3.5	
	Nd2	1.3317	~4000								18	
$CaGd_4(SiO_4)_3O$	Nd1	~1.06	~2000	-	-	-	P	300	-	Xe	-	39
$Ca_4La(PO_4)_3O$	Nd1	1.0613	~2000	-	-	~30	P	300	6.4×76	Xe	22	40

Table 2.1.1.2 (continued)
LASER CRYSTALS

NEODYMIUM (Nd^{3+}, $4f^3$) (Continued)

Host	Laser transition	Laser wavelength (μm)	Terminal level energy (cm⁻¹)	Laser ion concentration (%)	Fluorescence lifetime (msec)	Fluorescent linewidth (cm⁻¹)	Mode of operation	Temperature (K)	Crystal size (mm)	Optical pump	Threshold	Ref.
$CaLa_4(SiO_4)_3O$	Nd1	1.0610	~2000	~2	0.24	53	P	300	6.4×46	Xe	5.5	40,41
							CW		3×50	W	525	
$CaMoO_4$	Nd1	1.0573	~2000	1.3	0.12	12	P	300	6.5×75	Xe	10	10,42,43
		1.061		1.8		—	P		2.5×25		1	
							CW				1200	
		1.0673				—	P	77	—		7	
							CW				1200	
$Ca(NbO_3)_2$	Nd1	1.0615	1949	1—2	0.12	10	P	300	3.5×11	Xe	1	2,44
							CW		4×30		350	45
			1952			4.5	P	77	3.5×11	Xe	0.8	
							CW		4×30	Xe	400	
		1.0615→1.0625	1949			10→	P	300→650	5×20			46
	Nd2	1.3380	3898	1		13	P	300	6×15	Xe	4.5	47
		1.3425	3923			16					6	
		1.3370	3896			7		77			8	
		1.3415	3922			10					20	

Crystal	Nd (transition)	λ (μm)	(cm⁻¹)				Mode	T (K)	Dimensions (mm)	Pump	Threshold (J)	Ref
Ca$_5$(PO$_4$)$_3$F	Nd1 (IIa)	1.0630	1900	1	0.25	6	P	300	6×17	Xe	0.2-1	48,49
	Nd2 (IIc)	1.3347	3819			4	CW		2×28	W	145	50
		1.3345	3820			2	P	77	6×17	Xe	2	
CaSc$_2$O$_4$	Nd1	1.0720	~2100	1	0.105	~22	P	300	5×25	Xe	2.3	51,52
		1.0755				~25					2.7	
		1.0868	~2250			3.5					1	
		1.0730	~2100			4.5		77			0.7	
		1.0867	~2250								0.4	
	Nd2	1.3565	~4050			~30		300			1.5	
		1.3870	~4200			~12		77			6	
Ca$_3$(VO$_4$)$_2$	Nd1	1.067	~2000	2	0.15	160	P	300	3.2-6.3×52	Xe	3000	53
CaWO$_4$	$^4F_{3/2} \to {}^4I_{9/2}$	0.9145	471	0.14-3	0.18	15	P	77	3×50	Xe	4.6	54
	Nd1	1.0582	2016			~20		300	5×30	Xe	0.5	55,56,57,58
		1.0652				~20	CW		2×50	Hg	1200	
		1.0582 → 1.0597	—			20 →	P	300 → 700	4.5×45	Xe	1.3	28
		1.0587	—			7	P	77	5×30		1	
		1.0601	—								1	
		1.0634	—				P				80	
		1.0649	2016			7	CW	85	3×50	Hg	400	

Table 2.1.1.2 (continued)
LASER CRYSTALS

NEODYMIUM (Nd^{3+}, $4f^3$) (Continued)

Host	Laser transition	Laser wavelength (μm)	Terminal level energy (cm⁻¹)	Laser ion concentration (%)	Fluorescence lifetime (msec)	Fluorescent linewidth (cm⁻¹)	Mode of operation	Temperature (K)	Crystal size (mm)	Optical pump	Threshold	Ref.
	Nd2	1.3340	~3970			20	P	300	5 × 45	Xe	4	54, 59, 60
		1.3475				25					10	
		1.3370	~3925			23					1	
							CW				1500	
		1.3390	~4000			25	P		5 × 30		1	
		1.3345	3975			8		77			1	
		1.3372	3928			–					2	
		1.3459	3975			7					2	
		1.3880	4202			10					5	
$Ca_2Y_5F_{19}$	Nd1	1.0498	~2000	1	0.36	~100	P	300–600	5 × 34	Xe	15	28
		1.3190	~4000			~80			5 × 21	Xe	150	50
		1.3525	~4200			~100					180	
		1.3200	~4000			25					250	
$CaY_2Mg_2Ge_3O_{12}$	$^4F_{3/2} \rightarrow {}^4I_{9/2}$	0.941	819	–	–	–	CW	300	–	ArL	0.4	61
	Nd1	1.05896	2008	2–6	0.305	37	P		5 × 25	Xe	65	62
			–	–	–	–	CW		–	ArL	0.09	61

Material	Ion	λ (μm)										Ref.
CaY$_4$(SiO$_4$)$_3$O	Nd1	1.0672	~2000	–	0.2	53	P	300	5.2 × 24	Xe	50	40
CdF$_2$ – LaF$_3$	Nd1	~1.0665	~2000	~1	~0.3	~100	P	300	4.5 × ~20	Xe	500	63
CdF$_2$ – YF$_3$ (15%)	Nd1	1.0656 → 1.0629	~2000	1-2	~0.3	~100	P	300 → 600	6 × 23	Xe	12	46, 63
	Nd2	1.3425	~4000	~2		~120	P	300			40	36
		1.3165				55		77			150	
CeCl$_3$	Nd1 (IIa)	1.0647	2055	2.8	0.175		P	300	5 × 2.85	Xe	0.06	64
				~2.5			CW		1.5 × 8.9	ArL	8 mW	65
CeF$_3$	Nd1 (⊥c)	1.0410	1983	~4	~0.27	~25	P	300	5 × 25	Xe	25	66, 67
		1.0638	2189			~35					8.5	
		1.0404	1996		~0.25	6.5		77			3.5	
		1.0639	2199			15					6	
	Nd1 (~IIc)	1.0410	1983		~0.27	~25		300			8.5	
		1.0638	2189			~35					7.5	
		1.0404	1996		~0.25	6.5		77			3.5	
	Nd2 (⊥c)	1.3320	~4070	2	~0.27	30		300			20	36
		1.3240	~4040			8		77			20	
		1.3310	~4080			10					10	
		1.3675	~4280			15					20	
CeP$_5$O$_{14}$	Nd1	1.051	~2000	10	0.25-0.28	–	P	300		ArL		68

Table 2.1.1.2 (continued)
LASER CRYSTALS

NEODYMIUM (Nd³⁺, 4f³) (Continued)

Host	Laser transition	Laser wavelength (μm)	Terminal level energy (cm⁻¹)	Laser ion concentration (%)	Fluorescence lifetime (msec)	Fluorescent linewidth (cm⁻¹)	Mode of operation	Temperature (K)	Crystal size (mm)	Optical pump	Threshold	Ref.
$GdAlO_3$	Nd1	1.0690	2095	1	0.1	15	P	300	2.5 × 20	Xe	38	69,70
		1.0760	~2166						3 × 3 × 5		210	
		1.0689	2094					77	2.5 × 20		24	
		1.0759	2155								41	
$Gd_3Ga_5O_{12}$	Nd1	1.0621	2064	1	0.27	7.2	P	300	5 × 25	Xe	4	71,72
		1.0600	~2007		(0.3%)	–		~120			1.5	
		1.0615	~2063			–		~120			1.5	
		1.0584	1994			~2.3		77			1	
		1.0599	2007			~2.3		77			2	
	Nd2	1.3077	3927			~2.5		77			18	
$Gd_2(MoO_4)_3$	Nd1 (IIc)	1.0606	~1926	3	0.15	~35	P	300	5 × 10	Xe	–	73,74
		1.0701	2010			~40					2.5	
Gd_2O_3	Nd1	1.0741	~2000	–	0.12	–	P	300	2 × 6	Xe	36	75
		1.0789				20–30					9	
		1.0776				–		77			12	
		1.0789				8					3	

Crystal	Ion	λ (μm)	E (cm⁻¹)				Mode	T (K)	Dimensions (mm)	Pump		Ref.
GdP₅O₁₄	Nd1	1.051	~2000	10	0.25–0.28	10	P	300	–	ArL	3	68
Gd₃Sc₂Al₃O₁₂	Nd1	1.05995	1978	1	0.275	11.5	P	300	6×25	Xe	700	76,77
		1.06	1978			15	CW		2.5×29	W	–	
	Nd2	1.0620	2111			5	P	77	6×25	Xe	3	
	Nd1	1.05915	1979			12		300			7	
		1.3370	3931			14					7	78
Gd₃Sc₂Ga₃O₁₂	Nd2	1.0612	2070	1	0.26	65	P	300	5×22	Xe	3	
	Nd1	1.05755	1978		(0.3%)	–		77			12	
		1.0580	–			8.5					10	
		1.06045	2069			~18					3.5	
GdScO₃	Nd1	1.08515	~2065	1.5	0.135	–	P	300	4×12	Xe	7–20	79
HfO₂–Y₂O₃	Nd1	1.0604	~2000	0.5–0.7	0.45	130	P	300	(3–6)×(15–32)	Xe	48	80,81
		1.3305	~3900	0.5		~90	P	300			85	
K₅Bi(MoO₄)₄	Nd1	1.0660	~2000	5–10	0.25		P	300	7×11	Xe	0.4	82
KGd(WO₄)₂	Nd1	1.0672	1945	2.5	0.11	48	P	300	2.5×27	Xe	≈400	83
	Nd2	1.3510	~3900			22	CW				0.9	
KLa(MoO₄)₂	Nd1	1.0587	1940	~2	0.2	~65	P	300	6×40	Xe	3.5	84
		1.0585	1942		(0.5%)	12	P	77			5	
	Nd2	1.3350	3895			35		300			8	
KY(MoO₄)₂	Nd1 (IIa)	1.0669	1983	2	0.13	12	P	300	6.5	Xe	10	85,59
	Nd2 (IIa)	1.3485	~3940			35			5		20	

Table 2.1.1.2 (continued)
LASER CRYSTALS

NEODYMIUM (Nd^{3+}, 4f^3) (Continued)

Host	Laser transition	Laser wavelength (μm)	Terminal level energy (cm⁻¹)	Laser ion concentration (%)	Fluorescence lifetime (msec)	Fluorescent linewidth (cm⁻¹)	Mode of operation	Temperature (K)	Crystal size (mm)	Optical pump	Threshold	Ref.
KY(WO$_4$)$_2$	$^4F_{3/2} \rightarrow {}^4I_{9/2}$	0.9137	355	0.3–2.5	0.11	25	P	77	4 × 14	Xe	5.5	86, 87, 88
	Nd1	1.0688	1944		(0.3 %)	20	P	300	(4–5) × (14–25)	Xe	0.3	
							CW		5 × 25	Xe	400	
		1.0687	1943			3.3	P	77			0.5	
		1.0687 → 1.0690				3.3 →	P	77 → 600			—	
	Nd2	1.3525	~3900			13	P	300		Xe	0.9	
							CW				1100	
		1.3545	~3916			14	P				3	
		1.3515	3902			65		77			1	
		1.3545	3916			7					1	
LaAlO$_3$	Nd1	1.0804	2327	2	~0.24	~20	P	300	3 × 4 × 8	Xe	120	69
La$_2$Be$_2$O$_5$	Nd1 (IIb)	1.0698	1962	1	0.15	~27	P	300	5.5 × 25	Xe	1.8	89, 90
							CW		5 × 50	W		
	Nd1 (IIa)	1.079	2041			~40	P		5 × 50	Xe	10	
	Nd2 (IIb)	1.3510	3908			~32	P		—		2.5	91

LaF₃											
Nd1 (c–20°)	1.04065	1983	1–2	0.6–0.7	25	P	300	(5–6)×(25–40)	Xe	2.8	10, 92, 93
	1.06335	2188			35	CW		5×38		4000	
	1.0400	1980			6.5	P	77	(5–6)×(25–40)		8	
	1.0451	2069			6.5					7.5	
	1.0523	2092			6.5					15	
	1.0583	2188			6.5					—	
	1.06305	2188			15					—	
	1.0670	2223			18					5	
	1.04065 → 1.0410	1980			25→		300 → 430	5×38		—	28
	1.0595 → 1.0613	2190			—		380 → 820			—	
	1.0632 → 1.0642	2220			—		400 → 700			—	
	1.06335 → 1.0638	2190			35→		300 → 650			—	
Nd2 (c–20°)	1.3675	4270	4		25		300	5×25		40	22
	1.3235	4038			8		77			60	
	1.3670	4276			13					40	
Nd2 (c–73°)	1.3310	4070	2		21		300			7	
	1.3125	3974			6					20	
	1.3305	4077			10					5	

Table 2.1.1.2 (continued)
LASER CRYSTALS

NEODYMIUM (Nd^{3+}, 4f^3) (Continued)

Host	Laser transition	Laser wavelength (μm)	Terminal level energy (cm^{-1})	Laser ion concentration (%)	Fluorescence lifetime (msec)	Fluorescent linewidth (cm^{-1})	Mode of operation	Temperature (K)	Crystal size (mm)	Optical pump	Threshold	Ref.
LaF$_3$ – SrF$_2$	Nd1 ($\perp$c)	1.0486	~2000	2	0.22	~65	P	300	5×26	Xe	5.5	94
		1.0635				~65					11	
	Nd2 ($\perp$c)	1.3315	~4000			55					13	59
		1.3170				30		77			60	
		1.3275				30					40	
		1.3325				30					50	
LaNbO$_4$	Nd1	1.0624	1973	1–2	0.12	15	P	300	5×30	Xe	~10	95
La$_2$O$_3$	Nd1	1.079	~2000	1	0.125	11	P	77	4×8	Xe	230	96
La$_2$O$_2$S	Nd1 (π)	1.075	1910	1	0.095	10	P	300	10×75	Xe	0.3	97
							CW			W		
LiGd(MoO$_4$)$_2$	Nd1 ($\perp$c)	1.0599	2005	~2	0.14	~80	P	300	4×20	Xe	1.3	98,88
					(0.3%)		CW		5×25		1150	
	Nd2 ($\perp$c)	1.3400	3980	2		65	P	300	5×25		13	99
		1.3400	~3980			25		77			30	
LiLa(MoO$_4$)$_2$	Nd1 ($\perp$c)	1.0585	2000	2	0.15	100	P	300	5×25	Xe	3	100
		1.0658	2003		(0.3%)	30		77			15	

Material	Transition	λ (μm)	cm⁻¹			Threshold	Mode	Temp (K)	Dimensions	Pump		Ref.
LiNbO₃		1.3370	3909			75		300			20	99
		1.3375	3945			30		77			35	
	Nd1 (⊥c)	1.0846	2034	1.5	0.085	20	P	300	4 × 20	Xe	1.5	101
	(∥c)	1.0933	2107	0.5		30			3 × 3 × 10	KrL	75 μJ	102
	(∥c)	1.0933	2107	1.5		30			4 × 20	Xe	2.0	101
		1.0933 → 1.0922	2107			30→		300 → 620	3.5 × 12	Xe	—	46
		1.0787 → 1.0782	1990			—		450 → 590			—	
	(⊥c)	1.0846	2034			20→		300 → 600	5 × 15		—	
	Nd2 (⊥c)	1.3745	~3970	6		20		300	4 × 20		6	36, 101
	(∥c)	1.3870	~4035	3.5		20					3.5	
LiYF₄	Nd1 (π)	1.0471	2042	—	0.5	12		300	5 × 50	Xe	2	103, 104
	(σ)	1.0530		2		12.5					—	
LuAlO₃	Nd1 (~∥c)	1.0675	2026	1	0.16	7.5	P	300	5 × 18	Xe	—	105
		1.0759	2099		(~0.2%)	8.5					—	
		1.0832	2026			11.5		300			8	
		1.0831	2160			2.8		120			55	
		1.0671	2022			1.7		77			15	
	Nd2	1.3437	3950	~0.6	0.245	7.5	P	300	5 × 25	Xe	40	106
Lu₃Al₅O₁₂	$^4F_{3/2} \rightarrow {}^4I_{9/2}$	0.9473	878			6	P	77			1	
	Nd1	1.06425	2099			5.3	CW	300			950	

Table 2.1.1.2 (continued)
LASER CRYSTALS

NEODYMIUM (Nd^{3+}, 4f^3) (Continued)

Host	Laser transition	Laser wavelength (μm)	Terminal Laser level energy (cm^{-1})	Laser ion concentration (%)	Fluorescence lifetime (msec)	Fluorescent linewidth (cm^{-1})	Mode of operation	Temperature (K)	Crystal size (mm)	Optical pump	Threshold	Ref.
		1.0605	2004			1.4	P	77			0.5	
		1.06375 → 1.0672	2100			2 →		120 → 900			2.5	
	Nd2	1.3387	4025			9.5	CW	300			2000	
		1.3532	4038			10.5	P				4.5	
		1.3525	4040			5.8		77			0.9	
Lu$_3$Ga$_5$O$_{12}$	Nd1	1.0609	2005	1	0.27	9	P	300	5 × 23	Xe	8	107
		1.0623	2060			9					1.5	
		1.0587	1996			4					8	
		1.06025	2010			4		77			4.5	
		1.0615	2065			4.5					10	
	Nd2	1.3315	3930			13					6	
Lu$_3$Sc$_2$Al$_3$O$_{12}$	Nd1	1.0599	1978	1	0.275	10	P	300	5 × 25	Xe	3	77
		1.0591	1979			5		77			3	
	Nd2	1.3370	3931			12		300			5	
LuScO$_3$	Nd1	1.0785	~1900	1	0.22	~30	P	300	4.3 × 11	Xe	35	79

α — NaCaCeF₆	Nd1	1.0653	~2000	1	0.44	~100	P	300	6×20	Xe	7	108
		1.0653 → 1.0633				~100→		300→920				28
	Nd2	1.3190	~4000			95		300			40	36
		1.3165				40		77			140	
α — NaCaYF₆	Nd1	1.0539	~2000	0.5-4	0.36	70	P	300	6×29	Xe	11	110
		1.0629			(0.5%)	90					17	28
		1.0539 → 1.0549				70→		300→550			—	
		1.0629 → 1.0597				90→		300→1000				
	Nd2	1.3285	~4100			70		300			20	36
		1.3375				80					18	
		1.3600				120					35	
		1.3260				30		77			45	
		1.3390				45					70	
5NaF · 9YF₃	Nd1	1.0506	~2000	1	0.96	100	P	300	6×15	Xe	45	110
		1.0505				—					83	
	Nd2	1.3070	~4000	1-10	0.15	200					250	22
NaLa(MoO₄)₂	Nd1	1.0595	~2000			—	P	300	4×35	Xe	5.5	111, 112
		1.0653	1953			50	CW		4.8×24		1.6	113
		1.0653 → 1.0665				50→	P	300→750			2600	28

Table 2.1.1.2 (continued)
LASER CRYSTALS

NEODYMIUM (Nd³⁺, 4f³) (Continued)

Host	Laser transition	Laser wavelength (μm)	Terminal level energy (cm⁻¹)	Laser ion concentration (%)	Fluorescence lifetime (msec)	Fluorescent linewidth (cm⁻¹)	Mode of operation	Temperature (K)	Crystal size (mm)	Optical pump	Threshold	Ref.
	Nd2	1.3380	3900	1.5		45		300	5 × 21		4.5	99
		1.3440	3940			50					10	
		1.3380	3900			15		77			4	
		1.3430	3937			18					25	
NaLa(WO$_4$)$_2$	Nd1 (IIc)	1.0635	~2000	1	–	–	P	300	6 × 90	Xe	–	114
		1.3355	~4000			~50			6 × 21			99
PbMoO$_4$	Nd1	1.0586	~2000	–	0.13	–	P	295	3 × 52	Xe	60	10, 115
	Nd2 (⊥c)	1.3340	3950	1.5		~35		300	5 × 21	Xe	50	99
		1.3320	3947			5		77			100	
Pb$_5$(PO$_4$)$_3$F	Nd1	1.0551	1955	1	0.225	~180	P	300	(3.3–5.5) × 15	Xe	12	116
SrI$_4$O$_7$	Nd1	1.0576	~2100	0.7	0.34	20	P	300	6 × 20	Xe	28	23
		1.0828				30					25	
		1.0566				1.5		77			4.5	
		1.0568				1.6					15	
		1.0627				1.8					18	
	Nd2	1.3345	~4100			20		300			80	22

Material	Ion	λ (μm)										Ref.
$SrAl_{12}O_{19}$	Nd1	1.3665				33					120	
		1.3320				1.7		77			25	
		1.3530				2					31	
SrF_2	Nd1	1.0491	~2000	1.5	0.4	13	P	300	10	Xe	14	23,52
		1.3065	~4000			15					30	
		1.0370	2008	0.8	1.1	20	P	300	5.8×30	Xe	40	10,117
		1.0445				25					95	
		1.0437	2008	0.1	1.25	3		77	3×52		150	
		1.0370→1.0395				20→		300→530	5.8×30		–	
		1.0446	2045			–		500→550			–	
$SrF_2 - CeF_3$	Nd1	1.0590	~2000	2	–	~110	P	300	5×21	Xe	14	118
	Nd2	1.3255	~4000			~130					35	
$SrF_2 - CeF_3 - GdF_3$	Nd1	1.0589	~2000	2	0.46	100	P	300	5×45	Xe	54	119
$SrF_2 - GdF_3$ (10%)	Nd1	1.0528	~2000	2	0.4	100	P	300	5×30	Xe	7	119
	Nd2	1.3260	~4000			75					10	
		1.3250	~4000			25		77			15	
$SrF_2 - LaF_3$ (30%)	Nd1	1.0597→1.0583	~2000	0.5–1	~0.8	~80→	P	300→800	5×45	Xe	18→	28
		1.3250	~4000	1		90		300			45	22
		1.3160	~4100			30		77			150	
		1.3235				30					180	
		1.3355				40					250	

Table 2.1.1.2 (continued)
LASER CRYSTALS

NEODYMIUM (Nd^{3+}, $4f^3$) (Continued)

Host	Laser transition	Laser wavelength (μm)	Terminal level energy (cm⁻¹)	Laser ion concentration (%)	Fluorescence lifetime (msec)	Fluorescent linewidth (cm⁻¹)	Mode of operation	Temperature (K)	Crystal size (mm)	Optical pump	Threshold	Ref.
SrF₃ – LuF₃ (5%)	Nd1	1.0556	~2000	2	0.25	75	P	300	5 × 25	Xe	11	11
	Nd2	1.3200	~4000			70			6 × 20		35	31
SrF₂ – YF₃ (10%)	Nd1	1.0567	~2000	2	0.36	~65	P	300	6 × 20	Xe	15	120
	Nd2	1.3225	~4000			70					140	22
		1.3300				75					160	
		1.3320	~2000			25		77			200	
SrMoO₄	Nd1	1.0576	~2000	1–3	~0.17	30–40	P	295	3 × 52	Xe	10	10, 43
		1.0643				5					125	
		1.059				—		77			150	
		1.0611				—					500	
		1.0627				—					170	
		1.0640				—					17	
		1.0652				3					70	
	Nd2 (IIc)	1.3325	~3960	1		20		300	5 × 21		35	99
		1.3440	3962			6		77			50	
Sr(PO₄)₃F	Nd1 (IIa)	1.0585	~2000	2	0.31	~10	P	300	6.4 × 33	Xe	12	40

Material	Ion	λ (μm)					Mode	T (K)	Dimensions	Pump	Threshold	Ref.
SrWO₄	Nd1	1.063	~2000	–	0.2	–		295	3 × 52	Xe	180	10
		1.0574				3		77			4.7	
		1.0607				–					7.6	
		1.0627				–					5.1	
SrY₅F₁₉	Nd1	1.0493	~2000	2	~0.4	~110	P	300	5 × 16	Xe	40	121
	Nd2	1.3190	~4000			~130					75	
YAlO₃	⁴F₃/₂ → ⁴I₉/₂ (∥c)	0.930	670	1	0.18 (0.3%)	~30	P	300	5 × 8, 3 × 16	XeL	0.15	122
	Nd1	1.0795	2157	1–3		11	P		3.5–5 × 30–50	Xe	~1.5	123, 124
							CW		6.35 × 75	Kr	1200	123, 125, 126
		1.0645	2026			9.5	P		5 × 50	Xe	–	124
							CW			Xe	1750	125
		1.0729	2097			8.3	P		6.35 × 75	Kr	–	
		1.0909	~2300			–	P	300	6.3 × 75	Kr	–	125
		1.0989	2322			11					–	
		1.06405	2023			~1.5		77	(5–6) × (15–43)	Xe	<1	127
		1.07255	2097			~2					<1	
		1.06405 → 1.0654	2030			~1.5 →		77 → 500	4 × 14		–	46, 127
		1.0652 → 1.0659	2156			–		310 → 500			–	
		1.07255 → 1.0730	2097			2 → 19		77 → 490			–	
		1.07955 → 1.0802	2156			3 →		77 → 600			–	
		1.0796 → 1.0803	2270			–		600 → 700			–	

Table 2.1.1.2 (continued)
LASER CRYSTALS

NEODYMIUM (Nd^{3+}, $4f^3$) (Continued)

Host	Laser transition	Laser wavelength (μm)	Terminal level energy (cm⁻¹)	Laser ion concentration (%)	Fluorescence lifetime (msec)	Fluorescent linewidth (cm⁻¹)	Mode of operation	Temperature (K)	Crystal size (mm)	Optical pump	Threshold	Ref.
		1.0847	2318			—		~530			—	
		1.0913	2373			—					—	
		1.0991	2318			18		~500			—	
	Nd2	1.3400	3958			~20		300	6.3×76	Kr	10	59, 60, 125
							CW				—	
	(IIb)	1.3416				~20	P		5×25	Xe	1	
							CW		6.3×76	Kr	—	
		1.3391	3953			4.5	P	77	5×25	Xe	0.5	
		1.3514	4021								1	
		1.3644	4092			6					5	
$YAlO_3$+	Nd1 (IIc)	1.0645	2023	0.9	0.16		P	295	5×50	Xe	1.2	128
0.3% Cr							CW			Kr	910	126
$Y_3Al_5O_{12}$	$^4F_{3/2} \rightarrow {}^4I_{9/2}$	0.8910	200	1	0.255		P	300	13	ArL	0.1	129
		0.8999	311									
		0.9385	852									
		0.9460	852			~9			3×75	Xe	38	130

λ (μm)	(cm⁻¹)			Mode	Temp (K)	Size (mm)	Lamp	Output	Ref
1.06415	2110	~1	6.5	P	300	2.5×14-20	Xe	1	131
1.0521	2002		4.5	CW		2.5×28	W	594	132
1.0615	2002		3.6					330	
1.06415	2110		6.5					100	
1.0646	2029		–			4×76	Kr	–	133
1.0682	2146		~10	P		5×50	Xe	50	134
1.0737	2110		4.6	CW		2.5×28	Kr	348	132
1.0780	2146		~9			4×76		~1700	133
1.1054	2461		–					~3000	
1.1119	2514		10.2			2.5×28	W	623	132
1.1158	2461		10.6					646	
1.1225	2514		9.9					676	
1.0612	2002		~1	P	77	2.5×14-20	Xe	0.2	131
1.0610 → 1.0627	2002		1→16	P	77→600	5×40		–	28
1.0637 → 1.0670	2110		3.5→35		170→900	5×40		–	28
Nd2 1.3188	3922		6.5	CW	300	4×76	Kr	~1900	133
1.3200	3933		–					~2000	
1.3338	3922		7					~2800	
1.3350	3933		7.5					~2100	
1.3382	4034		7.7					~1600	
1.3410	~4000		–					~2000	
1.3533	4034		6	P		6×52	Xe	–	135

Table 2.1.1.2 (continued)
LASER CRYSTALS

NEODYMIUM (Nd^{3+}, 4f^3) (Continued)

Host	Laser transition	Laser wavelength (μm)	Terminal level energy (cm^{-1})	Laser ion concentration (%)	Fluorescence lifetime (msec)	Fluorescent linewidth (cm^{-1})	Mode of operation	Temperature (K)	Crystal size (mm)	Optical pump	Threshold	Ref.
		1.3564	~4000			—	CW		4 × 76	Kr	~2000	133
		1.3572	4055			6.5	P		6 × 52	Xe	1.9	135
		1.4140	~4000			—	CW		4 × 76	Kr	~3500	133
		1.4440	~4000			—					~5000	
	$^4F_{3/2} \rightarrow {}^4I_{15/2}$	1.833	5965			—	P	293	3 × 50	Kr	9	136
								230			7.5	
Y$_3$Al$_5$O$_{12}$ + 1% Cr	Nd1	1.0641	2110	1.3	0.21		P	300	3 × 30	Xe	2.1	137
							CW			Hg	750	
										W	>800	
		1.0612	2001		0.23			77		Hg	180	
										W	440	
Y$_3$Ga$_5$O$_{12}$	Nd1	1.0589	1990	1	0.26	6	P	300	5 × 20	Xe	5	72, 138
		1.0603	2005			6					4	
		1.0625	2060			6.5					2	
		1.0583	1990			2		77			1	
		1.05975	2005			2					7	

Material		λ (µm)										Ref.
(Y,Lu)$_3$Al$_5$O$_{12}$	Nd2	1.0614	2060			3						13	
	Nd1	1.3305	3930			8						12.5	
	Nd1	1.0642	~2100	0.3	0.2	—	P	300	4×4×25	Xe	—	139
		1.0608				—		77			—	
		1.0636				—					—	
		1.0726				—					—	
Y$_2$O$_3$	Nd1	~1.0746	1895	1.5	0.34	~8	CW	300	0.1×0.25	KrL	0.57 mW	140,141
		1.073	1892	1	0.26	2	P	77	2×2	Xe	260	
		1.078	1928			2					350	
	Nd2	~1.358	3837	1.5	0.34	~12	CW	300	0.1×0.25	KrL	0.7 mW	
Y$_2$O$_3$ – ThO$_2$ – Nd$_2$O$_3$	Nd1	~1.074	~2000	1		22	P	300	4.6×76	Xe	7.8	142
						(5%ThO$_2$)						
YP$_5$O$_{14}$	Nd1	~1.051	~2000	10	0.22	—	P	300	—	ArL	—	68
Y$_3$Sc$_2$Al$_3$O$_{12}$	Nd1	1.0595	~2000	1	—	2	P	300	5×20	Xe	5	31,51
		1.0622				6					7	
		1.0587				8		77			3	
	Nd2	1.3360	~4000			13		300			10	
Y$_3$Sc$_2$Ga$_3$O$_{12}$	Nd1	1.0583	~2000	1	—	8	P	300	5×10	Xe	1.5	31
		1.0615				~20					3	
		1.0575				2		77			1	
	Nd2	1.3310	~4000			~14		300			15	

Table 2.1.1.2 (continued)
LASER CRYSTALS

NEODYMIUM (Nd^{3+}, 4f^3) (Continued)

Host	Laser transition	Laser wavelength (μm)	Terminal laser level energy (cm⁻¹)	Laser ion concentration (%)	Fluorescence lifetime (msec)	Fluorescent linewidth (cm⁻¹)	Mode of operation	Temperature (K)	Crystal size (mm)	Optical pump	Threshold	Ref.
YScO₃	Nd1	1.0843	1950	0.5	0.24	—	P	300	6 × 18	Xe	55	79
		1.0770	1850			~40		77			70	
Y₂SiO₅	Nd1	1.0715	~2100	1–2	0.24	13	P	300	5 × 20	Xe	4	143
		1.0742				17					5	
	O	1.0782				~20					2	
	O	1.0710				2		77			1.5	
		1.0781				4					0.5	
	Nd2	1.3585	~4100			~20		300			4	
YVO₄	Nd1	1.0625	1964	~1	0.092	—	P	300	5 × 20	Xe	2.6	144
		1.0634	~2000			7	CW		4.8	ArL	0.11	145
		1.0641	1964			7	P		5 × 20	Xe	60	144
		1.0648	1985			—					50	
		1.0664	1985			8					1.3	
		1.069	~2000	0.1	0.033	—		~90	5 × 40		~1	146
		1.0664 → 1.0672	1985			8 →		300 → 690			—	46
	Nd2	1.3425	3913			8	P	300	5 × 20		9	22

Material	Transition	λ (μm)						T (K)		Pump		Ref.
ZrO_2–Y_2O_3		1.34	~4000	1–2		–	CW	77	4.8	ArL	~0.5	145
	Nd1	1.3415	3912	~1		~3	P	300	5 × 20	Xe	15	22
	Nd2	1.0608	~2000	0.5–0.7	0.45	–	P		3–6 × 24–32	Xe	7–20	80
		1.3320	~4000	0.5		~130	P		5 × 17		42	81
SAMARIUM (Sm^{3+}, $4f^5$)												
TbF_3	$^4G_{5/2} \rightarrow {}^6H_{7/2}$	0.5932	1039	0.1	2	~5	P	116	5 × 50	Xe	165	307
EUROPIUM (Eu^{3+}, $4f^6$)												
Y_2O_3	$^5D_0 \rightarrow {}^7F_2$	0.6113	859	5	0.87		P	220	1 × 12	Xe	85	147
YVO_4	$^5D_0 \rightarrow {}^7F_2$	0.6193	~1000				P	90	6.3 × 16	Xe	60	148
GADOLINIUM (Gd^{3+}, $4f^7$)												
$Y_3Al_5O_{12}$	$^6P_{7/2} \rightarrow {}^8S_{7/2}$	0.3146	0	~2	0.008	~20	P	300	3 × 20	Xe	2400	149
TERBIUM (Tb^{3+}, $4f^8$)												
$LiYF_4$ (+10% Gd)	$^5D_4 \rightarrow {}^7F_5$	0.5445	~2000	10	5	–	P	300	4 × 40	Xe	75	150
DYSPROSIUM (Dy^{3+}, $4f^9$)												
$Ba(Y_{1.26}Er_{0.7})F_8$	$^6H_{13/2} \rightarrow {}^6H_{15/2}$	3.022	216	2	7	60	P	77	–	Xe	510	151
HOLMIUM (Ho^{3+}, $4f^{10}$)												
$Ba(Y_{1.8}Ho_{0.2})F_8$	$^5F_5 \rightarrow {}^5I_5$	2.362	11260	10	–	–	P	77	–	Xe	555	152
		2.375	11317								755	
		2.363	11260					20			302	

Table 2.1.1.2 (continued)
LASER CRYSTALS

Host	Laser transition	Laser wavelength (μm)	Terminal level energy (cm⁻¹)	Laser ion concentration (%)	Fluorescence lifetime (msec)	Fluorescent linewidth (cm⁻¹)	Mode of operation	Temperature (K)	Crystal size (mm)	Optical pump	Threshold	Ref.
					HOLMIUM (Ho^{3+}, $4f^{10}$) (Continued)							
		3.377	11293								295	
$Ba(Y,Yb)_2F_8$	$^5S_2 \rightarrow {}^5I_8$	0.5515	385	0.5	—	—	P	77	—	Xe	635*	153
$BaY_{1.64}Er_{0.03}Tm_{0.03}F_8$	Ho	2.171	V	—	—	—	P	295	—	Xe	450	152
		2.0644	352				CW	85		W	—	
		2.065					P	77		Xe	46	
		2.0746						85			30	
		2.0746 → 2.076					CW	85		W	50	
		2.0866	381				P	77		Xe	30	
							CW	85		W	50	
		2.0555	310				P	20		Xe	9	
		2.074	352				P			Xe	55	
CaF_2	$^5S_2 \rightarrow {}^5I_8$	0.5512	270	0.4–0.8	0.7	~6	P	77	6.5 × 75	Xe	1200	154
	Ho	2.092	~230						3 × 52	Xe	260	10
$CaF_2 - ErF_3$ (2%)	Ho	2.030	~250	0.5	~10	~70	P	77	6 × 45	Xe	~140	155
$CaF_2 - ErF_3$ (3%)	Ho	2.06	~250	0.5–1	—	—	P	298	—	Xe	100	156
$- YbF_3$ (3%)		2.05	—			~6		100			16	

Material	Ion	λ (μm)					Mode	Temp (K)	Dimensions	Lamp		Ref
– TmF₃ (3%)		2.1				–	CW	77	5 × 80		90	
CaF₂ – YF₃	Ho	2.0318	~80	3	–	~90	P	65	5.5 × 45	Xe	280 / 145	157
											3000	
CaF₂ – YF₃ (+ Er, Tm, Yb)	Ho	2.06	–	0.5	–	–	P	90	–	Xe	100	158
		2.05						77			16	
CaMoO₄	Ho	2.0556	–	0.5	1.3	–	P	298	–	Xe	310	159
(+ 0.75% Er)		2.0707						100			170	
		2.0740	~250								107	
								20			45	
Ca(NbO₃)₂	Ho	2.047	–	0.5	2.2	–	P	77	3 × 30	Xe	90	2
Ca₅(PO₄)₃F (+ 0.3% Cr)	Ho	~2.079	355	3	0.5	~60	P	77	6 × 22	Xe	25	40
CaWO₄	Ho	2.046	230	0.5	–	–	P	77	3 × 52	Xe	80	10, 160
		2.059									250	
CaY₄(SiO₄)₃O (+ 37.5% Er, 3.8% Tm)	Ho	2.060	~300	2.5	2.5	55	P	77	7 × 32	Xe	15	40
ErAlO₃	Ho	2.1205	~570	2	–	~10	P	77	4 × 15	Xe	80	42
(Er, Lu) AlO₃	Ho	2.0010	~200	2-5	–	~6	P	77	5 × 25	Xe	75	161
		2.1205	~570								30	

* Infrared excitation.

Table 2.1.1.2 (continued)
LASER CRYSTALS

HOLMIUM (Ho^{3+}, $4f^{10}$) (Continued)

Host	Laser transition	Laser wavelength (μm)	Terminal level energy (cm⁻¹)	Laser ion concentration (%)	Fluorescence lifetime (msec)	Fluorescent linewidth (cm⁻¹)	Mode of operation	Temperature (K)	Crystal size (mm)	Optical pump	Threshold	Ref.
Er₂O₃	Ho1	2.121	–	–	10	–	P	145	12	Xe	20	162
								77			5	
							CW	77		W	200	
Er₃Sc₂Al₃O₁₂ (+ Tm)	Ho1	2.0985	~470	2	–	~22	P	77	5.5 × 20	Xe	10	77
Er₂SiO₅	Ho1	2.085	~500	1	0.08	–	P	77	–	Xe	~10	163
(Er, Th, Yb)₃Al₅O₁₂	Ho1	2.1010	~500	6	–	~12	P	77	6 × 31	Xe	10	164
ErVO₄	Ho1	2.0416	~230	1	–	–	P	77	5 × 7	Xe	–	165
(+ 1% Tm)	Ho1						CW			ArL	–	
GdAlO₃	Ho1	1.9925	91	3	–	–	P	90	2 × 20	Xe	240	166
Gd₃Ga₅O₁₂	$^5I_6 \rightarrow {}^5I_8$	1.2085	~500	1	–	–	P	~100	5.5 × 40	Xe	47	167
	$^5S_2 \rightarrow {}^5I_5$	1.4040	~11350		0.06			~110			42	
	Ho1	2.0885	~500		8.8			~110			50	
KGd(WO₄)₂	Ho2	2.9342	~5300	3–5	–	–	P	300	4 × 48	Xe	25	168
KY(WO₄)₂	Ho1	2.0720	~300	3	–	–	P	110	5.5 × 55	Xe	55	169
(+ 30% Er, 3% Tm for Ho1 only)								220			200	
for Ho1 only	Ho2	2.9395	~5300					300	5 × 30	Xe	14	170

Material	Transition	Wavelength (μm)					Mode	Temp (K)	Dimensions	Pump	Output	Ref
LaNbO$_4$	Ho1	2.07	~300	0.9	~20	8	P	90	6×20	Xe	20	171
(+ 0.5% Er for Ho1 only)	Ho2	2.8510	~5300	2	~0.1		P		6×22	Xe	50	172
LiNbO$_3$	Ho1	2.0786	310				P	77		Xe	425	24
LiYF$_4$	$^5S_2 - {}^5I_7$	0.7505	~5300	2	0.09	12	P	300	3×22	Xe	12	173
		0.7498			0.05	-		90	6×30		13.5	174
	$F_5 - {}^5I_7$	0.9794	~8800	2	0.1			300			18	
	$^5S_2 - {}^5I_6$	1.0143	~11300		0.05						18	
	$^5S_2 - {}^5I_5$	1.3960									78	
	$^5I_5 - {}^5I_7$	1.392		-	-		P		8×8×10	DNGL	0.001	175
		1.673	~5300		~10			115		Xe	24/cm	176
	Ho1	2.0672	~350	2			P	90	6×30	Xe	75	174
	$^5I_5 - {}^5I_6$	3.914	~8800	-	-		P	300	8×8×10	DNGL	0.0018	175
LiYF$_4$ (+ 5% Er)	Ho1	2.066	~300	2			P	77	4.5×24	Xe	40	177
Li(Y, Er) F$_4$ (+ 6.7% Tm)	Ho1	2.0654	~300	1.7	12	-	P	300	3.2×32	Xe	20	178
LuAlO$_3$	Ho1	2.1348	~470	-	-		P	90	-	Xe	3	179
Lu$_3$Al$_5$O$_{12}$	Ho1	2.1020	~480	2	-	11	P	77	6×23	Xe	50	180
	Ho2	2.9460	~5400	~5	-		P	300	6×37	Xe	60	181
Lu$_3$Al$_5$O$_{12}$ (+ 2% Er, 2% Tm)	Ho1	2.1020	~480	2	-	11	P	77	6×23	Xe	15	180
							CW			W	65	

Table 2.1.1.2 (continued)
LASER CRYSTALS

HOLMIUM (Ho^{3+}, 4f^{10}) (Continued)

Host	Laser transition	Laser wavelength (μm)	Terminal level energy (cm^{-1})	Laser ion concentration (%)	Fluorescence lifetime (msec)	Fluorescent linewidth (cm^{-1})	Mode of operation	Temperature (K)	Crystal size (mm)	Optical pump	Threshold	Ref.
Lu$_3$Al$_5$O$_{12}$ (+ 10% Yb, 0.3% Cr)	Ho2	2.9460	~5400	10	—	—	P	300	5×40	Xe	10	182
α−NaCaErF$_6$	Ho1	2.0345	~120	~1	—	~120	P	150	6×20	Xe	300	183
		2.0312				~100		77			180	
		2.0377									210	
NaLa(MoO$_4$)$_2$	Ho1	2.050	~250	1	8	~24	P	90	6×20	Xe	25	171
(+ 2.8% Er)	Ho1	2.050	~250	3	10	~24	P	90	6×20	Xe	32	171
SrY$_4$(SiO$_4$)$_3$O (+ 37.5% Er, 3.8% Tm)	Ho1	~2	~300	—	—	—	P	77	6×24	Xe	77	40
YAlO$_3$	Ho2	2.9180	~5300	2	—	—	P	300	5×15	Xe	7.5	181
		3.0132	5600	10	—	9			5.5×47		50	184
(+ 30−50%Er, 3−6.7% Tm)	Ho1	2.123	474	~2	4.8	17		300	2−6×50	Xe	~300	185
		2.119			6.6	9.8		77			1	
Y$_3$Al$_5$O$_{12}$	Ho1	2.0914	532	~4	4-5	—	P	77	~25	Xe	1760	128
		2.0975	462								44	
		2.1223	518								410	

Dopant	Ion	λ (μm)	λ_p (nm)				Mode	T (K)	Dim	Lamp	Power	Ref.
	Ho2	2.9403	~5400			9		300	5 × 25	Xe	25	181
(+ 0.5% Cr)	Ho1	2.0975	462	10	–	–	P	77	~25	Xe	25	186
			518	–			CW	85		W	210	
		2.1223					P	77		Xe	25	
							CW	85		Hg	1300	
										W	250	
(+ ~50% Er)	Ho1	2.0917	532	2	4	–	P	77	~25	Xe	390	186
		2.0979	462								11	
							CW	85		W	47	
										Hg	680	
(+ 50% Er, 6–7% Tm)	Ho1	2.123	518				P	77		Xe	3800	
		2.0982	462	2	~5	–	P	77	32	Xe	–	187
		2.0990	~462		5	–	P		50		3	188
							CW			W	45	
		2.1227	518				P		32		–	187
							CW			W	30	187
		2.1285	~538	2			P		50	Xe	14	188
		~2.13	532	1.65			P	300	3 × 50	Xe	60	189
									3 × 30		98	190
											0.6	
							CW	77		W	5	

Table 2.1.1.2 (continued)
LASER CRYSTALS

Host	Laser transition	Laser wavelength (μm)	Terminal level energy (cm⁻¹)	Laser ion concentration (%)	Fluorescence lifetime (msec)	Fluorescent linewidth (cm⁻¹)	Mode of operation	Temperature (K)	Crystal size (mm)	Optical pump	Threshold	Ref.
HOLMIUM (Ho³⁺, 4f¹⁰) (Continued)												
$Y_3Fe_5O_{12}$ (+ 5% Er, 5% Tm)	Ho1	2.086	402	2	3.6	~16	P	77	5	Xe	30	191
			413									
		2.089	413			~17					30	
			~425									
		2.107	461			~20					50	
			487									
$Y_3Ga_5O_{12}$	Ho1	2.086	418	–	–	–	P	77	5	Xe	70	191
		2.114	481								250	
YVO_4 (+ 25% Er, 7% Tm)	Ho1	2.0412	229	1	–	–	P / CW	77	5×7	Xe / ArL	50	165
$Yb_3Al_5O_{12}$	Ho1	2.0960	~100	1	–	–	P	77	2.5×25	Xe	143	192
$(Yb,Er,Tm)_3Al_5O_{12}$	Ho1	2.1010	~500	~6	–	~12	P	77	6×31	Xe	10	164
ZrO_2 (+ 12% ErO_3)	Ho1	2.115	~300	1	6.5	–	P	77	6×33	Xe	400	193
ERBIUM (Er³⁺, 4f¹¹)												
$Ba(Y_{1-x}Er_x)_2F8$	$^4S_{3/2} \rightarrow {}^4I_{15/2}$	0.5540	406	$x = .005 \rightarrow 0.2$	0.83	–	P	77	5	Xe	–	194

Material	Transition	λ (µm)	(cm⁻¹)				T (K)	Size (mm)	Pump		Ref.
$BaY_{1.2}Yb_{0.75}Er_{0.05}F_8$	$^2H_{9/2} \to ^4I_{13/2}$	0.5617	6740 x = 0.12, 0.2 (x = 0.005)								
	$^4F_{9/2} \to ^4I_{15/2}$	0.6709	~400 x = 0.2								
	$^2H_{9/2} \to ^4I_{11/2}$	0.7037	10336								
	$^4F_{9/2} \to ^4I_{15/2}$	0.6700	410	5	—	P	77	—	Xe	170	153
		0.6709		6					Xe	165	
		0.6709							Xe (IR)	195	
$CaAl_4O_7$ Er1		1.5500	~400	0.7–1	6.8	P	77	6 × 32	Xe	180	21
		1.5815	~500		~8					250	
Cc^{-}_2	$^4S_{3/2} \to ^4I_{13/2}$	0.8546	~7000	4–8	6–8	P	77	5 × 53		100	195
		0.8548								—	
	$^4S_{3/2} \to ^4I_{11/2}$	1.26	11000	0.1	0.3			8 × 60		2000	196
Er1		1.5298	~30	0.05	0.8–1.3		4	5 × 38		8.2	197
		1.5308								—	
	$^4S_{3/2} \to ^4I_{9/2}$	1.617	~400	0.1	20	P	77	9.5 × 51		1000	198
		1.696	12413					8 × 60		2000	196
		1.715	~13000		0.3					1000	
		1.726								1000	
Er2		2.7307	~6500	4	—	P	300	9 × 120		700	199
$CaF_2 - ErF_3$ $- TmF_3$ (0.5%) Er2		2.69	~6200	12.5	4.5		298	3 × 25	Xe	10	200

Table 2.1.1.2 (continued)
LASER CRYSTALS

ERBIUM (Er^{3+}, $4f^{11}$) (Continued)

Host	Laser transition	Laser wavelength (μm)	Terminal level energy (cm⁻¹)	Laser ion concentration (%)	Fluorescence lifetime (msec)	Fluorescent linewidth (cm⁻¹)	Mode of operation	Temperature (K)	Crystal size (mm)	Optical pump	Threshold	Ref.
$CaF_2 - YF_3$	$^4S_{3/2} \rightarrow {}^4I_{13/2}$	0.8430	6711	0.5–5	0.5	9.5	P	77	5.5 × 45	Xe	200	201, 202
		0.8456			(0.5%)	8.5					40	
$Ca(NbO_3)_2$	Er1	1.547	107	—	~10	~12	P	77	3 × 30	Xe	250	2
	Er1	1.61	~400								800	203
$CaWO_4$	Er1	1.612	375	1		~10	P	77	6 × 6 × 25.4	Xe	800	203
$GdAlO_3$	Er1	1.5646	224	3			P	77	2.5 × 20	Xe	18	204
	$^4S_{3/2} \rightarrow {}^4I_{9/2}$	1.6714	12454	0.5	0.09		P				85	205
$Gd_3Al_5O_{12}$	Er2	2.8128	~6500	10			P	300	—	Xe	125	206
$KY(WO_4)_2$	$^4S_{3/2} \rightarrow {}^4I_{13/2}$	0.8621	~6700	3			P	300	3.5 × 26	Xe	7.5	207
	$^4S_{3/2} \rightarrow {}^4I_{9/2}$	1.7178	~12000	3–10					5.5 × 23.5		19	
		1.7355		3–50					(3.5–6) × (17–47)		~5	
	Er2	2.8070	~6700						(3.5–6) × (23.5–35)		30–83	
(+Ho, Tm)	Er2	2.6887	~6500	30					3.5 × 55		10	169
$K(Y, Er)(WO_4)_2$	Er2	2.8070	~6700	50			P	300	6 × 35	Xe	30	207
$KGd(WO_4)_2$	$^4S_{3/2} \rightarrow {}^4I_{13/2}$	0.8468	6517	3					6 × 25	Xe	34	168
	$^4S_{3/2} \rightarrow {}^4I_{9/2}$	1.7155	12498				P	300			30	

Material	Transition	λ (μm)	cm⁻¹				Mode	T (K)	Geometry	Lamp		Ref
KLu(WO$_4$)$_2$		1.7325	12556	3–10					(5–6) × (25–40)		10	
	Er2	2.7222	6520	3					6 × 25		50	
		2.7990	6710	5–25					(4–5) × (24–25)		15	
LiYF$_4$	Er2	2.8092	~6700	3–25	–		P	300	–	Xe	15	206
	Er1	1.6113	400	0.05	–		P	77	5.6 × 25	Xe	500	208
	$^4S_{3/2} \rightarrow {}^4I_{13/2}$	0.85	6714	2	0.2	10	P	300	3–5 × 30	Xe	10	209
		0.8503	6673		–				5 × 50		32.5	210
					0.7			110			10	
	$^4S_{3/2} \rightarrow {}^4I_{11/2}$	0.8535	6725		0.2			300			25	
		1.2308	10313		0.7			110			40	
											7.5	
	$^4S_{3/2} \rightarrow {}^4I_{9/2}$	1.6470	12572		0.2			300	–		60	211
		1.7320			0.7			110	5 × 50		<10	
	Er2	2.81	~6700		10			300	–		55	212
		2.870	6738		10				5 × 50		130	
								110			90	
LuAlO$_3$	$^4S_{3/2} \rightarrow {}^4I_{9/2}$	1.6675	~12400	1	–		P	90	5 × 25	Xe	8	179
Lu$_3$Al$_5$O$_{12}$	$^4S_{3/2} \rightarrow {}^4I_{13/2}$	0.86325	6818	1.5	0.13	8	P	300	7 × 43	Xe	45	213
		0.86325	6824	8	(0.1%)	5.5		300	5.5 × 35		100	
	Er1	1.6525	534	2–5	6–7	6.5		77	6 × 23		100	180

Table 2.1.1.2 (continued)
LASER CRYSTALS

ERBIUM (Er^{3+}, $4f^{11}$) (Continued)

Host	Laser transition	Laser wavelength (μm)	Terminal level energy (cm⁻¹)	Laser ion concentration (%)	Fluorescence lifetime (msec)	Fluorescent linewidth (cm⁻¹)	Mode of operation	Temperature (K)	Crystal size (mm)	Optical pump	Threshold	Ref.
	$^4S_{3/2} \rightarrow {}^4I_{9/2}$ Er2	1.6630	578			8.5					40	
		1.7762	12772	1.5	0.11	–		300	7×43		30	213
		2.8298	6885	≈33	0.12	–			5×40		11	182
		2.9408		1.5					7×43		55	213
		2.9395		≈33					5×31		12	
(+ Tm, 3%)		2.6990	6560	10					5.5×31		18	
(+ Tm, 3%, Ho, 3%)		2.6990		9					6×29		25	
(+ Yb, 5%)		2.8298	~6885	5–50					5×40		5.5	182
		2.9395									10	
		2.9405		5							15	
(+ Yb, 5%, Cr 0.3%)		2.8298		33							6	
		2.9395									10.5	
YAlO₃	$^4S_{3/2} \rightarrow {}^4I_{13/2}$	0.84975	6841	0.5–2	0.12	6.5	P	300	5×50	Xe	100	181, 185
		0.8594	6774		(0.3%)	13					65	
		0.84965	6643			~1.5		77			20	
		0.85165	6671			~1.5					35	

$Y_3Al_5O_{12}$	Er1	1.5554	260	10	–	–	300		3.5 × 15		120	205
	$^4S_{3/2} \to {}^4I_{9/2}$	1.6632	12397	2	0.12	~50			5–6 × 50		20	181,214
		2.7309	6642	10	1	–			5 × 50		40	181
	$^4S_{3/2} \to {}^4I_{13/2}$	2.9155	6871	10	0.15	–	77		3.5 × 15		150	205
		0.8627	~6800	2.5	0.12	~10	300	P	(4–6) × (25–46)	Xe	45	181
	Er1	0.8627	6805	20	(0.3%)	6.5	77				80	37
		1.632	426	1.2	6.5	–	300		5 × 57		320	215
		1.6452	525	1	~8		77		~25		470	91
		1.6602									80	
	$^4S_{3/2} \to {}^4I_{9/2}$	1.7757	12766	1.2	0.12		300		5 × 57		22	216
	Er2	2.8302	~6880	2.5–10					(4–6) × (25–46)		35	37
		2.9365		33	0.09				(4–6) × 40		11	217
											~12	
(+ Yb)	Er1	1.6459	459	1	9.1	~20	295		6.3 × 51	Xe	75	218
	Er2	2.6975	6544	33	–	–	300		–	Xe	8	206
		2.8302	~6880		–						12	
(+ Yb, Tm)	Er2	2.9365	6880	33	–	–	300	P	–	Xe	35	206
(+ Tm)		2.6975	6544								7.5	
		2.8302	~6880								9.5	
+ (Tm, Ho)			6544								10	
$Y_{0.8}Gd_{0.2}AlO_3$	Er1	1.5542	267	5	5	–	77	P	3 × 20	Xe	90	205

Table 2.1.1.2 (continued)
LASER CRYSTALS

Host	Laser transition	Laser wavelength (μm)	Terminal level energy (cm⁻¹)	Laser ion concentration (%)	Fluorescence lifetime (msec)	Fluorescent linewidth (cm⁻¹)	Mode of operation	Temperature (K)	Crystal size (mm)	Optical pump	Threshold	Ref.
				ERBIUM (Er³⁺, 4f¹¹) (Continued)								
	$^4S_{3/2} \rightarrow {}^4I_{9/2}$	1.6600	12386	0.5	0.1				2.5 × 20		65	
$Y_{0.8}Gd_{0.2}ScO_3$	Er1	1.6437	400	3	6	–	P	77	2.5 × 15	Xe	70	205
	$^4S_{3/2} \rightarrow {}^4I_{9/2}$	1.6682	12364	0.5	0.15						60	
	Er2	2.8637	6737	10	2						120	
$Yb_3Al_5O_{12}$	Er1	1.6615	~500	3		~12	P	77	5 × 20	Xe	75	164
$ZrO_2 - Er_2O_3$	Er1	1.620	~300	12	–	–	P	77	6 × 33	Xe	800	193
				THULIUM (Tm³⁺, 4f¹²)								
CaF_2	Tm1	~1.9	–	–	–	–	P	77	–	–	–	197
(+ ErF₃, 2%)		1.894	~400	0.5	11.5	~100	P	77	5.5 × 45	Xe	~45	219
(+ ErF₃, 10%)					6		CW	65	5 × 80		3000	156
(+ ErF₃, 12.5%)					2.9		P	100	3 × 25		5	200
Ca (NbO₃)₂	Tm1	1.91	–	–	–	–	P	77	3 × 20	Xe	125	2
CaMoO₄ (+ Er, 0.75%)	Tm1	1.9060	~325	0.5	0.9	–	P	77	–	Xe	20	159
		1.9115									19	
CaWO₄	Tm1	1.911	325	0.3	–	–	P	77	3 × 52	Xe	60	10, 220
		1.916		0.5							73	

Material	Transition	Wavelength					Mode	Temp	Dimensions	Pump		Ref
ErAlO$_3$	Tml	1.872	~200	-	5.5	-	P	150	4×12	Xe	300	221
					3.5			77			15	
Er$_3$Al$_5$O$_{12}$	Tml	1.9-2.0	-	-	-	-	P	77	-	-	-	222
(Er, Lu) AlO$_3$	Tml	1.8845	~400	2-5	-	5.5	P	77	5×25	Xe	45	119
Er$_2$O$_3$	Tml	1.934	-	0.5	2.9	-	P	77	10	Xe	3	223
	W						CW			W	1500	
GdAlO$_3$	Tml	1.8529	~226	3	-	-	P	77	2×20	Xe	216	224
LiNbO$_3$	Tml	1.8532	~271	0.25-1	-	-	P	77	-	Xe	220	24
LiYF$_4$	$^1D_2 \rightarrow {}^3F_4$	0.4526	5969	10	0.001	~100	P	300	5×5×23	XeFL	0.02	225
Lu$_3$Al$_5$O$_{12}$	Tml	1.8855	~225	2	-	6	P	77	6×23	Xe	400	180
		2.0240	~580			10.5					160	
α − NaCaErF$_6$	Tml	1.8580	~470	~1	-	~75	P CW	150	6×20	Xe W	95	183
		1.8885	~570			~50					120	
		1.8580	~470			~70		77			65	
		1.8885	~540			~50					100	
SrF$_2$	Tml	1.972	-	-	-	-	P	77	3×52	Xe	1600	10
YAlO$_3$	Tml	1.856	~200	1	~10	-	P	90	5×20	Xe	-	226
(+ 0.1% Cr)		1.883	~300								8	
		1.9335	~420								7	
	$^3F_4 \rightarrow {}^3H_5$	~2.34	~8250		~0.1	-		300			12.5	
								90			5	

Table 2.1.1.2 (continued)
LASER CRYSTALS

THULIUM (Tm^{3+}, 4f^{12}) (Continued)

Host	Laser transition	Laser wavelength (μm)	Terminal level energy (cm^{-1})	Laser ion concentration (%)	Fluorescence lifetime (msec)	Fluorescent linewidth (cm^{-1})	Mode of operation	Temperature (K)	Crystal size (mm)	Optical pump	Threshold	Ref.
		2.348		0.8				300	5 × 55		–	227
		2.349									–	
(+ 0.25% Cr)		2.274		1					3 × 50	Kr	31	228
		2.318								Xe	132	
		2.353									110	
(+ 30% Er)	Tm1	1.861	~240	1	4	5.5		77	5 × 50		25	185
$Y_3Al_5O_{12}$	Tm1	1.8834	240	–	15	–	P	77	~25	Xe	590	186
		2.0132	582								208	
	$^3F_4 \rightarrow {}^3H_5$						CW	85		W	315	
(+ 0.1% Cr)		2.324	~8300	1	–	–	CW	300	3 × 50	W	200	228
(+ 0.5% Cr)	Tm1	2.019	600	–	–	–			~25		640	186
		2.0132	582					77			30	
								85			160	
										Hg	800	
(+ 50% Er)		1.880	228	2	15	–	P	77	~25	Xe	264	
		1.884	240								180	

Material	Ion	λ (µm)	(cm⁻¹)				Mode	T (K)	Pump (mm)	Source	Output	Ref
YVO_4	Tm1	2.014	582	–	–	–	P	77	–	W	170	229
$(Yb,Er)_3Al_5O_{12}$	Tm1	~2.0	–	–	–	–	CW	85	–	–	520	164
$ZrO_2 - Er_2O_3$	Tm1	1.8850	~240	3	–	~7	P	77	6×20	Xe	48	193
		2.0195	~580			~12					25	
		~1.896	~300	1	2.5	–	P	77	6×33	Xe	420	193
YTTERBIUM (Yb^{3+}, $4f^{13}$)												
CaF_2 (+ 2% Nd)	Yb1	1.0336	–	5–10	–	–	P	120	3×25	Xe	48	230
$Gd_3Ga_5O_{12}$ (+ 2% Nd)	Yb1	1.0232	–	2	–	~8	P	77	5×15	Xe	2.5	231
$Gd_3Sc_2Al_3O_{12}$ (+ 1.5% Nd)	Yb1	1.0299	~615	2	1.4	~13	P	77	5.5×20	Xe	7	77
$Lu_3Al_5O_{12}$	Yb1	1.0297	~621	2	~1.1	~21	P	175	6×30	Xe	35	231
	Yb1	1.0294	621	1	1	~7		77			1.5	
(+ 1% Nd)	Yb1	1.0294	–	2	–				5×15		1	
(+ 0.5% Nd, 0.1% Cr)	Yb1	1.0294	–	5	–				5×31		1.5	
$Lu_3Ga_5O_{12}$ (+ 1.5% Nd)	Yb1	1.0230	–	2	–	~8	P	77	5×25	Xe	1.5	231
$Lu_3Sc_2Al_3O_{12}$ (+ 1.5% Nd)	Yb1	1.0299	~615	2	1.4	~13	P	77	6×25	Xe	5	77
$Y_3Al_5O_{12}$	Yb1	1.0293	612	0.7	1.1	~7	P	77	6×24	Xe	9	231
	Yb1	1.0296	623	–	~1	~10			25		325	186
(+ 0.8% Nd)	Yb1	1.0297	~612	2	1.2	~25		200	6×24		145	231

Table 2.1.1.2 (continued)
LASER CRYSTALS

Host	Laser transition	Laser wavelength (μm)	Terminal level energy (cm⁻¹)	Laser ion concentration (%)	Fluorescence lifetime (msec)	Fluorescent linewidth (cm⁻¹)	Mode of operation	Temperature (K)	Crystal size (mm)	Optical pump	Threshold	Ref.
YTTERBIUM (Yb^{3+}, $4f^{13}$) (Continued)												
		1.0293	612		1.1	~7		77			4.5	
(+ 0.8% Nd,		1.0298	~612	2	–	~25		210	6×35		115	
0.5% Cr)					1	~7		77			2	
$Y_3Ga_5O_{12}$ (+ 1.5% Nd) Yb1		1.0233	–	2	–	~8	P	77	5×15	Xe	2	231
DIVALENT RARE-EARTH LASERS												
SAMARIUM (Sm^{2+}, $4f^6$)												
CaF_2	$5d \rightarrow {}^7F_1$	0.7085	~263	~0.01	.002	1.6	P	20	3×20	Xe	< 0.1	232, 233, 234
		0.7083	~263	< 0.01		–	P	65–90	8×20	RL		235
		0.7207	~506					85–90				
		0.7287	~658					110–130				
		0.7310	~700					155				
		0.745	~958					210				
SrF_2	${}^5D_o \rightarrow {}^7F_1$	0.6969	~270	0.1	14	~1	P	4.2		Xe	~4	236
DYSPROSIUM (Dy^{2+}, $4f^{10}$)												
CaF_2	${}^5I_7 \rightarrow {}^5I_8$	2.35867	90	0.03	12		P	77	3×6×25	Hg	~1	237
		(4.2K)		0.02			CW			Xe	80	238

		−2.358	28.8					4.2		W	15	239
		(77K)		0.05	10			27	3.2 × 6.3 × 25.4	Sun	50	240
			29.0	0.01–0.2	50–15	0.25 (77K)		4.2–77	5 × 75	W	20 (27K)	241
(polycrystalline, hot pressed)							P	77	3.2 × 22	Xe	25	242
								4			~4	
SrF_2	$^5I_7 \rightarrow {}^5I_8$	2.3659	23.4	0.03	50	0.5–0.6	P	77	55	Xe	100	243
					THULIUM (Tm^{2+}, $4f^{13}$)							
CaF_2	$^2F_{5/2} \rightarrow {}^2F_{7/2}$	1.116	556	0.05	4	0.03	P	4.2	6 × 25	Xe	50	244
							CW	2.7	6.2 × 25	Hg	1000	245
								4.2			600	
IRON GROUP LASERS												
					CHROMIUM (Cr^{3+}, ed^3)							
Al_2O_3	$^2E(\bar{E}) \rightarrow {}^4A_2$	0.6943(R_1)	0	~0.05	3	~10	P	300	28	Xe	>1000	246, 247
	$^2E(2\bar{A}) \rightarrow {}^4A_2$	0.6929(R_2)				7.5			28	Xe	>1000	248
	$^2E(\bar{E}) \rightarrow {}^4A_2$	0.6943				~10			2 × 25.4	Hg	2.5	249
							CW			Hg	700	249, 250
				0.02					0.04 × 0.25	KrL	0.005	251
		0.6934		0.05	4.3	0.1	P	77		Xe	~800	252
							CW		0.6 × 11.5	Hg	850	253
				0.04					6.4 × 12.7	ArL	0.053	254
		0.6943 → 0.6952	0			~10 →	P	300 → 550	6 × 28	Xe	–	28

Table 2.1.1.2 (continued)
LASER CRYSTALS

Host	Laser transition	Laser wavelength (μm)	Terminal level energy (cm⁻¹)	Laser ion concentration (%)	Fluorescence lifetime (msec)	Fluorescent linewidth (cm⁻¹)	Mode of operation	Temperature (K)	Crystal size (mm)	Optical pump	Threshold	Ref.
	$(Cr^{3+}$ pairs)	$0.7009(N_2)$	35	0.5—0.7	1.1	~2		300	3 × 40		3000	255
		$0.7041(N_1)$	33		1.2						—	
$BeAl_2O_4$	$^2E(\bar{E}) \rightarrow {}^4A_2$	0.6804	0	0.043	0.26	—	P	300	6.4 × 76	Xe	~250	256, 257
		~0.680		—	2.8			77	3 × 4.6 × 6.4		—	258
	$^4T_2 \rightarrow {}^4A_2$	0.701 → 0.818	>		0.26	—	P	300	6.4 × 76		~20	256, 259
		0.744 → 0.788	>	0.27			CW			Hg	~1500	260
$Y_3Al_5O_{12}$	$^2E(\bar{E}) \rightarrow {}^4A_2$	0.6874	0	0.5	1.49	1.4	P	~77	10 × 83	Xe	300	261
	VANADIUM (V^{2+}, $3d^3$)											
MgF_2	$^4T_2 \rightarrow {}^4A_2$	1.1213	>	0.2	2.3		P	77	—	Xe	1070	262
	COBALT (Co^{2+}, $3d^7$)											
KMg_gF_3	$^4T_2 \rightarrow {}^4T_1$	1.821	>	—	3.1	—	P	77	—	Xe	530	263
MgF_2	$^4T_2 \rightarrow {}^4T_1$	1.750	1087	1—5	1.3	—	P	77	—	Xe	690	263, 264
		1.8035	1256								730	
		1.99	>								660	
		2.05	>								700	
		1.63 → 2.11	>	0.9			CW	80	3 × 5 × 12	NdYL	0.4	265

NICKEL (Ni^{2+}, 3d^8)

Crystal	Transition	λ (μm)	Pol.				Mode	T (K)	Dim.	Pump	P	Ref.
ZnF$_2$	$^4T_2 \rightarrow\ ^4T_1$	2.165	V	—	0.4	—	P	77	—	Xe	430	263, 264
MgF$_2$	$^3T_2 \rightarrow\ ^3A_2$	1.623	V	1.5	11.5 (77K)	—	P	20→77	—	Xe	150	263, 266
		1.636						77→82			160	
		1.674→1.676						82→100			160→170	
		1.731→1.756						100→192			170→570	
		1.785→1.797						198→240			570→1650	
		1.731		—			CW	85	50	W	65	
		1.61→1.74		1				80	3×5×12	NdYL	0.02	265
		1.75						200	5×5×20		0.18	267
MgO	$^3T_2 \rightarrow\ ^3A_2$	1.3144	V	—	3.8 (77K)	—	P	77	—	Xe	230	263
		1.316		0.5	(77K)	—	CW	82	4×5×18.5	NdYL	0.13	268
		1.328						131			—	
		1.369						153			—	
		1.409						235			0.8	
MnF$_2$	$^3T_2 \rightarrow\ ^3A_2$	1.915	V	—	11.1 (77K)	—	P	77	—	Xe	840	263
		1.922									210	
		1.929					CW	85		W	270	
		1.939									240	
		1.865					P	20		Xe	740	

Table 2.1.1.2 (continued)
LASER CRYSTALS

ACTINIDE LASERS

URANIUM (U^{3+}, $5f^3$)

Host	Laser transition	Laser wavelength (μm)	Terminal level energy (cm^{-1})	Laser ion concentration (%)	Fluorescence lifetime (msec)	Fluorescent linewidth (cm^{-1})	Mode of operation	Temperature (K)	Crystal size (mm)	Optical pump	Threshold	Ref.
BaF_2	$^4I_{11/2} \rightarrow\ ^4I_{9/2}$	2.556	107	—	0.15	—	P	20	—	Xe	12	269
CaF_2	$^4I_{11/2} \rightarrow\ ^4I_{9/2}$	2.234	0	—	—	—	P	77	—	Xe	—	270
		2.439	398								10	271
		2.511	470			~30					5	271,272
		2.571	609			—		300			—	272
		2.613			<0.015						1200	273
					0.095			90			4.35	
					0.130			77			3.78	
							CW			Hg	2000	
							P	20		Xe	2	
SrF_2	$^4I_{11/2} -\ ^4I_{9/2}$	2.407	334	0.1	0.06		P	90		Xe	38	274
					0.08			77			32	
					0.11			20			8	

Table 2.1.1.3
SENSITIZED LASERS

Improved laser efficiency can occur by energy exchange from sensitizer ions or complexes to the laser-active ions. Table 2.1.1.3 lists all reported lasers using sensitizers. The convention used in Table 2.1.1.2 for ionic valence state and laser transition designation applies to this table also. Energy level diagrams for some of the sensitization processes are shown in Figure 2.1.1.12.

Table 2.1.1.3
SENSITIZED LASERS

Active ion	Laser transition	Sensitizer ion or group	Host crystal	Figure 12-	Ref.
Nd^{3+}	Nd1	Ce^{3+}	$BaF_2 - CeF_3$	—	12
	Nd1, Nd2		$CaF_2 - CeF_3$		28, 22
	Nd1		$CaF_2 - CeO_2$		30
			$CeCl_3$		64
	Nd1, Nd2		CeF_3		66, 67, 36
	Nd1, Nd2		$\alpha - NaCaCeF_6$		28, 108, 36
	Nd1		$SrF_2 - CeF_3$		119
	Nd1, Nd2		$SrF_2 - CeF_3 - GdF_3$		119
	Nd1	Cr^{3+}	$YAlO_3$	A	126, 128
			$Y_3Al_5O_{12}$		137
	Nd1	$(VO_4)^{3-}$	$Ca_3(VO_4)_2$	—	53
	Nd1, Nd2		YVO_4		144, 145, 229
Sm^{3+}	$^4G_{5/2} \rightarrow {}^6H_{7/2}$	Tb^{3+}	TbF_3		307
Tb^{3+}	$^5D_4 \rightarrow {}^7F_6$	Gd^{3+}	$LiYF_4$	—	150
Dy^{3+}	$^6H_{13/2} \rightarrow {}^6H_{15/2}$	Er^{3+}	$Ba(Y_{1.26}E_{0.7}) F_8$	—	151
Ho^{3+}	Ho1	Cr^{3+}	$Ca_5(PO_4)_3 F$	A	40
		Cr^{3+}	$Y_3Al_5O_{12}$		186
	Ho2	Cr^{3+}, Yb^{3+}	$Lu_3Al_5O_{12}$	A, C	182
	Ho1	Er^{3+}	$CaF_2 - ErF_3$	D	155
			$CaMoO_4$		159
			$ErAlO_3$		42
			$(Er, Lu) AlO_3$		161
			Er_2O_3		162
			$ErSiO_5$		163
			$LaNbO_4$		171
			$LiYF_4$		177

Table 2.1.1.3
SENSITIZED LASERS (continued)

Active ion	Laser transition	Sensitizer ion or group	Host crystal	Figure 12-	Ref.
			$\alpha - NaCaErF_6$		183
			$NaLa(MoO_4)_2$		171
			$Y_3Al_5O_{12}$		186
			ZrO_2		193
	Ho1	Er^{3+}, Tm^{3+}	$BaYF_8$	D	152
			$CaY_4(SiO_4)_3\,O$		40
			$Er_3Sc_2Al_3O_{12}$		77
			$ErVO_4$		165
			$KY(WO_4)_2$		169
Ho (cont'd)	Ho1 (cont'd)	Er^{3+}, Tm^{3+} (cont'd)	$Li(Y, Er)\,F_4$		178
			$Lu_3Al_5O_{12}$		180
			$SrY_4(SiO_4)_3\,O$		40
			$YAlO_3$		185
			$Y_3Al_5O_{12}$		187 – 190
			$Y_3Fe_5O_{12}$		191
			YVO_4		165
	Ho1	$Er^{3+}, Tm^{3+}, Yb^{3+}$	$CaF_2 - (Er, Yb, Tm)\,F_3$	C, D	156
			$CaF_2 - YF_3 - (Er, Yb, Tm)F_3$		158
			$(Yb, Er, Tm)_3\,Al_5O_{12}$		164
Er^{3+}	$^5S_2 \to {}^5I_8$	Yb^{3+}	$Ba(Y, Yb)_2\,F_8$	E	153
	Er1	color center	CaF_2	–	197
	$^4F_{9/2} \to {}^4I_{15/2}$	Yb^{3+}	$BaY_{1.2}Yb_{0.75}Er_{0.05}F_8$	E	153
	Er2	Yb^{3+}	$Lu_3Al_5O_{12}$	–	182
	Er1, Er2		$Y_3Al_5O_{12}$	C	218, 206
	Er1		$Yb_3Al_5O_{12}$		164
	Er2	Cr^{3+}, Yb^{3+}	$Lu_3Al_5O_{12}$	C	182
Tm^{3+}	Tm1, $^3F_4 \to {}^3H_5$	Cr^{3+}	$YAlO_3$	A	226, 228
			$Y_3Al_5O_{12}$		186, 228
	Tm1	Er^{3+}	$CaF_2 - ErF_3$	D	156, 200, 201
			$CaMoO_4$		159
			$ErAlO_3$		221

Table 2.1.1.3
SENSITIZED LASERS (continued)

Active ion	Laser transition	Sensitizer ion or group	Host crystal	Figure 12-	Ref.
			$Er_3Al_5O_{12}$		222
			$(Er, Lu) AlO_3$		119
			Er_2O_3		223
			$\alpha - NaCaErF_6$		183
			$(Y, Er) Al_5O_{12}$		186
	Tm1	Er^{3+}, Yb^{3+}	$(Yb, Er) Al_5O_{12}$	C, D	164
Yb^{3+}	Yb1	Nd^{3+}	CaF_2	B	230
			$Gd_3Ga_5O_{12}$		231
			$Gd_3Sc_2Al_3O_{12}$		77
			$Lu_3Al_5O_{12}$		231
			$Lu_3Ga_5O_{12}$		231
			$Lu_3Sc_2Al_3O_{12}$		77
			$Y_3Al_5O_{12}$		231
			$Y_3Ga_5O_{12}$		231
		Cr^{3+}, Nd^{3+}	$Lu_3Al_5O_{12}$	A, B	231
			$Y_3Al_5O_{12}$		231
Ni^{2+}	$^3T_2 \rightarrow {}^3A_2$	Mn^{2+}	MnF_2	—	263

Table 2.1.1.4
CASCADE LASERS

As the name implies, cascade lasers operate on two different, successive transitions of the active ions. In this table all reported cascade lasers are listed. Figure 1.1.1.13 diagrams the energy levels for each of the cascade lasers. All the ions listed are trivalent.

Table 2.1.1.4
CASCADE LASERS

Active ion	Cascade transitions	Host crystal	Figure 13-	Ref.
Ho^{3+}	$^5S_2 \rightarrow {}^5I_5$ $^5I_6 \rightarrow {}^5I_8$	$Gd_3Ga_5O_{12}$	A	167
Ho^{3+}	$^5S_2 \rightarrow {}^5I_5$ $^5I_5 \rightarrow {}^5I_7$	$LiYF_4$	A	175
Ho^{3+}	$^5S_2 \rightarrow {}^5I_5$ $^5I_5 \rightarrow {}^5I_6$	$LiYF_4$	A	175
Er^{3+}	$^4S_{3/2} \rightarrow {}^4I_{13/2}$ $^4I_{13/2} \rightarrow {}^4I_{15/2}$	$CaF_2 - YF_3$	B	201
Er^{3+}	$^4S_{3/2} \rightarrow {}^4I_{11/2}$ $^4I_{11/2} \rightarrow {}^4I_{13/2}$	$LiYF_4$	B	210
Tm^{3+}	$^3F_4 \rightarrow {}^3H_5$ $^3H_4 \rightarrow {}^3H_6$	$YAlO_3$ $+ 0.1\%$ Cr	C	226
Er^{3+} Ho^{3+}	$^4S_{3/2} \rightarrow {}^4I_{13/2}$ $^5I_7 \rightarrow {}^5I_8$	CaF_2	D	201
Er^{3+} Tm^{3+}	$^4S_{3/2} \rightarrow {}^4I_{13/2}$ $^3H_4 \rightarrow {}^3H_6$	CaF_2	D	201

Table 2.1.1.5
CW LASERS

The properties of reported CW lasers are listed in Table 2.1.1.5. The data found under columns identical to those in Table 2.1.1.2 are explained in the discussion of Table 2.1.1.2. Laser efficiency is, for lamp-pumped systems, laser output power divided by lamp electrical power and, for laser-pumped systems, output power divided by absorbed pump power. Both efficiency and output power depend upon a number of factors such as laser crystal quality, pump cavity efficiency, and output coupling efficiency; the original references should be consulted when comparing values. Very few of the systems listed have been operated under conditions yielding optimum efficiency or output.

Table 2.1.1.5
CW LASERS

Laser ion	Host crystal	Wavelength (μm)	Temp. (K)	Pump	Threshold (W)	Output power (W)	Efficiency (%)	Ref.
Nd^{3+}	$CaLa_4(SiO_4)_3 O$	1.0610	300	W	525	–	–	41
	$CaMoO_4$	1.061	300	Hg	1200	~0.001	–	42
		1.067	77		850	–	–	
	$Ca(NbO_3)_2$	1.0615	300	Xe	350	1	0.1	45
			77		400	0.1	0.05	
	$Ca_5(PO_4)_3 F$	1.0629	300	W	330	1.3	0.2	275
	$CaWO_4$	1.0584	300	Hg	–	~0.01	~0.006	55
				Xe	2600	<0.1	<0.003	57
		1.0649	85	Hg	600	0.5	0.03	56
		1.3370	300	Xe	1500	–	–	60
	$CaY_2Mg_2Ge_3O_{12}$	0.941	300	ArL	0.4	–	–	61
		1.059			0.09	0.1	20	
	$CeCl_3$	1.0647	300	Arl	0.006	–	–	65
	$Gd_3Sc_2Al_3O_{12}$	1.060	300	W	700	0.55	0.04	76
	$KGd(WO_4)_2$	1.0672	300	Xe	400	–	–	83
	$KY(WO_4)_2$	1.0688	300	Xe	~400	0.55	~0.03	88
		1.3525			1100	–	–	60
	$La_2Be_2O_5$	1.0698	300	W	–	9	0.3	90
	LaF_3	1.04065	300	Xe	4000	–	–	93
	La_2O_2S	1.075	300	W	<300	–	–	97
	$LiGd(MoO_4)_2$	1.0599	300	Xe	1150	–	–	88
	$Lu_3Al_5O_{12}$	1.06425	300	Xe	950	0.35	~0.02	88
		1.3387			2000	–	–	60
	$NaLa(MoO_4)_2$	1.0653	300	Xe	2600	–	–	113

Table 2.1.1.5 (continued)
CW LASERS

Laser ion	Host crystal	Wavelength (μm)	Temp. (K)	Pump	Threshold (W)	Output power (W)	Efficiency (%)	Ref.
	YAlO$_3$	1.0644	300	Kr	1700	35	0.8	125
		1.0795			1200	90	1.8	126
		1.3400*				14	0.45	125
		1.3416*						
	Y$_3$Al$_5$O$_{12}$	1.0521	300	Kr	~1200	33	0.7	133
		1.0615			~1500	66	1.3	
		1.06415		W	—	25	1	276
				Kr	1000	202	3.4	277
					~20000	1100†	2.0	278
Nd^{3+} (cont'd)	Y$_3$Al$_5$O$_{12}$ (cont'd)	1.0646			~1200	~36	~0.7	133
		1.0737			~1700	47	0.9	
		1.0780			~1500	24	0.5	
		1.0780			~1500	24	0.5	
		1.1054			~3000	6	0.1	
		1.1119			~1500	35	0.7	
		1.1158			~1600	33	0.7	
		1.1225			~1600	29	0.6	
		1.3188			~1900	24	0.5	
		1.3200			~2000	6	0.1	
		1.3338			~2800	9	0.2	
		1.3350			~2100	11	0.2	
		1.3382			~1600	17	0.3	
		1.3410			~2000	6	0.1	
		1.3564			~2000	10	0.2	
		1.4140			~3500	0.7	0.01	
		1.4440			~5000	0.1	0.002	
	Y$_3$Al$_5$O$_{12}$ + Cr	1.0641	300	Hg	750	10	0.4	137, 279
				W	800	—	—	137
				Sun	—	4.8	—	280
		1.0612	77	Hg	180	—	—	137

* Simultaneous operation on two wavelengths.

† Multiple laser rods in series inside one optical resonator.

Table 2.1.1.5 (continued)
CW LASERS

Laser ion	Host crystal	Wavelength (μm)	Temp. (K)	Pump	Threshold (W)	Output power (W)	Efficiency (%)	Ref.
				W	440			
	Y_2O_3	~1.07	300	KrL	0.57 mW	–	–	141
		~1.35			0.7 mW	–	–	
	YVO_4	1.0643	300	ArL	0.11	1	–	146
		1.34			0.5	0.35	7	
Ho^{3+}	$BaY_{1.64}Er_{0.3}Tm_{0.03}F_8$	2.0644	~85	W	–	–	–	152
		2.0746*			50	0.56	0.2	
		→2.076						
		2.0866*						
	$CaF_2 - ErF_3$ $- YbF_3 - TmF_3$	~2.1	~77	Xe	3000	–	–	156
	ErO_3	2.121	77	W	200	–	–	162
Ho^{3+} (cont'd)	$ErVO_4$ + Tm	2.0416	77	ArL	–	–	–	165
	$Lu_3Al_5O_{12}$ + Er, Tm	2.1020	77	W	65	–	–	180
	$Y_3Al_5O_{12}$ + Cr	2.0975	85	W	120	–	–	186
		2.1223			250	–	–	
				Hg	1300	–	–	
	$Y_3Al_5O_{12}$ (+ Er)	2.0979	85	W	47	–	–	186
				Hg	680	–	–	
	$Y_3Al_5O_{12}$ (+ Er, Tm)	2.0978	~77	W	~40	50	5	281
		2.0990	77	W	45	–	–	188
		2.123	85	W	30	15	5	187
		~2.13	~77	W	5	–	–	190
	YVO_4 (+ Er, Tm)	2.0412	~77	ArL	–	–	–	165
Tm^{3+}	$CaF_2 + ErF_3$	1.894	77	Xe	3000	–	–	156
	ErO_3	1.934	77	W	1500	–	–	223
	$Y_3Al_5O_{12}$	2.0132	85	W	315	–	–	186

* Simultaneous operation in two wavelength regions.

Table 2.1.1.5 (continued)
CW LASERS

Laser ion	Host crystal	Wavelength (μm)	Temp. (K)	Pump	Threshold (W)	Output power (W)	Efficiency (%)	Ref.
	(+ Cr)		85	W	160	–	–	186
				Hg	800			
	(+ Er)	2.014	85	W	520	–	–	186
Dy^{2+}	CaF_2	~2.358	77	Xe	80	–	–	238
				W	–	1.2	0.06	241
				Xe	–	200	1.3	282
			27	Sun	–	–	–	240
				W	20	–	–	241
Tm^{2+}	CaF_2	1.116	27	Hg	1000	–	–	245
			4.2		600			
Cr^{3+}	Al_2O_3	0.6943	300	Hg	700	2.4	~0.1	249, 2!
		~0.6944		KrL	0.005	2×10^{-4}	3	251
		0.6934	77	Hg	850	–	–	253
				ArL	0.053	0.042	16	254
	$BeAl_2O_4$	0.744 →	300	Hg	~1500	6.5	0.2	260
		0.788						
Co^{2+}	MgF_2	1.63 → 2.11	80	NdYL	0.4	~0.1	8	265
Ni^{2+}	MgF_2	1.731	85	W	65	1	0.2	263
		1.61–1.74	80	NdYL	0.02	~0.1	6	265
		1.75	200		0.18	–	–	267
	MgO	1.316	82	NdYL	0.13	9	36	268
		1.328	131		–	–	–	
		1.369	153					
		1.409	235		0.8			
	MnF_2	1.929	85	W	270	–	–	263
		1.939			240			

Table 2.1.1.6
Q-SWITCHED LASERS

Unlike the previous tables, Table 2.1.1.6 is intended to be representative, not exhaustive. The discussion of Table 2.1.1.2 should be referred to for columns with identical headings. For the most commonly used laser systems, Nd:Y₃Al₅O₁₂ (YAG) and Cr:Al₂O₃ (ruby) the data given reflects current state-of-the-art.

Table 2.1.1.6
Q-SWITCHED LASERS

Ion	Host crystal	Wavelength (μm)	Pump	Mode	Pulse energy (mJ)	Pulse width (nsec)	Pulse rate (Hz)	Crystal size (mm)	Comments	Ref.
Nd^{3+}	$CaLa_4(SiO_4)_3 O$	1.0610	Xe	P	585	—	10	6.3×76		283
	$La_2Be_2O_5$	~1.07	Xe	P	238	—	—	6.3×76	z–axis rod	89
		~1.08			400				x–axis rod	
	$YAlO_3$	1.064	Xe	P	225	—	1	5×50	c–axis rod,	126
									Cr–doped	
		1.0795	Kr	CW	2.9	250	2.5×10^4	6.3×76	b–axis rod	125
	$Y_3Al_5O_{12}$	1.06415	Xe	P	125		10	5×75	stable resonator, TEM_{oo} output	284
					200	10–12	10–20	6.3×75	unstable resonator	285, 286
					10	<20	1500	6.3×63		287
			Kr	CW	1.6	40	<1000	5×75	TEM_{oo} mode	288
		1.318	Xe	P	50	50	20	6×51	—	289

Table 2.1.1.6
Q-SWITCHED LASERS (continued)

Ion	Host crystal	Wavelength (μm)	Pump	Mode	Pulse energy (mJ)	Pulse width (nsec)	Pulse rate (Hz)	Crystal size (mm)	Comments	Ref.
Ho^{3+}	YVO_4	1.833	Kr	P	2	200	50	3 × 50	—	136
		1.0634	Xe	P	60	—	—	3 × 30	—	290
	$CaY_4(SiO_4)_3O$ + Er, Tm	2.060	Xe	P	320	—	—	7 × 32	77K crystal temp.	40
	$LiYF_4$ + Er, Tm	2.065	Xe	P	~100	<100	—	3 × 20	—	178
	$YAlO_3$ + Er, Tm	2.123	Xe	P	136	<200	—	2 × 50	77K crystal temp.	185
	$Y_3Al_5O_{12}$ + Er, Tm	2.098	W	CW	42	100–300	—	3 × 50	77K crystal temp.	291
Er^{3+}	CaF_2	~2.73	Xe	P	300	30	1	9 × 120	—	199
	$Y_3Al_5O_{12}$	2.936	Xe	P	8	80	20	4 × 70	—	292
Dy^{2+}	CaF_2	2.36	Xe	P	300	30–40	1	140	77K crystal temps.	293
			Xe	CW	~30	~30	500	8 × 280	temps.	294
Cr^{3+}	Al_2O_3	0.6943	Xe	P	2300	10	—	10 × 100	multimode	295
					200				TEM_{oo} mode	296
				CW	0.9	30	200	2 × 50	—	297

BeAl$_2$O$_4$	0.6804	Xe	P	500	20	–	6.3 × 76	R line	257
Ni^{2+}	~0.75			70	120			vibronic	260
MgF$_2$	1.67	NdYL	CW	0.2	1100	100	4 × 5 × 12	TEM$_{\infty}$ mode	265

Table 2.1.1.7
MODE-LOCKED LASERS

This table, like Table 2.1.1.6, is not a complete listing. The most common systems are included, however. Measurement of pulse width is often a subject of controversy and the original references cited should be checked to determine the method used. Pulse period is the spacing between individual pulses. For pulsed systems the output shown is an estimate of the energy of a single pulse while for CW systems the average power output is given. There are two types of active locking techniques listed, acoustooptic loss modulation (AO loss) and electrooptic phase modulation (EO phase). Intracavity etalons were used in one case to vary the pulse width. Passive mode locking (dye) was achieved with an intracavity dye cell.

Table 2.1.1.7
MODE-LOCKED LASERS

Active ion	Host crystal	Wavelength (μm)	Mode	Pulse width (ps)	Pulse period (nsec)	Output (W or mJ)	Locking technique	Ref.
Nd^{3+}	$CaLa(SiO_4)_3O$	~1.061	P	5–15	6.6	0.01	dye	298
	$La_2Be_2O_5$	1.070	P	11–15	5.5	~0.25	dye	299
		1.079						
	$LiYF_4$	1.047	P	35	7.3	—	dye	104
		1.053						
	$Y_3Al_5O_{12}$	1.06415	CW	80	3.3	0.1	AO loss	300
				30	2.6	1.0	EO phase	301
			P	20–25	3.6	—	dye	302
				35	6.6	11	dye	303
				100–1000	—	0.2 –0.28	AO loss + etalons	304
Cr^{3+}	Al_2O_3	0.6943	P	5	10	—	dye	305
		0.6934		~35	—	200	dye, 77K laser rod	306

Table 2.1.1.8
BROADLY TUNABLE LASERS

Table 2.1.1.8 includes all reported lasers which have been tuned over a broad wavelength range. An asterisk on the turning range indicates that turning was not continuous over the wavelength span. The maximum output is that observed at the peak of the tuning curve. Temperature tuning is achieved by varying the crystal temperature. Other techniques use the various elements listed as part of the laser cavity. Other column headings are discussed in Table 2.1.1.2.

Table 2.1.1.8
BROADLY TUNABLE LASERS

Active ion	Host crystal	Transition	Tuning range (μm)	Mode	Maximum output (W or mJ)	Tuning technique	Ref.
Ce^+	$LiYF_4$	$5d \to {}^2F_{5/2}$	$0.306 \to 0.315$	P	~0.6	grating	2
		$5d \to {}^2F_{7/2}$	$0.323 \to 0.328$		~0.06		
Sm^{2+}	CaF_2	$5d \to {}^7F_1$	$0.7083 \to 0.745^*$	P	—	temperature (65 → 210K)	235
Cr^{3+}	$BeAl_2O_4$	${}^4T_2 \to {}^4A_2$	$0.701 \to 0.818$	P	2000	birefringent filter	259
			$0.744 \to 0.788$	CW	3.5		260
Co^{2+}	MgF_2	${}^4T_2 \to {}^4T_1$	$1.63 \to 2.11$	CW	0.1	birefringent filter	265
Ni^{2+}	MgF_2	${}^3T_2 \to {}^3A_2$	$1.623 \to 1.797^*$	P	—	temperature (20 → 240K)	263
			$1.62 \to 1.84^*$	P	—	prism	
			$1.61 \to 1.74$	CW	0.1	birefringent filter	265
Ni^{2+}	MgO	${}^3T_2 \to {}^3A_2$	$1.316 \to 1.409^*$	CW	9	temperature (82 → 235K)	268

*Tuning not continuous over entire range indicated.

REFERENCES TO TABLES

Note: Titles of Russian papers may differ slightly from those appearing in the translation journals.

1. Ehrlich, D. J., Moulton, P. F., and Osgood, R. M., Jr., Optically pumped Ce:LaF₃ laser at 286 nm, *Opt. Lett.,* 5, 339, 1980.
2. Ehrlich, D. J., Moulton, P. F., and Osgood, R. M., Jr., Ultraviolet solid-state Ce:YLF laser at 325 nm, *Opt. Lett.,* 4, 184, 1978, and unpublished data.
3. Ballman, A. A., Porto, S. P. S., and Yariv, A., Calcium niobate Ca(NbO₃)₂ — A new laser host crystal, *J. Appl. Phys.,* 34, 3155, 1963.
4. Yariv, A., Porto, S. P. S., and Nassau, K., Optical laser emission from trivalent praseodymium in calcium tungstate, *J. Appl. Phys.,* 33, 2519, 1962.
5. German, K. R., Kiel, A., and Guggenheim, H. J., Radiative and nonradiative transitions of Pr³⁺ in trichloride and tribromide hosts, *Phys. Rev. B,* B11, 2436, 1975.
6. German, K. R., Kiel, A., and Guggenheim, H., Stimulated emission from PrCl₃, *Appl. Phys. Lett.,* 22, 87, 1973.
7. Solomon, R., and Mueller, L., Stimulated emission at 5985 Å from Pr³⁺ in LaF₃, *Appl. Phys. Lett.,* 3, 135, 1963.
8. Esterowitz, L., Allen, R., Kruer, M., Bartoli, F., Goldberg, L. S., Jensen, H. P., Linz, A., and Nicolai, V. O., Blue light emission by Pr:LiYF₄-laser operated at room temperature, *J. Appl. Phys.,* 48, 650, 1977.
9. Johnson, L. F., Optically pumped pulsed crystal lasers other than ruby, in *Lasers, A Series of Advances,* Vol. 1, Levine, A. K., Ed., Marcel Dekker, New York, 1966, 137.
10. Johnson, L. F. and Ballman, A., Coherent emission from rare-earth ions in electro-optic crystals, *J. Appl. Phys.,* 40, 297, 1969.
11. Johnson, L. F., Optical maser characteristics of rare-earth ions, *J. Appl. Phys.,* 34, 897, 1963.
12. Kaminskii, A. A., Sobolev, B. P., Babdasarov, Kh. S., et al., Investigation of stimulated emission from crystals with Nd³⁺ ions, *Phys. Stat. Solidi,* 23a, K135, 1974.
13. Kaminskii, A. A., On the possibility of investigation of the 'Stark' structure of Tr³⁺ ion spectra in disordered fluoride crystal systems, *Sov. Phys.-JETP,* 31, 216, 1970.
14. Kaminskii, A. A. and Sarkisov, S. E., Stimulated emission by Nd³⁺ ions in crystals due to the ⁴F₃/₂ → ⁴I₁₃/₂ transition. I, *Inorg. Mater.,* 9, 453, 1973.
15. Alam, M., Gooen, K. H., DiBartolo, B., Linz, A., Sharp, E., Gillespie, and Janney, G., Optical spectra and laser action of neodymium in a crystal Ba₂MgGe₂O₇, *J. Appl. Phys.,* 39, 4728, 1968.
16. Miller, J. E., Sharp, F. J., and Horowitz, D. J., Optical spectra and laser action of neodymium in a crystal Ba₀.₂₅Mg₂.₇₅Y₂Ge₃O₁₂, *J. Appl. Phys.,* 43, 462, 1972.
17. Kaminskii, A. A., Koptsik, V. A., Maskaek, Yu A., et al., Stimulated emission from Nd³⁺ ions in Ferroelectric Ba₂NaNb₅O₁₅ crystals (bananas), *Phys. Stat. Solidi,* 28a, K5, 1975.
18. Horowitz, D. J., Gillespie, L. F., Miller, J. E., and Sharp, E. J., Laser action of Nd³⁺ in a crystal Ba₂ZnGe₂O₇, *J. Appl. Phys.,* 43, 3527, 1972.
19. Kaminskii, A. A., Schultze, D., Hermoneit, B., et al., Spectroscopic properties and stimulated emission in the ⁴F₃/₂ → ⁴I₁₁/₂ and ⁴F₃/₂ → ⁴I₁₃/₂ transitions of Nd³⁺ ions from cubic bi₄Ge₃O₁₂ crystals, *Phys. Stat. Solidi,* 33a, 737, 1976.
20. Kaminskii, A. A., Sarkisov, S. E., Maier, A. A., et al., Eulytine with TR³⁺ ions as a laser medium, *Sov. Tech. Phys. Lett.,* 2, 59, 1976.
21. Sarkisov, S. E., Lomonov, V. A., Kaminskii, A. A., et al., Spectroscopic investigation of the stable crystalline mixed system Bi₄(Ge₁₋ₓSiₓ)₅O₁₂, in *Abstracts of Papers of Fifth All-Union Symposium on Spectroscopy of Crystals, Kazan,* 1976, 195 (in Russian).
22. Kevorkov, A. M., Kaminskii, A. A., Bagdasarov, Kh. S., Tevosyan, T. A., and Sarkisov, S. E., Spectroscopic properties of CaAl₄O₇:Nd³⁺ crystals, *Inorg. Mater.,* 9, 146, 1973.
23. Kaminskii, A. A., Sarkisov, S. E., and Bagdasarv, Kh. S., Study of stimulated emission from Nd³⁺ ions in crystals at the ⁴F₃/₂ → ⁴I₁₃/₂ transition. P 2, *Inorg. Mater.,* 9, 457, 1973.
24. Kevorkov, A. M., Kaminskii, A. A., Bagdasarov, Kh. S., Tevosyan, T. A., and Sarkisov, S. E., Spectroscopic properties of SrAl₄O₇:Nd³⁺ crystals, *Inorg. Mater.,* 9, 1637, 1973.
25. Robinson, M. and Asawa, C. K., Stimulated emission from Nd³⁺ and Yb³⁺ in noncubic sites of neodymium- and ytterbium-doped CaF₂, *J. Appl. Phys.,* 38, 4495, 1967.
26. Johnson, L. F., Optical maser characteristics of Nd³⁺ in CaF₂, *J. Appl. Phys.,* 33, 756, 1962.
27. Kaminskii, A. A., Korniyenko, L. S., and Prokhorov, A. M., Spectral study of stimulated emission from Nd³⁺ in CaF₂, *Sov. Phys.-JETP,* 21, 318, 1965.
28. Kaminskii, A., High-temperature spectroscopic investigation of stimulated emission from lasers based on crystals and glass activated with Nd³⁺ ions, *Sov. Phys.-JETP,* 27, 388, 1968.

29. Voron'ko, Yu. K., Kaminskii, A. A., Korniyenko, L. S., Osiko, V. V., Prokhorov, A. M., and Udoven'chik, V. T., Investigation of stimulated emission from $CaF_2:Nd^{3+}$ (Type II) crystals at room temperature, *JETP Lett.*, 1, 39, 1965.

30. Kaminskii, A. A., Osiko, V. V., and Voron'ko, Yu. K., Mixed systems on the basis of fluorides as new laser materials for quantum electronics. The optical and emission parameters, *Phys. Stat. Sol.*, 21, 17, 1967.

31. Kaminskii, A. A., Sobolev, B. P., Bagdasarov, Kh. S., et al., Investigation of stimulated emission of the $^4F_{3/2} \rightarrow ^4I_{13/2}$ transition of Nd^{3+} ions in crystals, *Phys. Stat. Sol.*, 26a, K63, 1974.

32. Kaminskii, A. A., Mikaelyan, R. G., and Zigler, I. N., Room-temperature induced emission of CaF_2-SrF_2 crystals containing Nd^{3+}, *Phys. Stat. Sol.*, 31, K85, 1969.

33. Kaminskii, A. A., Dsiko, V. V., and Voron'ko, Yu. K., Five-component fluoride: a new laser material, *Sov. Phys.-Crystallogr.*, 13, 267, 1968.

34. Bagdasarov, Kh. S., Voron'ko, Yu. K., Kaminskii, A. A., Osiko, V. V., and Prokhorov, A. M., Stimulated emission in yttro-fluorite:Nd^{3+} crystals at room temperature, *Sov. Phys.-Crystallogr.*, 10, 626, 1966.

35. Kaminskii, A. A., Osiko, V. V., Prokhorov, A. M., and Voron'ko, Yu. K., Spectral investigation of the stimulated radiation of Nd^{3+} in CaF_2-YF_3, *Phys. Lett.*, 22, 419, 1966.

36. Kaminskii, A. A. and Sarkisov, S. E., Study of stimulated emission from Nd^{3+} ions in crystals emitting at the $^4F_{3/2} \rightarrow ^4I_{13/2}$ Transition. 1, *Inorg. Mater.*, 9, 453, 1973.

37. Kaminskii, A. A., Butaeva, T. I., Kevorkov, A. M., et al. New data on stimulated emission by crystals with high concentrations of Ln^{3+} ions, *Inorg. Mater.*, 1238, 1976.

38. Es'kov, N. A., Osiko, V. V., Sobol, A. A., et al., A new laser garnet $Ca_3Ga_2Ge_3O_{12}$-Nd^{3+}, *Inorg. Mater.*, 14, 1764, 1978.

39. Aivea, A. F.,Westinghouse, unpublished.

40. Steinbruegge, K. B., Hennigsen, R. H., Hopkins, R., Mazelsky, R., Melamed, N. T., Riedel, E. P., and Roland, G. W., Laser properties of Nd^{3+} and Ho^{3+} doped crystals with the apatite structure, *Appl. Optics*, 11, 999, 1972.

41. Steinbruegge, K. B., High average power characteristics of CaLaSOAP:Nd laser materials, in *Digest of Technical Papers CLEA 1973 IEEE/OSA*, Washington, D.C., 1973, 49.

42. Duncan, R. C., Continuous room-temperature Nd^{3+}:$CaMoO_4$ laser, *J. Appl. Phys.*, 36, 874, 1965.

43. Flournoy, P. A. and Brixner, L. H., Laser characteristics of niobium compensated $CaMoO_4$ and $SrMoO_4$, *J. Electrochem. Soc.*, 112, 779, 1965.

44. Kaminskii, A. A. and Li, L., Spectroscopic and stimulated emission studies of crystal compounds in a CaO-Nb_2O_5 system, $Ca(NbO_3)2:Nd^{3+}$ crystals, *Inorg. Mater.*, 6, 254, 1970.

45. Bagdasarov, Kh. S., Gritsenko, M. M., Zubkova, F. M., Kaminskii, A. A., Kevorkov, A. M., and Li, L., CW $Ca(NbO_3)_2:Nd^{3+}$ crystal laser, *Sov. Phys.-Crystallogr.*, 15, 323, 1970.

46. Kaminskii, A. A., High-temperature spectroscopic investigation of stimulated emission from lasers based on crystals activated with Nd^{3+} ions, *Phys. Stat. Sol.*, 1a, 573, 1970.

47. Kaminskii, A. A., Sarkisov, S. E., and Li, L., Investigation of stimulated emission in the $^4F_{3/2}$ $^4I_{13/2}$ transition of Nd^{3+} ions in crystals (III), *Phys. Stat. Sol.*, 15a, K141, 1973.

48. Ohlmann, R. C., Steinbruegge, K. B., and Mazelsky, R., Spectroscopic and laser characteristics of neodymium-doped calcium fluorophosphate, *Appl. Optics*, 7, 905, 1968.

49. Bruk, Z. M., Voron'ko, Yu. K., Maksimova, G. V., Osiko, V. V., Prokhorov, A. M., Shipilov, K. F., and Schcherbakov, I. A., Optical properties of a stimulated emission from Nd^{3+} in fluorapatite crystals, *JETP Lett.*, 8, 221, 1968.

50. Aleksandrov, V. I., Kaminskii, A. A., Maksimova, G. V., Prokhorov, A. M., Sarkisov, S. E., Sobol', A. A., and Tatarintsev, V. M., Study of stimulated emission from Nd^{3+} ions in crystals emitting at the $^4F_{3/2}$ $^4I_{13/2}$ transition, *Sov. Phys.-Dokl.*, 18, 495, 1974.

51. Bagdasarov, Kh. S., Kaminskii, A. A., Kevorkov, A. M., and Prokhorov, A. M., Investigation of the stimulated radiation emitted by Nd^{3+} ions in $CaSc_2O_4$ crystals, *Sov. J. Quantum Electron.*, 4, 927, 1975.

52. Bagdasarov, Kh. S., Kaminskii, A. A., Kevorkov, A. M., et al., Stimulated emission of Nd^{3+} ions in an $SrAl_{12}O_{19}$ crystal at the transitions $^4F_{3/2}$ $^4I_{11/2}$ and $^4F_{3/2}$ $^4I_{13/2}$, *Sov. Phys.-Dokl.*, 19, 350, 1974.

53. Brixner, L. H. and Flournoy, A. P., Calcium orthovanadate $Ca_3(VO_4)_2$ — a new laser host crystal, *J. Electrochem. Soc.*, 112, 303, 1965.

54. Johnson, L. F. and Thomas, R. A., Maser oscillations at 0.9 and 1.35 microns in $CaWO_4:Nd^{3+}$, *Phys. Rev.*, 131, 2038, 1963.

55. Johnson, L. F., Boyd, G. D., Nassau, K., and Soden, R. R., Continuous operation of a solid-state optical maser, *Phys. Rev.*, 126, 1406, 1962.

56. Johnson, L. F., Characteristics of the $CaWO_4:Nd^{+3}$ optical maser, in *Quantum Electronics Proceedings of the Third International Congress*, Grivet, P. and Bloembergen, N., Eds., Columbia University Press, New York, 1964, 1021.

57. **Kaminskiii, A. A., Korniyenko, L. S., Maksimova, G. V., Osiko, V. V., Prokhorov, A. M., and Shipulo, G. P.**, CW CaWO$_4$:Nd^{3+} laser operating at room temperature, *Sov. Phys.-JETP*, 22, 22, 1966.

58. **Kaminskii, A. A.**, Spectral composition of stimulated emission from a CaWO$_4$:Nd^{3+} laser, *Inorg. Mater.*, 6, 347, 1970.

59. **Kaminskii, A. A., Sarkisov, S. E., and Li, L.**, Investigation of stimulated emission in the $^4F_{3/2} \rightarrow$ $^4I_{13/2}$ Transition of Nd^{3+} ions in crystals. III, *Phys. Stat. Sol.*, 15a, K141, 1973.

60. **Kaminskii, A. A., Sarkisov, S. E., Klevtsov, P. V., Bagdasarov, Kh. S., Pavlyuk, A. A., and Petrosyan, A. G.**, Investigation of stimulated emission in the $^4F_{13/12}$ transition of Nd^{3+} ions in crystals. V, *Phys. Stat. Sol.*, 17a, K75, 1973.

61. **Birnbaum, M., Tucker, A. W., and Fincher, C. L.**, CW room-temperature laser operation of Nd:CAMGAR at 0.941 and 1.059 μ, *J. Appl. Phys.*, 49, 2984, 1978.

62. **Sharp, E. J., Miller, J. E., Horowitz, D. J., et al.**, Optical spectra and laser action in Nd^{3+}-doped CaY$_2$M$_{g2}$Ge$_3$O$_{12}$ *J. Appl. Phys.*, 45, 4974, 1974.

63. **Bagdasarov, Kh. S., Izotova, O. Ye., Kaminskii, A. A., Li, L., and Sobolev, B. P.**, Optical and laser properties of mixed CdF$_2$-YF$_3$:Nd^{3+} crystals, *Sov. Phys.-Dokl.*, 14, 939, 1970.

64. **Singh, S., Van Uitert, L. G., Potopowicz, J. R., and Grodkiewitz, W. H.**, Laser emission at 1.065 μm from neodymium-doped anhydrous cerium trichloride at room temperature, *Appl. Phys. Lett.*, 24, 10, 1974.

65. **Singh, S., Chesler, R. B., Grodkiewicz, W. H., et al.**, Room temperature CW Nd^{3+}:CdCl$_3$ laser, *J. Appl. Phys.*, 46, 436, 1975.

66. **O'Connor, J. R., and Hargreaves, W. A.**, Lattice energy transfer and stimulated emission from CeF$_3$:Nd^{3+}, *Appl. Phys. Lett.*, 4, 208, 1964.

67. **Dmitruk, M. V., Kaminskii, A. A., and Shcherbakov, I. A.**, Spectroscopic studies of stimulated emission from a CeF$_3$:Nd^{3+} laser, *Sov. Phys.-JETP*, 27, 900, 1968.

68. **Gualtieri, J. G. and Aucoin, T. R.**, Laser performance of large Nd-pentaphosphate crystals, *Appl. Phys. Lett.*, 28, 189, 1976.

69. **Bagdasarov, Kh. S., Bogomolova, G. A., Gritsenko, M. M., Kaminskii, A. A., and Kervorkov, A. M.**, Spectroscopic study of the LaAlO$_3$:Nd^{3+} laser crystal, *Sov. Phys.-Crystallogr.*, 17, 357, 1972.

70. **Arsen'yev, P. A. and Bienert, K. E.**, Synthesis and optical properties of neodymium-doped gadolinium aluminate (GdAlO$_3$) single crystals, *J. Appl. Spectrosc.*, (USSR), 17, 1623, 1972.

71. **Geusic, J. E., Marcos, H. M., and Van Uitert, L. G.**, Laser oscillations in Nd-doped yttrium aluminum, yttrium gallium and gadolinium garnets, *Appl. Phys. Lett.*, 4, 182, 1964.

72. **Bagdasarov, Kh. S., Bogomolova, G. A., Gritsenko, M. M., Kaminskii, A. A., Kevorkov, A. M., Prokhorov, A. M., and Sarkisov, S. E.**, Spectroscopy of stimulated emission from Gd$_3$Ga$_5$O$_{12}$:Nd^{3+} crystals, *Sov. Phys.-Dokl.*, 19, 353, 1974.

73. **Borchardt, H. J., and Bierstedt, P. E.**, Gd$_2$(MoO$_4$)$_3$: A ferro-electric laser host, *Appl. Phys. Lett.*, 8, 50, 1966.

74. **Kaminskii, A. A.**, Laser and spectroscopic properties of activated ferroelectrics, *Sov. Phys.-Crystallogr.*, 17, 194, 1972.

75. **Soffer, B. H. and Hoskins, R. H.**, Fluorescence and stimulated emission from Gd$_2$O$_3$:Nd^{3+} at room temperature and 77°K, *Appl. Phys. Lett.*, 4, 113, 1964.

76. **Brandle, C. D. and Vanderleeden, J. C.**, Growth, optical properties and CW laser action of neodymium-doped gadolinium scandium aluminum garnet, *IEEE J. Quantum Electron.*, QE-10, 67, 1974.

77. **Bagdasarov, Kh. S., Kaminskii, A. A., Kevorkov, A. M., and Prokhorov, A. M.** Rare earth scandium-aluminum garnets with impurity of TR^{3+} ions as active media for solid state lasers, *Sov. Phys.-Dokl.*, 19, 671, 1975.

78. **Kaminskii, A. A., Bagdasarov, Kh. S., Bogomolova, G. A., et al.**, Luminescence and stimulated emission of Nd^{3+} ions in Gd$_3$Sc$_2$Ga$_3$O$_{12}$ crystals, *Phys. Stat. Sol.*, 34a, K109, 1976.

79. **Bagdasarov, Kh. S., Kaminskii, A. A., Kevorkov, A. M., et al.**, Investigation of the stimulated emission of cubic crystals of YScO$_3$ with Nd^{3+} ions, *Sov. Phys.-Dokl.*, 20, 681, 1975.

80. **Aleksandrov, V. I., Voron'ko, Yu. K., Mikhalevich, V. G., et al.**, Spectroscopic properties and emission of Nd^{3+} in ZnO$_2$ and HfO$_2$ crystals, *Sov. Phys.-Dokl.*, 16, 657, 1972.

81. **Aleksandrov, V. I., Kaminskii, A. A., Maksimova, G. V., et al.**, Stimulated radiation of Nd^{3+} ions in crystals for the $^4F_{3/2} \rightarrow$ $^4I_{13/2}$ transition, *Sov. Phys.-Dokl.*, 18, 495, 1974.

82. **Kaminskii, A. A., Sarkisov, S. E., Bohm, J., et al.**, Growth, spectroscopic and laser properties of crystals in the K$_5$Bi$_{1-x}$Nd$_x$ (MoO$_4$)$_4$ system, *Phys. Stat. Sol.*, 43a, 71, 1977.

83. **Kaminskii, A. A., Pavlyuk, A. A., Klevtsov, P. V., et al.**, Stimulated radiation of monoclinic crystals of KY(WO$_4$)$_2$ and KGd(WO$_4$)$_2$ with Ln^{3+} ions, *Inorg. Mater.*, 13, 482, 1977.

84. **Kaminskii, A. A., Klevtsov, P. V., Li, L., et al.**, Stimulated emission of radiation by crystals of KLa(MoO$_4$)$_2$ with Nd^{3+} ions, *Inorg. Mater.*, 9, 1824, 1973.

85. **Kaminskii, A. A., Klevtsov, P. V., and Pavlyuk, A. A.**, Stimulated emission from KY(MoO$_4$)$_2$:d^{3+} crystal laser, *Phys. Stat. Sol.*, 1a, K91, 1970.

86. **Kaminskii, A. A., Klevtsov, P. V., Li, L., and Pavlyuk, A. A.,** Spectroscopic and stimulated emission studies of the new $KY(WO_4)_2$:d^{3+} laser crystal, *Inorg. Mater.,* 8, 1896, 1972.

87. **Kaminskii, A. A., Klevtsov, P. V., Li, L., and Pavlyuk, A. A.,** Laser $^4F_{3/2} \rightarrow {}^4I_{9/2}$ and $^4F_{3/2} \rightarrow {}^4I_{13/2}$ transitions in $KY(WO_4)_2$:d^{3+}, *IEEE J. Quantum Electron.,* QE-8, 457, 1972.

88. **Kaminskii, A. A., Klevtsov, P. V., Bagdasarov, Kh. S., Mayyer, A. A., Pavlyuk, A. A., Petrosyan, A. G., and Provotorov, M. V.,** New cw crystal lasers, *JETP Lett.,* 16, 387, 1972.

89. **Morris, R. C., Cline, C. F., and Begley, R. F.,** Lanthanum beryllate: a new rare-earth ion host, *Appl. Phys. Lett.,* 27, 444, 1975.

90. **Jenssen, H. P., Begley, R. F., Webb, R., and Morris, R. C.,** Spectroscopic properties and laser performance of Nd^{+3} in lanthanum beryllate, *J. Appl. Phys.,* 47, 1496, 1976.

91. **Matrosov, V. I., Timosheckin, M. I., Tsvetkov, E. I., et al.,** Investigation of the conditions of crystallization of lanthanum beryllate in *Abstracts of the Fifth All-Union Conference on Crystal Growth, Proc. Acad. Sci. Georgian Sov. Soc. Rep.,* Tiflis, 1977, 167 (in Russian).

92. **Vylegzhanin, D. N. and Kaminskii, A. A.,** Study of electron-phonon interaction in LaF_3:Nd^{3+} crystals, *Sov. Phys.-JETP,* 35, 361, 1972.

93. **Voron'ko, Yu. K., Dmitruk, M. V., Kaminskii, A. A., Osiko, V. V., and Shpakov, V. N.,** CW stimulated emission in an LaF_3:Nd^{3+} laser at room temperature, *Sov. Phys.-JETP,* 27, 400, 1968.

94. **Dmitruk, M. V., Kaminskii, A. A., Osiko, V. V., and Tevosyan, T. A.,** Induced emission of hexagonal LaF_3-SrF_2:Nd^{3+} crystals at room-temperature, *Phys. Stat. Sol.,* 25, K75, 1968.

95. **Bakhsheyeva, G. F., Karapetyan, V. Ye., Morozov, A. M., Morozova, L. G., Tolstoy, M. N., and Feofilov, P. P.,** Optical constants, fluorescence and stimulated emission of neodymium-doped lanthanum niobate single crystals, *Opt. Spectrosc.,* 28, 38, 1970.

96. **Hoskins, R. H. and Soffer, B. H.,** Fluorescence and stimulated emission from La_2O_3:Nd^{3+}, *J. Appl. Phys.,* 36, 323, 1965.

97. **Alves, R. V., Buchanan, R. A., Wickersheim, K. A., and Yates, E. A.,** Neodymium-activated lanthanum oxysulfide: a new high-gain laser material, *J. Appl. Phys.,* 42, 3043, 1971.

98. **Kaminskii, A. A., Mayer, A. A., Nikonova, N. S., Provotorov, M. V., and Sarkisov, S. E.,** Stimulated emission from the new $LiGd(MoO_4)_2$:Nd^{3+} crystal laser, *Phys. Stat. Sol.,* 12a, K73, 1972.

99. **Kaminskii, A. A. and Sarkisov, S. E.,** Study of stimulated emission from Nd^{3+} ions in crystals at the $^4F_{3/2} \rightarrow {}^4I_{13/2}$ transition. 4, *Sov. J. Quantum Electron.,* 3, 248, 1973.

100. **Kaminskii, A. A., Mayer, A. A., Provotorov, M. V., and Sarkisov, S. E.,** Investigation of stimulated emission from $LiLa(MoO_4)_2$:Nd^{3+} crystal laser, *Phys. Stat. Sol.,* 17a, K115, 1973.

101. **Belabaev, K. G., Kaminskii, A. A., and Sarkisov, S. E.,** Stimulated emission from ferroelectric $LiNbO_3$ crystals containing Nd^{3+} and Mg^{2+} ions, *Phys. Stat. Sol.,* 28a, K17, 1975.

102. **Kaminow, I. P., Stulz, L. W.,** Nd:$LiNbO_3$ laser, *IEEE J. Quantum Electron.,* QE-11, 306, 1975.

103. **Harmer, A. L., Linz, A., and Gabbe, D. R.,** Fluorescence of Nd^{3+} in lithium yttrium fluoride, *J. Phys. Chem. Sol.,* 30, 1483, 1969.

104. **Le Goff, D., Bettinger, A., and Labadens, A.,** Etude d'un oscillateur a blocage de modes utilisant un cristal de $LiYF_4$ dope au neodyme, *Opt. Commun.,* 26, 108, 1978, in French.

105. **Kaminskii, A. A., Ivanov, A. O., Sarkisov, S. E., et al.,** Comprehensive investigations of the spectral and lasing characteristics of the $LuAlO_3$ crystal doped with Nd^{3+}, *Sov. Phys.-JETP,* 44, 516, 1976.

106. **Kaminskii, A. A., Bogomolova, G. A., Bagdasarov, Kh. S., and Petrosyan, A. G.,** Luminescence, absorption and stimulated emission of $Lu_3Al_5O_{12}$-Nd^{3+} crystals, *Opt. Spectrosc.,* 39, 643, 1975.

107. **Bagdasarov, Kh. S., Bogomolova, G. A., Kaminskii, A. A., et al.,** Study of the stimulated emission of $Lu_3Ga_5O_{12}$ crystals containing Nd^{3+} ions at the transitions $^4F_{3/2} \rightarrow {}^4I_{11/2}$ and $^4F_{3/2} \rightarrow {}^4I_{13/2}$, *Sov. Phys.-Dokl.,* 19, 584, 1975.

108. **Kaminskii, A. A., Lapsker, Ya. Ye., and Sobolev, B. P.,** Induced emission of $NaCaCeF_6$:Nd^{3+} at room temperature, *Phys. Stat. Sol.,* 23, K5, 1967.

109. **Bagdasarov, Kh. S., Kaminskii, A. A., Lapsker, Ya. Ye., and Sobolev, B. P.,** Neodymium-doped α-gagarinite laser, *JETP Lett.,* 5, 175, 1967.

110. **Bagdasarov, Kh. S., Kaminskii, A. A., and Sobolev, B. P.,** Laser based on $5NaF \cdot 9YF_3$:Nd^{3+} cubic crystals, *Sov. Phys.-Crystallogr.,* 13, 779, 1969.

111. **Morozov, A. M., Tolstoy, M. N., Feofilov, P. P., and Shapovalov, V. N.,** Fluorescence and stimulated emission in neodymium in lanthium molybdate-sodium crystals, *Opt. Spectrosc.,* 22, 224, 1967.

112. **Zverev, G. M., Kolodnyy, G. Ya.,** Stimulated emission and spectroscopic studies of neodymium-doped binary lanthanum molybdate-sodium single crystals, *Sov. Phys.-JETP,* 25, 217, 1967.

113. **Kaminskii, A. A., Kolodnyy, G. Ya., and Sergeyeva, N. I.,** CW $NaLa(MoO_4)_2$:Nd^{3+} crystal laser operating at 300°K, *J. Appl. Spectrosc. (USSR),* 9, 1275, 1968.

114. **Belokrinitskiy, N. S., Belousov, N. D., Bonchkovskiy, V. I., Kobzar'-Zlenko, V. A., Skorobogatov, B. S., and Soskin, M. S.,** Study of stimulated emission from Nd^{3+}-doped $NaLa(WO_4)_2$ single crystals, *Ukrainskiy Fizicheskiy Zhurnal,* 14, 1400, 1969 (in Russian).

115. **Kariss, Ya. E., Tolstoy, M. N., and Feofilov, P. P.,** Stimulated emission from neodymium in lead molybdate single crystals, *Opt. Spectrosc.,* 18, 99, 1965.

116. **Morozov, A. M., Morozova, L. G., Fedorov, V. A., and Feofilov, P. P.**, Spontaneous and stimulated emission of neodymium in lead fluorophosphate crystals, *Opt. Spectrosc.,* 39, 343, 1975.

117. **Kaminskii, A. A. and Li, L.**, Spectroscopic studies of stimulated emission in an $SrF_2:Nd^{3+}$ (Type I) crystal laser, *J. Appl. Spectrosc., (USSR),* 12, 29, 1970.

118. **Kaminskii, A. A., Sarkisov, S. E., and Sobolev, B. P.**, unpublished.

119. **Kaminskii, A. A., Sarkisov, S. E., Seyranyan, K. B., and Sobolev, B. P.**, Stimulated emission from Nd^{3+} ions in SrF_2-GdF_3 crystals, *Inorg. Mater.,* 9, 310, 1973.

120. **Garashina, L. S., Kaminskii, A. A., Li, L., and Sobolev, B. P.**, Laser based on $SrF_2-YF_3:Nd^{3+}$ cubic crystals, *Sov. Phys.-Crystallogr.,* 14, 799, 1970.

121. **Kaminskii, A. A., Sarkisov, S. E., Seyranyan, K. B., and Sobolev, B. P.**, Study of stimulated emission in $Sr_2Y_5F_{19}$ crystals with Nd^{3+} ions, *Sov. J. Quantum Electron.,* 4, 112, 1974.

122. **Birnbaum, M. and Tucker, A. W.**, Nd-YALO Oscillation at 0.95 µm at 300°K, *IEEE J. Quantum Electron.,* QE-9, 46, 1973.

123. **Bagdasarov, Kh. S., and Kaminskii, A. A.**, RE^{3+}-doped $YAlO_3$ as an active medium for lasers, *JETP Lett.,* 9, 303, 1969.

124. **Weber, M. J., Bass, M., Andringa, K., Monchamp, R. R., and Comperchio, E.**, Czochralski growth and properties of $YAlO_3$ laser crystals, *Appl. Phys. Lett.,* 15, 342, 1969.

125. **Massey, G. A. and Yarborough, J. M.**, High average power operation and nonlinear optical generation with the $Nd:YAlO_3$ Laser, *Appl. Phys. Lett.,* 18, 576, 1971.

126. **Bass, M. and Weber, M. J.**, YALO:Robust at age 2, *Laser Focus,* No. 9, 34, 1971.

127. **Kaminskii, A. A.**, Temperature pulsations and multi-frequency laser action in $YAlO_3:Nd^{3+}$, *JETP Lett.,* 14, 222, 1971.

128. **Bass, M. and Weber, M. J.**, Nd, $Cr:YAlO_3$ laser tailored for high-energy Q-switched operation, *Appl. Phys. Lett.,* 17, 395, 1970.

129. **Birnbaum, M., Tucker, A. W., and Pomphrey, P. J.**, New Nd:YAG laser transition $^4F_{3/2} \rightarrow {}^4I_{9/2}$, *IEEE J. Quantum Electron.,* QE-8, 502, 1972.

130. **Wallace, R. W. and Harris, S. E.**, Oscillation and doubling of the 0.946 µ line in $Nd^{3+}:YAG$, *Appl. Phys. Lett.,* 15, 111, 1969.

131. **Geusic, J. E., Marcos, H. M., and Van Uitert, L. G.**, Laser oscillations in Nd-doped yttrium aluminum, yttrium gallium and gadolinium garnets, *Appl. Phys. Lett.,* 4, 182, 1964.

132. **Smith, R. G.**, New room temperature CW laser transition in YAlG:Nd, *IEEE J. Quantum Electron.,* QE-4, 505, 1968.

133. **Marling, J. B.**, 1.05-1.44 µm tunability and performance of the CW $Nd^{3+}:YAG$ laser, *IEEE J. Quantum Electron.,* QE-14, 56, 1978.

134. **Marin, V. I., Nikitin, V. I., Soskin, M. S., and Khizhnyak, A. I.**, Superluminescence emitted by $YAG:Nd^{3+}$ crystals and stimulated emission due to weak transitions, *Sov. J. Quantum Electron.,* 5, 732, 1975.

135. **DeSerno, U., Ross, D., and Zeidler, G.**, Quasicontinuous giant pulse emission of $^4F_{3/2} \rightarrow {}^4I_{13/2}$ transition at 1.32 µm in $YAG:Nd^{3+}$, *Phys. Lett.,* 28A, 422, 1968.

136. **Wallace, R. W.**, Oscillation of the 1.833 µ line in $Nd^{3+}:YAG$, *IEEE J. Quantum Electron.,* QE-7, 203, 1971.

137. **Kiss, Z. J. and Duncan, R. C.**, Cross-pumped $Cr^{3+}-Nd^{3+}:YAG$ laser system, *Appl. Phys. Lett.,* 5, 200, 1964.

138. **Kaminskii, A. A.**, *Laser Crystals,* Springer-Verlag, New York, 1980.

139. **Voron'ko, Yu. K., Maksimova, G. V., Mikhalevich, V. G., Osiko, V. V., Sobol', A. A., Timosheckin, M. I., and Shipulo, G. P.**, Spectroscopic properties of and stimulated emission from yttrium-lutecium-aluminum garnet crystals, *Opt. Spectrosc.,* 33, 376, 1972.

140. **Hoskins, R. H. and Soffer, B. H.**, Stimulated emission from $Y_2O_3:Nd^{3+}$, *Appl. Phys. Lett.,* 4, 22, 1964.

141. **Stone, J. and Burrus, C. A.**, $Nd:Y_2O_3$-single-crystal fiber laser: room-temperature CW operation at 1.07- and 1.35-µm wavelength, *J. Appl. Phys.,* 49, 2281, 1978.

142. **Greskovich, C. and Chernoch, J. P.**, Improved polycrystalline ceramic lasers, *J. Appl. Phys.,* 45, 4495, 1974.

143. **Bagdasarov, Kh. S., Kaminskii, A. A., Kevorkov, A. M., et al.**, Laser properties of $Y_2SiO_5-Nd^{3+}$ crystals irradiated at the $^4F_{3/2} \rightarrow {}^4I_{11/2}$ and $^4F_{3/2} \rightarrow {}^4I_{13/2}$ transitions, *Sov. Phys.-Dokl.,* 18, 664, 1974.

144. **Kaminskii, A. A., Bogomolova, G. A., and Li, L.**, Absorption, fluorescence, stimulated emission and splitting of the Nd^{3+} levels in a YVO_4 crystal, *Inorg. Mater.,* 5, 573, 1969.

145. **Tucker, A. W., Birnbaum, M., Fincher, C. L., and DeShazer, L. G.**, Continuous-wave operation of $Nd:YVO_4$ at 1.06 and 1.34 µ, *J. Appl. Phys.,* 47, 232, 1976.

146. **O'Connor, J. R.**, Unusual crystal-field energy levels and efficient laser properties of $YVO_4:Nd^{3+}$, *Appl. Phys. Lett.,* 9, 407, 1966.

147. **Chang, N. C.**, Fluorescence and stimulated emission from trivalent europium in yttrium oxide, *J. Appl. Phys.,* 34, 3500, 1963.

148. O'Connor, J. R., Optical and laser properties of Nd^{3+}- and Eu^{3+}-doped YVO_4, *Trans. Metallurg. Soc. AIME,* 239, 362, 1967.

149. Azamatov, Z. T., Arsen'yev, P. A., and Chukichev, M. V., Spectra of gadolinium in YAG single crystals, *Opt. Spectrosc.,* 28, 156, 1970.

150. Jenssen, H. P., Castleberry, D., Gabbe, D., and Linz, A., Stimulated emission at 5445 A in Tb^{3+}:YLF, in *Digest of Technical Papers CLEA 1973 IEEE/OSA,* Washington, D.C., 1973, 47.

151. Johnson, L. F. and Guggenheim, H. J., Laser emission at 3 μ from Dy^{3+} in BaY_2F_8, *Appl. Phys. Lett.,* 23, 96, 1973.

152. Johnson, L. F. and Guggenheim, H. J., Electronic- and phonon-terminated laser emission from Ho^{3+} in BaY_2F_8, *IEEE J. Quantum Electron.,* QE-10, 442, 1974.

153. Johnson, L. F. and Guggenheim, H. J., Infrared-pumped visible laser, *Appl. Phys. Lett.,* 19, 44, 1971.

154. Voron'ko, Yu. K., Kaminskii, A. A., Osiko, V. V., and Prokhorov, A. M., Stimulated emission from Ho^{3+} in CaF_2 at 5512 Å, *JETP Lett.,* 1, 3, 1965.

155. Dmitruk, M. V., Kaminskii, A. A., Osiko, V. V., and Fursikov, M. M., Sensitization in CaF_2-ErF_3:Ho^{3+} lasers, *Inorg. Mater.,* 3, 516, 1967.

156. Voron'ko, Yu. K., Dmitruk, M. V., Murina, T. M., and Osiko, V. V., CW lasers based on mixed yttrofluorine-type crystals, *Inorg. Mater.,* 5, 422, 1969.

157. Dmitruk, M. V. and Kaminskii, A. A., Stimulated emission in a laser based on CaF_2-YF_3 crystals with Ho^{3+} and Er^{3+} ions, *Sov. Phys.-Crystallogr.,* 14, 620, 1970.

158. Robinson, M. and Devor, D. P., Thermal switching of laser emission of Er^{3+} at 2.69 μ and Tm^{3+} at 1.86 μ in mixed crystals of CaF_2:ErF_3:TmF_3, *Appl. Phys. Lett.,* 10, 167, 1967.

159. Johnson, L. F., Van Uitert, L. G., Rubin, J. J., and Thomas, R. A., Energy transfer from Er^{3+} to Tm^{3+} and Ho^{3+} ions in crystals, *Phys. Rev.,* 133A, 494, 1964.

160. Johnson, L. F., Boyd, G. D., Nassau, K., Optical maser characteristics of Ho^{3+} in $CaWO_4$, *Proc. IRE,* 50, 87, 1962.

161. Bagdasarov, Kh. S., Kaminskii, A. A., Kevorkov, A. M., Sarkisov, S. E., and Tevosyan, T. A., Stimulated emission from (Er,Lu)$A1O_3$ crystals with Ho^{3+} and Tm^{3+} ions, *Sov. Phys.-Crystallogr.,* 18, 681, 1974.

162. Hoskins, R. H. and Soffer, B. H., Energy transfer and CW laser action in Ho^{3+}:Er_2O_3, *IEEE J. Quantum Electron.,* QE-2, 253, 1966.

163. Morozov, A. M., Petrov, M. V., Startsev, V. R., et al., Luminescence and stimulated emission of holmium in yttrium- and erbium-oxyortho-silicate single crystals, *Opt. Spectrosc.,* 41, 641, 1976.

164. Bagdasarov, Kh. S., Kaminskii, A. A., Kevorkov, A. M., Prokhorov, A. M., Sarkisov, S. E., and Tevosyan, T. A., Stimulated emission from RE^{3+} ions in YAG crystals, *Sov. Phys.-Dokl.,* 19, 592, 1975.

165. Wunderlich, J. A., Sliney, J. G., and DeShazer, L. G., Stimulated emission at 2.04 μm in Ho^{3+}-doped $ErVO_4$ and YVO_4, *IEEE J. Quantum Electron.,* QE-13, 69, 1977.

166. Arsen'yev, P. A. and Bienert, K. E., Spectral properties of Ho^{3+} in $GdA1O_3$ crystals, *Phys. Stat. Sol.* 13a, K129, 1972.

167. Kaminskii, A. A., Fedorov, V. A., Sarkisov, S. E., et al., Stimulated emission of Ho^{3+} and Er^{3+} ions in $Gd_3Ga_5O_{12}$ crystals and cascade laser action of Ho^{3+} ions over the $^5S_2 \rightarrow {}^5I_5 \rightarrow {}^5I_6 \rightarrow {}^5I_8$ scheme, *Phys. Stat. Sol.,* 53a, K219, 1979.

168. Kaminskii, A. A., Pavlyuk, A. A., Butaeva, T. I., et al., Stimulated emission by subsidiary transitions of Ho^{3+} and Er^{3+} ions in $KGd(WO_4)_2$ crystals, *Inorg. Mater.,* 13, 1251, 1977.

169. Kaminskii, A. A., Pavlyuk, A. A., Klevtsov, P. V., et al., Stimulated radiation of monoclinic crystals of $KY(WO_4)_2$ and $KGd(WO_4)_2$ with Ln^{3+} ions, *Inorg. Mater.,* 13, 482, 1977.

170. Kaminskii, A. A., Pavlyuk, A. A., Chan, Ng., et al., 3μ stimulated emission by Ho^{3+} ions in $KY(WO_4)_2$ crystals at 300 K, *Sov. Phys.-Dokl.,* 24, 201, 1979.

171. Korovkin, A. M., Morozov, A. M., Tkachuk, A. M., Fedorov, A. A., Fedorov, V. A., and Feofilov, P. P., Spontaneous and stimulated emission from holmium in $NaLa(MoO_4)_2$ and $LaNbO_4$ crystals, in *Sbornik. Spektroskopiya Kristallov,* Nauka, Moskva, 1975.

172. Kaminskii, A. A., Fedorov, V. A., and Chan, Ng., Three-micron stimulated emission by Ho^{3+} ions in an $LaNbO_4$ crystal, *Inorg. Mater.,* 14, 1061, 1978.

173. Chicklis, E. P., Naiman, C. S., Esterowitz, L., and Allen, R., Deep red laser emission in Ho:YLF, *IEEE J. Quantum Electron.,* QE-13, 893, 1977.

174. Podkolzina, I. G., Tkachuk, A. M., Fedorov, V. A., and Feofilov, P. P., Multifrequency generation of stimulated emission of Ho^{3+} ion in $LiYF_4$ crystals, *Opt. Spectrosc.,* 40, 111, 1976.

175. Esterowitz, L., Eckardt, R. C., and Allen, R. E., Long-wavelength stimulated emission via cascade laser action in Ho:YLF, *Appl. Phys. Lett.,* 35, 236, 1979.

176. Gifeisman, Sh. N., Tkachuk, A. M., and Prizmak, V. V., Optical spectra of Ho^{3+} ion in $LiYF_4$ crystals, *Opt. Spectrosc.,* 44, 68, 1978.

177. **Remski, R. L., James, L. T., Gooen, K. H., DiBartolo, B., and Linz, A.,** Pulsed laser action in LiYF₄:Er³⁺, Ho³⁺ at 77°K, *IEEE J. Quantum Electron.,* QE-5, 212, 1969.

178. **Chicklis, E. P., Naiman, C. S., Folweiller, R. C., Gabbe, D. R., Jenssen, H. P., and Linz, A.,** High efficiency room-temperature 2.06-μm laser using sensitized Ho³⁺:YLF, *Appl. Phys. Lett.,* 19, 119, 1971.

179. **Ivanov, A. O., Mochalov, I. V., Petrov, M. V.,** et al., Spectroscopic properties of single crystals of rare-earth aluminum garnet and rare-earth-ortho-aluminate activated by the ions Ho³⁺, Er³⁺ and Th³⁺, in *Abstracts of Papers of Fifth All-Union Symposium on Spectroscopy of Crystals,* Kazan, 1976, 195 (in Russian).

180. **Kaminskii, A. A., Bagdasarov, Kh. S., Petrosyan, A. G., and Sarkisov, S. E.,** Investigation of stimulated emission from Lu₃Al₅O₁₂ crystal with Ho³⁺, Er³⁺ and Tm³⁺ ions, *Phys. Status Solidi,* 18a, K31, 1973.

181. **Kaminskii, A. A., Butaeva, T. I., Ivanov, A. O.,** et al., New data on stimulated emission of crystals containing Er³⁺ and Ho³⁺ ions, *Sov. Tech. Phys. Lett.,* 2, 308, 1976.

182. **Kaminskii, A. A. and Petrosyan, A. G.,** Sensitized stimulated emission from self-saturating 3-μm transitions of Ho³⁺ and Er³⁺ ions in Lu₃Al₅O₁₂ crystals, *Inorg. Mater.,* 15, 425, 1979.

183. **Bagdasarov, Kh. S., Kaminskii, A. A., and Sobolev, B. P.,** Stimulated emission in lasers based on α-NaCaErF₆:Ho³⁺ and α-NaCaErF₆:Tm³⁺ crystals, *Inorg. Mater.,* 5, 527, 1969.

184. **Kaminskii, A. A.,** Moderne Problem der Laserkristallphysik, *Physikalische Gesellschaft der DDR,* Dresden, 1977.

185. **Weber, M., Bass, M., Varitimos, T., and Bua, D.,** Laser action from Ho³⁺, Er³⁺ and Tm³⁺ in YAlO₃, *IEEE J. Quantum Electron.,* QE-9, 1079, 1973.

186. **Johnson, L. F., Geusic, J. E., and Van Uitert, L. G.,** Coherent oscillations from Tm³⁺, Ho³⁺, Yb³⁺ and Er³⁺ ions in yttrium aluminum garnet, *Appl. Phys. Lett.,* 7, 127, 1965.

187. **Johnson, L. F., Geusic, J. E., and Van Uitert, L. G.,** Efficient, high-power coherent emission from Ho³⁺ ions in yttrium aluminum garnet, assisted by energy transfer, *Appl. Phys. Lett.,* 8, 200, 1966.

188. **Bakradze, R. V., Zverev, G. M., Kolodnyi, G. Ya.,** et al., Sensitized luminescence and stimulated radiation from yttrium-aluminum garnet crystals, *Sov. Phys.-JETP,* 26, 323, 1968.

189. **Remski, R. L. and Smith, D. J.,** Temperature dependence of pulsed laser threshold in YAG: Er³⁺, Tm³⁺, Ho³⁺, *IEEE J. Quantum Electron.,* QE-6, 750, 1970.

190. **Hopkins, R. H., Melamed, N. T., Henningsen, T., and Roland, G. W.,** Technical Rpt. AFAL-TR-70-103, 1970. Air Force Avionics Laboratory.

191. **Johnson, L. F., Dillon, J. F., and Remeika, J. P.,** Optical properties of Ho³⁺ ions in yttrium gallium garnet and yttrium iron garnet, *Phys. Rev.,* B1, 1935, 1970.

192. **Arsen'yev, P. A.,** Spectral parameters of trivalent holmium in a YAG lattice, *Ikrainskiy Fizicheskiy Zhurnal,* 15, 689, 1970 (in Russian).

193. **Aleksandrov, V. I., Murina, T. M., Zhekov, V. K., and Tatarintsev, V. M.,** Stimulated emission from Tm³⁺ and Ho³⁺ in zirconium dioxide crystals, in *Sbornik. Kratkiye Soobshcheniya po Fizike,* An SSSR Fizicheskiy Institut im P. N. Lebedeva, 1973, No. 2, 17—21 (in Russian).

194. **Johnson, L. F. and Guggenheim, H. J.,** New laser lines in the visible from Er³⁺ ions in BaY₂F₈, *Appl. Phys. Lett.,* 20, 474, 1972.

195. **Voron'ko, Yu. K. and Sychugov, V. I.,** The stimulated emission of Er³⁺ ions in CaF₂ at λ₁ = 8456 Å and λ₂ = 8548 Å, *Phys. Stat. Sol.,* 25, K119, 1968.

196. **Voron'ko, Yu. K., Zverev, G. M., and Prokhorov, A. M.,** Stimulated emission from Er³⁺ ions in CaF₂, *Sov. Phys.-JETP,* 21, 1023, 1964.

197. **Forrester, P. A. and Sampson, D. F.,** A new laser line due to energy transfer from colour centers to erbium ions in CaF₂, *Proc. Phys. Soc.,* 88, 199, 1966.

198. **Pollack, S. A.,** Stimulated emission in CaF₂:Er³⁺, *Proc. IEEE,* 51, 1793, 1963.

199. **Batygov, S. Kh., Kulevskii, L. A., Prokhorov, A. M.,** et al., Erbium-doped CaF₂ crystal laser operating at room temperature, *Sov. J. Quantum Electron.,* 4, 1469, 1975.

200. **Robinson, M. and Devor, D. P.,** Thermal switching of laser emission of Er³⁺ at 2.69 μ and Tm³⁺ at 1.86 μ in mixed crystals of CaF₂:ErF₃:TmF₃, *Appl. Phys. Lett.,* 10, 167, 1967.

201. **Kaminskii, A. A.,** Cascading lasers based on activated crystals, *Inorg. Mater.,* 7, 802, 1971.

202. **Kaminskii, A. A.,** Spectroscopic studies of stimulated emission from Er³⁺ ions in CaF₂-YF₃ crystals, *Opt. Spectrosc.,* 31, 507, 1971.

203. **Kiss, Z. J. and Duncan, R. C.,** Optical maser action in CaWO₄:Er³⁺, *Proc. IRE.,* 50, 1531, 1962.

204. **Arsen'yev, P. A. and Bienert, K. E.,** Absorption, luminescence, and stimulated emission spectra of Er³⁺ ions in GdAlO₃ crystals, *Phys. Stat. Sol.,* 10a, K85, 1972.

205. **Arsen'yev, P. A., Potemkin, A. V., Fenin, V. V., and Senff, I.,** Investigation of stimulated emission of Er³⁺ ions in mixed crystals with perovskite structure, *Phys. Stat. Sol.,* 43a, K15, 1977.

206. **Kaminskii, A. A.,** Inorganic materials with Ln³⁺ ions for producing stimulated radiation in the 3-μm band, *Inorg. Mater.,* 15, 809, 1979.

207. **Kaminskii, A. A., Pavlyuk, A. A., Balashov, I. F.,** et al., Stimulated emission by $KY(WO_4)_2$-Er^{3+} crystals at 0.85, 1.73 and 2.8 μm at 300 K, *Inorg. Mater.,* 14, 1765, 1978.

208. **Krupke, W. F. and Gruber, J. B.,** Energy levels of Er^{3+} in LaF_3 and coherent emission at 1.61 μm, *J. Chem. Phys.,* 41, 1225, 1964.

209. **Chicklis, E. P., Naiman, C. S., and Linz, A.,** Stimulated emission at 0.85 μm in Er^{3+}:YLF, in *Digest of Technical Papers VII International Quantum Electronics Conference,* Montreal, 1972, 17.

210. **Petrov, M. V. and Tkachuk, A. M.,** Optical spectra and multifrequency stimulated emission of $LiYF_4$-Er^{3+} crystals, *Opt. Spectrosc.,* 45, 81, 1978.

211. **Chicklis, E. P. and Naiman, C. S.,** A review of near-infrared optically pumped solid-state lasers, in *Proceedings of the First European Electro-Optics Markets and Technology Conference,* IPC Science and Technology Press, 1973, 77.

212. **Chicklis, E. P., Esterowitz, L., Allen, R., and Kruer, M.,** Stimulated emission at 2.81 μm in Er:YLF, in *Proceedings of LASERS '78,* Orlando, Fla., 1978.

213. **Kaminskii, A. A., Butaeva, T. I., Fedorov, V. A., Bagdasarov, Kh. S., and Petrosyan, A. G.,** Absorption, luminescence and stimulated emission investigations in $Lu_3Al_5O_{12}$-Er^{3+} crystals, *Phys. Stat. Sol.,* 39a, 541, 1977.

214. **Weber, M. J., Bass, M., and Demars, G. A.,** Laser action and spectroscopic properties of Er^{3+} in $YAlO_3$, *J. Appl. Phys.,* 42, 301, 1971.

215. **White, K. O. and Schlenser, S. A.,** Coincidence of Er:YAG laser emission with methane absorption at 1645.1 nm, *Appl. Phys. Lett.,* 21, 419, 1972.

216. **Zverev, G. M., Garmash, V. M., Onischenko, A. M.,** et al., Induced emission by trivalent erbium ions in crystals of yttrium-aluminum garnet, *J. Appl. Spectrosc. (USSR),* 21, 1467, 1974.

217. **Prokhorov, A. M., Kaminskii, A. A., Osiko, V. V.,** et al., Investigations of the 3 μm stimulated emission from Er^{3+} ions in aluminum garnets at room temperature, *Phys. Stat. Sol.,* 40a, K69, 1977.

218. **Thornton, J. R., Rushworth, P. M., Kelly, E. A., McMillan, R. W., and Harper, L. L.,** in *Proceedings 4th Conference Laser Technology,* Vol. II, University of Michigan, Ann Arbor, 1970, 1249.

219. **Kaminskii, A. A. and Osiko, V. V.,** Sensitization in a CaF_2-ErF_3:Tm^{3+} laser, *Inorg. Mater.,* 3, 519, 1967.

220. **Johnson, L. F., Boyd, G. D., and Nassau, K.,** Optical maser characteristics of Tm^{3+} in $CaWO_4$, *Proc. IRE,* 50, 86, 1962.

221. **Giorbachov, V. A., Zhekov, V. I., Murina, T. M.,** et al., Spectroscopic and growth properties of erbium aluminate with Tr^{3+} impurity ions, *Short Communications in Physics,* 4, 16, 1973.

222. **Van Uitert, L. G., Grodkiewicz, W. H., and Dearborn, E. F.,** Growth of large optical-quality yttrium and rare-earth aluminum garnets, *J. Am. Ceram. Soc.,* 48, 105, 1965.

223. **Soffer, B. H. and Hoskins, R. H.,** Energy transfer and CW laser action in Tm^{3+}:Er_2O_3, *Appl. Phys. Lett.,* 6, 200, 1965.

224. **Arsenev, P. A. and Bienert, K. E.,** Absorption, luminescence and stimulated emission spectra of Tm^{3+} in $GdAlO_3$ Crystals, *Phys. Stat. Sol.,* 13a, K125, 1972.

225. **Baer, J. W., Knights, M. G., Chicklis, E. P., and Jenssen, H. P.,** XeF-pumped laser operation of Tm:YLF at 452 nm, in *Proceedings Topical Meeting on Excimer Lasers, IEEE-OSA,* Charleston, S.C., 1979.

226. **Ivanov, A. O., Mochalov, I. V., Tkachuk, A. M., Fedorov, V. A., and Feofilov, P P.,** Spectral characteristics of the thulium ion and cascade generation of stimulated radiation in a $YAlO_3$:Tu^{3+}:Cr^{3+} crystal, *Sov. J. Quantum Electron.,* 5, 117, 1975.

227. **Hobrock, L. M., DeShazer, L. G., Krupke, W. F., Keig, G. A., and Witter, D. E.,** Four-level operation of Tm:Cr:$YAlO_3$ laser at 2.35 μ, in *Digest of Technical Papers VII International Quantum Electronics Conference,* Montreal, 1972, 15.

228. **Caird, J. A., DeShazer, L. G. and Nella, J.,** Characteristics of room-temperature 2.3-μm laser emission from Tm^{3+} In YAG and $YAlO_3$, *IEEE J. Quantum Electron.,* QE-11, 874, 1975.

229. **Rubin, J. J. and Van Uitert, L. G.,** Growth of large yttrium vanadate single crystals for optical maser studies, *J. Appl. Phys.,* 37, 2920, 1966.

230. **Robinson, M. and Asawa, C. K.,** Stimulated emission from Nd^{3+} and Yb^{3+} in noncubic sites of neodymium- and ytterbium-doped CaF_2, *J. Appl. Phys.,* 38, 4495, 1967.

231. **Bogomolova, G. A., Vylegzhanin, D. N., and Kaminskii, A. A.,** Spectral and lasing investigations of garnets with Yb^{3+} ions, *Sov. Phys.-JETP,* 42, 440, 1976.

232. **Sorokin, P. P. and Stevenson, M. J.,** Solid-state optical maser using divalent samarium in calcium fluoride, *IBM J. Res. Dev.,* 5, 56, 1961.

233. **Kaiser, W., Garrett, C. G. B., and Wood, D. L.,** Fluorescence and optical maser effects in CaF_2:Sm^{2+}, *Phys. Rev.,* 123, 766, 1961.

234. **Anan'yev, Yu. A., Grezin, A. K., Mak, A. A., Sedov, B. M., and Yudina, Ye. N.,** A fluorite:samarium laser, *Sov. J. Opt. Technol.,* 35, 313, 1968.

235. **Vagin, Yu. S., Marchenko, V. M., and Prokhorov, A. M.,** Spectrum of a laser based on electron-vibrational transitions in a CaF_2:Sm^{2+} crystal, *Sov. Phys.-JETP,* 28, 904, 1969.

236. Sorokin, P. P., Stevenson, M. J., Lankard, J. R., and Pettit, G. D., Spectroscopy and optical maser action in $SrF_2:Sm^{2+}$, *Phys. Rev.*, 127, 503, 1962.

237. Kiss, Z. J. and Duncan, R. C., Pulsed and continuous optical maser action in $CaF_2:Dy^{2+}$, *Proc. IRE*, 50, 1531, 1962.

238. Yariv, A., Continuous operation of a $CaF_2:Dy^{2+}$ optical maser, *Proc. IRE*, 50, 1699, 1962.

239. Kiss, Z. J., The CaF_2-Tm^{2+} and the CaF_2-Dy^{2+} optical maser systems, in *Quantum Electronics Proceedings of the Third International Congress,* Grivet, P. and Bloembergen, N., Eds., Columbia University Press, New York, 1964, 805.

240. Kiss, Z. J., Lewis, H. R., and Duncan, R. C., Sun pumped continuous optical maser, *Appl. Phys. Lett.*, 2, 93, 1963.

241. Pressley, R. J. and Wittke, J. P., $CaF_2:Dy^{2+}$ lasers, *IEEE J. Quantum Electron.*, QE-3, 116, 1967.

242. Hatch, S. E., Parsons, W. F., and Weagley, J. R., Hot-pressed polycrystalline $CaF_2:Dy^{3+}$ laser, *Appl. Phys. Lett.*, 5, 153, 1964.

243. Zolotov, Ye. M., Osiko, V. V., Prokhorov, A. M., and Shipulo, G. P., Study of fluorescence and laser properties of $SrF_2:Dy^{2+}$ crystals, *J. Appl. Spectrosc. (USSR)*, 8, 627, 1968.

244. Kiss, Z. J. and Duncan, R. C., Optical maser action in $CaF_2:Tm^{2+}$, *Proc. IRE*, 1532, 1962.

245. Duncan, R. C. and Kiss, Z. J., Continuously operating $CaF_2:Tm^{2+}$ Optical Maser, *Appl. Phys. Lett.*, 3, 23, 1963.

246. Maiman, T. H., Stimulated optical radiation in ruby, *Nature*, 187, 493, 1960.

247. Maiman, T. H., Optical maser action in ruby, *Br. Commun. Electron.*, 7, 674, 1960.

248. McClung, F. J., Schwarz, S. E., and Meyers, F. J., R_2 line optical maser action in ruby, *J. Appl. Phys.*, 33, 3139, 1962.

249. Roess, D., Analysis of room temperature CW ruby lasers, *IEEE J. Quantum Electron.*, QE-2, 208, 1966.

250. Evtuhov, V. and Kneeland, J. K., Power output and efficiency of continuous ruby laser, *J. Appl. Phys.*, 38, 4051, 1967.

251. Burrus, C. A. and Stone, J., Room-temperature continuous operation of a ruby fiber laser, *J. Appl. Phys.*, 49, 3118, 1978.

252. Collins, R. J., Nelson, D. F., Schawlow, A. L., Bond, W., Garrett, C. G. B., and Kaiser, W., Coherence, narrowing, directionality, and relaxation oscillations in the light emission from ruby, *Phys. Rev. Lett.*, 5, 303, 1960.

253. Nelson, D. F. and Boyle, W. S., A continuously operating ruby optical maser, *Appl. Optics*, 1, 181, 1962.

254. Birnbaum, M., Tucker, A. W., and Fincher, C. L., CW ruby laser pumped by an argon ion laser, *IEEE J. Quantum Electron.*, QE-13, 808, 1977.

255. Schawlow, A. L. and Devlin, G. E., Simultaneous optical maser action in two ruby satellite lines, *Phys. Rev. Lett.*, 6, 96, 1961.

256. Walling, J. C., Jenssen, H. P., Morris, R. C., O'Dell, E. W., and Peterson, O. G., Tunable-laser performance in $BeAl_2O_4:Cr^{3+}$, *Opt. Lett.*, 4, 182, 1979.

257. Walling, J. C. and Peterson, O. G., High gain laser performance in alexandrite, *IEEE J. Quantum Electron.*, QE-16, 119, 1980.

258. Bukin, G. V., Volkov, S. Yu., Matrosov, V. N., Sevast'yanov, B. K., and Timoshechkin, M. I., Stimulated emission from alexandrite ($BeAl_2O_4:Cr^{3+}$), *Sov. J. Quantum Electron.*, 8, 671, 1978.

259. Walling, J. C. and Sam, C. L., unpublished data, 1980.

260. Walling, J. C., Peterson, D. G., and Morris, R. C., Tunable CW alexandrite laser, *IEEE J. Quantum Electron.*, QE-16, 120, 1980.

261. Sevast'yanov, B. K., Bagdasarov, Kh. S., Pasternak, L. B., Volkov, S. Yu., and Drekhova, V. P., Stimulated emission from Cr^{3+} ions in YAG crystals, *JETP Lett.*, 17, 47, 1973.

262. Johnson, L. F. and Guggenheim, H. J., Phonon-terminated coherent emission from V^{2+} ions in MgF_2, *J. Appl. Phys.*, 38, 4837, 1967.

263. Johnson, L. F., Guggenheim, H. J., and Thomas, R. A., Phonon-terminated optical masers, *Phys. Rev.*, 149, 179, 1966.

264. Johnson, L. F., Dietz, R. E., and Guggenheim, H. J., Spontaneous and stimulated emission from Co^{3+} ions in MgF_2 and ZnF_2, *Appl. Phys. Lett.*, 5, 21, 1964.

265. Moulton, P. F. and Mooradian, A., Broadly tunable CW operation of $Ni:MgF_2$ and $CoMgF_2$ lasers, *Appl. Phys. Lett.*, 35, 838, 1979. (Unpublished results which improve upon those indicated in this reference have been included.)

266. Johnson, L. F., Dietz, R. E., and Guggenheim, H. J., Optical maser oscillation from Ni^{2+} in MgF_2 involving simultaneous emission of phonons, *Phys. Rev. Lett.*, 11, 318, 1963.

267. Moulton, P. F., Mooradian, A., and Reed, T. B., Efficient CW optically pumped $Ni:MgF_2$ laser, *Opt. Lett.*, 3, 164, 1978.

268. Moulton, P. F., Mooradian, A., Chen, Y., and Abraham, M. M., unpublished.

269. Porto, S. P. S. and Yariv, A., Optical maser characteristics $BaF_2:U^{3+}$, *Proc. IRE*, 50, 1542, 1962.

270. Porto, S. P. S. and Yariv, A., Trigonal sites and 2.24 micron coherent emission of U^{3+} in CaF_2, *J. Appl. Phys.*, 33, 1620, 1962.

271. Porto, S. P. S. and Yariv, A., Low lying energy levels and comparison of laser action of U^{3+} in CaF_2 in *Proceedings 3rd International Conference Quantum Electronics*, Grivet, P. and Bloembergen, N., Eds., Columbia University Press, New York, 1964, 717.

272. Wittke, J. P., Kiss, Z. J., Duncan, R. C., and McCormick, J. J., Uranium-doped calcium fluoride as a laser material, *Proc. IEEE*, 51, 56, 1963.

273. Boyd, G. D., Collins, R. J., Porto, S. P. S., Yariv, A., and Hargreaves, W. A., Excitation, relaxation and continuous maser action in 2.613 μm transition of CaF_2:U^{3+}, *Phys. Rev. Lett.*, 8, 269, 1962.

274. Porto, S. P. S. and Yariv, A., Excitation, relaxation and optical maser action at 2.407 microns in SrF_2:U^{3+}, *Proc. IRE*, 50, 1543, 1962.

275. Hopkins, R. H., Steinbruegge, K. B., Melamed, N. T., et al., Technical RPT. AFAL-TR-69-239, Air Force Avionics Laboratory, 1969.

276. Guesic, J. E., Advances in cw solid state lasers, *NEREM Rec.*, 8, 192, 1966.

277. Foster, J. D. and Osterink, L. M., Thermal effects in Nd:YAG laser, *J. Appl. Phys.*, 41, 3656, 1970.

278. Boyden, J. M. and Erickson, E. G., Second harmonic generation, Report AD729681, Holobeam, 1971.

279. Pressley, R. J., Collard, J. R., Goerdertier, P. V., et al., Technical RPT. AFAL-TR-66-129, Air Force Avionics Laboratory, 1966.

280. Huff, L., Sun-pumped laser, in *Digest of Technical Papers CLEA 1973 IEEE/OSA*, Washington, D.C., 1973, 48.

281. Beck, R. and Gurs, K., Ho laser with 50-W output and 6.5% slope efficiency, *J. Appl. Phys.*, 46, 5224, 1975.

282. Kostin, V. V., Thesis for Candidate's Degree, Lebedev Physics Institute, Academy of Sciences of the USSR, Moscow, 1973 (in Russian).

283. Steinbruegge, K. B. and Baldwin, G. D., High average power characteristics of CaLa SOAP:Nd laser materials, in *Digest of Technical Papers CLEA 1973 IEEE/OSA*, Washington, D.C., 1973, 48.

284. Kogan, R. M. and Crow, T. G., A high-brightness, one joule, frequency-doubled Nd:YAG laser, in *Digest of Technical Papers CLEA 1977 IEEE/OSA*, Washington, D.C., 1977, 70.

285. Herbst, R. L., Komine, H., and Byer, R. L., A 200 mJ unstable resonator Nd:YAG oscillator, *Opt. Commun.*, 21, 5, 1977.

286. Andreou, D., Construction and performance of a 20-pps unstable Nd:YAG oscillator, *Rev. Sci. Instrum.*, 49, 586, 1978.

287. Corcoran, V. J., McMillan, R. W., and Barnoske, S. K., Flashlamp-pumped YAG:Nd^{+3} laser action at kilohertz rates, *IEEE J. Quantum Electron.*, QE-10, 618, 1974.

288. Koechner, W., *Solid-State Laser Engineering*, Springer-Verlag, New York, 1976, 408.

289. Bethea, C. G., Megawatt power at 1.318 μ in Nd^{2+}:YAG and simultaneous oscillation at both 1.06 and 1.318 μ, *IEEE J. Quantum Electron.*, QE-9, 254, 1973.

290. Bass, M., Electrooptic Q-switching of the Nd:YVO_4 laser without an intracavity polarizer, *IEEE J. Quantum Electron.*, QE-11, 938, 1975.

291. Devor, D. P. and Soffer, B. H., 2.1 μm Laser of 20 W output power and 4-percent efficiency from Ho^{3+} in sensitized YAG, *IEEE J. Quantum Electron.*, QE-8, 231, 1972.

292. Bagdasarov, Kh. S., Danilov, V. P., Zhekov, V. I., et al., Pulse-periodic $Y_3Al_5O_{12}$:Er^{3+} laser with high activator concentration, *Sov. J. Quantum Electron.*, 8, 83, 1978.

293. Davydov, A. A., Kulevskii, L. A., Prokhorov, A. M., et al., Parametric generation with CdSe crystal pumped by CaF_2:Dy^{2+} laser, *JETP Lett.*, 15, 513, 1972.

294. Kostin, V. V., Kulevsky, L. A., Murina, T. M., et al., CaF_2:Dy^{2+} giant pulse laser with high repetition rate, *IEEE J. Quantum Electron.*, QE-2, 611, 1966.

295. Koechner, W., *Solid State Laser Engineering*, Springer-Verlag, New York, 1976, 425.

296. Koechner, W., *Solid State Laser Engineering*, Springer-Verlag, New York, 1976, 203.

297. Evtuhov, V. and Neeland, J. K., A continuously pumped repetitively Q-switched ruby laser and applications to frequency-conversion experiments, *IEEE J. Quantum Electron.*, QE-5, 207, 1969.

298. Eckhardt, R. C., DeRosa, J. L., and Letellier, J. P., Characteristics of an Nd:CaLaSOAP mode-locked oscillator, *IEEE J. Quantum Electron.*, QE-10, 620, 1974.

299. Goldberg, L. S. and Bradford, J. N., Passive mode locking and picosecond pulse generation in Nd:lanthanum beryllate, *Appl. Phys. Lett.*, 29, 585, 1976.

300. DiDomenico, M., Jr., Geusic, J. E., Marcos, H. M., and Smith, R. G., Generation of ultrashort optical pulses by mode locking the YAℓG:Nd laser, *Appl. Phys. Lett.*, 8, 180, 1966.

301. Osterink, L. M., and Foster, J. D., A mode-locked Nd:YAG laster, *J. Appl. Phys.*, 39, 4163, 1968.

302. Clobes, A. R. and Brienza, M. J., Passive mode locking of a pulsed Nd:YAG laser, *Appl. Phys. Lett.*, 14, 287, 1969.

303. Dewhurst, R. J. and Jacoby, D., A mode-locked unstable Nd:YAG laser, *Opt. Commun.*, 28, 107, 1979.

304. **Kuizenga, D. J.,** Development of an actively modelocked and Q-switched oscillator for laser fusion program at LLL, in *Picosecond Phenomena,* Shank, C. V., Ippen, E. P., and Shapiro, S. L., Eds., Springer-Verlag, New York, 1978, 302.

305. **Carman, R. L., Fleck, J., and James, L.,** Self-focusing and self-phase modulation in picosecond pulse oscillators, *IEEE J. Quantum Electron.,* EQ-8, 586, 1972.

306. **Kirkin, A. N., Leontovich, A. M., and Mozharovskii, A. M.,** Generation of high power ultrashort pulses in a low temperature ruby laser with a small active volume, *Sov. J. Quantum Electron,* 8, 1489, 1978.

307. **Kazakov, B. N., Orlov, M. S., Petrov, M. V., Stolov, A. L., and Takachuk, A. M.,** Induced emission of Sm^{3+} ions in the visible region of the spectrum, *Opt. Spectrosc.,* 47, 676, 1979.

2.1.2 STOICHIOMETRIC LASERS*,**

S. R. Chinn

INTRODUCTION

A stoichiometric crystal laser is by definition a laser whose gain medium contains the lasing ion as an intrinsic constituent of the insulating crystal lattice. In such laser crystals the active ion may be partially replaced by other ions; however, for the purpose of limiting the content of this section, the pure or truly stoichiometric form of such mixed crystals must have demonstrated laser action. Although not exactly synonymous, the term "high-concentration" is often used to describe such lasers. In the Soviet literature these lasers are frequently referred to as "self-activated". The major distinction to be made between this type of laser material and the more common solid state laser crystals developed earlier is that the active ions in the latter case occur in the lattice as impurities with concentrations generally less than a few percent. A review of this type of solid state insulating crystal laser is given in Section 2.1.1, "Paramagnetic Ion Lasers".

The first reported stoichiometric laser was HoF_3,[1] closely followed by $PrCl_3$.[2] Interest in this field was stimulated in 1972-73 by the achievement of lasing in NdP_5O_{14} (neodymium pentaphosphate).[3,4] The significance of this development lay in the utilization of Nd^{3+}, a lasing ion of great practical importance but whose concentration in earlier hosts had been severely limited. Since then many other stoichiometric laser crystals have been synthesized, and the potential for future development seems very promising.

Several review articles describing the characteristics and properties of stoichiometric laser materials and lasers have been published,[5-11] and should be referred to for a general summary. The purpose of this section is to provide a brief outline of the unique properties of stoichiometric lasers, and a tabulation of reported laser results with a guide to the literature. Except for a listing of some crystal parameters, the emphasis is on lasing behavior, so that the references do not provide a complete list of sources pertaining to other aspects of stoichiometric laser crystals. Furthermore, materials such as $Na_{0.5}Gd_{0.5-x}Nd_xWO_4$[12] which may have a high lasing ion concentration but have not exhibited lasing in the limiting stoichiometric case will not be included. Crystals that are potential stoichiometric lasers, such as EuP_5O_{14},[13] but have not yet demonstrated laser action are similarly excluded. References to amorphous forms of stoichiometric laser materials or to high-concentration laser glasses are found in Section 2.3, "Glass Lasers".

OPTICAL EXCITATION

The materials described in this section are all insulating crystals, requiring excitation by optical absorption of radiation from an emitting source. It may be possible to use electron beam excitation as an alternate means of pumping, but the penetration depths would be quite small (on the order of a few micrometers); no laser action in stoichiometric crystals has yet been reported by this means. Another possibility may be to incorporate the rare earth ion in a semiconductor, with the lasing transition occurring within the conduction band gap. This scheme would allow direct electrical excitation of the rare earth, but has not yet been achieved.

* This work was sponsored by the Department of the Air Force.

** At the time this article was prepared, the author was a member of the technical staff, Lincoln Laboratory, Massachusetts Institute of Technology, Lexington, Massachusetts.

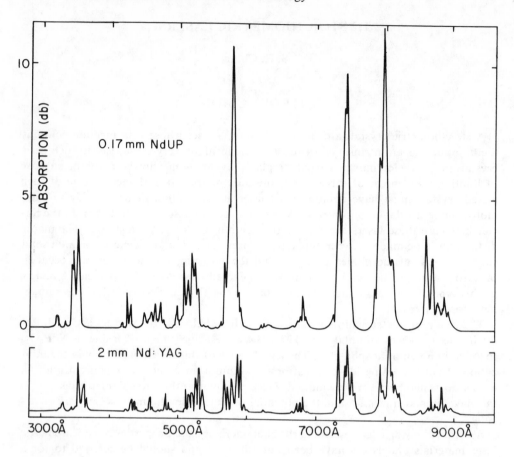

FIGURE 2.1.2.1.   Comparison of the absorption spectra of NdP_5O_{14} (here abbreviated NdUP) and Nd:YAG. Vertical scales (in db) are the same for both spectra. Note the difference in sample thicknesses.[3]

The necessity for optical pumping of insulating crystal lasers was one of the prime motivations for developing stoichiometric hosts. The increase in the concentration of the active ions leads to a proportional increase in the optical absorption coefficient, whose value reaches several hundred cm^{-1} in some typical Nd^{3+} absorption bands. This greatly enhanced absorption not only allows more efficient utilization of the pump radiation, but gives high gain densities in small volumes, enabling a reduction in the size of the crystal. A comparison of the absorption spectra for Nd^{3+} in NdP_5O_{14} and Nd:YAG is shown in Figure 2.1.2.1. The absorption coefficients in the pentaphosphate are 30 to 50 times larger than in Nd:YAG. One particular geometry which takes advantage of this feature is the slab waveguide configuration, which uses a highly absorbing thin layer of the active medium as an optical waveguide.[14-17] The useful size of stoichiometric crystals is most likely to be determined by the absorption length in the material, which depends on the particular type of absorbing ion, pump transition, and source being used. The ability to vary the lasing ion concentration over a wide range allows further optimization of a particular laser design.

The potential for reducing the stoichiometric laser size has other consequences which may affect fundamental aspects of the laser operation. The laser crystal itself may be used as a mode-selecting etalon, and a small active volume may reduce cavity loading and dispersive effects from the gain medium. Smaller transverse laser dimensions should also facilitate heat removal from the crystal.

CONCENTRATION QUENCHING

The laser transitions in the stoichiometric crystals described in this section are between $4f^n$ electronic levels of trivalent rare earth ions. Compared to the far less localized ionic d states of transition-metal-doped lasers, these f states are relatively well-shielded from the crystal field environment and from each other. However, when the rare earth concentration is sufficiently high to cause a large probability of interionic separations less than a few angstroms, a metastable fluorescent level may decay non-radiatively by cross-relaxation involving a nearby, initially unexcited ion. For example, this quenching in Nd:YAG occurs by the $^4F_{3/2}$-$^4I_{15/2}$, $^4I_{9/2}$-$^4I_{15/2}$ pair process.[18] The dominant mechanism for this interaction is thought to be a dipole-dipole Coulomb interaction.[19,20]

The strength of such concentration quenching, which determines whether a stoichiometric crystal will lase, depends on the rare earth ion and crystal. Ions such as Nd^{3+} or Pr^{3+}, which have states of approximately half the energy of the metastable level, are more susceptible to quenching than Eu^{3+} or Ho^{3+}, which do not. The crystal lattice also affects the amount of quenching through different factors. First, the crystal field determines the perturbations of the intermediate coupled electronic states. Increased resonance overlap enhances the cross-relaxation rate if the sum of the energies of two such crystal field levels or twice the energy of a single level (lying below the initially populated level) nearly equals the initial energy. Second, the lattice constrains the minimum separation between rare earth ions. In many of the stoichiometric crystals listed below, the lasing ions are separated by complexes such as $(PO_4)^{3-}$ which give a near-neighbor lanthanide separation about 5 to 6 Å. This relatively large distance decreases the strength of multipole interactions between ions, and direct- and super-exchange interactions are also reduced. The lattice also affects interionic Coulomb interactions through the related electronic oscillator strengths, whose dominant electric dipole terms are induced by the odd-parity crystal field components. Another postulated effect which may increase the fluorescence quenching at high concentrations is the increased resonant spatial migration of the optical excitation to nonradiative quenching centers or surfaces.[21]

A complete understanding of the physical processes of optical energy diffusion and nonradiative cross-relaxation in stoichiometric rare earth laser crystals is still lacking. Nevertheless, there is a large amount of experimental data for many materials which leads to two conclusions concerning Nd^{3+}:

First, the measured fluorescence decay rate, $1/\tau$, from the $^4F_{3/2}$ manifold is the sum of a constant contribution, $1/\tau_0$, and a nonradiative part, $1/\tau_q$, which increases linearly with the mole-fraction of Nd,x. (The multipole Coulomb interaction theory,[19,20] which appears to hold for Nd:YAG, predicts a dependence of $1/\tau_q \sim x^2$.)

The low-concentration decay rate, $1/\tau_0$, is dominated by the radiative rate, but also includes nonradiative multiphonon decay from $^4F_{3/2}$ to $^4I_{15/2}$ ($\Delta E \sim 5000$ cm^{-1}). This latter rate is crystal dependent and in the stoichiometric laser materials has been measured only for ErP_5O_{14} and NdP_5O_{14},[22] for which it is 8×10^2 sec^{-1}, much less than the low-concentration rate, 2.85×10^3 sec^{-1}, and the implied radiative rate, 2.05×10^3 sec^{-1}. Since this multiphonon rate is not known for other stoichiometric laser crystals, it is presumed to be relatively small as well, and the radiative rate has been equated to $1/\tau_0$ in the following discussion.

The magnitude of the quenching rate also is roughly proportional to the radiative rate in going from one material to another. If we define $q_0 = 1/\tau_0$ and $q = 1/\tau_q$, then $1/\tau = q + q_0$ and the quenching rate is $q = 1/\tau - q_0$. The normalized quenching rate is found by dividing q/q_0, giving $q_n = (\tau_0-\tau)/\tau$. This concentration quenching behavior is illustrated in Figure 2.1.2.2, where the normalized quenching is plotted as

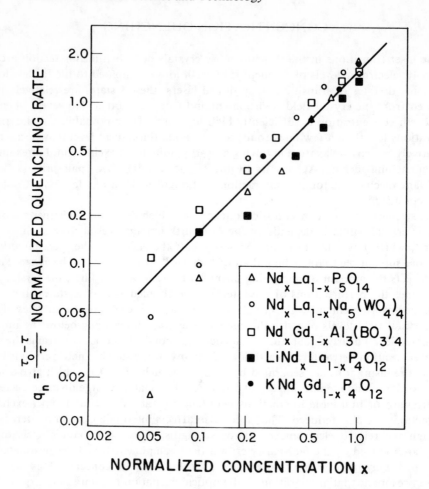

FIGURE 2.1.2.2.   The normalized fluorescence quenching rate $q_n = (\tau_o - \tau)/\tau$ as a function of normalized concentration (mole fraction) of Nd for several stoichiometric laser crystals.[7]

a function of mole-fraction of Nd, x, for several materials. The absolute values of radiative and quenched lifetimes may be found in the tables below.

Secondly, the measured fluorescence decays over the entire range of Nd dilution in the stoichiometric crystals are characterized by a single exponential.

Recent theoretical work suggests that both facts may be explained by resonant diffusion of the $^4F_{3/2}$ excitation which is rapid compared to the decay processes.[23-26] This same type of quenching behavior has also been observed for Nd → Eu, Nd → Tb, and Eu → Yb nonradiative cross-relaxation in those mixed rare earth pentaphosphate crystals.[25]

Nd LASERS

The Nd^{3+} ion has played an important role in the technology of conventional solid state lasers, and the same holds true for stoichiometric laser development. Several reasons for this emphasis can be seen from Figure 2.1.2.3, where the energy levels and lasing transitions for trivalent rare earths in stoichiometric crystals are indicated. First, Nd^{3+} has a four-level system, making population inversion relatively easy. Second, there are many absorbing transitions which decay rapidly and efficiently to the meta-

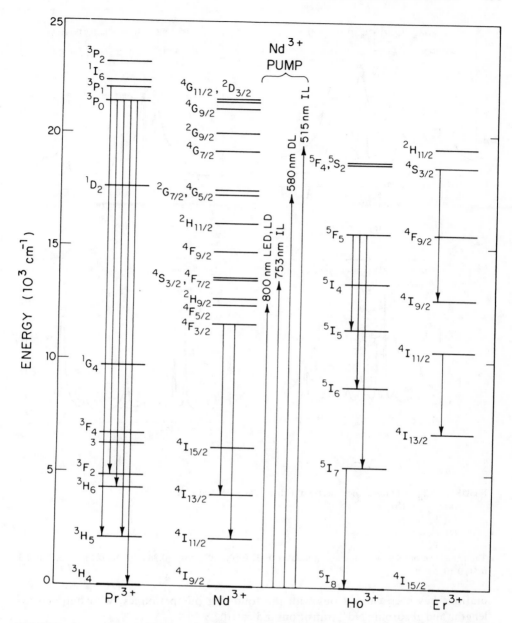

FIGURE 2.1.2.3. Partial energy level diagram for trivalent rare earth ions which have shown laser action in stoichiometric crystals. The lasing transitions are shown, along with the lower energy pumping transitions for Nd^{3+}. Crystal field splittings within the spin-orbit manifolds are not shown.

stable $^4F_{3/2}$ manifold. Third, there is a large gap between this manifold and the next lower one, giving a negligibly small multiphonon nonradiative decay rate. Finally, there is a favorable radiative branching ratio which gives large cross-sections from $^4F_{3/2}$ to $^4I_{11/2}$, the most widely used transitions in the 1040 to 1080 nm wavelength region. (The weaker 1300 nm $^4F_{3/2}$-$^4I_{13/2}$ and, with cooling, 940 nm $^4F_{3/2}$-$^4I_{9/2}$ transitions will also lase.)

For these reasons, most of the research in stoichiometric crystal lasers has dealt with Nd. In particular, the energies of the absorbing transitions in Nd have made optical pumping quite convenient. Laboratory evaluation of new Nd laser materials or config-

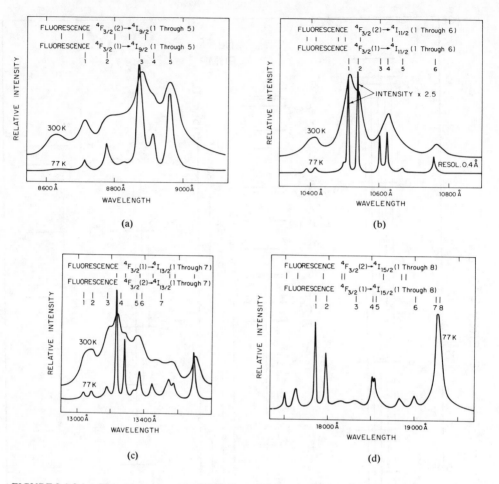

FIGURE 2.1.2.4.   Fluorescence spectra of NdP_5O_{14} at 300 K and 77 K.[40]
  (a) $^4F_{3/2} \rightarrow {}^4I_{9/2}$
  (b) $^4F_{3/2} \rightarrow {}^4I_{11/2}$
  (c) $^4F_{3/2} \rightarrow {}^4I_{13/2}$
  (d) $^4F_{3/2} \rightarrow {}^4I_{15/2}$

The respective branching ratios are quoted as 0.43, 0.50, 0.07, ∼0;[41] 0.362, 0.502, 0.132, ∼0;[42] or 0.38, 0.51, 0.10, ∼0.[43]

urations has been carried out with the following pump sources, with output wavelengths and absorbing Nd^{3+} transitions indicated:*

1.	Dye laser (Rhodamine 6G)	580 nm	$^2G_{7/2}$, $^4G_{5/2}$
2.	Ion laser (Ar)	514.5 nm	$^2G_{9/2}$, $^4G_{7/2}$
3.	Ion laser (Kr)	800 nm	$^2H_{9/2}$, $^4F_{5/2}$
4.	Semiconductor diode laser light-emitting diode	800, 870 nm	$^3H_{9/2}$, $^4F_{5/2}$, $^4F_{3/2}$
5.	Doubled ruby laser	347 nm ⎫	$^4D_{3/2}$, $^4D_{5/2}$, $^4D_{1/2}$
6.	Excimer laser	353 nm ⎭	

More recently, practical stoichiometric laser devices no longer restricted to the laboratory bench have been made using pulsed Xe flash lamps[28,29] (with broad-band emission) and cw diode lasers.[30] Pumping by cw incandescent or gas discharge lamps should prove feasible with the use of forced cooling systems for lamp and laser rod. Details

---

*   The identification of the transitions is from correspondence with these levels in other crystals.

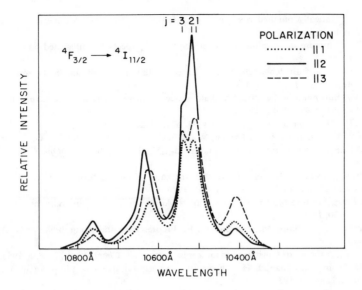

FIGURE 2.1.2.5.   Polarization dependence of the $^4F_{3/2}-^4I_{11/2}$ spectra in NdP$_5$O$_{14}$ at 300 K. Axis 2 is parallel to b, and axes 1 and 3 are the indicatrix axes rotated 9° from a and c in the a-c plane.[41]

of the design of miniature neodymium stoichiometric lasers pumped by diodes are discussed in References 9 and 31.

At this relatively early stage of development, it is not certain where or to what extent stoichiometric lasers will find their widest application. Initial progress in two areas has been reported, the use of NdP$_5$O$_{14}$ as a compact source with potential for laser range-finding[29] and of LiNdP$_4$O$_{12}$ as a carrier generator in a high-data-rate optical communications system.[32] The communications applications of (Nd,La)P$_5$O$_{14}$ are also promising, following demonstration of a short-pulse, cw mode-locked laser.[33] Waveguide KNdP$_4$O$_{12}$[16] and LiNdP$_4$O$_{12}$[34-36] and fiber (Nd,La)P$_5$O$_{14}$[37] lasers have been fabricated, and enhancement of the optical absorption by co-doping NdAl$_3$(BO$_3$)$_4$ with Cr^{3+} has been achieved,[38,39] all with potential use as miniature sources for optical communications systems.

REFERENCES

1. **Devor, D. P., Soffer, B. H., and Robinson, M.,** Stimulated emission from Ho^{3+} at 2 μm in HoF$_3$, *Appl. Phys. Lett.,* 18, 122, 1971.
2. **Varsanyi, F.,** Surface lasers, *Appl. Phys. Lett.,* 19, 169, 1971.
3. **Danielmeyer, H. G. and Weber, H. P.,** Fluorescence in neodymium ultraphosphate, *IEEE J. Quantum Electron.,* QE-8, 805, 1972.
4. **Weber, H. P., Damen, T. C., Danielmeyer, H. G., and Tofield, B. C.,** Nd-ultraphosphate laser, *Appl. Phys. Lett.,* 22, 534, 1973.
5. **Danielmeyer, H. G.,** Stoichiometric laser materials, *Festkörperprobleme,* 15, 253, 1975.
6. **Weber, H. P.,** Review: Nd pentaphosphate lasers, *Opt. Quant. Electron.,* 7, 431, 1975.
7. **Chinn, S. R., Hong, H. Y-P., and Pierce, J. W.,** Minilasers of neodymium compounds, *Laser Focus,* 12, 64, 1976.
8. **Hong, H. Y-P. and Chinn, S. R.,** Influence of local-site symmetry on fluorescence lifetime in high-Nd-concentration laser materials, *Mater. Res. Bull.,* 11, 461, 1976.
9. **Budin, J.-P., Neubauer, M., and Rondot, M.,** On the design of neodymium miniature lasers, *IEEE J. Quantum Electron.,* QE-14, 831, 1978.

10. Möckel, P., Optically pumped miniature solid-state lasers from stoichiometric neodymium compounds, *Frequenz*, 32, 85, 1978.
11. Huber, G., Miniature neodymium lasers: principles and aspects for integrated optics, *Proc. SPIE*, 164, 2, 1978.
12. Peterson, G. E. and Bridenbaugh, P. M., Laser oscillation at 1.06 μm in the series $Na_{0.5}Gd_{0.5-x}Nd_xWO_4$, *Appl. Phys. Lett.*, 4, 173, 1964.
13. Brecher, C., Europium in the ultraphosphate lattice: polarized spectra and structure of EuP_5O_{14}, *J. Chem. Phys.*, 61, 2297, 1974.
14. Wittke, J. P., Thin-film lasers, *RCA Review*, 33, 674, 1972.
15. Krühler, W. W., Plättner, R. D., Fabian, W., Möckel, P., and Grabmaier, J. G., Laser oscillation of $Nd_{0.14}Y_{0.86}P_5O_{14}$ layers epitaxially grown on $Gd_{0.33}Y_{0.67}P_5O_{14}$ substrates, *Opt. Commun.*, 20, 354, 1977.
16. Kubodera, K., Miyazawa, S., Nakano, J., and Otsuka, K., Laser performance of an epitaxially grown $KNdP_4O_{12}$ waveguide, *Opt. Commun.*, 27, 345, 1978.
17. Lutz, F., Leiss, M., and Müller, J., Epitaxy of $NdAl_3(BO_3)_4$ for thin film miniature lasers, *J. Cryst. Growth*, 47, 130, 1979.
18. Danielmeyer, H. G., Blätte, M., and Balmer, P., Fluorescence quenching in Nd:YAG, *Appl. Phys.*, 1, 269, 1973.
19. Dexter, D. L., A theory of sensitized luminescence in solids, *J. Chem. Phys.*, 21, 836, 1953.
20. Dexter, D. L. and Schulman, J. H., Theory of concentration quenching in inorganic phosphors, *J. Chem. Phys.*, 22, 1063, 1954.
21. Lempicki, A., Concentration quenching in Nd^{3+} stoichiometric materials, *Opt. Commun.*, 23, 376, 1977.
22. Mazurak, Z., Ryba-Romanowski, W., and Jezowska-Trzebiatowska, B., Radiative and non-radiative transitions in ErP_5O_{14} single crystals, *J. Lumin.*, 17, 401, 1978.
23. Voron'ko, Yu. K., Mamedov, T. G., Osiko, V. V., Prokhorov, A. M., Sakun, V. P., and Shcherbakov, I. A., Nature of nonradiative excitation-energy relaxation in condensed media with high activator concentrations, *Sov. Phys. JETP*, 44, 251, 1976.
24. Bondar', I. A., Denker, B. I., Domanskii, A. I., Mamedov, T. G., Mezentseva, L. P., Osiko, V. V., and Shcherbakov, I. A., Investigation of anomalously weak quenching of Nd^{3+} ion luminescence in $La_{1-x}Nd_xP_5O_{14}$, *Sov. J. Quantum Electron.*, 7, 167, 1977.
25. Fay, D., Huber, G., and Lenth, W., Linear concentration quenching of luminescence in rare earth laser materials, *Opt. Commun.*, 28, 117, 1979.
26. Nettel, S. J. and Lempicki, A., Resonant energy transfer in stoichiometric rare earth compounds, *Opt. Commun.*, 30, 387, 1979.
27. Dieke, G. H., *Spectra and Energy Levels of Rare Earth Ions in Crystals*, John Wiley & Sons, New York, 1968.
28. Chinn, S. R. and Zwicker, W. K., Flash-lamp-excited NdP_5O_{14} laser, *Appl. Phys. Lett.*, 31, 178, 1977.
29. Chinn, S. R., Research Studies on Neodymium Pentaphosphate Miniature Lasers, Final Report ESD-TR-78-392, DDC No. AD-A073140, Lincoln Laboratory (MIT), Lexington, Mass., 1978.
30. Kubodera, K. and Otsuka, K., Efficient $LiNdP_4O_{12}$ lasers pumped with a laser diode, *Appl. Opt.*, 18, 3882, 1979.
31. Kubodera, K. and Otsuka, K., Diode-pumped miniature solid-state laser: design considerations, *Appl. Opt.*, 16, 2747, 1977.
32. Kimura, T., Saruwatari, M., Yamada, J., Uehara, S., and Miyashita, T., Optical fiber (800-Mbit/sec) transmission experiment at 1.05 μm, *Appl. Opt.*, 17, 2420, 1978.
33. Chinn, S. R. and Zwicker, W. K., FM mode-locked $Nd_{0.5}La_{0.5}P_5O_{14}$ laser, *Appl. Phys. Lett.*, 34, 847, 1979.
34. Nakano, J., Kubodera, K., Miyazawa, S., Kondo, S., and Koizumi, H., $LiBi_xNd_{1-x}P_4O_{12}$ waveguide laser layer epitaxially grown on $LiNdP_4O_{12}$ substrate, *J. Appl. Phys.*, 50, 6546, 1979.
35. Kubodera, K. and Otsuka, K., Single-transverse-mode-$LiNdP_4O_{12}$ slab waveguide laser, *J. Appl. Phys.*, 50, 653, 1979.
36. Kubodera, K. and Otsuka, K., Laser performance of a glass-clad $LiNdP_4O_{12}$ rectangular waveguide, *J. Appl. Phys.*, 50, 6707, 1979.
37. Weber, H. P., Liao, P. F., Tofield, B. C. and Bridenbaugh, P. M., CW fiber laser of NdLa pentaphosphate, *Appl. Phys. Lett.*, 26, 692, 1975.
38. Hattendorff, H.-D., Huber, G., and Danielmeyer, H. G., Efficient cross pumping of Nd^{3+} by Cr^{3+} in $Nd(Al,Cr)_3(BO_3)_4$ lasers, *J. Phys. C*, 11, 2399, 1978.
39. Hattendorff, H.-D., Huber, G., and Lutz, F., CW laser action in $Nd(Al,Cr)_3(BO_3)_4$, *Appl. Phys. Lett.*, 34, 437, 1979.

40. **Blätte, M., Danielmeyer, H. G., and Ulrich, R.,** Energy transfer and the complete level system of NdUP, *Appl. Phys.,* 1, 275, 1973.

41. **Huber, G., Krühler, W. W., Bludau, W., and Danielmeyer, H. G.,** Anisotropy in the laser performance of NdP_5O_{14}, *J. Appl Phys.,* 46, 3580, 1975.

42. **Lomheim, J. S. and DeShazer, L. G.,** New procedure of determining neodymium fluorescence branching ratios as applied to 25 crystal and glass hosts, *Opt. Commun.,* 24, 89, 1978.

43. **Mazurak, Z., Strek, W., and Jezowska-Trzebiatowska, B.,** Analysis of spectral line intensities of Nd^{3+} ions in neodymium pentaphosphate, *Acta Phys. Pol.,* A53, 415, 1978.

Table 2.1.2.1
STOICHIOMETRIC LASER CRYSTAL PROPERTIES

Table Notes

1. Compounds are listed in both Tables 2.1.2.1 and 2.1.2.2 in the sequence of the lasing rare earth elements in the periodic table. The compounds of a given rare earth are listed in groups of similar composition i.e., phosphates, and in approximate chronological order within similar groups and among different groups.
2. Space group and lasing ion site symmetries are given in international notation.
3. Lasing ion concentrations are calculated from lattice constants.
4. Densities are calculated from lattice constants and atomic weights. Where data are available, measured values of density are given in parentheses.
5. Notation for the orientation of refractive index axes follows that of the quoted references. Subscripts "o" and "e" refer to ordinary and extraordinary polarizations, whereas "a", "b", "c" refer to crystal axes. Where indicatrix axes differ from crystallographic axes they are denoted by α, β, and γ.

Table 2.1.2.1
STOICHIOMETRIC LASER CRYSTAL PROPERTIES

Material	Symmetry	Space group	Lasing ion site symmetry	Lattice constant (Å)	Lasing ion concentration (cm⁻³)	Density (g/cm³)	Refractive index	Growth method	Ref.
$PrCl_3$	Hexagonal	$P6_3/m$ $Z = 2$	$\bar{6}$	a = 7.422 c = 4.275	9.81×10^{21}	4.025			1, 2
$PrBr_3$	Hexagonal	$P6_3/m$ $Z = 2$	$\bar{6}$	a = 7.92 c = 4.38	8.41×10^{21}	5.31			1
PrP_5O_{14}	Monoclinic(I)	$P2_1/c$ $Z = 4$	1	a = 8.787 b = 9.041 c = 13.08 $\beta = 89.58°$	3.85×10^{21}	3.32		Flux	3, 4
				a = 8.78 b = 9.02 c = 13.02 $\beta = 89.5°$	3.88×10^{21}	3.35 (3.3)	1.65		5
NdP_5O_{14}[a] $(Nd_xLa_{1-x}P_5O_{14})$	Monoclinic (I)	$P2_1/c$ $Z = 4$	1	a = 8.771 b = 9.012	3.88×10^{21}	3.37 (3.5)	$n_o = 1.592$ $n_\beta = 1.587$	Flux	3, 4, 6—16

Material	y range	Crystal system	Space group, Z	Nd site	Lattice constants (Å, °)	Density	N_{Nd}	Refractive index	Growth	Ref.
$(Nd_yY_{1-y}P_5O_{14})$	0.17 <y<1				c = 13.057, β = 89.58°			n_r = 1.610		
$Nd_yY_{1-y}P_5O_{14}$, 0<y<.17		Monoclinic (II) or Orthorhombic	C2/c, Z = 4; Pnma, Z = 4	1 (2 sites); 1	a = 8.939, b = 12.730, c = 8.718		4.03×10^{21} y		Flux	9
$LiNdP_4O_{12}$		Monoclinic	C2/c, Z = 4	2	a = 16.408, b = 7.035, c = 9.729, β = 126.38°	3.43	4.42×10^{21}	n_a = 1.590, n_b = 1.607, n_c = 1.600	Flux; Kyropoulos; Flux	17; 18, 19
					a = 16.45, b = 7.07, c = 13.25, β = 143.62° Pseudo-orthorhomb.	3.39 (3.4)				
					a = 9.844, b = 7.008, c = 13.25 Pseudo-orthorhomb.			nα = 1.6065 (c), $n_β$ = 1.6125 (b), n = 1.6195 (⊥100)	Flux	20—23
$NaNdP_4O_{12}$		Monoclinic	P2₁/c, Z = 4	1	a = 9.907, b = 13.10, c = 7.201, β = 90.51°	3.43 (3.45)	4.28×10^{21}		Kyropoulos	24, 25
$KNdP_4O_{12}$		Monoclinic	P2₁, Z = 2	1	a = 7.266, b = 8.436, c = 8.007, β = 91.97°	3.38	4.08×10^{21}	1.60	Flux	26, 27
$K_3Nd(PO_4)_2$		Monoclinic	P2₁/m, Z = 2	m	a = 9.532, b = 5.631, c = 7.444, β = 90.95°	3.75	5.01×10^{21}		Flux; Melt	28, 29
$Na_3Nd(PO_4)_2$		Orthorhombic	Pbc2₁, Z = 24		a = 15.874, b = 13.952, c = 18.470	3.93 (3.87)	5.87×10^{21}		Flux	28, 30
$NdAl_3(BO_3)_4$		Rhombohedral (Huntite)	R32, Z = 3	32	a = 9.3416, b = 7.3066	4.15	5.43×10^{21}	n_o = 1.79, n_e = 1.72	Flux	31—35

[a] Thermal Conductivity (W/cm·K): $K_a = 2.10 \times 10^{-2}$, $K_b = 9.66 \times 10^{-3}$, $K_c = 1.40 \times 10^{-2}$. Specific heat: 0.14 Cal/g, Hardness (Moh): 6—7.

Table 2.1.2.1 (continued)
STOICHIOMETRIC LASER CRYSTAL PROPERTIES

Material	Symmetry	Space group	Lasing ion site symmetry	Lattice constant (Å)	Lasing ion concentration (cm⁻³)	Density (g/cm³)	Refractive index	Growth method	Ref.
$NdNa_5(WO_4)_4$	Tetragonal	$I4_1/a$ $Z = 4$	4	$a = 11.559$ $b = 11.453$	2.61×10^{21}	5.43	1.75	Flux	36
$K_5Nd(MoO_4)_4$	Trigonal (Palmierite)	$R\bar{3}m$	1	$a = 7.753$ $\alpha = 45°25'$ Rhombohed. or $a = 5.97$ $c = 20.86$ Hexagonal	2.38×10^{21}	3.87	$n_a = 1.713$ $n_c = 1.788$	Melt Hydrothermal	37—39
Structurally: $[K_2(K_{0.5}Nd_{0.5})(MoO_4)_3]$				$a = 5.96$ $c = 20.49$			$n_a = 1.62$ $n_c = 1.787$		40
$Na_3Nd_2Pb_6(PO_4)_6Cl_2$	Hexagonal (Apatite)	$P6_3/m$ $Z = 1$		$a = 9.946$ $c = 7.287$	3.20×10^{21}	5.90	1.98	Melt	41—43
$K_5NdLi_2F_{10}$	Orthorhombic	Pnma $Z = 4$	m	$a = 20.65$ $b = 7.779$ $c = 6.902$	3.61×10^{21}	3.26	1.40	Melt	44—46
HoF_3	Orthorhombic	Pnma $Z = 4$	m	$a = 6.404$ $b = 6.875$ $c = 4.379$	2.08×10^{22}	7.64	$n_a = 1.566$ $n_f = 1.598$	Czochralski	47—49
$HoLiF_4$	Tetragonal (Scheelite)	$I4_1/a$ $Z = 4$	4	$a = 5.164$ $c = 10.78$	1.39×10^{22}	5.70 (5.72)	$n_o = 1.464$ $n_r = 1.498$	Czochralski	50, 51
$Ho_3Al_5O_{12}$	Cubic (Garnet)	Ia3d $Z = 8$	222	$a = 12.01$	4.62×10^{21}	6.30 (6.30)		Czochralski	52
$ErLiF_4$	Tetragonal (Scheelite)	$I4_1/a$ $Z = 4$	4	$a = 5.150$ $c = 10.68$	1.41×10^{22}	5.86 (5.83)	$n_o = 1.464$ $n_r = 1.497$	Czochralski	50, 51
$Er_3Al_5O_{12}$	Cubic (Garnet)	Ia3d $Z = 8$	222	$a = 11.98$	4.65×10^{21}	6.40			53
$KEr(WO_4)_2$	Monoclinic	C2/c $Z = 4$		$a = 8.05$ $b = 10.35$ $c = 7.59$ $\alpha = 94°$ $[KY(WO_4)_2]$	6.5×10^{21}	7.4		Melt	54, 55

Table 2.1.2.2
STOICHIOMETRIC LASER OPERATION

Table Notes and Abbreviations

1. Fluorescent linewidths are the approximate full-width, half-maxima of individual transitions between two crystal field levels at 300 K, unless a different temperature is noted. Generally the fluorescence band is composed of several overlapping transitions and is much broader than the individual component linewidth. When the full half-bandwidth is given, it is indicated by (t).

2. Laser ion concentration (x) is the mole fraction of lasing ion; i.e., 1.0 means 100%.

3. Mode of operation:

   P   Pulsed
   cw   Quasicontinuous (chopped)
   CW   Continuous

4. Optical pump, wavelength, configuration:

   (A) Pump source
       DL   Dye laser
       IL   Ion laser
       EL   Excimer laser
       SL   Solid state laser
       FL   Flash lamp
       LD   Laser diode
       LED Light-emitting diode
       *   Second-harmonic

   (B) Configuration
       L   Longitudinal
       T   Transverse

5. Threshold input (in parenthesis): Threshold input refers to optical power or energy, except for the case of flash lamp excitation, where electrical energy is given, and one case of LED pumping, where electrical power is denoted by subscript "e".

Table 2.1.2.2
STOICHIOMETRIC LASER OPERATION

Material	Laser transition	Laser wavelength (nm)	Fluorescent lifetime (μsec) x<.01	x=1	Laser cross section (10⁻¹⁹ cm²)	Fluorescent linewidth (cm⁻¹)	Laser ion concentration (x)	Mode of operation	Temperature (K)	Optical pump wavelength (nm) configuration	Output (and threshold input)	Ref.
$Pr_xLa_{1-x}Cl_3$	$^3P_0-^3F_2$	645.1	14.7			4	1.0	P	300	DL(488) L,T no external mirrors	(1 μJ)	56—58
	$^3P_0-^3H_6$	616.4			0.37	2			65		(100 nJ)	
						0.8			8			
	$^3P_1-^3H_5$	529.8							12		(125 nJ)	
	$^3P_0-^3F_2$	645.1					0.01		65		(125 nJ)	
	$^3P_0-^3H_6$	616.4							65		(125 nJ)	
	$^3P_1-^3H_5$	529.8							12		(40 nJ)	
	$^3P_0-^3H_4$	489.2							14			
$PrBr_3$	$^3P_0-^3F_2$	645.1	0.158	0.088		1.0		P	300	DL(488) L,T	(1 μJ)	56
$Pr_xLa_{1-x}P_5O_{14}$ $(Pr_xY_{1-x}P_5O_{14})$	$^3P_0-^3F_2$	637.4			0.5	12	0.9, 1.0	P	300	DL(347) L	60 μJ (1.1—1.5 mJ)	5,59,60
$Nd_xLa_{1-x}P_5O_{14}$	$^4F_{3/2}-^4I_{11/2}$	1052	350	135	0.7 (a) 1.2 (b) 0.7 (c)	25						61,62
							1.0	P	300	DL(580) L	1.8 μJ (40 μJ)	63
							0.5	P	300	DL(580) L	(20 μJ)	64
							0.5	CW	300	IL(514.5) L	7mW (25 mW)	64
							1.0,Sc doped	CW	300	IL(514.5) L	3mW (4 mW)	65
							1.0	CW,cw	300	DL(580) T	3mW (4 mW)	66
							0.5	CW,cw	300	IL(752.5) L fiber	(10 mW)	67
		1051.1 1051.2 1050.5 1053 1063			1.1 (a) 2.0 (b) 1.6 (c)		1.0	cw	300	IL(568.2) L	(1.7—4 mW)	68
							1.0	CW,cw	300	IL(514.5) L Integral mirrors	19 mW (90 mW)	69
							1.0	P	300	FL T	80 mJ (400 mJ)	62,70
							0.5	P	300	SL*(347) L SL (694) L	(50 mJ) (50 mJ)	71
							1.0	P	300	EL (353) T	80 μJ (3 mJ)	72
							1.0	P	300	SL*(532) L	2.5 μJ (420 μJ)	73
							1.0	CW	300	IL(752.5) L	(1.6 mW)	74
							1.0	cw and gain-switched	300	LD(800) T	3.5 mW (7 mW)	75,76
		1052					1.0	cw, intra-cavity SHG	300	DL(580) L	3.0 mW (7 mW) 1.0mW	77
		526*					0.5 0.25	CW mode-locked	300	DL(580) L	50 mW (20 mW)	78

Material	Transition	λ					Mode	T (K)	Pump	L/T	Power (threshold / output)	Ref.
$Nd,Y_{1-x}P_5O_{14}$	$^4F_{3/2} - {}^4I_{13/2}$	1323			60	0.75	CW,cw	224,260	LED(800)	T	(5W,16 W),	79,80 81
	$^4F_{3/2} - {}^4I_{11/2}$	1052		0.28		1.0	P	300	FL	T	2.7 mJ (200 mJ)	82
				0.24		1.0	CW	300	IL(530.9)	L	7 mW (70 mW)	83
				1.43 $(C2/c)$	30 $(C2/c)$							84
						0.16 $(C2/c)$	P	300	DL(580)	L		9
						0.99 $(P2_1/c)$	P	300	DL(580)	L	300 W (400 kW)	85
						0.90 $(P2_1/c)$	cw	300	IL(476.5–514.5)	L	200 mW (70 mW)	86
						0.14 $(C2/c)$	CW	300	IL(514.5) Epitaxial layer	L	140 μW (3 mW)	87
$LiNd,La_{1-x}P_4O_{12}$	$^4F_{3/2} - {}^4I_{13/2}$	1319 – 1322		0.2	15	0.14	cw	300	IL(514.5)	L	1.4 mW (2.1 mW)	84
	$^4F_{3/2} - {}^4I_{11/2}$	1048	325 135	0.88 (a)	20	0.14	cw	300	IL(514.5)	L	1.5 mW (12 mW)	84
				1.7 (b)								
				0.73 (c)								
				1.5 (a)	19: 17 strong, 25 weak							18,34,88
				1.3 (b)								
				3.2 (c)								
						1.0	cw	300	IL(514.5)	L / T	(38 mW) / 4 mW (95 mW)	89
						1.0	cw,CW	300	DL(582)	L	1.1 mW (360 μW)	90
						1.0	cw	300	IL(514.5)	L	4.7 mW (12.7 mW)	91
						1.0	P	300	DL(596.5)	L	7 mW (140 μJ)	92
						1.0	CW	300	IL(514.5)	L	(200 μW)	93
						1.0	CW single-mode	300	IL(514.5)	L	3.8 mW (1.2 mW)	94
						1.0	CW mode-stabilized	300	IL(514.5)	L	(40-50 mW)	95
						1.0	CW	300	IL(514.5)	L	(140-200 mW)	96
						0.5	CW, stable dual polar	300	IL(514.5)	L / T	28 mW (40 mW)	97
						1.0,0.1	CW	300	IL(514.5)	L	0.65 mW (8-14 mW)	98,99
						1.0	CW spont. phase-locked	300	IL(514.5)	L	(14 mW)	100
						1.0	P	300	LD(870)	L	60 μW (3.5 mW) / (0.6-1 μJ)	101
						1.0	cw	300	DL(595)	L	5mW (20 mW)	102
						1.0	cw, TE∞	300	IL(514.5) slab wvgd. / Slab waveguide	L / T	3 mW (7 mW) / 1.2 mW (80 mW)	103

Table 2.1.2.2 (continued)
STOICHIOMETRIC LASER OPERATION

Material	Laser transition	Laser wavelength (nm)	Fluorescent lifetime (μsec) x<.01, x=1	Laser cross section (10⁻¹⁹ cm²)	Fluorescent linewidth (cm⁻¹)	Laser ion concentration (x)	Mode of operation	Temperature (K)	Optical pump wavelength (nm) configuration	Pump power	Output (and threshold input)	Ref.
						1.0	CW single-mode	300	IL(514.5) L	4 mW	(7.2 mW)	104
						1.0	CW	300	LD(805) L	2 mW	(0.6 mW)	105
						1.0	cw	238	LED(800) L		(60 mW)	106
						1.0	cw	238	LED(800) T		(60 mW)	107
						1.0	cw, CW	300	IL(514.5) L rect. wvgd.	1.8 mW	(400 mW) (42.8 mW)	108
						1.0	cw, CW	300	IL(514.5) L Ext. mirror Int. mirror	1.6 mW 1 mW	(1.7 mW) (1 mW)	109
LiNd$_{.5}$(La,Gd)$_{.5}$P$_5$O$_{12}$			179 Gd$_{.5}$ 192 La$_{.5}$	2.1 (c) 2.0 (c)		0.5 0.5	CW	300	IL(514.5) L		(3.3 mW) (4.2 mW)	110
LiBi$_{1-x}$Nd$_x$P$_4$O$_{12}$						0.99	cw	300	IL(514.5) L Epitaxial wvgd. on LNP		(3.4-5 mW)	111
LiNd$_x$La$_{1-x}$P$_5$O$_{12}$	$^4F_{3/2} - {}^4I_{13/2}$	1317		0.67 (c) 0.71 (c)	23	1.0	CW	300	IL(514.5) L	11 mW	(17 mW)	112
						1.0	cw	300	IL(514.5) L	20 mW	(6.5 mW)	113
						1.0	CW	300	LD(805) L	0.5 mW	(2.7 mW)	105
						1.0	cw, CW	300	IL(514.5) L	1.5 mW	(77 mW)	108
						1.0	cw, CW	300	IL(514.5) L	1.5 mW		109
NaNdP$_4$O$_{12}$	$^4F_{3/2} - {}^4I_{11/2}$	1051	110	1.2 (a) 1.7 (b) 2.1 (b) 1.4 (a) 0.8 (b) 1.7 (c)	25	1.0	CW	300	IL(514.5) L Ext. mirror Int. mirror	1.5 mW 1 mW	(8 mW) (30 mW)	24
											(2 mW)	88
KNd$_x$Gd$_{1-x}$P$_4$O$_{12}$	$^4F_{3/2} - {}^4I_{13/2}$	1320	275 100	0.8 (1) 0.9 (2) 0.9 (3)	23	1.0	CW	300	IL(514.5) L		(5.5 mW)	112
	$^4F_{3/2} - {}^4I_{11/2}$	1052		0.47 (1) 0.62 (2) 0.39 (3)	32	1.0	CW	300	IL(514.5) L		(47.7 mW)	112
												88
												114
						1.0	cw	300	DL(580) L	6.6 mW	(450 μW)	115
						1.0	CW	300	IL(514.5) L		(10.9 mW)	112

Material	Transition	λ (nm)	(a)	(b)	(c)	x	Mode	T (K)	Pump	L/T	Output	Output	Ref.
$K_3Nd_xLa_{1-x}(PO_4)_2$	$^4F_{3/2}-^4I_{13/2}$	1320		0.15(1) 0.30(2) 0.14(3)	42	1.0	cw	300	IL(514.5) Epitax. wvgd.	L	1.8 mW	(40 mW)	27,116, 117, 114
	$^4F_{3/2}-^4I_{11/2}$	1055	460 21	0.7		1.0	CW	300	IL(514.5)	L		(43.4 mW)	112
						1.0	cw	300	DL(585)	L	8 mW	(7.5 mW)	29,118
$Na_5Nd_xLa_{1-x}(PO_4)_4$	$^4F_{3/2}-^4I_{11/2}$		359 23			0.5	cw	300	DL(585)	L	6 mW; Lasing obsvd. sample qlty. poor	(3.5 mW)	118
$Nd_xGd_{1-x}Al_3(BO_3)_4$	$^4F_{3/2}-^4I_{11/2}$	1065	50 20	8.0 (a)	30	1.0	cw,CW	300	DL(580)	L	7 mW	(1.2 mW)	115
				10		1.0	cw,CW	300	IL(530.9)	L		(0.87 mW)	83
						1.0	cw	300	IL(514.5) Integral mirrors	L	5 mW	(130 mW)	119
$Nd(Al,Cr)_3(BO_3)_4$	$^4F_{3/2}-^4I_{13/2}$	1345				1.0	cw gain-switched	300	LD(800)	T	0.4 mW	(16.5 mW)	76
	$^4F_{3/2}-^4I_{11/2}$	1063	15			1.0		300	Epitaxial layers	L	Not yet reported		35
			12	1.7	50	1.0	cw,CW	300	IL(530.9)	L	9 mW	(50 mW)	83
						1.0	cw; Cr³⁺ to Nd³⁺ Energy transfer	300	DL(585) (615)	L	7 mW	(11 mW)	120,121
$Nd(Ga,Cr)_3(BO_3)_4$	$^4F_{3/2}-^4I_{11/2}$	1066	18			1.0	P superradiant	300	DL(590) epitaxial waveguide	L	1.2 mW	(23 mW)	122
$Nd_xLa_{1-x}Na_5(WO_4)_4$	$^4F_{3/2}-^4I_{11/2}$	1063	220 90	5-10	33	1.0	cw	300	DL(580)	L	7 mW	(330 μW)	36,118
$K_5Bi_{1-x}Nd_x(MoO_4)_4$	$^4F_{3/2}-^4I_{11/2}$	1066	215 70	0.7	90(t)	0.1	P	300	FL	T		(60 J)	40
						1.0	P	300	SL-Raman (745,805)	L		(25 mJ)	
$Na_2(Nd,La_{1-x})_3$	$^4F_{3/2}-^4I_{11/2}$					1.0	cw, duty cycle 1:1 1:1000	300	IL(514.5)	L			123
$Pb_5(PO_4)_3Cl$	$^4F_{3/2}-^4I_{11/2}$	1059 1068	200 110	0.57	80(t)	1.0	CW	300	IL(514.5)	L	0.3 mW 3.5 mW	(22 mW) (15 mW)	41,43
$K_5Nd,Ce_{1-x}Li_2F_{10}$	$^4F_{3/2}-^4I_{11/2}$	1048 1052	520 350	0.38	23	1.0	cw	300	DL(586)	L	3.5 mW	(4.5 mW) (1.4 mW)	46
$Ho,Er_{1-x}F_3$ (Ho³⁺)	$^5I_7-^5I_8$	2090	9.0×10³ 2.6×10⁴ (77 K) 7.0×10³ 1.9×10⁴ (298 K)		4(77 K)	1.0	P	77	FL	T		(1 J)	49
$HoLiF_4$	$^5F_5-^5I_8$ $^5F_5-^5I_6$	2352 1486				1.0	P	90	FL	T		(4-5 J/cm)	51,124

Table 2.1.2.2 (continued)
STOICHIOMETRIC LASER OPERATION

Material	Laser transition	Laser wavelength (nm)	Fluorescent lifetime (μsec) x<.01, x=1	Laser cross section (10^{-19} cm²)	Fluorescent linewidth (cm⁻¹)	Laser ion concentration (x)	Mode of operation	Temperature (K)	Optical pump wavelength (nm) configuration	Output (and threshold input)	Ref.
Ho₃Al₅O₁₂	⁵F₅–⁵I₇ ⁵I₇–⁵I₈	979 2122 2129	10³			1.0	P	90	FL T	(9 J) (14 J)	52
(Ho,Y₁₋ₓ)₃Al₅O₁₂		2123 2097	8.5×10³, 0.5×10³ (77 K)			1.0	P	77	FL T	(50 J) (400 J)	125
ErLiF₄	⁴S₃/₂ – ⁴I₉/₂	1732	110 70	0.26		1.0	P	90	FL T	(5 J/cm)	51,124
(Er,Lu₁₋ₓ)₃Al₅O₁₂	⁴I₁₁/₂ – ⁴I₁₃/₂	2937			9	1.0	P	300	FL	(18 J)	53,126, 127
(Er,Y₁₋ₓ)₃Al₅O₁₂ Er₃(Al,Ga)₅O₁₂		2830	80			0.33	P	300	FL Epitaxial wave-guide	(12 J)	128
KEr(WO₄)₂	⁴I₁₁/₂ – ⁴I₁₃/₂	2807				1.0	P	300	FL T	(12 J)	55

REFERENCES

1. **Zachariasen, W. H.,** Crystal chemical studies of the 5f-series of elements. I. New structure types, *Acta Crystallogr.,* 1, 265, 1948.
2. **Templeton, D. H. and Dauben, C. H.,** Lattice parameters of some rare earth compounds and a set of crystal radii, *J. Am. Chem. Soc.,* 76, 5237, 1954.
3. **Beucher, M.,** Données cristallographiques sur les ultraphosphates de terres rares du type TP_5O_{14}, *Colloque International sur les Terres Rares,* Paris et Grenoble, 1969, Centre National de Recherche Scientifique, Paris, 1970.
4. **Bagieu-Beucher, M. and Tranqui, D.,** Les ultraphosphates de terres rares et d'yttrium du type TP_5O_{14}, *Bull. Soc. Fr. Minéral. Cristallogr.,* 93, 505, 1970.
5. **Borkowski, B., Grzesiak, E., Kaczmarek, F., Kaluski, Z., Karolczak, J., and Szymanski, M.,** Chemical synthesis and crystal growth of laser quality praseodymium pentaphosphate, *J. Crystal Growth,* 44, 320, 1978.
6. **Hong, H. Y-P.,** Crystal structures of neodymium metaphosphate (NdP_3O_9) and ultraphosphate (NdP_5O_{14}), *Acta Crystallogr. Sect. B,* B30(2), 468, 1974.
7. **Albrand, K.-R., Attig, R., Fenner, J., Jeser, J. P., and Mootz, D.,** Crystal structure of the laser material NdP_5O_{14}, *Mater. Res. Bull.,* 9, 129, 1974.
8. **Danielmeyer, H. G. and Weber, H. P.,** Fluorescence in neodymium ultraphosphate, *IEEE J. Quantum Electron.,* QE-8, 805, 1972.
9. **Krühler, W. W., Huber, G., and Danielmeyer, H. G.,** Correlations between site geometries and level energies in the laser system $Nd_{1-x}Y_xP_5O_{14}$, *Appl. Phys.,* 8, 261, 1975.
10. **Danielmeyer, H. G., Jeser, J. P., Schönherr, E., and Stetter, W.,** The growth of laser quality NdP_5O_{14} crystals, *J. Cryst. Growth,* 22, 298, 1974.
11. **Miller, D. C., Shick, L. K., and Brandle, C. D.,** Growth of rare earth pentaphosphates in phosphoric acid, *J. Cryst. Growth,* 23, 313, 1974.
12. **Tofield, B. C., Weber, H. P., Damen, T. C., and Pasteur, G. A.,** On the growth of neodymium pentaphosphate crystals for laser action, *Mater. Res. Bull.,* 9, 435, 1974.
13. **Marais, M., Chinh, N. D., Savary H., and Budin, J. P.,** Croissance cristalline des pentaphosphates de terres rares, NdP_5O_{14} et $Nd_xLa_{1-x}P_5O_{14}$, *J. Cryst. Growth,* 35, 329, 1976.
14. **Kasano, H. and Furuhata, Y.,** Morphology control of NdP_5O_{14} single crystals grown from polyphosphoric acids, *J. Electrochem. Soc.,* 126, 1567, 1979.
15. **Plättner, R. D., Krühler, W. W., Zwicker, W. K., Kovats, T., and Chinn, S. R.,** The growth of large, laser quality $Nd_xRE_{1-x}P_5O_{14}$ crystals, *J. Cryst. Growth,* 49, 274, 1980.
16. **Chinn, S. R. and Zwicker, W. K.,** Thermal conductivity and specific heat of NdP_5O_{14}, *J. Appl. Phys.,* 49, 5892, 1978.
17. **Hong, H. Y-P.,** Crystal structure of $NdLiP_4O_{12}$, *Mater. Res. Bull.,* 10, 635, 1975.
18. **Yamada, T., Otsuka, K., and Nakano, J.,** Fluorescence in lithium neodymium ultraphosphate single crystals, *J. Appl. Phys.,* 45, 5096, 1974.
19. **Otsuka, K., Yamada, T., and Nakano, J.,** Crystal growth and optical properties of solid state laser material $LiNdP_4O_{12}$, *Rev. Electr. Commun. Lab.(Jpn.),* 26, 1129, 1978.
20. **Koizumi, H.,** An efficient laser material, lithium neodymium phosphate $LiNdP_4O_{12}$, *Acta Crystallogr. Sect. B,* 32, 266, 1976.
21. **Bohm, J., Schlage, R., Schultze, D., and Waligora, C.,** Zur kenntnis des lithium-neodym-polyphosphates $LiNd(PO_3)_4$, *Krist. Tech.,* 13, 423, 1978.
22. **Nakano, J., Miyazawa, S., and Yamada, T.,** Flux growth of $LiNdP_4O_{12}$ single crystals, *Mater. Res. Bull.,* 14, 21, 1979.
23. **Nakano, J., Yamada, T., and Miyazawa, S.,** Phase diagram for a portion of the system Li_2O-Nd_2O_3-P_2O_5, *J. Am. Ceram. Soc.,* 62, 465, 1979.
24. **Nakano, J., Otsuka, K., and Yamada, T.,** Fluorescence and laser-emission cross sections in $NaNdP_4O_{12}$, *J. Appl. Phys.,* 47, 2749, 1976.
25. **Koizumi, H.,** Sodium neodymium metaphosphate $NaNdP_4O_{12}$, *Acta Crystallogr. Sect. B,* 32, 2254, 1976.
26. **Hong, H. Y-P.,** Crystal structure of potassium neodymium metaphosphate, $KNdP_4O_{12}$, a new acentric laser material, *Mater. Res. Bull.,* 10, 1105, 1975.
27. **Miyazawa, S., Koizumi, H., Kubodera, K., and Iwasaki, H.,** Epitaxial growth of $KNdP_4O_{12}$ laser waveguides, *J. Cryst. Growth,* 47, 351, 1979.
28. **Apinitis, S. K., Vitinya, I. A., and Sedmalis, U. Ya.,** Synthesis, crystallographic and X-ray investigation of $Na_3PO_4 \cdot NdPO_4$ and $K_3PO_4 \cdot NdPO_4$, *Izv. Akad. Nauk Latv. SSR, Kim. Ser.,* 6, 676, 1974.
29. **Hong, H. Y-P. and Chinn, S. R.,** Crystal structure and fluorescence lifetime of potassium neodymium orthophosphate, $K_3Nd(PO_4)_2$, a new laser material, *Mater. Res. Bull.,* 11, 421, 1976.
30. **Salmon, R., Parent, C., Vlasse, M., and LeFlem, G.,** The crystal structure of a new high-Nd-concentration laser material: $Na_3Nd(PO_4)_2$, *Mater. Res. Bull.,* 13, 439, 1978.

31. **Hong, H. Y-P. and Dwight, K.,** Crystal structure and fluorescence lifetime of NdAl$_3$(BO$_3$)$_4$, a promising laser material, *Mater. Res. Bull.,* 9, 1661, 1974. [Recent evidence indicates that the space group R32 may be an average pseudo-symmetry of two microscopically mixed monoclinic phases (H. G. Danielmeyer and G. Huber, private communication).]

32. **Filimonov, A. A., Leonyuk, N. I., Meissner, L. B., Timchenko, T. I., and Rez, I. S.,** Nonlinear optical properties of isomorphic family of crystals with yttrium-aluminum-borate (YAB) structure,. *Krist Tech.,* 9, 63, 1974.

33. **Leonyuk, N. I., Pashkova, A. V., and Semenova, T. D.,** Preparation and morphology of crystals of aluminum rare earth borates, *Inorg. Mater., USSR,* 11, 154, 1975.

34. **Hong, H. Y-P. and Chinn, S. R.,** Influence of local-site symmetry on fluorescence lifetime in high-Nd-concentration laser materials, *Mater. Res. Bull.,* 11, 461, 1976.

35. **Lutz, F., Leiss, M., and Müller, J.,** Epitaxy of NdAl$_3$(BO$_3$)$_4$ for thin film miniature lasers, *J. Cryst. Growth,* 47, 130, 1979.

36. **Hong, H. Y-P. and Dwight, K.,** Crystal structure and fluorescence lifetime of a laser material NdNa$_5$(WO$_4$)$_4$ *Mater. Res. Bull.,* 9, 775, 1974.

37. **Efremov, V. A. and Trunov, V. K.,** Double molybdates with palmierite structure, *Sov. Phys. Crystallogr.,* 19, 613, 1975.

38. **Klevtsov, P. V., Kozeev, L. P., Protasova, V. I., Kharchenko, L. Yu., Glinskaya, L. A., Klevtsova, R. F., and Bakakin, V. V.,** Synthesis of crystals and X-ray diffraction investigation of double molybdates K$_5$Ln(MoO$_4$)$_4$, Ln = La-Tb, *Sov. Phys. Crystallogr.,* 20, 31, 1975.

39. **Maeda, M., Sakiyama, K., and Ikeda, T.** Dielectric and optical properties of K$_5$Nd(MoO$_4$)$_4$, *Jpn. J. Appl. Phys.,* 18, 25, 1979.

40. **Kaminskii, A. A., Sarkisov, S. E., Bohm, J., Reiche, P., Schultze, D., and Uecker, R.,** Growth, spectroscopic and laser properties of crystals in the K$_5$Bi$_{1-x}$Nd$_x$(MoO$_4$)$_4$ system, *Phys. Status Solid., A,* 43, 71, 1977.

41. **Michel, J.-C., Morin, D., and Auzel, F.,** Intensité de fluorescence et durée de vie du niveau ^{4}F$_{3/2}$ de Nd^{3+} dans une chlorapatite fortement dopée. Comparison avec d'autres matériaux, *C. R. Acad. Sci. Ser. B,* 281, 445, 1975.

42. **Joukoff, B., Fadly, M., Ostorero, J., and Makram, H.,** Flux growth of lead neodymium chlorapatite, *J. Crystal Growth,* 43, 81, 1978.

43. **Budin, J.-P., Michel, J.-C., and Auzel, F.,** Oscillator strengths and laser effect in Na$_2$Nd$_2$Pb$_6$(PO$_4$)$_6$Cl$_2$ (chloroapatite), a new high-Nd-concentration laser material, *J. Appl. Phys.,* 50, 641, 1979.

44. **McCollum, B. C. and Lempicki, A.,** A new, high yield luminescing compound: K$_5$NdLi$_2$F$_{10}$, *Mater. Res. Bull.,* 13, 883, 1978.

45. **Hong, H. Y-P. and McCollum, B. C.,** Crystal structure of K$_5$NdLi$_2$F$_{10}$, *Mater. Res. Bull.,* 14, 137, 1979.

46. **Lempicki, A., McCollum, B. C., and Chinn, S. R.,** Spectroscopy and lasing in K$_5$NdLi$_2$F$_{10}$ (KNLF), *IEEE J. Quantum Electron.,* QE-15, 896, 1979.

47. **Zalkin, A. and Templeton, D. H.,** The crystal structures of YF$_3$ and related compounds, *J. Am. Chem. Soc.,* 75, 2453, 1953.

48. **Brunton et al.,** Oak Ridge National Laboratory Report, ORNL-3761, 37, 1965; as quoted in *Crystal Data, Determinative Tables,* 3rd ed., Vol. 2, Donnay, J. D. H. and Ondik, H. M., Eds., U. S. Department of Commerce, Washington, D.C., 1973.

49. **Devor, D. P., Soffer, B. H., and Robinson, M.,** Stimulated emission from Ho^{3+} at 2 μm in HoF$_3$, *Appl. Phys. Lett.,* 18, 122, 1971.

50. **Keller, C. and Schmutz, H.,** Die reaktion von lithiumfluorid mit den trifluoriden der lanthaniden und einiger actiniden, *J. Inorg. Nucl. Chem.,* 27, 900, 1965.

51. **Morozov, A. M., Podkolzina, I. G., Tkachuk, A. M., Fedorov, V. A., and Feofilov, P. P.,** Luminescence and induced emission lithium-erbium and lithium-holmium binary fluorides, *Opt. Spectrosc.,* 39, 338, 1975.

52. **Ivanov, A. O., Mochalov, I. V., Tkachuk, A. M., Fedorov, V. A., and Feofilov, P. P.,** Emission of λ = 2μ stimulated radiation by holmium in aluminum holmium garnet crystals, *Sov. J. Quantum Electron.,* 5, 115, 1975.

53. **Prokhorov, A. M., Kaminskii, A. A., Osiko, V. V., Timoshechkin, M. I., Zharikov, E. V., Butaeva, T. I., Sarkisov, S. E., Petrosyan, A. G., and Fedorov, V. A.,** Investigations of the 3 μm stimulated emission from Er^{3+} ions in aluminum garnets at room temperature, *Phys. Status Solid A,* 40, K69, 1977.

54. **Klevtsov, P. V., Kozeeva, L. P., and Klevtsova, R. F.,** Crystallographic study of potassium-yttrium tungstate and molybdate, *Inorg. Mater. USSR,* 4, 1004, 1968.

55. **Kaminskii, A. A., Pavlyuk, A. A., Butaeva, T. I., Bobovich, L. I., and Lyubchenko, V. V.,** Stimulated emission in the 2.8 μm band by a self-activated crystal of KEr(WO$_4$)$_2$, *Inorg. Mater., USSR,* 15, 424, 1979.

56. Varsanyi, F., Surface lasers, *Appl. Phys. Lett.,* 19, 169, 1971.
57. German, K. R., Kiel, A., and Guggenheim, H., Stimulated emission from $PrCl_3$, *Appl. Phys. Lett.,* 22, 87, 1973.
58. German, K. R. and Kiel, A., Radiative and nonradiative transitions in $LaCl_3$: Pr and $PrCl_3$, *Phys. Rev. B,* 8, 1846, 1973.
59. Szymanski, M., Karolczak, J., and Kaczmarek, F., Laser properties of praseodymium pentaphosphate single crystals, *Appl. Phys.,* 19, 345, 1979.
60. Dornauf, H. and Heber, J., Fluorescence of Pr^{3+}-ions in $La_{1-x}Pr_xP_5O_{14}$, *J. Lumin.,* 20, 271, 1979.
61. Singh, S., Miller, D. C., Potopowicz, J. R., and Shick, L. K., Emission cross section and fluorescence quenching of Nd^{3+} lanthanum pentaphosphate, *J. Appl. Phys.,* 46, 1191, 1975.
62. Chinn, S. R. and Zwicker, W. K., Flash-lamp-excited NdP_5O_{14} laser, *Appl. Phys. Lett.,* 31, 178, 1977.
63. Weber, H. P., Damen, T. C., Danielmeyer, H. G., and Tofield, B. C., Nd-ultraphosphate laser, *Appl. Phys. Lett.,* 22, 534, 1973.
64. Damen, T. C., Weber, H. P., and Tofield, B. C., NdLa pentaphosphate laser performance, *Appl. Phys. Lett.,* 23, 519, 1973.
65. Danielmeyer, H. G., Huber, G., Krühler, W. W., and Jeser, J. P., Continuous oscillation of a (Sc,Nd) pentaphosphate laser with 4 milliwatts pump threshold, *Appl. Phys.,* 2, 335, 1973.
66. Chinn, S. R., Pierce, J. W., and Heckscher, H., Low-threshold, transversely excited NdP_5O_{14} laser, *IEEE J. Quantum Electron.,* QE-11, 747, 1975.
67. Weber, H. P., Liao, P. F., Tofield, B. C., and Bridenbaugh, P. M., CW fiber laser of NdLa pentaphosphate, *Appl. Phys. Lett.,* 26, 692, 1975.
68. Huber, G., Krühler, W. W., Bludau, W., and Danielmeyer, H. G., Anisotropy in the laser performance of NdP_5O_{14}, *J. Appl. Phys.,* 46, 3580, 1975.
69. Winzer, G., Möckel, P. G., Oberbacher, R., and Vité, L., Laser emission from polished NdP_5O_{14} crystals with directly applied mirrors, *Appl. Phys.,* 11, 121, 1976.
70. Chinn, S. R., Research studies on neodymium pentaphosphate miniature lasers, Final Report ESD-TR-78-392, (DDC Number AD-A073140, Lincoln Laboratory, M. I. T., Lexington, Mass, 1978.
71. Kaczmarek, F. and Szymanski, M., Performance of NdLa pentaphosphate laser pumped by nanosecond pulses, *Appl. Phys.,* 13, 55, 1977.
72. Wilson, J., Brown, D. C., and Zwicker, W. K., XeF excimer pumping of NdP_5O_{14}, *Appl. Phys. Lett.,* 33, 614, 1978.
73. Gaiduk, M. I., Grigor'yants, V. V., Zhabotinskii, M. E., Makovestskii, A. A., and Tishchenko, R. P., Neodymium pentaphosphate microlaser pumped by the second harmonic of a YAG:Nd^{3+} laser, *Sov. J. Quantum Electron.,* 9, 250, 1979.
74. Weber, H. P. and Tofield, B. C., Heating in a cw Nd-pentaphosphate laser, *IEEE J. Quantum Electron.,* QE-11, 368, 1975.
75. Chinn, S. R., Pierce, J. W., and Heckscher H., Low-threshold transversely excited NdP_5O_{14} laser, *Appl. Opt.,* 15, 1444, 1976.
76. Chinn, S. R., Hong, H. Y-P., and Pierce, J. W., Spiking oscillations in diode-pumped NdP_5O_{14} and $NdAl_3(BO_3)_4$ lasers, *IEEE J. Quantum Electron.,* QE-12, 189, 1976.
77. Chinn, S. R., Intracavity second-harmonic generation in a Nd pentaphosphate laser, *Appl. Phys. Lett.,* 29, 176, 1976.
78. Chinn, S. R. and Zwicker, W. K., FM mode-locked $Nd_{0.5}La_{0.5}P_5O_{14}$ laser, *Appl. Phys. Lett.,* 34, 847, 1979.
79. Budin, J. -P. Neubauer, M., and Rondot, M., Miniature Nd-pentaphosphate laser with bonded mirrors side pumped with low-current-density LED's, *Appl. Phys. Lett.,* 33, 309, 1978.
80. Budin, J. -P., Neubauer, M., and Rondot, H., On the design of neodymium miniature lasers, *IEEE J. Quantum Electron.,* QE-14, 831, 1978.
81. Blätte, M., Danielmeyer, H. G., and Ulrich R., Energy transfer and the complete level system of NdUP, *Appl. Phys.,* 1, 275, 1973.
82. Choy, M. M., Zwicker, W. K., and Chinn, S. R., Emission cross section and flashlamp-excited NdP_5O_{14} laser at 1.32 μm, *Appl. Phys. Lett.,* 34, 387, 1979.
83. Huber, G. and Danielmeyer, H. G., NdP_5O_{14} and $NdAl_3(BO_3)_4$ lasers at 1.3 μm, *Appl. Phys.,* 18, 77, 1979.
84. Krühler, W. W. and Plättner, R. D., Laser emission of (Nd,Y)-pentaphosphate at 1.32 μm, *Opt. Commun.,* 28, 217, 1979.
85. Krühler, W. W., Jeser, J. P., and Danielmeyer, H. G., Properties and laser oscillation of the (Nd,Y) pentaphosphate system, *Appl. Phys.,* 2, 329, 1973.
86. Gualtieri, J. G. and Aucoin, T. R., Laser performance of large Nd-pentaphosphate crystals, *Appl. Phys. Lett.,* 28, 189, 1976.

87. Krühler, W. W., Plättner, R. D., Fabian, W., Mockel, P., and Grabmaier, J. G., Laser oscillation of $Nd_{0.14}Y_{0.86}P_5O_{14}$ layers epitaxially grown on $Gd_{0.33}Y_{0.67}P_5O_{14}$ substrates, *Opt. Commun.*, 20, 354, 1977.

88. Nakano, J., Kubodera, K., Yamada, T., and Miyazawa, S., Laser-emission cross sections of $MeNdP_4O_{12}$ (Me = Li,Na,K) crystals, *J. Appl. Phys.*, 50, 6492, 1979.

89. Otsuka, K. and Yamada, T., Transversely pumped LNP laser performance, *Appl. Phys. Lett.*, 26, 311, 1975.

90. Chinn, S. R. and Hong, H. Y-P., Low-threshold cw $LiNdP_4O_{12}$ laser, *Appl. Phys. Lett.*, 26, 649, 1975.

91. Otsuka, K., Yamada, T., Saruwatari, M., and Kimura, T., Spectroscopy and laser oscillation properties of lithium neodymium tetraphosphate, *IEEE J. Quantum Electron.*, QE-11, 330, 1975.

92. Otsuka, K., Yamada, T., Nakano, J., Kimura, T., and Saruwatari, M., Lithium neodymium tetraphosphate laser, *J. Appl. Phys.*, 46, 4600, 1975.

93. Otsuka, K. and Yamada, T., Continuous oscillation of a lithium neodymium tetraphosphate laser with 200-μW pump threshold, *IEEE J. Quantum Electron.*, QE-11, 845, 1975.

94. Otsuka, K. and Yamada, T., Single-longitudinal-mode $LiNdP_4O_{12}$ laser, *Proc. IEEE*, 63, 1621, 1975.

95. Otsuka, K. and Iwasaki, H., Stabilization of oscillating modes in a $LiNdP_4O_{12}$ laser, *IEEE J. Quantum Electron.*, QE-12, 214, 1976.

96. Otsuka, K. and Yamada, T., Resonant absorption in $LiNdP_4O_{12}$ lasers, *Opt. Commun.*, 17, 24, 1976.

97. Otsuka, K., Kubodera, K., and Nakano, J., Stabilized dual-polarization oscillation in a $LiNd_{.5}La_{.5}P_4O_{12}$ laser, *IEEE J. Quantum Electron.*, QE-13, 398, 1977.

98. Otsuka, K., Oscillation properties of anisotropic lasers, *IEEE J. Quantum Electron.*, QE-14, 49, 1978.

99. Otsuka, K., Simultaneous oscillations of different transitions in lasers, *IEEE J. Quantum Electron.*, QE-14, 1007, 1978.

100. Otsuka, K., Observations of spontaneous phase locking of $LiNdP_4O_{12}$ lasers, *IEEE J. Quantum Electron.*, QE-14, 639, 1978.

101. Saruwatari, M., Kimura, T., and Otsuka, K., Miniaturized cw $LiNdP_4O_{12}$ laser pumped with a semiconductor laser, *Appl. Phys. Lett.*, 29, 291, 1976.

102. Kubodera, K., Nakano, J., Otsuka, K., and Miyazawa, S., A slab waveguide laser formed of glass-clad $LiNdP_4O_{12}$, *J. Appl. Phys.*, 49, 65, 1978.

103. Kubodera, K. and Otsuka, K., Single-transverse-mode $LiNdP_4O_{12}$ slab waveguide laser, *J. Appl. Phys.*, 50, 653, 1979.

104. Kubodera, K., Otsuka, K., and Miyazawa, S., Stable $LiNdP_4O_{12}$ miniature laser, *Appl. Opt.*, 18, 844, 1979.

105. Kubodera, K. and Otsuka, K., Efficient $LiNdP_4O_{12}$ lasers pumped with a laser diode, *Appl. Opt.*, 18, 3882, 1979.

106. Saruwatari, M., Kimura, T., Yamada, T., and Nakano, J., $LiNdP_4O_{12}$ laser pumped with an Al_xGa_{1-x} As electroluminescent diode, *Appl. Phys. Lett.*, 27, 682, 1975.

107. Saruwatari, M. and Kimura, T., LED pumped lithium neodymium tetraphosphate lasers, *IEEE J. Quantum Electron.*, QE-12, 584, 1976.

108. Kubodera, K. and Otsuka, K., Laser performance of a glass-clad $LiNdP_4O_{12}$ rectangular waveguide, *J. Appl. Phys.*, 50, 6707, 1979.

109. Krühler, W. W., Plättner, R. D., and Stetter, W., CW oscillation at 1.05 μm and 1.32 μm of $LiNd(PO_3)_4$ lasers in external resonator and in resonator with directly applied mirrors, *Appl. Phys.*, 20, 329, 1979.

110. Otsuka, K., Nakano, J., and Yamada, T., Laser emission cross section of the system $LiNd_{0.5}M_{0.5}P_4O_{12}$ (M = Gd,La), *J. Appl. Phys.*, 46, 5297, 1975.

111. Nakano, J., Kubodera, K., Miyazawa, S., Kondo, S., and Koizumi, H., $LiBi_xNd_{1-x}P_4O_{12}$ waveguide laser layer epitaxially grown on $LiNdP_4O_{12}$ substrate, *J. Appl. Phys.*, 50, 6546, 1979.

112. Otsuka, K., Miyazawa, S., Yamada, T., Iwasaki, H., and Nakano, J., CW laser oscillations in $MeNdP_4O_{12}$ (Me = Li,Na,K) at 1.32 μm, *J. Appl. Phys.*, 48, 2099, 1977.

113. Saruwatari, M., Otsuka, K., Miyazawa, S., Yamada, T. and Kimura, T., Fluorescence and oscillation characteristics of $LiNdP_4O_{12}$ lasers at 1.317 μm, *IEEE J. Quantum Electron.*, QE-13, 836, 1977.

114. Gueugnon, C. and Budin, J. P., Determination of fluorescence quantum efficiency and laser emission cross sections of neodymium crystals: application to $KNdP_4O_{12}$, *IEEE J. Quantum Electron.*, QE-16, 94, 1980.

115. Chinn, S. R. and Hong, H. Y-P., CW laser action in acentric $NdAl_3(BO_3)_4$ and $KNdP_4O_{12}$, *Opt. Commun.*, 15, 345, 1975.

116. Kubodera, K., Miyazawa, S., Nakano, J., and Otsuka, K., Laser performance of an epitaxially grown $KNdP_4O_{12}$ waveguide, *Opt. Commun.*, 27, 345, 1978.

117. Miyazawa, S. and Kubodera, K., Fabrication of $KNdP_4O_{12}$ laser epitaxial waveguide, *J. Appl. Phys.*, 49, 6197, 1978.

118. **Chinn, S. R. and Hong, H. Y-P.**, Fluorescence and lasing properties of $NdNa_5(WO_4)_4$, $K_3Nd(PO_4)_2$ and $Na_3Nd(PO_4)_2$, *Opt. Commun.*, 18, 87, 1976.

119. **Winzer G., Möckel, P. G., and Krühler, W.**, Laser emission from miniaturized $NdAl_3(BO_3)_4$ crystals with directly applied mirrors, *IEEE J. Quantum Electron.*, QE-14, 840, 1978.

120. **Hattendorff, H.-D., Huber, G., and Danielmeyer, H. G.**, Efficient cross pumping of Nd^{3+} by Cr^{3+} in $Nd(Al, Cr)_3(BO_3)_4$ lasers, *J. Phys. C*, 11, 2399, 1978.

121. **Hattendorff, H.-D., Huber, G., and Lutz, F.**, CW laser action in $Nd(Al,Cr)_3(BO_3)_4$, *Appl. Phys. Lett.*, 34, 437, 1979.

122. **Lutz, F., Rüppel, D., and Leiss, M.**, Epitaxial layers of the laser material $Nd(Ga,Cr)_3(BO_3)_4$, *J. Cryst. Growth*, 48, 41, 1980.

123. **Lenth, W., Hattendorff, H.-D., Huber, G., and Lutz, F.**, Quasi-cw laser action in $K_5Nd(MoO_4)_4$, *Appl. Phys.*, 17, 367, 1978.

124. **Christensen, H. P.**, Spectroscopic analysis of $LiHoF_4$ and $LiErF_4$, *Phys. Rev. B*, 19, 6564, 1979.

125. **Ashurov, M. Kh., Voron'ko, Yu. K., Zharikov, E. V., Kaminskii, A. A., Osiko, V. V., Sobol', A. A., Timoshechkin, M. I., Fedorov, V. A., and Shabaltai, A. A.**, Structure, spectroscopy, and stimulated emission of crystals of yttrium holmium aluminum garnets, *Inorg. Mater. USSR*, 15, 979, 1979.

126. **Basiev, T. T., Zharikov, E. V., Zhekov, V. I., Murina, T. M., Osiko, V. V., Prokhorov, A. M., Starikov, B. P., Timoshechkin, M. I., and Shcherbakov, I. A.**, Radiative and nonradiative transitions exhibited by Er^{3+} ions in mixed yttrium-erbium aluminum garnets, *Sov. J. Quantum Electron.*, 6, 796, 1976.

127. **Zharikov, E. V., Zhekov, V. I., Murina, T. M., Osiko, V. V., Timoshechkin, M. I., and Schcherbakov, I. A.**, Cross section of the $^4I_{11/2}$-$^4I_{13/2}$ laser transition in Er^{3+} ions in yttrium-erbium-aluminum garnet crystals, *Sov. J. Quantum Electron*, 7, 117, 1977.

128. **Dmitruk, M. V., Zhekov, V. I., Prokhorov, A. M., and Timoshechkin, M. I.**, Spectroscopic properties of $Er_3Al_{5-x}Ga_xO_{12}$ films obtained by liquid -phase epitaxy, *Inorg. Mater. USSR*, 15, 976, 1979.

2.1.3 COLOR CENTER LASERS

Linn F. Mollenauer

INTRODUCTION

Performance of Color Center Lasers

Certain color centers in the alkali halides can be used to create optically pumped, broadly tunable lasers[1-3] for the near infrared: F_2^+ and F_2^+-like centers, incorporated into a mere handful of hosts, now yield cw and mode-locked laser action tunable over almost the entire range 0.82 to 2.5 μm; $F_A(II)$ and $F_B(II)$ centers, in just two hosts, cover the smaller but important range 2.25 to 3.3 μm. Performance of significant representatives of these is summarized in Table 2.1.3.1.

A typical cw color center laser is shown schematically in Figure 2.1.3.1. It has much in common with a dye laser. Mirrors M_0, M_1, and M_2 constitute the usual folded, astigmatically compensated cavity. But instead of the dye jet stream, it has a 1 to 2mm thick slab of crystal containing the laser active color centers on a cold finger at 77 K. A vacuum provides the necessary thermal isolation for the cold finger, and helps to preserve the polished surfaces of the alkali-halide crystal. To prevent atmospheric absorption from interfering with laser action, the remainder of the laser cavity beam path is usually purged with dry N_2. For further constructional and operational details of color center lasers, see References 2, 6, and 13 through 16.

Basic Physics of Laser-Active Color Centers

As already indicated, of the well-known laser-active color centers, there are just two basic types: the F_2^+ and F_2^+-like centers on the one hand, and the very closely related $F_A(II)$ and $F_B(II)$ on the other. A complete description of the physics of these is beyond the scope of this handbook article, and the interested reader is referred to the rather extensive literature.[17-19] Discussion here will be limited largely to description of configurations and energy levels.

The F_2^+ center is shown in Figure 2.1.3.2. It consists of a single electron trapped by a pair of adjacent anion (halide ion) vacancies. Its configuration is suggestive of an H_2^+ molecular ion, with the two vacancies (as attractive centers) playing the role of the protons. In fact, the model of an H_2^+ molecular ion embedded in a dielectric continuum works amazingly well.[18,19] According to that model, energy levels are to be calculated as

$$E_{F_2^+} = k_o^{-2} \, E_{H_2^+}(r_{12})$$ (1)

where the proton separation r_{12} is given by

$$r_{12} = k_o^{-1} \, R_{12}$$ (2)

Here k_o is the dielectric constant and R_{12} the actual vacancy pair separation. The very large tuning range possible with F_2^+ centers (see Figure 2.1.3.10 below) is largely a consequence of Equation 1 and the fact that k_o^2 ranges over a greater than 2.5 to 1 ratio when all possible hosts are included.

Figure 2.1.3.3 shows an empirically determined energy level diagram of the F_2^+ center. There the levels are named after their molecular ion counterparts. The lasers use the $1s\sigma_g \rightarrow 2p\sigma_u$ transition. It has nearly ideal properties for laser action: The oscillator strength is high (f $\sim$ 0.2); the quantum efficiency of the pumping-emission cycle is

Table 2.1.3.1
PERFORMANCE OF COLOR CENTER LASERS: TUNING RANGES AND OUTPUT POWERS

Host	LiF	NaF	KF	NaCl	KCl:Na	KCl:Li	KCl:Na	KCl:Li	RbCl:Li
Center	F_2^+	$(F_2^+)^*$	F_2^+	F_2^+	$(F_2^+)_A$	$(F_2^+)_A$	$F_B(II)$	$F_A(II)$	$F_A(II)$
Pump μm/pwr*	0.647 4 W	0.87	1.064	1.064	1.34	1.34	0.6	0.647	0.647
		1 W	5 W	5 W	0.1 W	0.15 W	1.5 W	2.6 W	2.2 W
Tuning range μm	0.82— 1.05	0.99 1.22	1.22 1.50	1.40 1.75	1.62 1.91	2.0 2.5	2.25 2.65	2.5 2.9	2.6— 3.3
Max. pwr output, cw	1.8 W	400 mW	2.7 W	1 W	12 mW	10 mW	35 mW	240 mW	55 mW
Ref.	5, 6	12	6	7	8	9	10, 11	11	11

* The pump powers listed here correspond to the maximum output powers listed below.

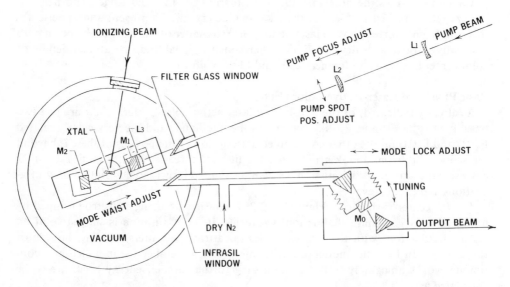

FIGURE 2.1.3.1.   Schematic of a cw color center laser (see text).

100%, independently of temperature; the Stokes shift is just large enough to allow complete resolution of the absorption and emission bands; finally, there are no excited state absorptions at transition energies even close to that of the emission.

It is important to note that the laser transition is polarized along the axis of the center (i.e., the line joining the vacancy pair); this is always one of the six [110] axes of the crystal. Optical pumping to a level higher than the $2p\sigma_u$, if accompanied by sufficient thermal excitation, can lead to a reorientation of the center's axis. In this way, multiphoton excitation by the laser pump source itself can lead to an orientational bleaching of the crystal. Thus, the principal reason for cooling the crystals is to freeze out, or at least to slow down considerably, this orientational bleaching. Techniques for dynamically maintaining the desired center orientation are also known.[6,19]

F_2^+ centers can be associated with certain impurities. For example, if the impurity is a foreign alkali metal ion, then the center[8] is called $(F_2^+)_A$. The $(F_2^+)^*$ center listed in

F AND F_2^+ CENTERS

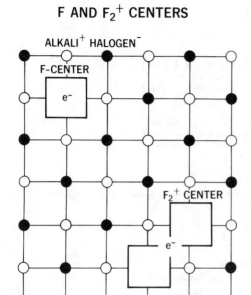

FIGURE 2.1.3.2.   The F_2^+ center and its progenitor, the ordinary F center (see text).

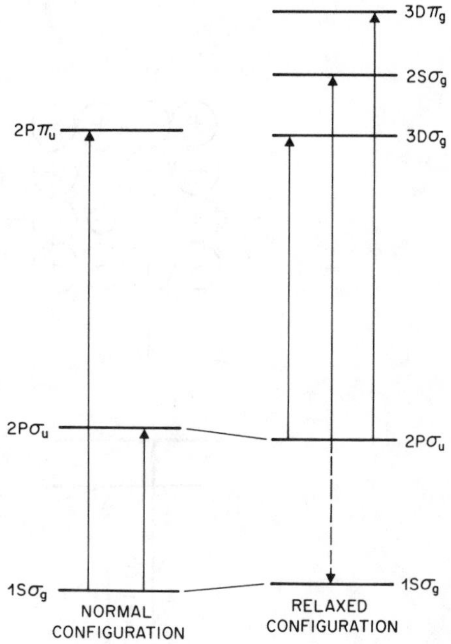

FIGURE 2.1.3.3.   Empirically determined energy level diagram of the F_2^+ center. Here the energy levels are named after their molecular-ion counterparts.

Table 2.1.3.1 represents an F_2^+ center associated with still another (but as yet unknown) impurity.[12] Such association with impurities usually has a number of benefits:

1.   New and useful tuning ranges are created.
2.   Orientational bleaching is made more difficult or excluded altogether.
3.   Most important of all, the room-temperature shelf life of the centers is greatly increased. (See remarks at the end of the section entitled ''Methods for Creating Color Centers.'')

Configurations and energy levels of an $F_A(II)$ center are shown in Figure 2.1.3.4. In its normal configuration, the $F_A(II)$ center consists of an ordinary F center (an anion vacancy containing a trapped electron) adjacent to a foreign metal ion. (The foreign ion is usually Li^+.) Absorption is from the s-like ground state to the p-like excited states of the normal configuration, and appears as a double-peaked band in the visible. (As one would expect, the excited state orbital overlapping the foreign ion has a somewhat different energy from the other two — this accounts for the double peak.)

Immediately following excitation, the system relaxes to the double-well configuration shown on the right-hand side of Figure 2.1.3.4. The energy levels of this relaxed configuration are completely unrelated to those of the normal configuration, and emission takes place at an energy four to five times smaller than that of absorption. As in the case of the F_2^+ center, the gain cross-section is high. On the other hand in contrast with the F_2^+ center, the quantum efficiency of luminescence is always less than unity, and is a monotonically decreasing function of temperature.[20] Thus, the threshold for laser action is considerably lowered by cooling of the crystal to liquid N_2 temperature. It should be pointed out that since at least 80% of the energy of every pump photon must be dissipated as heat, even when the laser is operated well above threshold, the power handling capability of the $F_A(II)$ center lasers is somewhat limited (see Table 2.1.3.1.)

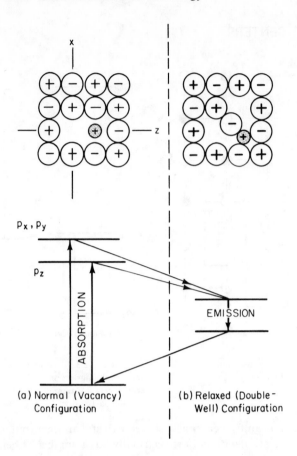

FIGURE 2.1.3.4.   Configurations and energy levels of an $F_A(II)$ center (see text). (For clarity, the electron of this center has not been shown.)

It can be seen from Figure 2.1.3.4 that with every completed optical pumping cycle, the $F_A(II)$ center has a certain probability of returning to a new orientation. In general, the resultant flopping of the center about the position of the foreign ion leads to an (essentially instantaneous) orientational bleaching, unless the orientation of the crystal axes (with respect to the pump and laser polarization directions) has been chosen carefully. The proper choice of orientation has been well discussed in the literature.[1,2,10]

The normal configuration of an $F_B(II)$ center is shown in Figure 2.1.3.5. Except for the fact that now two foreign ions are involved, the $F_B(II)$ center is very similar to the $F_A(II)$ center, and most of the behavior described above for $F_A(II)$ centers applies to the $F_B(II)$ centers as well.

Methods for Creating Color Centers

Creation of any of the laser-active color centers involves two fundamental steps: the generation of F centers, and the aggregation of these, either into pairs to form F_2^+ centers or with foreign metal ions, to form $F_A(II)$ or $F_B(II)$ centers. However, the choice of particular technique for the F center generation is based on a number of considerations, such as the requirement of electron traps in the crystals with F_2^+ centers.

When $F_A(II)$ or $F_B(II)$ centers are desired, additive coloration is most suitable. In that process, the crystal is heated to a temperature within $\sim 100°C$ of its melting point (to facilitate the diffusion of F centers) and brought to equilibrium with an excess of

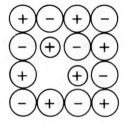
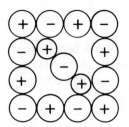

NORMAL

CONFIGURATION

RELAXED

CONFIGURATION

FIGURE 2.1.3.5.   Normal and relaxed configurations of an $F_B(II)$ center. (For clarity, the electron of this center has not been shown.)

the alkali metal vapor. In practice, this is best accomplished with an apparatus based upon the heat pipe, and utilizing a simple air lock for the introduction and extraction of crystals.[21]

When the resultant F centers are optically excited at room temperature, they become ionized. The empty vacancies formed are highly mobile, and soon find and become attached to a foreign metal ion. Recapture of an electron then completes the $F_A(II)$ center formation.

When F_2^+ or F_2^+-like centers are desired, it is usually necessary to incorporate divalent transition metal ion impurities, such as Mn^{++}, Ni^{++}, Cr^{++}, or Pd^{++}, to serve as electron traps. In this case the additive coloration cannot be used, as it chemically reduces the divalent ions to the monovalent state (thereby ruining their usefulness as electron traps). Instead, F centers and anion vacancies are created by radiation damage.[3] The most effective radiation source is an electron beam of 1 to 2 MeV energy and several μamp/cm² current density. To prevent undue aggregation, the crystals are cooled to $\sim -100°C$ during the irradiation. The irradiated crystals are then usually stored under liquid N_2 until needed, at which time they are quickly warmed to room temperature and briefly (10 to 15 min) held there. During this period, the F centers and vacancies aggregate to form F_2^+ centers. They are then cooled once again to operating temperature, usually 77 K; otherwise, the ordinary F_2^+ centers tend to disappear, in a matter of hours or days.

Often, after the above steps have been followed, the majority of centers are deionized, i.e., one has mostly F_2 rather than F_2^+ centers. This situation can usually be rectified through use of a special two-step ionization process.[22]

It is important to note that electron traps can also be created (as an alternative to the transition metal ions mentioned above) out of the debris of radiation damaged OH^- or SH^-. The effect of OH^- in enhancing the generation of electron-deficient centers was first described a number of years ago.[23] Recently, studies on crystals containing rather large and controlled additions of OH^- and SH^- have yielded rather promising results.[24]

The rather limited shelf-life of the F_2^+ centers at room temperature has been a serious barrier to their widespread use. However, the association of F_2^+ centers with stabilizing impurities, as in the $(F_2^+)_A$ or $(F_2^+)^*$ centers, appears to be the solution to this problem.[8,12] For example, the $(F_2^+)^*$ center in NaF has a shelf life at room temperature of at least several months. Unfortunately, work on these centers is too new to allow for extensive comment here.

NOTE ADDED IN PROOF

Recently the F⁺ center in CaO (a single electron trapped at an oxygen ion vacancy) has been used[26] to make a stable, cw tunable laser for the range 357 - 420 nm. The pump source was an Ar⁺ ion laser operating at 351 nm.

ABSORPTION AND EMISSION BANDS OF F_A(II) and F_B(II) CENTERS

(This section consists of graphs only.)

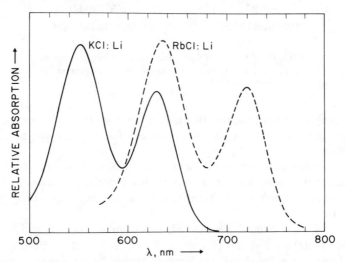

FIGURE 2.1.3.6.   Absorption bands of F_A(II) centers in the two hosts of greatest technical importance. Absorption bands of the F_B(II) centers in KCl:Na are not shown, since accurate shapes are difficult to determine for F_B centers. However, the peaks are thought to occur at 520 nm and 600 nm.

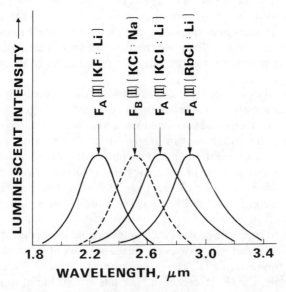

FIGURE 2.1.3.7.   Emission bands of type (II) centers in various hosts. Of the bands shown here, only that of KF:Li is not of technical importance, due to the low quantum efficiency in that particular host.[25]

ABSORPTION AND EMISSION BANDS OF F_2^+ AND F_2^+-LIKE CENTERS

(This section consists of graphs only.)

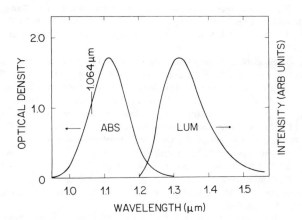

FIGURE 2.1.3.8.   Absorption and emission bands of the $1s\sigma_g \rightarrow 2p\sigma_u$ transition (laser active transition) of the F_2^+ center in the particular host KF, chosen to illustrate typical band shapes and typical Stokes shift.

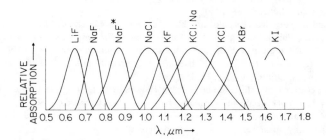

FIGURE 2.1.3.9.   Absorption bands of the $1s\sigma_g \rightarrow 2p\sigma_u$ transition (laser pump bands) of F_2^+ and F_2^+-like centers in various hosts. NaF* refers to the $(F_2^+)^*$ center in NaF, and KCl:Na refers to the $(F_2^+)_A$ center; all others refer to the isolated F_2^+ centers. Only the peak is shown for KI, since the full curve has not been reported in the literature. KCl:Li is not shown; the absorption peaks at 1.35 μm and has a width comparable to that shown for KCl:Na.

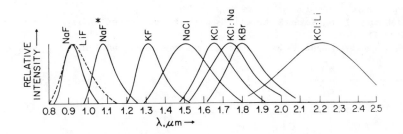

FIGURE 2.1.3.10.   Emission bands of the $1s\sigma_g \rightarrow 2p\sigma_u$ transition (laser tuning bands) of F_2^+ and F_2^+-like centers in various hosts. The LiF band has been shown in dashed outline, in order clearly to distinguish it from the closely overlapping curve of NaF. Note that due to a considerable variation in Stokes shifts, the peaks here do no follow exactly the same order as above in Figure 2.1.3.9.

REFERENCES

1. **Mollenauer, L. F. and Olson, D. H.**, Broadly tunable lasers using color centers, *J. Appl. Phys.*, 46, 3109, 1975.
2. **Mollenauer, L. F.**, Color center lasers, in *Quantum Electronics*, Part B, Tang, C. L., Ed., Ch. 6, Vol. 15B of *Methods of Experimental Physics*, Academic Press, New York, 1979.
3. **Mollenauer, L. F.**, Dyelike lasers for the 0.9-2μm region using F_2^+ centers in alkali halides, *Opt. Lett.*, 1, 164, 1977.
4. **Gusev, Yu L., Marennikov, S. I., and Chebataev, V. P.**, "Laser Action on F_2^+ and F_2^- Color Centers in a LiF Crystal in the Range 0.88-1.2 μm," *Sov. Tech. Phys. Lett.*, 3, 124, 1977. This very brief paper describes pulsed laser action in several different bands of the same crystal. However, the laser action in the 1.1-1.2μm band ascribed there to F_2^- centers most probably has another origin; F_2^- centers are too volatile under optical pumping to allow for sustained laser action.
5. **Mollenauer, L. F., Bloom, D. M., and DelGaudio, A. M.**, Broadly tunable cw lasers using F_2^+ centers for the 1.26-1.48 and 0.82-1.07 μm bands, *Opt. Lett.*, 3, 48, 1978.
6. **Mollenauer, L. F. and Bloom, D. M.**, Color-center laser generates picosecond pulses and several watts cw over the 1.24-1.45 μm range, *Opt. Lett.*, 4, 247, 1979.
7. **Mollenauer, L. F.**, unpublished.
8. **Schneider, I. and Marrone, M. J.**, Continuous-wave laser action of $(F_2^+)_A$ centers in sodium-doped KCl crystals, *Opt. Lett.*, 4, 390, 1979.
9. **Schneider, I. and Marquardt, C.**, Tunable, cw laser action using $(F_2^+)_A$ centers in Li-doped KCl, *Opt. Lett.*, 214, 1980.
10. **Liftin, G., Beigang, R., and Welling, H.**, Tunable cw laser operation in F_B(II)-type color center crystals, *Appl. Phys. Lett.*, 31, 381, 1977.
11. As reported in Table 2 of **Koch, K. P., Litfin, G., and Welling, H.**, Continuous-wave laser oscillation with extended tuning range in F_A(II)-F_B(II) color center crystals, *Opt. Lett.*, 4, 387, 1979.
12. **Mollenauer, L. F.**, Room temperature stable, F_2^+-like center yields cw laser tunable over the 0.99-1.22μm range, *Opt. Lett.*, 5, 188, 1980.
13. **Litfin, G. and Beigang, R.**, Design of tunable cw colour centre lasers, *J. Phys. E*, 11, 1978.
14. **German, K. R.**, Polarization bean splitters for pumping of FII-center lasers, *Opt. Lett.*, 4, 68, 1979.
15. **German, K. R.**, Diffraction grating tuners for cw lasers, *Appl. Opt.*, 18, 2348, 1979.
16. **Jackson, D. J., Lawler, J. E., and Hänsch, T. W.**, Broadly tunable pulsed laser for the infrared using color centers, *Opt. Commun.*, 29, 357, 1979.
17. **Lüty, F.**, F_A centers in alkali halide crystals, in *Physics of Color Centers*, Fowler, W. B., Ed., Academic Press, New York, 1968, and the many shorter papers referenced therein.
18. **Aegerter, M. A. and Lüty, F.**, The F_2^+ Center in KCl Crystals, *Phys. Stat. Sol.*, b 43, 227f and 245f, 1971, and the many earlier references contained therein.
19. **Mollenauer, L. F.**, Excited-state absorption spectrum of F_2^+ centers and the H_2^+ Molecular-ion model, *Phys. Rev. Lett.*, 43, 1524, 1979.
20. See Figure 2.1.3.5, Ref. 2.
21. **Mollenauer, L. F.**, Apparatus for the coloration of laser-quality alkali halide crystals, *Rev. Sci. Instr.*, 49, 809, 1978.
22. **Mollenauer, L. F. and Bloom, D. M.**, Simple two-step photoionization yields high densities of laser-active F_2^+ centers, *Appl. Phys. Lett.*, 33, 506, 1978.
23. **Chandra,A.**, Impurity effects on the ionization states of F-aggregate color centers in sodium fluoride, *J. Chem. Phys.*, 51, 1499, 1969.
24. **Gellermann, W., Lüty, F., Koch, K. T., and Litfin, G.**, F_2^+ center stabilization and tunable laser opertion in OH$^-$ doped alkali halides, *Phys. Stat. Sol.*, 57, 111, 1980.
25. **Mollenauer, L. F., Hatch, B. A., and Olson, D. H.**, Discovery of F_A(II) centers in KF, *Phys. Rev. B*, 12, 731, 1975.
26. **Henderson, B.**, Tunable visible lasers using F$^+$ centers in oxides, *Opt. Lett.*, 6, 437, 1981.

2.2 SEMICONDUCTOR LASERS

Henry Kressel and Michael Ettenberg

INTRODUCTION

Semiconductor lasers consist of *injection lasers*, where a p-n junction or heterojunction is used to inject excess carriers into the active region, *optically pumped lasers,* where an external light source produces excess carriers, and *electron-beam pumped lasers* which use high energy electrons to produce the excess carriers. Injection lasers, which are the most practical devices, are discussed at length in this review. The other types of lasers are used for experimental purposes in the study of semiconductors which cannot be used for the fabrication of injection devices. Examples of such materials are wide bandgap energy II — VI compounds, such as ZnO, which can only be doped to one conductivity type, thus making p-n junction fabrication impossible.

OPERATING PRINCIPLES

In this section, we review a few of the key concepts concerning laser action in semiconductors. Extensive theoretical treatments of this subject can be found elsewhere.[1]

Direct and Indirect Bandgap Semiconductors

In direct bandgap semiconductors (the only ones in which stimulated emission has been observed), both photon emission and absorption can occur without the need for a phonon to conserve momentum. This is because the lowest conduction band minimum and highest valence band maximum are at the same wave vector ($\bar{k}$) in the Brillouin zone. Figure 2.2.1 shows the schematic diagram of electron energy vs. $\bar{k}$ in a semiconductor, such as GaAs, where the smallest bandgap energy $E_g = E_c - E_v$ is at $\bar{k}$ = [000].

In indirect bandgap semiconductors the conduction band minimum and valence band maximum are not at the same $\bar{k}$ value. Hence, photon emission and absorption require the participation of phonons to conserve momentum. A schematic diagram of an indirect bandgap semiconductor such as GaP or AlAs is shown in Figure 2.2.2. In these semiconductors the lowest-lying conduction band minima are along $\bar{k}$ = <100>.

By mixing different bandgap compounds, a continuous variation of the band structure is possible. For example, in $GaAs_{1-x}P_x$ the change from direct to indirect bandgap transition occurs at about x $\simeq$ 0.45, with $E_g \simeq$ 1.96 eV. Another alloy system of great technological interest is $Al_xGa_{1-x}As$ in which the crossover occurs at x $\simeq$ 0.4. Figure 2.2.3 shows[1] the variation of the bandgap energy of $Al_xGa_{1-x}As$ as a function of x. The composition at which the <100> and [000] conduction band minima are at the same energy marks the crossover from a direct to an indirect bandgap semiconductor ($E_g \sim$ 1.92 eV).

Lasing in indirect bandgap semiconductors is improbable because the lowest-energy band-to-band transition probabilities are much smaller than in direct semiconductors. Thus, the radiative lifetime is long. Because of the relatively long lifetime of electrons in the indirect minima, there is time for nonradiative recombination processes to occur, thus yielding a low internal quantum efficiency. Furthermore, the stimulated recombination rate is related to the band-to-band absorption coefficient. Since this coefficient is lower for indirect than direct transitions, the potential laser gain is correspondingly reduced.

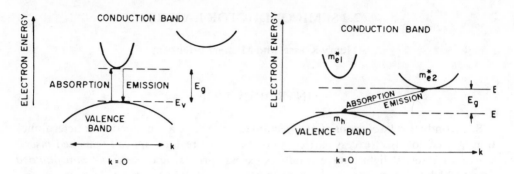

FIGURE 2.2.1. (Left) Photon absorption and emission in a direct bandgap semiconductor. FIGURE 2.2.2 (Right) Photon emission and absorption in an indirect bandgap semiconductor. The effective electron masses m_{e1}^* and m_{e2}^* refer to the two conduction band minima shown. In general, $m_{e2}^* > m_{e1}^*$. In the case of GaP and AlAs, the conduction band minima occur along $k = <100>$. m_h is the density-of-states effective hole mass.

Population Inversion and Lasing Criteria

Optical gain is possible when population inversion has been achieved; i.e., the probability of photon emission with energy $h\nu$ is greater than the inverse process of absorption at the same photon energy. Analysis of laser action in semiconductors is complicated by the fact that the density-of-states distribution affects the gain dependence on excess carrier density. We begin with the basic requirements for laser action.

In thermodynamic equilibrium, the probability of an electron occupying a state with energy E is given by the Fermi-Dirac distribution function

$$f = \left(1 + \exp \frac{E - F}{kT}\right)^{-1} \tag{1}$$

where F is the Fermi level, T is the temperature, and k is Boltzmann's constant.

When minority carriers are injected into a semiconductor (for example, electrons into p-type material), a steady state distribution of carriers in the conduction and valence bands, independently, can be assumed. Hence, a quasi-Fermi level for electrons, F_c, and for holes, F_v, is defined, where

$$f_c = \left(1 + \exp \frac{E - F_c}{kT}\right)^{-1} \tag{2}$$

Similarly, for the valence band, the probability for a state at a given energy to be empty (containing a hole) is

$$1 - f_v = 1 - \left(1 + \exp \frac{E - F_v}{kT}\right)^{-1} \tag{3}$$

It was shown by Bernard and Duraffourg[2] that the condition for net gain at a photon energy $h\nu$ is

$$F_c - F_v > h\nu \tag{4}$$

The carrier distribution must be degenerate for this condition to be satisfied. In order to obtain lasing in p-type material, for example, it is therefore necessary to inject a sufficient density of electrons to fill states in the conduction band until the quasi-Fermi level F_c has been raised (and F_v lowered) sufficiently for condition (4) to be satisfied.

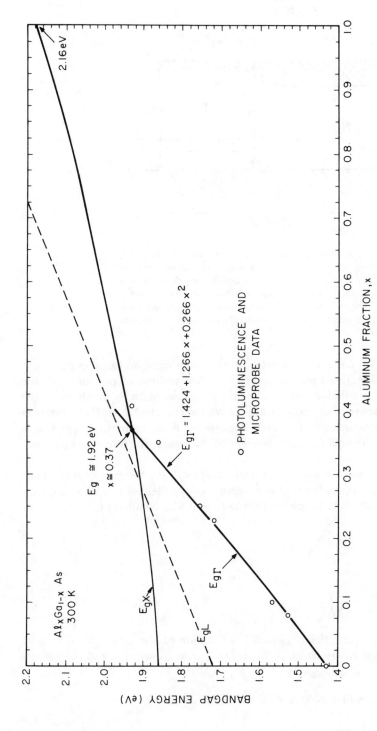

FIGURE 2.2.3.   Band structure of Al_xGa_{1-x} As at 300 K. (From Kressel, H. and Butler, J. K., *Semiconductor Lasers and Heterojunction LEDs*, Academic Press, New York, 1977. With permission.)

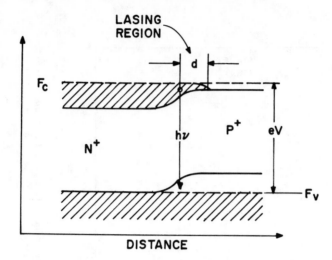

FIGURE 2.2.4.   Energy bands in a p-n junction under very high forward bias voltage V. Both the n and p sides of the junction were initially degenerate. As shown, a high density of electrons is injected into the p side, with negligible hole injection into the n side. Both the electron and hole populations are substantially degenerate over the active region d where the condition $F_c - F_v > h\nu$ is satisfied.

The lasing criterion (4) applies, of course, to both homogeneous samples (electron-beam or optically pumped) and p-n junctions. In a forward-biased p-n junction, the injected electrons are distributed over a distance d (a few microns) in which, if the injection level is high enough, the population is inverted (Figure 2.2.4). Hole injection into the n side is small in such diodes, partly because the bandgap in the p-type side is effectively reduced by the high acceptor concentration; hence, the injection efficiency of electrons is high.

The second laser condition is that the optical gain match the losses over one transit of the beam through the cavity. With a Fabry-Perot cavity of length L, with reflectivity R at the lasing wavelength (0.32 for cleaved GaAs facets), this condition is

$$R \exp (g - \tilde{\alpha}) = 1 \tag{5a}$$

or

$$g = \tilde{\alpha} + \frac{1}{L} \ln \left(\frac{1}{R}\right) \tag{5b}$$

where g is the optical gain coefficient (which is a function of the current density), and $\tilde{\alpha}$ is the optical loss in the cavity which includes free carrier absorption and absorption in the surrounding noninverted regions.

EXTERNALLY EXCITED LASERS

Electron-Beam Pumped Lasers

The utility of electron-beam pumped lasers is limited by the need for elaborate and bulky electronics, and owing to excessive thermal dissipation, the inability to operate

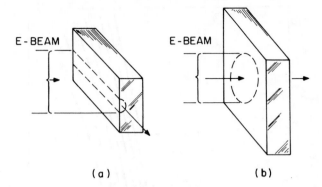

FIGURE 2.2.5.   Electron-beam pumper laser configurations. (a) "Stripe-pumped" and (b) central pumping.

at high duty cycles. The sample size is generally on the order of 1 mm or less. Figures 2.2.5 (a) and (b) show two types of sample configurations. In Figure 2.2.5a the beam is shaped in the transverse direction with the sample edges forming the Fabry-Perot walls. In the configuration shown in Figure 2.2.5b a circular beam is used and the back and front surfaces of the sample are the Fabry-Perot cavity walls.

The penetration depth of the electron beam (and hence the approximate active region width) increases with the electron energy in the beam. Energy values as low as 5 keV have been used, with upper values in the 200 keV range. Rapid radiation damage to the sample is a major consideration in discouraging the use of high energy beams.

Optically Pumped Lasers

The excess carriers in optically pumped lasers are produced by strongly absorbed radiation from an external laser. Gas, solid state, or injection lasers can be used for pumping. For example, a GaAs sample can be pumped using an injection laser made from a higher bandgap material. In general, the radiation is absorbed within about a micrometer of the surface. Heterojunction structures can also be optically pumped if the pumped region is covered with a higher bandgap, nonabsorbing region.

DIODE LASERS

Independent of the material, the design of a semiconductor laser is guided by several basic needs:

- A recombination region is needed where population inversion is produced by the injection of electron-hole pairs. In general, the lower the temperature, the lower the pair density needed to produce sufficient gain to sustain oscillations in the cavity. For example, a pair density of about 2×10^{18} cm^{-3} is typically needed to initiate lasing at room temperature in a GaAs laser diode. The thinner the recombination region (i.e., the smaller the volume of material in which population inversion is needed), the lower the threshold current density — within limits.
- An optical cavity is needed to produce feedback. A Fabry-Perot cavity is usually used which is produced by cleaving two parallel facets of the device a distance typically 300 μm apart. However, feedback can also be produced by periodic variations of the refractive index (in distributed feedback lasers), thus eliminating the need for the cavity mirrors.
- An internal waveguide must be provided to confine the stimulated radiation to a

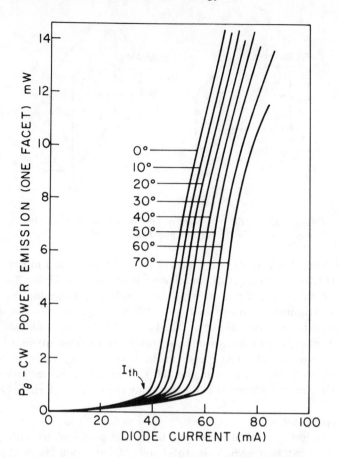

FIGURE 2.2.6.   Power output from a cw laser diode of AlGaAs as a function of current and heat sink temperature. The threshold current is I_{th} and the differential quantum efficiency above threshold is η_{ext}.

volume which includes the region producing gain. The degree of this radiation confinement depends on the dielectric steps bounding that region. In heterojunction lasers the dielectric steps at the heterojunctions provide a strong and controllable confinement. A much lesser degree of confinement exists in homojunction lasers.

Structural parameters and material quality are essential features of useful lasers. Although many materials are candidates for laser diode construction, few have so far been used because of the difficulty of producing certain semiconductors of sufficient quality.

Key injection laser parameters are the threshold current density, the differential quantum efficiency and the power level which can be reliably emitted. Figure 2.2.6 shows a typical curve of power emitted P_θ versus diode current for a cw double-heterojunction AlGaAs laser diode. We see the region of spontaneous emission below threshold and the steep increase in power emitted as the lasing threshold is passed. The threshold current density (and threshold current) as well as the operating range of the laser depend on the diode topology and the internal geometry.

Laser Topology

Figure 2.2.7 shows the two basic laser diode configurations, "broad-area" and

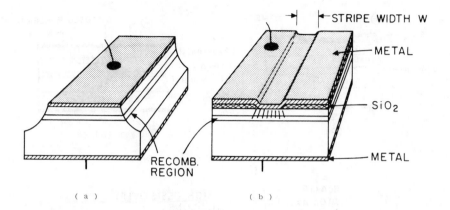

FIGURE 2.2.7.   (a) Conventional "broad-area" diode formed by sawing two sides of the diode. A reflecting film evaporated on one facet increases the emission from the other facet. The facets are formed by cleaving two parallel planes. (b) Stripe-contact diode used for cw operation. The active diode area if formed by confining the current flow to a narrow region of the diode surface. The diode is usually mounted with the p-side down on a heat sink. (From Kressel, H., Ettenberg, M., Wittke, J. P., and Ladany, I., *Semiconductor Devices for Optical Communication*, Kressel, H., Ed., Springer-Verlag, Heidelberg, 1980. With permission.)

"stripe-contact," which differ in the way the active junction area is defined.[3] In both structures the Fabry-Perot resonator is formed by cleaving two parallel facets. A reflecting film is sometimes placed over one facet to increase the output at the other end. Waves propagating in directions off the facet normal must be suppressed by the introduction of special loss mechanisms. In the broad-area devices this is achieved by sawing the sidewalls.

In stripe-contact lasers this is achieved by constricting the current to the region under the contract, the rest of the material thereby becoming highly absorbing. The simplest planar stripe-contact laser is the oxide-isolated one shown in Figure 2.2.8a. Other structures are shown in Figures 2.2.8b and 2.2.8c.

Vertical Geometry

Figure 2.2.9 shows the schematic cross section of the most important laser diode structures.[1] The energy diagram, the refractive index profile, the distribution of the optical energy ($\propto E^2$) and the position of the recombination region are shown for each structure.

- In the homojunction laser,[4-6] the recombination region width is essentially set by the minority carrier diffusion length. The small radiation confinement is the result of refractive index gradients resulting from dopant concentration gradients and carrier concentration differences. Typically a p$^+$-p-n configuration is used, where the p$^+$-p interface provides a small potential barrier.
- In the single-heterojunction (close-confinement) laser a p$^+$-p heterojunction forms one boundary of the waveguide and provides a potential barrier for carrier confinement within the p-type recombination region.[7-8] The refractive index step at the p$^+$-p heterojunction is typically a factor of five larger than that at the p-n homojunction. The AlGaAs/GaAs laser threshold current densities are typically 1/4 to 1/5 of the GaAs homojunction values at room temperature ($\sim$10,000 A/cm^2 vs. $\sim$50,000 A/cm^2).
- In the double-heterojunction (DH) laser, the recombination region is bound by two higher bandgap regions to confine the carriers and the radiation. Typical

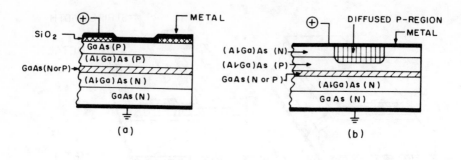

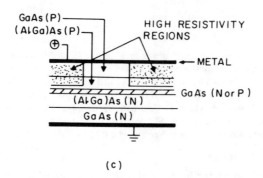

FIGURE 2.2.8.   Relatively simple techniques for forming planar stripe-contact laser diodes. (a) Oxide isolated stripe-contact. (b) Selective Zn diffusion into n-type surface layer. (c) The formation of high resistance regions adjoining the active region to constrain the current flow (proton and oxygen bombardment has been used).

AlGaAs/GaAs DH laser threshold current densities are 1000 to 2000 A/cm^2.[9-11]

- In the large optical-cavity (LOC) double-heterojunction laser, the waveguide region is deliberately made wider than the recombination region in order to reduce the optical power density and hence produce a device suited for higher peak operation than is possible with the DH laser.[12]

- The very-narrow-spaced double-heterojunction laser is a subclass of the basic DH device, but it is designed with an extremely thin recombination region, $d \approx 0.1$ μm, adjusted to permit the wave to partially spread outside the recombination region but still provide full carrier confinement. By this means the beam divergence is reduced while keeping $J_{th} < 2000$ A/cm^2.

- In the four-heterojunction (FH) laser, five regions are included.[13] A submicron-thick recombination region is bracketed by two heterojunctions which are further enclosed within two outer heterojunctions. This device combines the LOC and DH concepts. The recombination region is generally centered within the waveguide region. By adjusting the index step between the recombination region and the adjoining region, maximum design flexibility is achieved. High peak power operation is obtained by widening the optical field distribution.

The heterojunction laser design flexibility, made possible by sophisticated epitaxial technologies, offers ample opportunity for specific device designs consistent with applications. Thus, lasers can be designed for high peak power but low duty cycle operation with moderate threshold currents. For example, peak power values in excess of 10 W are easily obtained from single-heterojunction or LOC lasers at duty cycles of about 0.1% and much higher values can be achieved by stacking diodes, the formation of arrays or by widening the emitting region in the junction plane. On the other end

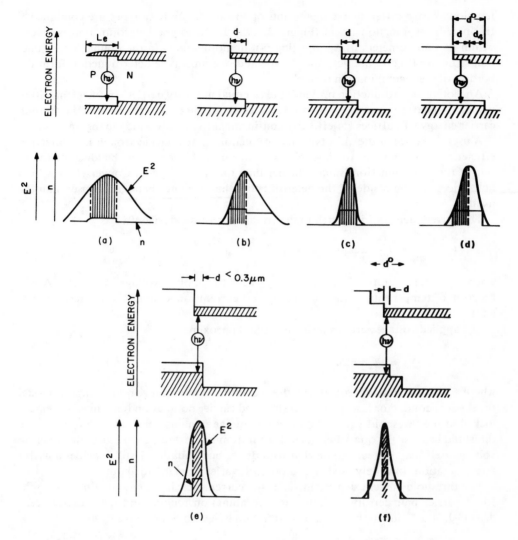

FIGURE 2.2.9.   Schematic cross section of 6 laser configurations. The shaded region in each case is the region of electron-hole recombination. The curve E^2 denotes the optical field intensity.

of the power scale are low threshold current lasers designed for cw operation at room temperature. Such lasers have small junction areas and are designed to emit continuous power levels of about 5 to 20 mW, although about 100 mW have been obtained from laboratory devices.

The diode geometry varies with the intended use of the device. Low threshold is not essential for low duty cycle pulsed-power lasers, but the ability to sustain high optical power levels without facet damage is required. For cw lasers, on the other hand, the lowest possible threshold current density is needed and this is generally achieved with narrow recombination region devices of the double-heterojunction class, although low threshold operation can also be obtained with FH or LOC lasers having a small heterojunction spacing.

Current-Voltage Characteristics

Heterojunctions consist of isotype heterojunctions, where the two adjoining regions have different bandgap energy but the same conductivity type, and anisotype heterojunctions where different bandgaps and opposite conductivity types are combined.[14] In single-heterojunction lasers (Figure 2.2.9b) an anisotype heterojunction is placed within two micrometers of an injecting p-n homojunction. In the double-heterojunction structure (Figure 2.2.9c), one isotype and one anisotype heterojunction form the borders of the recombination region.

An isotype heterojunction provides the essential minority carrier confinement within the recombination region. A potential barrier about ten times greater than the thermal carrier energy kT suffices for effective confinement to the low bandgap region.

A useful feature of the anisotype heterojunction is the high electron or hole injection efficiency. For example, for high electron injection efficiency, the bandgap of the n-side of the heterojunction is made higher than that of the p-side. Conversely, for high hole injection, the p-side of the heterojunction has a higher bandgap energy than the n-side.

The current density J is related to the voltage across the device V_a,

$$J = J_o \{ \exp [A (V_a - V_s)] - 1 \} \tag{6}$$

where J_o is the saturation current density, V_a is the applied voltage, $V_s = IR_s$, A is a function of temperature, junction quality and current range, and R_s is the series resistance.

The applied voltage across a lasing device is approximately

$$V_a \cong IR_s + E_g/e \tag{7}$$

where E_g is the bandgap energy for the recombination region. In the simplest model for the semiconductor laser, where the injected carrier population is assumed to remain locked at the threshold value, the junction voltage remains constant with drive above threshold because the quasi-Fermi level separation is constant. In practice, the junction voltage itself may increase somewhat with drive, but Equation 7 still provides a useful approximation. In widely used laser diodes, R_x varies from 0.1 to about 5 Ω.

The threshold current density includes the current density component needed to inject a carrier pair density N_{th} into the recombination region, and the excess current density J_e and J_h due to the loss of electrons and holes from the active region,

$$J_{th} = \frac{e \, N_{th} d}{\tau_s} + J_h + J_e \tag{8}$$

Here, τ_s is the carrier lifetime for spontaneous recombination in the active region of width d; in GaAs devices, $\tau_s = 2-3$ nsec near threshold.

The ratio of electron-to-hole flow across the high bandgap n-low bandgap p heterojunction is normally very large in typical DH lasers. Likewise, the loss of minority carriers at the isotype heterojunction is negligible at room temperature for reasonable barrier heights ($\geqslant 0.2$ eV). However, at elevated temperatures, carrier loss will unduly increase J_{th} unless $\Delta E_g \geqslant 0.4$ eV.

RADIATION PROPERTIES

The basic heterojunction laser can be viewed as a three-layer dielectric slab with constant refractive indices in each layer as shown in Figure 2.2.10, where the recombination or active region is shown bracketed by two higher bandgap regions. Maxwell's equations are solved with appropriate boundary conditions to determine the modal properties. This analysis yields the fraction of the radiation confined to the active

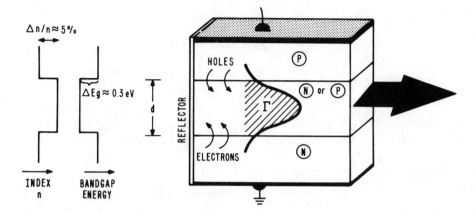

FIGURE 2.2.10.   Double-heterojunction laser diode showing a propagating wave with the fraction Γ within the recombination region. The bandgap energy steps and associated refractive index steps are shown on the left of the laser schematic.

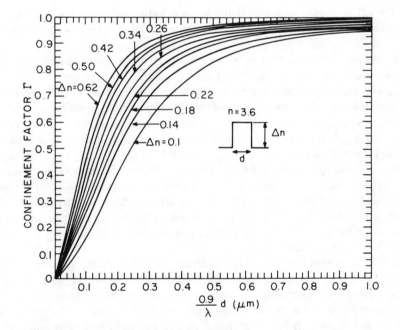

FIGURE 2.2.11.   Fraction of the radiation confined to the recombination region of a double-heterojunction as a function of the width of the region d (normalized for wavelength). The index steps Δn for the various curves are indicated. The index of the recombination region is n = 3.6. (After Kressel, H. and Butler, J. K., *Semiconductor Lasers and Heterojunction LEDs,* Academic Press, New York, 1977.)

region Γ and transverse modes capable of propagation in the waveguide. However, which of the modes actually dominates is not determined from such an analysis unless the losses for the various modes are estimated. It is assumed that the mode having the lowest propagation loss reaches threshold first.

There is no cut-off condition for a symmetric waveguide, i.e., one with equal index steps at the two walls, but as the waveguide width d is reduced, more of the wave propagates outside the guide. Figure 2.2.11 shows the fraction of the radiation con-

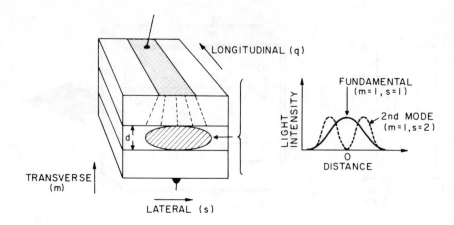

FIGURE 2.2.12.   Modes in laser diode. (After Kressel, H., Ettenberg, M., Wittke, J. P., and Ladany, I., *Semiconductors Devices for Optical Communication,* Kressel, H., Ed., Springer-Verlag, Heidelberg, 1980. With permission.)

fined to the active region of width d for various values of the index step Δn.[1] As we shall see shortly, Γ is a key quantity in determining the efficiency and threshold of a laser.

The distribution of the radiation in the vicinity of the recombination region of the double-heterojunction laser obviously affects its radiated beam shape.

The allowed electromagnetic modes of the laser cavity are separable into two independent sets, having either transverse electric (TE) or transverse magnetic (TM) polarization. In the present notation, the fundamental mode is denoted 1. The mode numbers m, s, and q give the number of antinodes in the optical field along the transverse, lateral, and longitudinal axes of the cavity, Figure 2.2.12.

The allowed longitudinal modes are determined by the effective index of refraction and the cavity length. The resulting Fabry-Perot mode spacing is several angstrom units in typical laser diodes. The lateral modes are dependent on the method used to define the two edges of the diode width and on the temperature and gain profile across the diode which impacts the refractive index profile. Generally, only low-order modes are excited in narrow stripe-contact lasers.

The transverse modes depend on the dielectric variation perpendicular to the junction plane. In typical DH lasers, only the fundamental transverse mode is excited, a condition achieved by restricting the width of the waveguiding recombination region d (i.e., heterojunction spacing) to a few tenths of a micrometer. Therefore, the far-field radiation pattern consists of a single lobe in the direction perpendicular to the junction. (Higher-order transverse modes give rise to "rabbit-ear" lobes.)

The far-field pattern in the direction perpendicular to the junction plane reflects the radiation distribution in the vicinity of the active region. This, in turn, depends on d and on the refractive index step Δn at the heterojunctions. Figure 2.2.13b shows the dependence of the beam width $\theta\perp$ (full angular width at the half-intensity points) on d/λ for various Δn values.[15,16] (For $Al_xGa_{1-x}As/GaAs$ heterojunctions $\Delta n \approx 0.62$ x.) The decrease in $\theta\perp$ for small d reflects the reduced radiation confinement as the waveguide becomes narrower.

Fundamental lateral mode operation can be obtained by sufficiently restricting the width of the active region. The lateral modes depend on the dielectric profile in the plane of the junction and thus, on the technique used for junction area definition. The simplest case occurs when there are two high (and equal) steps Δn perpendicular to the junction plane forming a rectangular box cavity. Consider a diode of width W. From

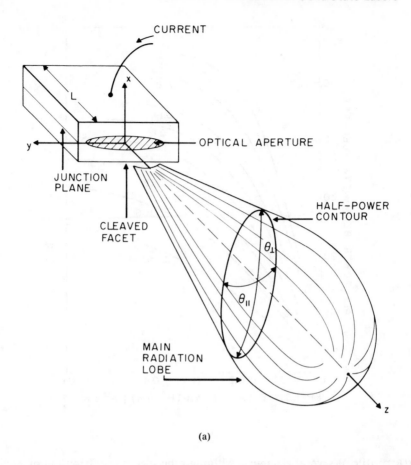

(a)

FIGURE 2.2.13.   (a) Far-field from stripe-contact laser operating in the fundamental lateral and transverse mode. (b) Half-power beamwidth for symmetric double-heterojunction lasers as a function of the normalized heterojunction spacing for various refractive index step values. The index in the active region is 3.6 (GaAs). (From Butler, J. K. and Kressel, H., *RCA Rev.*, 38, 542, 1977. With permission.)

a simple analysis of the critical angle for total internal reflection at the sidewalls, we find that the highest lateral mode number S_m capable of propagating in the structure is

$$S_m = \text{Integer} \left[1 + \left(\frac{2nW}{\lambda} \right) \left(\frac{2\Delta n}{n} \right)^{1/2} \right] \tag{9}$$

Therefore, to operate in the fundamental lateral mode, the index step condition is

$$\frac{\Delta n}{n} \lesssim \frac{1}{8} \left(\frac{\lambda}{nW} \right)^2 \tag{10}$$

For GaAs, $n = 3.6$ and $\lambda \approx 0.9$ μm; hence $\Delta n/n \lesssim 7.8 \times 10^{-3} W^{-2}$, where W is in micrometers.

This step-index, step-sidewall model is most appropriate for sawed-side or etched-side devices. Analysis of the planar stripe lasers is more difficult because the shallow dielectric profile is related to the current and gain distribution; hence, it is subject to change with current and optical power level.

Mode guiding in planar stripe lasers results from the combined contributions to the

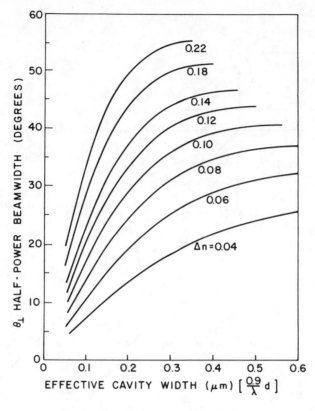

FIGURE 2.2.13b.

dielectric profile of several elements. Although the specific contribution of each is not readily calculated, the following are relevant.

1. Increasing gain near the stripe center produces a corresponding increase in the imaginary part of the dielectric constant
2. Local heating related to the current distribution and power dissipation increases the real part of the dielectric constant.
3. However, acting in the opposite direction to reduce the dielectric constant under the stripe is the distribution of free carriers, i.e., the higher the carrier density, the lower the index.

A shallow maximum in the dielectric profile under the stripe results from the combined effect of the above three factors. With changes in current and optical power density, the local heating and gain coefficient profile are likely to change, thus affecting the ability of the various lateral modes to propagate.

Fundamental lateral mode operation is desirable not only because it eases coupling into low numerical aperture fibers (including single-mode fibers), but also because mode changes with current are frequently accompanied by kinks in the power output vs. current curves. Furthermore, fundamental spatial mode operation is required for all critical optical applications such as the formation of small spots or parallel beams. The key principle which governs the restriction of laser operation to a single spatial mode is that the difference between the propagation losses of the fundamental and the higher-order modes be as large as possible. Restricting the stripe width of lasers to very small values (<8 μm) is one method of achieving this objective,[17,18] although at the expense of useful power from the device.

Forming the stripe contact in other configurations such as misaligning it with respect to the mirrors,[19] curving it,[20] or bending it[21] as well as altering its width[22] along the length have all been employed with some success, but at the sacrifice of the threshold current density. Furthermore, since the mode stability depends on the current flow, changes in temperature, duty cycle and threshold alter the spatial pattern.

The most successful structures for providing lateral mode stability incorporate a lateral waveguide. The many structures devised are limited by the ability to grow epitaxial crystal layers into the desired configurations and to provide materials with matching lattice parameter, index of refraction and bandgap which allow both optical and carrier confinement to the active portion of the device. Two basic concepts are employed. In one concept, the lateral mode is stabilized by introducing loss into the wings of the mode so that higher order modes, which distribute more power to the wings, have higher thresholds. In the other, a lateral index of refraction profile is provided such that a waveguide is formed which only supports the fundamental mode. In Figure 2.2.14 we illustrate some of these laterally guided structures.

The simplest structure is the buried heterojunction (BH) laser[23] in Figure 2.2.14a. Here, the lateral guide is provided by etching and regrowing a lower index material to form the lateral guide. However, the guide is very narrow limiting the power to 1 to 2 mW. To improve the power output, the basic structure can also be a large optical cavity (BH-LOC).[24] In a similar LOC type structure shown in Figure 2.2.14b, called a stripe-buried-heterostructure[25] (SBH), only the active region is etched and the waveguide is regrown around it while the rest of the optical cavity remains untouched. In the channel-substrate-planar device,[26] Figure 2.2.14c, absorption in the wings is employed. The transverse junction stripe[27] (TJS), Figure 2.2.14d, laser employs a homojunction in the lateral direction to stabilize the mode while in the transverse direction we have a conventional double-heterojunction structure without an electrical junction.

There are a number of etched substrate configurations other than the CSP structure. The configuration may be grown in a well,[28] Figure 2.2.14e; it may use a LOC structure in a well[29] or on the well shoulder[30] or between two wells as in the double-dovetail confined double-heterojunction laser[31] shown in Figure 2.2.14f. All rely on the non-planarity in the active (i.e., light-guiding) region to provide the lateral waveguide.

All the structures shown incorporate variations in the index of refraction due to alloy compositional changes at the heterojunction interfaces. The lasers are illustrated for AlGaAs where the addition of Al to the alloy lowers the index without a significant change in lattice parameter. Such structures can also be prepared in other alloy systems such as InGaAsP.

These lasers, in addition to providing a fundamental spatial mode, also tend to exhibit linear power output versus current characteristic as well as single longitudinal mode operation over a useful current range as illustrated for a CDH laser in Figure 2.2.15.[32] The linearity of the power curve is due to the presence of a fixed waveguide whose losses do not change with diode current. While not very well understood, a coincidence is often found between fundamental lateral mode and a preferential single frequency (longitudinal) mode operation.

Although single longitudinal mode operation can be obtained in some laterally guided structures, the mode shifts with temperature and power output.

The distributed feedback laser[33] provides a means of better restricting the laser operation to a single longitudinal mode which is relatively independent of device temperature or power. This structure does not affect the transverse or lateral modes. The theory of DFB lasers has been reviewed by Yariv.[34]

The distributed feedback laser (DFB) incorporates periodic variations of the refractive index along the direction of wave propagation. The energy of the wave propagation in one direction is continuously fed back in the opposite direction by Bragg scat-

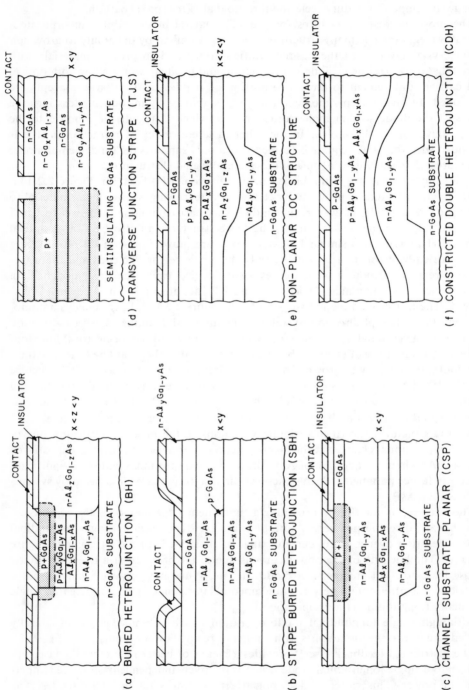

FIGURE 2.2.14.   Schematic cross section of some of the laser configurations designed for fundamental lateral mode operation. The shaded areas are post-growth Zn diffusions to form heavily-doped, p-type regions which help confine the injected carriers. All employ a fixed lateral waveguide.

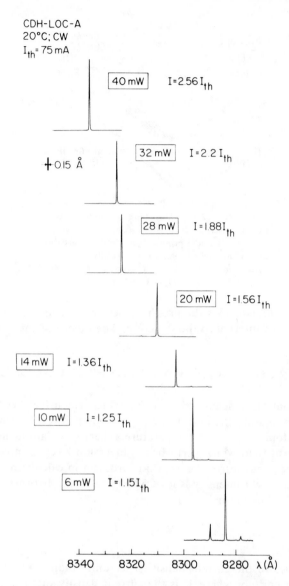

CDH-LOC-A
20°C; CW
$I_{th} = 75$ mA

40 mW $I = 2.56 I_{th}$

$+ 0.15$ Å

32 mW $I = 2.2 I_{th}$

28 mW $I = 1.88 I_{th}$

20 mW $I = 1.56 I_{th}$

14 mW $I = 1.36 I_{th}$

10 mW $I = 1.25 I_{th}$

6 mW $I = 1.15 I_{th}$

8340   8320   8300   8280 λ (Å)

FIGURE 2.2.15.   Typical CDH-LOC spectral characteristics, at various power levels, as the current is increased to 2.5 times threshold value (40 mW from one facet). (From Botez, D., *Appl. Phys. Lett.*, 36(3), 190-192, 1980. With permission.)

tering, producing resonance. The greatest interest has focused on injection DFB lasers using AlGaAs/GaAs heterojunctions where variations in the Al content along the laser axis produce the required dielectric perturbations.

Figure 2.2.16 shows a schematic of a DFB-AlGaAs/GaAs laser in which the corrugations are incorporated within the optical cavity defined by two outer heterojunctions, the GaAs recombination region being separate from the corrugated region.[35] The periodicity of the corrugations is about 0.4 μm. Devices of this type have operated at room temperature with threshold current densities of about 5000 A/cm², and have been shown capable of cw operation. Operation in a single longitudinal mode has been obtained, consistent with theory, although the power level which can be reached in a single longitudinal mode is uncertain.

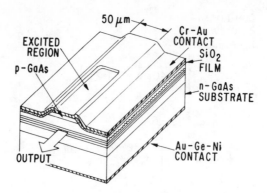

FIGURE 2.2.16. Distributed feedback AlGaAs/
GaAs laser diode in stripe-contact configuration.
The corrugations producing the feedback are shown.
Only the excited region is electrically active. (After
Nakamura, M., Aiki, K., Umeda, J., and Yariv, A.,
Appl. Phys. Lett., 27, 403, 1975.)

Other distributed feedback structures have been described in the literature where the corrugations are not internal to the structure. The theory of these devices is reviewed by Wang.[36]

GAIN COEFFICIENT AND THRESHOLD CONDITION

Gallium arsenide is a semiconductor for which extensive theoretical calculations have been made relating the current density to the gain coefficient as a function of bandstructure, doping level and temperature. These calculations neglect the presence of a strong optical field which perturbs the inversion level for a given injection rate. However, this does not introduce a serious problem in calculating the threshold condition because the field intensity is low. The relationship between the current density and gain (g) derived by Stern[37] is

$$g = \beta_s (J_{nom} - J_1)^b \qquad (11)$$

Here, J_{nom} and J_1 are the current densities per unit volume usually conveniently expressed in A per cm^2-μm where J_1 is the current density just necessary to invert the carrier population and J_{nom} is that necessary to produce useful gain. β_s is a parameter and b is a measure of the linearity of the relationship between gain and current; β_s, b and J_1 are all temperature, material and doping-dependent quantities. To relate J_{nom} in this basic expression to the more useful current density at threshold, J_{th}, we need to determine:

1.   The gain value at threshold g_{th} which must equal the prorated mode loss
2.   The radiation confinement Γ to the recombination region producing the gain
3.   The internal quantum efficiency for spontaneous recombination η_i
4.   The excess current components (Equation 8), neglected here.

The condition for the laser threshold taking into account the absorption coefficient in the three regions of the laser is easily derived. For simplicity, assume that the absorption coefficient in regions 3 and 1 of the double-heterojunction structure of Figure 2.2.10 is equal to α_{out} and that the free carrier absorption coefficient within the recom-

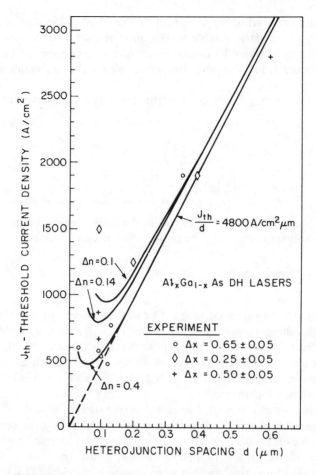

FIGURE 2.2.17.   Threshold current density as a function of double-heterojunction spacing d of $Al_xGa_{1-x}As/GaAs$ lasers for varying values of x and corresponding index steps Δn. The lines are calculated. The experimental data are indicated. (From Kressel, H. and Ettenberg, M., *J. Appl. Phys.*, 47, 3533, 1976. With permission.)

bination region is α_{fc}. Then the gain coefficient at threshold g_{th} is determined from the condition,

$$\Gamma\,(g_{th} - \alpha_{fc}) = \alpha_{out}\,(1-\Gamma) + \text{cavity loss at the mirrors} \qquad (12)$$

For a Fabry-Perot cavity of length L and facet reflectivity R, the cavity end loss = $1/L\ln(1/R)$; R = 0.32 in GaAs. Combining Equation 12 and the definition of J_{nom} we obtain an expression for the threshold current density taking b $\approx$ 1,

$$J_{th} \cong \frac{d}{\eta_i}\left(\frac{g_{th}}{\beta_s} + J_1\right) \qquad (13)$$

For lightly doped GaAs at 300 K, J_1 = 4100 A/(cm²-μm) and β_s = 0.044 (cm-μm)/A; at 80 K, J_1 = 600 A/(cm²-μm) and β_s = 0.16 (cm-μm)/ A.

It is evident that decreasing d reduces J_{th} until the eventual steep Γ decrease; for any heterojunction structure there is, therefore, a minimum achievable J_{th} which is determined by the magnitude of the index steps at the heterojunctions. Figure 2.2.17 shows experimental data comparing the variation of the threshold current density at room temperature for a symmetric double-heterojunction $GaAs/Al_xGa_{1-x}As$ laser where d and the index step were varied by changing the Al concentration x.[38] Except where the value of d is reduced below the optimum value, we see that J_{th} varies with d linearly, $J_{th}/d \cong 4800$ A/(cm²-μm).

Other important laser parameters are the differential quantum efficiency above threshold η_{ext},

$$\eta_{ext} = \eta_i' \; \frac{\frac{1}{L}\ln\left(\frac{1}{R}\right)}{\frac{1}{L}\ln\left(\frac{1}{R}\right) + \Gamma\left[\alpha_{fc} + \frac{(1-\Gamma)}{\Gamma}\,\alpha_{out}\right]} \tag{14}$$

where η_i is the internal quantum efficiency in the lasing regime (which may be larger than η_i in the spontaneous regime). The power conversion efficiency, η_p, is given by

$$\eta_p \cong \frac{P_\theta}{I^2 R_s + IE_g/e} \tag{15}$$

where P_θ is the emitter power at current I. The power conversion efficiency is evidently very low near threshold (a few percent) but increases rapidly with drive to reach a maximum at a drive current two to three times the threshold current, if thermal effects or facet damage are not limiting. Power conversion efficiency values up to $\sim$35% and differential quantum efficiency values of $\sim$85% have been obtained in the cw mode of operation at room temperature.[39]

The temperature dependence of threshold current density is derived from two factors: first, the increased carrier population needed to maintain the required inverted population in the bands with increasing temperature, and second, the leakage of carriers out of the active region in heterojunction structures. Other factors include changes in internal absorption and internal quantum efficiency.[40] Although each of these parameters has a complex variation, a simple exponential relationship[41] has been found useful over a moderate temperature range (−60°C to 120°C). The expression is of the form

$$J_{th} = J_{th}(O)\exp(\Delta T/T_o) \tag{16}$$

where $J_{th}(O)$ is the threshold current density at a reference temperature, ΔT is the temperature difference between the laser and the reference value and T_o is a parameter characteristic of the laser.

For AlGaAs double-heterojunction lasers around room temperature, the following expression[42] can be used to estimate T_o,

$$T_o \cong 300\,\Delta E_g \tag{17}$$

where T_o is in degrees Kelvin and ΔE_g is the bandgap step at the heterojunction in electron volts. With increasing duty cycle, the threshold increases due to Joule heating. Hence, the threshold current is

$$I_{th} \cong I_{th}(O)\exp\left\{(I_{th}E_g/e + I_{th}^2 R_s)\,D\phi/T_o\right\} \tag{18}$$

where $I_{th}(O)$ is the low duty cycle pulsed threshold, R_s is the diode electrical resistance, D is the duty cycle (1 for cw), and ϕ is the diode thermal resistance.

SPECTRAL EMISSION

The wavelength of the emission depends on the bandgap energy, the dopant concentration in the recombination region and the junction temperature. For commonly used DH lasers with lightly doped recombination regions, the lasing peak energy is 20 to 30 meV below the bandgap energy. The spectrum shifts toward lower energy with increasing temperature at a rate of ~0.5 meV/K (~2.5 Å/K at ~8500 Å).

The spectral width of the laser emission depends on the number of longitudinal modes excited. The evolution of the spectrum with increasing current from a representative cw laser diode is shown in Figure 2.2.18. Near threshold the emission is relatively wide, with many longitudinal modes. With increasing drive a few of the modes become dominant, and in some lasers a single mode dominates. Commonly, however, one finds three to four lines with a total spectral width of 5 to 10 Å within the half-intensity points. Lasers with very short cavities (under 100 μm) have greater longitudinal mode separations, and oscillate in a single longitudinal mode more frequently than longer lasers. The cw power emitted in a single mode in such short lasers sometimes exceeds 10 mW. Some special structures also exhibit good spectral purity (Figure 2.2.15). However, if the drive current of such a laser is widely varied, the output wavelength will shift, and the overall linewidth will broaden. Therefore, maintaining spectral purity in pulsed laser operation is generally difficult unless the current range is restricted.

MATERIALS

Semiconductor laser action has been obtained in many direct bandgap materials with external pumping (see Table 2.2.1) but laser diodes can only be produced if p- and n-type doping can be achieved in the same structure. In principle, it should be possible to combine semiconductors with different bandgaps to produce heterojunction lasers as long as the active region has a direct gap. In practice, the electrical resistance of the device is important, hence the use of materials with very low doping levels or low carrier mobilities leads to experimental difficulties. Furthermore, there are stringent requirements on the crystalline quality of the materials if useful devices are to be produced.

As discussed earlier, heterojunction lasers require confining heterojunction barriers with bandgap steps of several kT. It is essential, however, that the lattice parameter at the heterojunctions be matched as closely as possible in order to avoid misfit dislocation formation. The tolerable lattice mismatch for reasonable device performance is not generally established but values under 0.1% are desirable.

A simplified model provides insight into the impact of nonradiative interfacial recombination on the internal quantum efficiency. For thin double-heterojunction structures, the internal quantum efficiency for spontaneous emission is approximately,[43]

$$\eta_i \cong (1 + \tau_r/\tau_{nr} + 2S\tau_r/d)^{-1} \qquad (19)$$

where τ_r is the radiative carrier lifetime, τ_{nr} is the nonradiative carrier lifetime in the absence of interfaces, S is the interfacial recombination velocity (assumed equal at the heterojunctions), and the condition SL/D ≪ 1 is satisfied (L is the minority carrier diffusion length in the absence of interfaces and D is the minority carrier diffusion constant).

Consider a p-type recombination region; τ_r is typically 3 nsec near threshold, and with d = 0.2 μm we need S ≤ 3.3 × 10³ cm/sec to ensure an internal quantum efficiency of at least 50%. The impact on the threshold current density has been analyzed.[44]

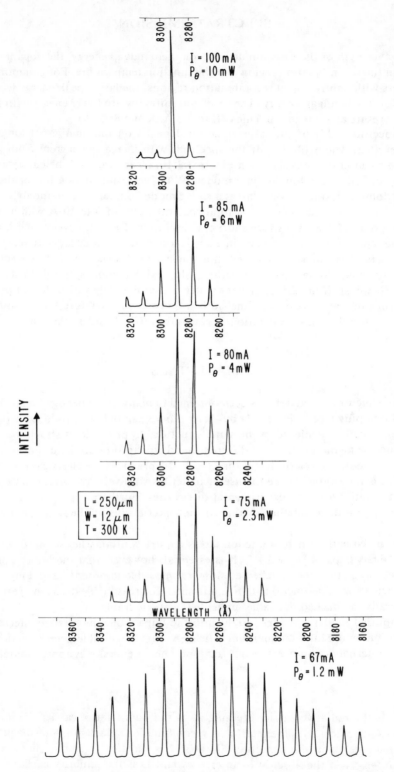

FIGURE 2.2.18.   Spectra from AlGaAs cw laser operating at room temperature at increasing current showing a relatively pure emission at high power levels.

Table 2.2.1
SEMICONDUCTORS USED FOR LASERS[a]

Material	Wavelength (μm)	References in which various types of excitations were studied			
		Injection	Optical	Electron beam	Breakdown
GaN	0.36		1		
GaAs	0.83—0.92	2—4	5	6—8	9, 10
InP	0.89—0.91	11, 12	13, 36		14
GaSb	1.50—1.60	15, 16	17	18	
InAs	3.00—3.20	19, 81	17—20	21	
InSb	4.80—5.30	22	23	24	
InGaP	0.59—0.90	25	26	41a	
AlGaAs	0.62—0.90	27—30, 79, 80		31	
GaAsP	0.63—0.90	32, 33		34	
InGaAs	0.85—3.20	35	36		
InAsP	0.90—3.20	37			
GaAsSb	0.95—1.60	38			
InAsSb	1.00—5.30	39			
AlGaSb	1.10—1.60			40	
InGaPAs	0.62—3.20	41		41a	
AlGaAsSb	0.62—1.60	41			
ZnO	0.37		41b	42	
ZnS	0.33		43	44, 45	
ZnSe	0.46			46	
ZnTe	0.53			47, 48	
CdS	0.49		49, 50	51, 52	
CdSe	0.69		53, 54	55	
CdTe	0.79			56, 57	
ZnCdS	0.33—0.49		84		
CdSSe	0.59—0.69		41b, 58	59	
CdHgTe	3.80—4.10		60		
PbS	4.30	61, 82		62	
PbSe	8.50	63		62	
PbTe	6.50	64		62	
PbSSe	4.70—5.50	65	66	67	
PbSnSe	8.50—32.0	68		69	
PbSnTe	6.50—32.0	68		69	
PbGeTe	4.40—6.50	70			
PbCdS	3.50	83			
Te	3.70			71	
GaSe	0.59—0.60		72	73	
InSe	0.97			74	
In_2Se	1.60			75	
Cd_3P_2	2.10		76		
$CdSnP_2$	1.10			77	
$CdSiAs_2$	0.77			78	

[a] Heterojunction structures are specifically described in Table 2.2.2 although some are included in this table as well. Much of the information in this table was collected by Pankove, J. I., and appeared in *Handbook of Lasers*, Pressley, R. J., Ed., CRC Press, Cleveland, 1971, chap. 12.

REFERENCES (Table 2.2.1)

1. Dingle, R., Shaklee, K. L., Leheny, R. F., and Zetterstrom, R. B., Stimulated emission and laser action in gallium nitride, *Appl. Phys. Lett.*, 19, 5, 1971.

Table 2.2.1 (continued)

2. Hall, R. N., Fenner, G. E., Kingsley, J. D., Soltys, T. J., and Carlson, R. O., Coherent light emission from GaAs junctions, *Phys. Rev. Lett.*, 9, 366, 1962.
3. Nathan, M. E., Dumke, W. P., Burns, C., Dill, Jr., F. H., and Lasher, G. J., Stimulated emission of radiation from GaAs p-n junction, *Appl. Phys. Lett.*, 1, 62, 1962.
4. Quist, T. M., Rediker, R. H., and Keyes, R., Semiconductor maser of GaAs, *Appl. Phys. Lett.*, 1, 91, 1962.
5. Basov, N. G., Grasyuk, A. Z., and Katulin, V. A., Induced radiation in optically excited gallium arsenide, *Sov. Phys. Dokl.*, 10, 343, 1965.
6. Hurwitz, C. E. and Keyes, R. J., Electron beam pumped GaAs laser, *Appl. Phys. Lett.*, 5, 139, 1964.
7. Kurbatov, L. N., Kabanov, A. N., and Sigriyanskii, B. B., Generation of coherent radiation in gallium arsenide by electron excitation, *Sov. Phys. Dokl.*, 10, 1059, 1966.
8. Cusano, D. A., Radiative recombination from GaAs directly excited by electron beams, *Solid State Commun.*, 2, 353, 1964.
9. Weiser, K. and Woods, J. F., Evidence for avalanche injection laser in p-type GaAs, *Appl. Phys. Lett.*, 7, 225, 1965.
10. Southgate, P. D., Stimulated emission from bulk field-ionized GaAs, *IEEE J. Quantum Electron.*, QE-4, 179, 1968.
11. Weiser, K. and Levitt, R. S., Stimulated light emission from indium phosphide, *Appl. Phys. Lett.*, 2, 178, 1963.
12. Basov, N. G., Eliseev, P. G., and Ismailov, I., Some properties of semiconductor lasers based on indium phosphide, *Sov. Phys. Solid State*, 8, 2087, 1967.
13. Eliseev, P. G., Ismailoz, I., and Mikhailina, L. I., Coherent emission of InP optically excited by an injection laser, *JETP Lett.*, 6, 15, 1967.
14. Southgate, F. D. and Mazzochi, R. T., Stimulated emission in field-ionized bulk InP , *Phys. Lett.*, 2BA, 216, 1968.
15. Chipaux, C. and Eymard, R., Study of the laser effect in GaSb alloys, *Phys. Stat. Sol.*, 10, 165, 1965; *Solid State Phys.*, 10, 165, 1965.
16. Kryukova, I. V., Karnaukhov, V. G., and Paduchikh, L. I., Stimulated radiation from diffused p-n junction in gallium antimonide, *Sov. Phys. Sol. Stat.*, 7, 2757, 1966.
17. Benoit-a-la Guillaume, C. and Laurant, J. M., Laser effect in InAs and GaSb by optical excitation, *Compt. Rend.*, 262, 275, 1966.
18. Benoit-a-la Guillaume, C. and Debever, J. M., Laser effect in gallium antimonide by electron bombardment, *Compt. Rend.*, 259, 2200, 1964.
19. Melngailis, I., Maser action in InAs diodes, *Appl. Phys. Lett.*, 2, 176, 1963.
20. Melngailis, I., Optically pumped InAs laser, *IEEE J. Quantum Electron.*, QE-1, 104, 1965.
21. Kryukova, I. V., Leskovich, V. I., and Matveenko, E. V., Mechanisms of laser action in epitaxial InAs subjected to electron beam excitation, *Sov. J. Quantum Electron.*, 9, 823, 1979.
22. Melngailis, I., Phelan, R. J., and Rediker, R. H., Luminescence and coherent emission in large-volume injection plasma in InSb, *Appl. Phys. Lett.*, 5, 99, 1964.
23. Phelan, R. J., and Rediker, R. H., Optically pumped semiconductor lasers, *Appl. Phys. Lett.*, 6, 70, 1965.
24. Benoit-a-la Guillaume, C. and Debever, J. M., Laser effect excited by an electron beam in *Radiative Recombination in Semiconductors*, Dunod, Paris, 1965, 255.
25. Macksey, H. M., Holonyak, Jr., N., Scifres, D. R., Dupuis, R. D., and Zack, G. W., In$_{1-x}$Ga$_x$P p-n junction lasers, *Appl. Phys. Lett.*, 19, 271, 1971.
26. Burnham, R. D., Holonyak, Jr., N., and Keune, D. L., Stimulated emission in In$_{1-x}$Ga$_x$P, *Appl. Phys. Lett.*, 17, 430, 1970.
27. Rupprecht, H., Woodall, J. M., and Pettit, G. D., Stimulated emission from Ga$_{1-x}$Al$_x$As diodes at 70 K, *IEEE J. Quantum Electron.*, QE-4, 35, 1968.
28. Susaki, W., Sogo, T., and Oku, T., Lasing action in (Ga$_{1-x}$Al$_x$)As diodes, *IEEE J. Quantum Electron.*, QE-4, 122, 1968.
29. Alferov, Zh. I., Andreev, V. M., Korolkov, V. I., Portnoi, E. L., and Tretyakov, D. N., Injection properties of n-Al$_x$Ga$_{1-x}$As-p-GaAs heterojunctions, *Sov. Phys. Semiconductors*, 2, 843, 1969.
30. Kressel, H. and Hawrylo, F. Z., Stimulated emission at 300°K and simultaneous lasing at two wavelength in epitaxial Al$_x$Ga$_{1-x}$ As injection lasers, *Proc. IEEE*, 56, 1598, 1968.
31. Dolginov, L. M., Druzhinina, L. V., and Kryukova, I. V., Parameters of electron-beam-pumped Al$_x$Ga$_{1-x}$ As lasers in the visible part of the spectrum, *Sov. J. Quantum Electron.*, 4, 104, 1975.
32. Holonyak, Jr., N. and Bevacqua, S. F., Coherent (visible) light emission from Ga(As$_{1-x}$P$_x$) junctions, *Appl. Phys. Lett.*, 1, 82, 1962.

Table 2.2.1 (continued)

33. **Tietjen, J. J., Pankove, J. I., Hegyi, I. J., and Nelson, H.,** Vapor-phase growth of $GaAs_{1-x}P_x$ room-temperature injection lasers, *Trans. AIME,* 239, 385, 1967.

34. **Basov, N. G., Bogdankevich, O. V., Eliseev, P. G., and Lavrushin, B. M.,** Electron beam excited lasers made from solid solutions of GaP_xAs_{1-x}, *Sov. Phys. Sol. Stat.,* 8, 1073, 1966.

35. **Melngailis, I., Strauss, A. J., and Rediker, R. H.,** Semiconductor diode masers of $In_xGa_{1-x}As$, *Proc. IEEE,* 51, 1154, 1963.

36. **Rossi, J. A. and Chinn, S. R.,** Efficient optically pumped InP and $In_xGa_{1-x}As$ lasers, *J. Appl. Phys.,* 43, 4806, 1972.

37. **Alexander, F. B., Bird, V. R., and Carpenter, D. B.,** Spontaneous and stimulated infrared emission from indium phosphide-arsenide diodes, *Appl. Phys. Lett.,* 4, 13, 1964.

38. **Sugiyama, K. and Saito, H.,** GaAsSb/AlGaAsSb double heterojunction lasers, *Jpn. J. Appl. Phys.,* 11, 1057, 1972.

39. **Basov, N. G., Dudenkova, A. V., and Krasil'nikov, A. I.,** Semiconductor p-n junction lasers in the $InAs_{1-x}Sb_x$ system, *Sov. Phys. Sol. Stat.,* 8, 847, 1966.

40. **Akimov, Yu. A., Burov, A. A., and Zagarinskii, E. A.,** Electron-beam-pumped $Al_xGa_{1-x}Sb$ semiconductor laser, *Sov. J. Quantum Electron.,* 5, 37, 1975.

41. **Bogatov, A. P., Dolginov, P. M., and Druzhinina, L. V.,** Heterojunction lasers made of $Ga_xIn_{1-x}As_yP_{1-y}$ and $Al_xGa_{1-x}Sb_yAs_{1-y}$ solid solutions, *Sov. J. Quantum Electron.,* 4, 1281, 1975.

41a. **Ermanov, O. N., Garba, L. S., Golvanov, Y. A., Sushov, V. P., and Chukichev, M. V.,** Yellow-green InGaP and InGaPAs LEDs and electron-beam-pumped lasers prepared by LPE and VPE, *IEEE Trans. Elect. Devices,* ED-26, 1190, 1979.

41b. **Johnston, Jr., W. D.,** Characteristics of optically pumped platelet lasers of ZnO, CdS, CdSe and $CdS_{0.6}Se_{0.4}$ between 300° and 80°K, *J. Appl. Phys.,* 42, 2731, 1971.

42. **Nicoll, F. H.,** Ultraviolet ZnO laser pumped by an electron beam, *Appl. Phys. Lett.,* 9, 13, 1966.

43. **Wang, S. and Chang, C. C.,** Coherent fluorescence from zinc sulphide excited by two-photon absorption, *Appl. Phys. Lett.,* 12, 193, 1968.

44. **Hurwitz, C. E.,** Efficient ultraviolet laser emission in electron-beam-excited ZnS, *Appl. Phys. Lett.,* 9, 116, 1966.

45. **Bogdankevich, O. V., Zverev M. M., Pechenov, A. N., and Sysoev, L. A.,** Recombination radiation of ZnS single crystals excited by a beam of fast electrons, *Sov. Phys. Sol. Stat.,* 8, 2039, 1967.

46. **Bogdankevich, O. V., Zverev, M. M., Krasilnikov, A. I., and Pechenov, A. N.,** Laser emission in electron-beam excited ZnSe, *Phys. St. Sol.,* 19, K5-6, 1967.

47. **Hurwitz, C. E.,** Laser emission from electron beam excited ZnTe, *IEEE J. Quantum Electron.,* QE-3, 333, 1967.

48. **Vlasov, A. N., Kozina, G. S., and Fedorova, O. B.** Stimulated emission from zinc telluride single crystals excited by fast electrons, *Sov. Phys. JETP,* 25, 283, 1967.

49. **Konyukhov, V. K., Kulevskii, L. A., and Prokhorov, A. M.,** Optical oscillation in CdS under the action of two-photon excitation by a ruby laser, *Sov. Phys. Dokl.,* 10, 943, 1966.

50. **Basov, N. G., Grasyuk, A. Z., Zubarev, I. G., and Katulin, V. A.,** Laser action in CdS induced by two-photon optical excitation from a ruby laser, *Sov. Phys. Sol. Stat.,* 7, 2932, 1966.

51. **Basov, N. G., Bogdankevich, O. V., and Devyatkov, A. G.,** Exciting a semiconductor quantum generator (laser) with a fast electron beam, *Sov. Phys. Dokl.,* 9, 288, 1964.

52. **Kurbatov, L. N., Mashchenko, V. E., and Mochalkin, N. N.,** Coherent radiation from cadmium sulfide single crystals excited by an electron beam, *Opt. Spectrosc.,* 22, 232, 1967.

53. **Grasyuk, A. Z., Efimkov, V. F., and Zubarev, I. G.,** Semiconductor CdSe laser with two-photon optical excitation, *Sov. Phys. Sol. Stat.,* 8, 1548, 1966.

54. **Holonyak, Jr., N., Sirkis, H. D., Stillman, G. E., and Johnson, M. R.,** Laser operation of CdSe pumped with a Ga(AsP) laser diode, *Proc. IEEE,* 54, 1068, 1966.

55. **Hurwitz, C. E.,** Electron beam pumped lasers of CdSe and CdS, *Appl. Phys. Lett.,* 8, 121, 1966.

56. **Vavilov, V. S. and Nolle, Z. L.,** Cadmium telluride laser with electron excitation, *Sov. Phys. Dokl.,* 10, 827, 1966.

57. **Vavilov, V. S., Molle, Z. L., and Egorov, V. D.,** New data on the electron-excited recombination radiation spectrum of cadmium telluride, *Sov. Phys. Sol. Stat.,* 7, 749, 1965.

58. **Brodin, M. S., Vitrikhovskii, N. I., Zakrevskii, S. V., and Reznichenko, V. Ya.,** Generation in mixed CdS_x-CdSe_{1-x} crystals excited with ruby laser radiation, *Sov. Phys. Sol. Stat.,* 8, 2461, 1967.

59. **Hurwitz, C. E.,** Efficient visible lasers of CdS_xSe_{1-x} by electron-beam excitation, *Appl. Phys. Lett.,* 8, 243, 1966.

60. **Melngailis, I. and Strauss, A. J.,** Spontaneous and coherent photoluminescence in $Cd_xHg_{1-x}Te$, *Appl. Phys. Lett.,* 8, 179, 1966.

61. **Butler, J. F. and Calawa, A. R.,** PbS diode laser, *J. Electrochem. Soc.,* 112, 1056, 1965.

Table 2.2.1 (continued)

62. **Hurwitz, C. E., Calawa, A. R., and Rediker, R. H.,** Electron beam pumped laser of PbS, PbSe and PbTe, *IEEE J. Quantum Electron.,* QE-1, 102, 1965.
63. **Butler, J. F., Calawa, A. R., Phelan, R. J., Strauss, A. J., and Rediker, R. H.,** PbSe diode laser, *Sol. State Commun.,* 2, 303, 1964.
64. **Butler, J. F., Calawa, R. A., Phelan, R. J., Jarman, T. C., and Strauss, A. J., Rediker, R. H.,** PbTe diode laser, *Appl. Phys. Lett.,* 5, 75, 1964.
65. **McLane, G. F. and Sleger, K. J.,** Vacuum deposited epitaxial layers of $PbS_{1-x}Se_x$ for laser devices, *J. Electron. Mat.,* 4, 465, 1975.
66. **Mooradian, A., Strauss, A. J., and Rossi, J. A.,** Broad band laser emission from optically pumped $PbS_{1-x}Se_x$, *IEEE J. Quantum Electron.,* QE-9, 347, 1973.
67. **Kurbatov, L. N., Mashchenko, V. E., and Mochalkin, N. N.,** Coherent radiation recombination of some semiconductors under high excitation, *Proc. 9th Int. Conf. Phys. Semiconductor,* Nauka, 621, 1969.
68. **Butler, J. F., Calawa, A. R., and Harman, T. C.,** Diode lasers of $Pb_{1-y}Sn_ySe$ and $Pb_{1-x}Sn_xTe$, *Appl. Phys. Lett.,* 9, 427, 1966.
69. **Kurbatov, L. N., Britov, A. D., and Dirochka, A. I.,** Stimulated radiation from solid solutions of chalcogenide of lead and tin in the range of 10 μm, *Quantum Electron.,* Basov, N. G., Ed., *Soviet Radio,* 97, 1972.
70. **Anticliffe, G. A., Parker, S. G., and Bate, R. T.,** CW operation and nitric oxide spectroscopy using diode laser of $Pb_{1-x}Ge_xTe$, *App Phys. Lett.,* 21, 505, 1972.
71. **Benoit-a-la Guillaume, C. and Debever, J. M.,** Spontaneous and stimulated emission in tellurium by electron bombardment, *Sol. State Commun.,* 3, 19, 1965.
72. **Abdullaev, G. B., Aliev, M. Kh., and Mirzoev, B. R.,** Laser emission by GaSe under two-photon optical excitation conditions, *Sov. Phys. Semiconductor,* 4, 1189, 1971.
73. **Basov, N. G., Bogdankevich, O. V., and Abdullaev, A. N.,** Radiation in GaSe single crystals induced by excitation with fast electrons, *Sov. Phys. Dokl.,* 10, 329, 1965.
74. **Kurbatov, L. N., Dirochka, A. I., and Britov, A. D.,** Stimulated emission of indium monoselenide subjected by electron bombardment, *Sov. Phys. Semiconductor,* 5, 494, 1971.
75. **Kurbatov, D. N., Dirochka, A. I., and Ogorodnik, A. D.,** Recombination radiation of In_2Se, *Sov. Phys. Semiconductor,* 4, 1195, 1971.
76. **Bishop, S. G., Moore, W. J., and Swiggard, E. M.,** Optically pumped Cd_3P_2 laser, *Appl. Phys. Lett.,* 16, 459, 1970.
77. **Berkovskii, F. M., Goryunova, N. A., and Ordov, V. M.,** $CdSnP_2$ laser excited with an electron beam, *Sov. Phys. Semiconductor,* 2, 1027, 1969.
78. **Averkieva, G. K., Goryunova, N. A., and Prochukhan, V. D.,** Stimulated recombination radiation emitted by $CdSiAs_2$, *Sov. Phys. Semiconductor,* 5, 151, 1971.
79. **Kressel, H. and Hawrylo, F. Z.,** Red light-emitting Al_xGa_{1-x} As heterojunction laser diodes, *J. Appl. Phys.,* 44, 4222, 1973.
80. **Kressel, H., Lockwood, H. F., and Nelson, H.,** Low-threshold Al_xGa_{1-x} As visible and IR-light emitting diode laser, *IEEE J. Quantum Electron.,* QE-6, 278, 1970.
81. **Patel, N. and Yariv, A.,** Electrical and optical characteristics of InAs junction lasers, *IEEE J. Quantum Electron.,* QE-6, 383, 1970.
82. **Ralston, R. W., Walpole, J. M., and Calawa, A. R.,** High CW output power in stripe-geometry PbS diode lasers, *J. Appl. Phys.,* 45, 1323, 1974.
83. **Nill, K. W. Strauss, A. J., and Blum, F. A.,** Tunable cw $Pb_{0.98}Cd_{0.02}S$ diode laser emitting at 3.5 μm, *Appl. Phys. Lett.,* 22, 677, 1973.
84. **Brodin, M. S., Budnick, P. I., Vitrikhovskii, B. L., and Zakrevskii, V. S.,** *Proc. Int. Conf. Phys. Semiconductor,* Moscow, 1968, 610.

The recombination velocity is related to defects at the heterojunction interface. A simple analysis, which yields satisfactory agreement with experiment, starts with an estimate of the density of centers introduced by misfit dislocations generated to relieve the misfit strain. Studies of heterojunction interfaces of InGaP/GaAs[45] show that the interfacial recombination velocity increases with the lattice misfit strain $\Delta a_o/a_o$ (in the range 10^{-4} to 10^{-2}) following an approximate relationship,

$$S = (3.8 \pm 1.2) \times 10^7 \ (\Delta a_o/a_o) \quad \text{cm/sec} \qquad (20)$$

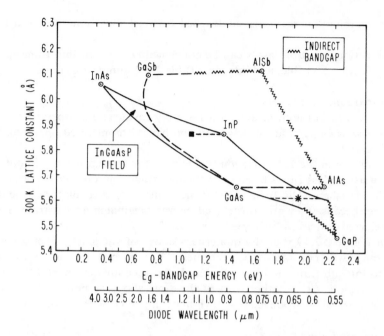

FIGURE 2.2.19.   Lattice constant for various III-V compounds and alloys related to the bandgap energy and the corresponding diode emission wavelength. The dotted line illustrates heterojunction lattice matching possibilities using InGaAsP on an InP substrate as well as a GaAsP substrate layer (usually deposited on GaAs). (From Kressel, H. and Butler, J. K., *Semiconductor Lasers and Heterojunction LEDs*, Academic Press, New York, 1977. With permission.)

However, even in heterojunction structures where the lattice misfit is very small (below 10^{-4}) there is still a substantial amount of interfacial recombination owing to the presence of nonradiative recombinatio.i centers of uncertain origin. In AlGaAs/GaAs, for example, S is between 10^2 and 10^4 cm/sec.

Heterojunction lasers require a direct bandgap material in the recombination region but not in the bounding layers. Figure 2.2.19 shows the lattice constant, bandgap energy, and corresponding diode emission wavelength of several III-V materials that can be doped both n- and p-type. The AlGaAs alloys are particularly useful because the small (about 0.1%) lattice constant change between AlAs and GaAs makes it possible to grow very nearly lattice-matched structures of AlGaAs on GaAs substrates.

Heterojunction structures can be epitaxially deposited on lattice-mismatched binary substrates such as InP, GaAs, GaP, InAs or InSb. However, to minimize the density of misfit dislocations in the epitaxial structures containing the active region of the device, it is essential to grade the composition between the substrate and the active region.

It is also noteworthy that mismatched layers can be grown without misfit dislocations if the layers are kept sufficiently thin because the strain energy in the layer is less than the energy needed to nucleate the strain-relieving misfit dislocations. The maximum thickness h_c of a dislocation-free layer in a III-V compound structure grown on the usual growth planes is approximately [46] $h_c \cong a^2_o/(2\sqrt{2}\,\Delta a_o)$.

Thus, very thin layers (usually well under one micrometer) can be produced without misfit dislocations even if the lattice constant differs between the layers. For example, if the lattice misfit is $\Delta a_o/a_o = 5 \times 10^{-3}$ and $a_o = 5.6$ Å, the calculated $h_c \cong 0.4$ μm. This effect is of great benefit in the fabrication of double-heterojunction lasers where the thin recombination region is sandwiched between slightly mismatched bounding layers.

Heterojunction injection lasers can be conveniently grouped into three spectral regions: near-infrared, visible, and infrared. Table 2.2.2 summarizes representative data.

Near-Infrared Emission Lasers

These devices cover the spectral region 0.8 to 0.9 μm at room temperature. Structures of $Al_xGa_{1-x}As/Al_yGa_{1-y}As$ are widely used and have demonstrated operating lifetimes of several years.[1,47]

Lasers emitting in the 1.2 to 1.5 μm region are of particular interest in optical communication using glass fibers because of the low attenuation in that spectral region in state-of-the-art fibers. The InGaAsP/InP heterojunction structures are the most important ones because they are constructed on a lattice-matching substrate. Thus, they are the preferred devices for this spectral region.[48,51]

Note in Figure 2.2.19 that the quaternary alloys of InGaAsP cover a wide direct bandgap range with constant lattice parameter. For example, this alloy can be lattice-matched to InP substrates and thus produce heterojunction devices emitting in the one micrometer region. At the other limit of the same alloy, we note that visible heterojunction devices can be produced by lattice-matching to GaAsP grown on GaAs.

Visible Emission Lasers

At room temperature, the laser diode emission extends into the red portion of the spectrum. Aluminum gallium arsenide lasers can operate in the pulsed mode to about 0.69 μm at room temperature and to 0.72 μm in the cw mode of operation.[52,53]

The use of materials with a more favorable band structure for visible emission permits an extension of the emission further into the red. These include InGaAsP/InGaP and GaAsP/InGaP structures which can be lattice-matched at the heterojunction. However, there is a mismatch to the GaAs substrate which results in a high defect density in the device itself.

Infrared Emission

Heterojunction lasers using IV-VI compounds can provide emission from 2.5 to about 30 μm, although a substantial lattice mismatch is present in the structure. These heterojunction lasers have operated cw at higher temperatures than previously possible with homojunction lasers because of the reduced threshold current density. For example, stripe-contact PbSnTe double-heterojunction lasers have operated cw at heat sink temperatures as high as 114 K. This device can be temperature tuned between 15.9 and 8.54 μm.[54]

MODULATION

A major advantage of laser diodes compared to other laser types is the ability to conveniently modulate their output at rates exceeding one GHz.[1] This is possible because the lifetime for stimulated carrier recombination is reduced compared to that for spontaneous recombination.

Although lasers can be modulated at GHz rates, they must remain current-biased to threshold in order to avoid a turn-on delay time t_d related to the spontaneous carrier lifetime τ,

$$t_d = \tau \ln [I/(I - I_{th})] \qquad (21a)$$

Table 2.2.2
HETEROJUNCTION LASER DIODES USING ALLOYS OTHER THAN AlAs-GaAs

Materials	Laser type	J_{th} (A/cm²)	d_3^a (μm)	λ(μm)	Operating temperature (K)	Growth technique	Ref.
GaAsP-GaAs	SH	1.4×10^5 8×10^4	1 to ~10	~0.9	300	VPE	1
GaAsSb-AlGaAsSb	DH	9×10^3	0.8	0.98	300	LPE	2
InGa$_{0.5}$P$_{0.5}$-GaAs	LOC	8×10^3	2.5	~0.89	300	VPE	3
InGaP-InGaAs	LOC	1.49×10^4	1.6	1.075	300	VPE	4
InGaAsP-GaAsP	SH	6.2×10^4	1	0.63	77	LPE + VPE	5
AlGaAsP-GaAsP	SH	6.6×10^4	1.24	~0.81	300	LPE	6
AlGaAsP-GaAsP	DH	1.6×10^4	1.25	0.845	300	LPE + VPE	7
PbSnTe-PbTe	SH	780	—	~8.9	77	Evaporation	8
PbSnTe-PbTe	DH	4.2×10^3	9	~8.35	77	LPE	9
PbSnTe-PbTe	DH	1.3×10^3	5	~8.4	77	LPE	10
InGa$_{0.5}$P$_{0.5}$-Al$_{0.69}$Ga$_{0.32}$As	SH	2.5×10^3	1	0.638	77	LPE	11
PbS$_{0.72}$Se$_{0.28}$ Pb$_{0.78}$Se$_{0.22}$	SH	~200	~3	4.78	12	Evaporation	12
In$_{0.88}$Ga$_{0.12}$As$_{0.23}$P$_{0.72}$-InP	DH	~2.8×10^3	0.6	1.1	300	LPE	13
GaAsP-InGaP	LOC	5.5×10^4	~4	0.675	273	VPE	14
	DH	3.37×10^3	0.2	0.703	300	VPE	15

^{a} Distance between the two heterojunctions: d_3 for DH structure, d for LOC structure.

REFERENCES (Table 2.2.2)

1. Craford, M. G., Groves, W. O., and Fox, M. J., GaAs-GaAsP heterostructure injection lasers, *J. Electrochem. Soc.*, 118, 355, 1971. The abrupt heterojunction barrier was formed using GaAs$_{0.9}$P$_{0.1}$ GaAs ($\Delta E_g \approx 0.1$ eV). Graded heterojunctions with 4%P/μm were also studied with similar device results.

2. Sugiyama, K. and Saito, H., GaAsSb-AlGaAsSb GaAsSb-AlGaAsSb double heterojunction lasers, *Jpn. J. Appl. Phys.*, 11, 1057, 1972.

3. Nuese, C. J., Ettenberg, M., and Olsen, G. H., Room-temperature heterojunction laser diodes from vapor-grown In$_{1-x}$Ga$_x$P/GaAs structures, *Appl. Phys. Lett.*, 25, 612, 1974.

4. Nuese, C. J. and Olsen, G. H., Room-temperature heterojunction laser diodes of In$_x$Ga$_{1-x}$AS/In$_y$Ga$_{1-y}$P with emission wavelength between 0.9 and 1.15 μm. *Appl. Phys. Lett.*, 26, 528, 1975. The heterojunction barrier consisted of the near-lattice matching compositions In$_{0.68}$Ga$_{0.32}$P-In$_{0.16}$Ga$_{0.84}$As. The highest $\Delta a_o/a_o$, value used was 0.2%.

5. Coleman, J. J., Hitchens, W. R., Holonyak, Jr., N, Ludowise, M. J., Groves, W. O., and Keune, D. L., Liquid phase epitaxial In$_{1-x}$Ga$_x$P$_{1-z}$As$_z$/GaAs$_{1-y}$P$_y$ quaternary (LPE)-ternary (VPE) heterojunction lasers (x~0.70, z~0.01, y~0.40; λ<6300 Å, 77°K), *Appl. Phys. Lett.*, 25, 725, 1974.

6. Burnham, R. D., Holonyak, Jr., N., and Scifres, D. R., Al$_x$Ga$_{1-x}$As$_{1-y}$P$_y$-GaAs$_{1-y}$P$_y$ heterostructure laser and lamp junctions, *Appl. Phys. Lett.*, 17, 455, 1970.

7. Burnham, R. D., Holonyak, Jr., N., Korb, H. W., Macksey, H. M., Scifres, D. R., and Woodhouse, J. B., Double heterojunction AlGaAsP quaternary lasers, *Appl. Phys. Lett.*, 19, 25, 1971.

8. Walpole, J. N., Calawa, A. R., Ralston, R. W., Harman, T. C., and McVittie, J. P., Single heterojunction Pb$_{1-x}$Sn$_x$Te diode lasers,, *Appl. Phys. Lett.*, 23, 620, 1973. The structure used consisted of Pb$_{0.88}$Sn$_{0.12}$Te-PbTe($\Delta a_o/a_o \cong 0.24$%). The J_{th} values obtained were reported to be ~⅓ of those measured for homojunction lasers.

9. Groves, S. H., Nill, K. W., and Strauss, A. J., Double heterostructure Pb$_{1-x}$Sn$_x$Te-PbTe lasers with cw operation at 77 K, *Appl. Phys. Lett.*, 25, 331, 1974.

10. Tomasetta, L. R. and Fonstad, C., Threshold reduction in Pb$_{1-x}$Sn$_x$Te laser diodes through the use of double heterojunction geometries, *Appl. Phys. Lett.*, 25, 440, 1974; see also 24, 567, 1974. A high dislocation density ($10^8 - 10^9$ cm^{-2}) in the substrate was noted.

11. Schul, G. and Mischel, P., Ga$_x$In$_{1-x}$P-Ga$_y$Al$_{1-y}$ As heterojunction close-confinement injection laser, *Appl. Phys. Lett.*, 26, 394, 1975.

Table 2.2.2 (continued)

12. Sleger, K. G., McLane, G. F., Strom, U., Bishop, S. G., and Mitchell, D., Single-heterostructure PbS$_{1-x}$Se, diode lasers, *J. Appl. Phys.*, 45, 5069, 1974; $\Delta a_o/a_o \cong 0.2\%$ at the n-n heterojunction.
13. Hsieh, J. J., Room-temperature operation of GaInAsP/InP double-heterostructure diode lasers emitting at 1.1 μ, *Appl. Phys. Lett.*, 28, 283, 1976.
14. Ladany, I., Kressel, H., and Nuese, C. J., unpublished.
15. Kressel, H., Olsen, G. H., and Nuese, C. J., Visible GaAs$_{0.7}$P$_{0.3}$ cw heterojunction lasers, *Appl. Phys. Lett.*, 30, 249, 1977.

The ultimate laser modulation rate is limited by the photon lifetime in the Fabry-Perot cavity,

$$\frac{1}{\tau_{ph}} = \frac{c}{n}\left[\tilde{\alpha} + \frac{1}{L}\ln\frac{1}{R}\right] \tag{21b}$$

where c is the velocity of light, n is the refractive index in the cavity at λ_L, R is the facet reflectivity and $\tilde{\alpha}$ is the averaged internal absorption coefficient. With the quantity in the bracket being typically $\sim$50 cm^{-1}, $\tau_{ph} \cong 2 \times 10^{-12}$ sec. Thus, the photon lifetime does not represent a practical limit to the modulation capability of the laser diode.

The analysis of the transient laser properties for semiconductors follows the rate equation approach for the carrier and photon population.[1]

Based on the assumptions that the cavity is fully and uniformly inverted and that only a single mode is excited, this analysis predicts that the injected carrier density in the lasing region is fixed at the threshold value. This means that the gain is assumed fixed at its threshold value because the gain coefficient is a function of the carrier density.

Fluctuations in the carrier and photon population about the steady state value will be dampened in a time dependent on the spontaneous carrier lifetime, the photon lifetime, and the injection level. These interactions give rise to fluctuations in the laser output, which we now discuss.

Suppose that we turn the laser diode on with a step current I. After an initial transient oscillatory effect with a time constant a_t, the photon density in the cavity and the carrier pair density will reach their steady state values (Figure 2.2.20). The oscillation frequency ω_c, and the time constant a_t can be calculated with simplifying assumptions from the rate equations which include the assumption that only small deviations occur from the equilibrium photon and carrier densities,

$$a_t \cong \frac{1}{2\tau_s}\left(\frac{J}{J_{th}} + 1\right) \tag{22}$$

$$\omega_c \cong 2\pi f_c \cong \left[\frac{1}{\tau_s\tau_{ph}}\left(\frac{J}{J_{th}} - 1\right)\right]^{1/2} \tag{23}$$

From Equation 22 we see that the oscillations disappear in a time of the order of the spontaneous carrier lifetime, or a few nanoseconds. From Equation 23 we see that the resonance frequency f_c increases with decreasing spontaneous carrier lifetime and photon lifetime, and with the ratio J/J_{th}. Therefore, the laser should be biased above threshold to minimize both the turn-on delay and oscillations. For example, with $J = 2J_{th}$, $\tau_s = 2 \times 10^{-9}$ sec and $\tau_{ph} = 10^{-12}$ sec, $f_c \cong 4$ GHz. Furthermore, if a modulating current density $J = J_o e^{i\omega t}$ is applied to the diode (with appropriate dc bias), the modulation efficiency will decrease rather steeply with frequency when $\omega > \omega_c$.

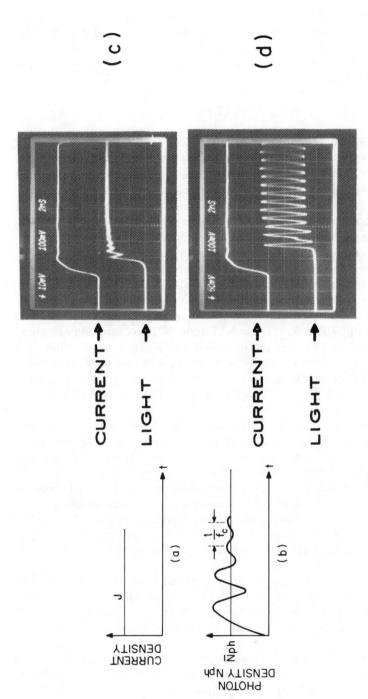

FIGURE 2.2.20.   Schematic of transient effect in laser diodes as it is turned on at t = 0 in (a). The damped oscillations in the emitted radiation intensity are shown in (b). The steady state photon density is N_{ph}. The turn-on delay is not shown.¹ (c) Actual damped oscillatory behavior as measured in a DH AlGaAs laser diode as it is turned on. The delay in laser turn-on after application of the current is evident. (d) Actual undamped oscillations which sometimes appears in lasers that have degraded.

In addition to the transient oscillations discussed above, continuous oscillations are also observed in lasers resulting from nonuniform inversion of the cavity and shot noise. Nonuniform inversion in a laser can produce oscillations because of saturable absorption.[55]

With regard to oscillations due to shot noise in the cavity, theory predicts that the noise intensity peaks at frequencies ranging from the MHz region well into the GHz region dependent on J/J_{th} and the temperature.[56] The noise intensity is predicted to decrease with increasing J/J_{th}, i.e., the maximum is reached near threshold with a rapid decrease above threshold as J increases.

RELIABILITY

Injection lasers operate under quite stressful conditions because they combine high current densities (order 10^4 A/cm^2), and high optical intensities approaching in some cases 10^6 W/cm^2 at the facets. As a result, these devices are subject to certain failure modes and it is essential that they be prudently operated in order to achieve long life.

Laser diodes may degrade[1] as a result of two basic processes — *facet damage* which changes the reflectivity of the facets and *internal damage* which is caused by the formation of defects in the recombination region. The degradation manifests itself primarily in an increase in threshold current although other parameters may also change.

Facet damage is subdivided into "catastrophic" degradation, where large mechanical damage occurs at the emitting facets, and facet "erosion" which is a milder form of damage. Both forms of facet damage are a function of the optical flux density. For example, catastrophic degradation occurs when the optical intensity at the facets reaches values approaching 10^6 W/cm^2 for pulse lengths of 1 μsec duration or longer. (The allowable pulse duration increases approximately with the square root of the pulse length for shorter pulses.) The damage limit may be increased by the use of antireflecting facet coatings.[57] Facet erosion is believed to be caused by a chemical attack of the laser facet by a process which is optically activated. Facet coatings of Al_2O_3 are used to prevent facet erosion.[58]

The internal laser damage (which causes "gradual" degradation) consists of the formation of nonradiative recombination centers in the recombination region. These lower the internal quantum efficiency and increase the internal radiation absorption. Thus, these centers produce (when their density is sufficiently high) an increased threshold current and decreased differential quantum efficiency. It is generally believed that these centers consist of lattice defects which migrate into the recombination region by a diffusion process which is enhanced by the energy released in nonradiative electron-hole recombination. The initial quality of the material is important in order to produce reliable devices because existing lattice defects act as nuclei for the growth of large defects.

The gradual degradation rate increases with temperature[59] and current density. The temperature dependence of the process has been characterized by an expression of the form $\tau_l \propto \exp(E/kT)$ where τl is the laser lifetime, E is an energy value of about 0.7 eV for AlGaAs lasers operating cw, k is Boltzmann's constant, and T is the Kelvin temperature.[60]

Injection laser failures follow a log-normal failure rate (a Gaussian distribution on a logarithmic time scale). For example, in Figure 2.2.21 we plot the time to failure on log-normal coordinates for a group of 24 cw AlGaAs lasers operating at 20°C.[60] The mean time to failure of these devices is about 100,000 hr.

The most studied changes in the laser parameters with operating times are the threshold current and the differential quantum efficiency. However, other changes may oc-

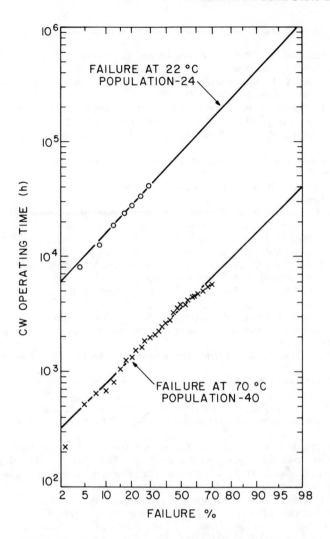

FIGURE 2.2.21.   Time to failure on log-normal coordinates for 24 AlGaAs lasers operating cw at a heat sink temperature of 22°C and 40 lasers operating at 70°C.[60] Both failure statistics follow a log-normal distribution with the difference in time between them representing an "activation energy" for degradation of about 0.7 eV.

cur. These include changes in the number of longitudinal and lateral modes. In addition, self-sustained oscillations are sometimes observed in devices after long-term operation as shown in Figure 2.2.20d.[61] It is believed that these oscillations are caused by the defects introduced during operation into the recombination region.

ACKNOWLEDGMENTS

We are grateful to Jacques I. Pankove for discussions and aid in preparing Table 2.2.1.

REFERENCES

1. Kressel, H. and Butler, J. K., *Semiconductor Lasers and Heterojunction LEDs,* Academic Press, New York, 1977.

2. **Bernard, M. G. A. and Duraffourg, G.**, Laser condition in semiconductors, *Phys. Stat. Sol.*, 1, 699, 1961.

3. **Kressel, H., Ettenberg, M., Wittke, J. P., and Ladany, I.**, *Semiconductor Devices for Optical Communication*, Kressel, H., Ed., Springer-Verlag, Heidelberg, 1979.

4. **Hall, R. N., Fenner, G. E., Kingsley, J. D., Soltys, T. J., and Carlson, R. O.**, Coherent light emission from GaAs junctions, *Phys. Rev. Lett.*, 9, 366, 1962.

5. **Nathan, M. E., Dumke, W. P., Burns, C., Diil, Jr., F. H., and Lasker, G. J.**, Stimulated emission of radiation from GaAs p-n junction, *Appl. Phys Lett.*, 1, 62, 1962.

6. **Quist, T. M., Rediker, R. H., and Keyes, R.**, Semiconductor maser of GaAs, *Appl. Phys. Lett.*, 1, 91, 1962.

7. **Kressel, H. and Nelson, H.**, Close-confinement gallium arsenide p-n junction lasers with reduced optical loss at room temperature, *RCA Rev.*, 30, 106, 1969.

8. **Hayashi, I., Panish, M. B., and Foy, P. W.**, A low-threshold room-temperature injection laser, *IEEE J. Quantum Electron.*, QE-5, 211, 1969.

9. **Alferov, Zh. I., Andreev, V. M., Portnoi, E. L., and Trukan, M. K.**, AlAs-GaAs heterojunction injection lasers with a low room temperature threshold, *Sov. Phys. Semicon.*, 3, 1107, 1970.

10. **Panish, M. B., Hayashi, I., and Sumski, S.**, Double-heterostructure injection lasers with room-temperature thresholds as low as 2300 A/cm², *Appl. Phys. Lett.*, 16, 326, 1970.

11. **Kressel, H. and Hawrylo, F. Z.**, Fabry-Perot structure $Al_xGa_{1-x}As$ injection lasers with room-temperature threshold current densities of 2530 A/cm², *Appl. Phys. Lett.*, 17, 169, 1970.

12. **Lockwood, H. F., Kressel, H., Sommers, Jr., H. S., and Hawrylo, F. Z.**, An efficient large optical cavity injection laser, *Appl. Phys. Lett.*, 17, 499, 1970.

13. **Thompson, G. H. B. and Kirby, P. A.**, (GaAl)As lasers with a heterostructure for optical confinement and additional heterojunctions for extreme carrier confinement, *IEEE J. Quantum Electron.*, QE-9, 311, 1973.

14. **Milnes, A. G. and Feucht, D. L.**, *Heterojunctions and Metal-Semiconductor Junctions*, Academic Press, New York, 1972.

15. **Butler, J. K. and Kressel, H.**, Design curves for double-heterojunction laser diodes, *RCA Rev.*, 38, 542, 1977.

16. For analytical expression see, **Botez, D. and Ettenberg, M.**, Beamwidth approximations for the fundamental mode in symmetric double-heterojunction lasers, *IEEE J. Quantum Electron.*, QE-14, 827, 1978.

17. **Dixon, R. W., Nash, F. R., Hartman, R. L., and Hepplewhite, T.**, Improved light output linearity in stripe geometry double heterostructure (AlGa)As lasers, *Appl. Phys. Lett.*, 29, 372, 1976.

18. **Asbeck, P. M., Cammack, D. A., and Daniele, J. J.**, Nongaussian fundamental mode patterns in narrow-stripe geometry lasers, *Appl. Phys. Lett.*, 33, 504, 1978.

19. **Frescura, B. L., Hwang, C. J., Luechinger, H., and Ripper, J. E.**, Suppression of output nonlinearities in double-heterostructure lasers by use of misaligned mirrors, *Appl. Phys. Lett.*, 31, 770, 1977.

20. **Scifres, D. R., Streifer, W., and Burnham, R. D.**, Curved stripe GaAs:GaAlAs diode lasers and waveguides, *Appl. Phys. Lett.*, 32, 231, 1978.

21. **Matsumoto, N. and Kawaguchi, H.**, Semiconductor lasers with bent guide of planar structure, *Jpn. J. Appl. Phys.*, 16, 1885, 1977.

22. **DeWaard, P. J.**, Strip geometry DH lasers with linear output/current characteristics, *Electron. Lett.*, 13, 400, 1977.

23. **Tsukada, T.**, GaAs-(GaAl)As buried heterojunction injection lasers, *J. Appl. Phys.*, 45, 4899, 1974.

24. **Chinone, N., Saito, K., Ito, R., Aiki, K., and Shige, N.**, Highly efficient (GaAl)As buried heterostructure lasers with buried optical guide, *Appl. Phys. Lett.*, 35, 513, 1979.

25. **Tsang, W. and Logan, R. A.**, $GaAs-Al_xGa_{1-x}As$ strip buried heterostructure lasers, *IEEE J. Quantum Electron.*, QE-15, 451, 1979.

26. **Aiki, K., Nakamura, N., Kuorda, T., Umeda, J., Ito, R., Chinone, N., and Maeda, M.**, Transverse mode stabilized $Al_xGa_{1-x}As$ injection lasers with channel-substrate-planar structures, *IEEE J. Quantum Electron.*, QE-14, 89, 1978.

27. **Kumabe, H., Tanaka, T., Namizaki, H., Ishii, M., and Susaki, W.**, High temperature single mode cw operation with a junction-up TJS laser, *Appl. Phys. Lett.*, 33, 38, 1978.

28. **Kirkby, P. A. and Thompson, G.H. B.**, Channeled substrate buried heterostructure GaAs-(GaAl)As injection laser, *J. Appl. Phys.*, 47, 4578, 1976.

29. **Burnham, R. D., Scifres, D. R., Striefer, W., and Peled, S.**, Nonplanar large optical cavity GaAs/(GaAl)As semiconductor laser, *Appl. Phys. Lett.*, 35, 734, 1979.

30. **Botez, D. and Zorey, D.**, Constricted double heterojunction (AlGa)As diode laser, *Appl. Phys. Lett.*, 32, 761, 1978.

31. **Botez, D.**, Single mode cw operation of double dove-tail constricted DH (AlGa)As diode lasers, *Appl. Phys. Lett.*, 33, 872, 1978.

32. **Botez, D.**, cw high-power single mode lasers using constricted double-heterostructures with a large optical cavity (CDH-LOC), *Appl. Phys. Lett.,* 36(3), 190—192, 1980.

33. **Kogelnik, H. and Shank, H. and Shank, C. V.**, Stimulated emission in a periodic structure, *Appl. Phys. Lett.,* 18, 152, 1971.

34. **Yariv, A.**, *Quantum Electronics,* 2nd ed., John Wiley & Sons, New York, 1975, 165.

35. **Nakamura, M., Aiki, K., Umeda, J., and Yariv, A.**, cw operation of distributed-feedback GaAs-GaAlAs diode lasers at temperatures up to 300 K, *Appl. Phys. Lett.,* 27, 403, 1975.

36. **Wang, S.**, Design considerations of the DBR injection lasers and the waveguiding structures for integrated optics, *IEEE J. Quantum Electron.,* QE-13, 176, 1977.

37. **Stern, F.**, Gain-current relation for GaAs lasers with n-type and undoped active layers, *IEEE J. Quantum Electron.,* QE-9, 290, 1973.

38. **Kressel, H. and Ettenberg, M.**, Low-threshold double-heterojunction AlGaAs/GaAs laser diodes: theory and experiment, *J. Appl. Phys.,* 47, 3533, 1976.

39. **Chinone, N., Saito, K., Ito, R., Aiki, K., and Shige, N.**, Highly efficient (GaAl)As buried-hetero-structure lasers with buried optical guide, *Appl. Phys. Lett.,* 35, 513, 1979.

40. **Goodwin, A. R., Peters, J. R., Pion, M., Thompson, G. H. B., and Whiteaway, J. E. A.**, Threshold temperature characteristics of double heterostructure $Ga_{1-x}Al_xAs$ lasers, *J. Appl. Phys.,* 46, 3146, 1975.

41. **Pankove, J. I.**, Temperature dependence of emission efficiency and lasing threshold in laser diodes, *IEEE J. Quantum Electron.,* QE-4, 119, 1968.

42. **Ettenberg, M., Nuese, C. J., and Kressel, H.**, The temperature dependence of threshold current for double heterojunction lasers, *J. Appl. Phys.,* 50, 2949, 1979.

43. **Ettenberg, M. and Kressel, H.**, Interfacial recombination at (AlGa)As/GaAs heterojunction structures, *J. Appl. Phys.,* 47, 1538, 1976.

44. **Van Opdorp, C. and Vennvliet, H.**, On the relation between threshold current and interface recombination velocities in double-heterojunction lasers, *IEEE J. Quantum Electron.,* QE-15, 817, 1979.

45. **Ettenberg, M. and Olsen, G. H.**, The recombination properties of lattice-mismatched $In_xGa_{1-x}P$/GaAs heterojunctions, *J. Appl. Phys.,* 48, 4275, 1977.

46. **Matthews, J. W.**, *Epitaxial Growth,* Matthews, J. W., Ed., Academic Press, New York, 1975, 562.

47. **Casey, Jr., H. C. and Panish, M. B.**, *Heterostructure Lasers,* Part A and B, Academic Press, New York, 1978.

48. **Hsieh, J. J.**, Room-temperature GaInAsP/InP double-heterostructure diode lasers emitting at 1.1 μm, *Appl. Phys. Lett.,* 28, 283, 1976.

49. **Yamamoto, T., Sakai, K., Akiba, S., and Suematsu, Y.**, $In_{1-x}Ga_xAs_yP_{1-y}$/InP DH lasers fabricated on InP (100) substrates, *IEEE J. Quantum Electron.,* QE-14, 95, 1978.

50. **Akiba, S., Sakai, K., and Yamamoto, T.**, $In_{0.53}Ga_{0.47}As/In_{1-x}Ga_xAs_yP_{1-y}$ double-heterostructure laser with emission wavelength of 1.67 μm at room temperature, *Jpn. J. Appl. Phys.,* 17, 1899, 1978.

51. **Olsen, G. H., Nuese, C. J., and Ettenberg, M.**, Reliability of vapor-grown InGaAs and InGaAsP heterojunction laser structures, *IEEE J. Quantum Electron.,* QE-15, 688, 1979.

52. **Kressel, H. and Hawrylo, F. Z.**, Red-light-emitting diodes operating cw at room temperature, *Appl. Phys. Lett.,* 28, 598, 1976.

53. **Ladany I. and Kressel, H.**, Visible cw (AlGa)As heterojunction laser diodes, *Int. Electron. Devices Meet. Tech. Dig.,* p129, 1976.

54. **Walpole, J. N., Calawa, A. R., Harman, T. C., and Groves, S. H.**, Double-heterostructure PbSnTe laser grown by molecular-beam epitaxy with cw operation up to 114 K, *Appl. Phys. Lett.,* 28, 552, 1976.

55. **Lasher, G. J.**, Analysis of a proposed bistable injection laser, *Solid-State Electron.,* 7, 707, 1964.

56. **Haug, H.**, Quantum mechanic rate equations for semiconductor lasers, *Phys. Rev.,* 184, 338, 1969.

57. **Ettenberg, M., Sommers, Jr., H. S., Kressel, H., and Lockwood, H. F.**, Control of facet damage in GaAs laser diodes, *Appl. Phys. Lett.,* 18, 571, 1971.

58. **Ladany, I., Ettenberg, M., Lockwood, H. F., and Kressel, H.**, Al_2O_3 half-wave films for long life cw lasers, *Appl. Phys. Lett.,* 30, 87, 1977.

59. **Hartman, R. L. and Dixon, R. W.** Reliability of DH GaAs lasers at elevated temperatures, *Appl. Phys. Lett.,* 26, 239, 1975.

60. For a review, see **Ettenberg, M. and Kressel, H.**, The reliability of AlGaAs cw laser diodes, *IEEE J. Quantum Electron.,* QE-16, 186, 1980.

61. **Channin, D. J., Ettenberg, M., and Kressel, H.**, Self-sustained oscillations in (AlGa)As oxide-defined stripe lasers, *J. Appl. Phys.,* 50, 6700, 1979.

2.3 GLASS LASERS

S. E. Stokowski

INTRODUCTION

Snitzer[1] first discovered glass lasers by observing laser action in a Nd-doped Ba-crown glass. Within a few years stimulated emission was also observed in glasses doped with other rare earth ions.[2-6] However, over the past two decades the development of glass lasers has concentrated primarily on Nd:glass systems, which presently are the world's most powerful lasers.[7]

Lasers made from vitreous and crystalline materials comprise the two classes of solid state lasers. Their different material properties are complementary for use in lasers. Because of their lower cross sections, glass lasers store energy well and thus make good short pulse lasers and amplifiers. On the other hand, crystalline materials are better for cw oscillators and amplifiers because of their higher gain and good thermal conductivity.

Glass has advantages over crystalline materials. It can be cast in a variety of forms and sizes, from small fibers to meter-sized pieces. Tremendous flexibility in choosing glass and laser properties is afforded by the ability to vary the glass composition over very large ranges. Glass is also relatively inexpensive because of the shorter time required for its manufacture and the use of inexpensive chemical components. Further, large pieces of laser glass can be made with excellent homogeneity, uniformly distributed rare earth concentrations, low birefringence, and can be finished easily, even in large sizes. The only major drawback of glass is its low thermal conductivity, which limits its applicability in high average power systems.

This chapter lists the properties of glass lasers and laser glass materials. The text is kept to the minimum necessary to describe the tables and figures and to discuss laser glass properties directly related to laser performance. The intent is to provide a convenient tabulation of information on glass lasers with a sufficiently extensive bibliography for the reader to find additional detailed information, if desired. Several books[8-11] have been written on solid state lasers with detailed descriptions of laser theory, operation, construction, and applications. In addition, the spectroscopy of rare-earth ions in solids is covered extensively in other texts.[13-17]

The following section is devoted to specific glass lasers and their performance characteristics. Some discussion of their design and operational techniques is included. The next section contains a description of rare-earth spectroscopic properties important to glass laser operation. The variation of these properties with glass composition is summarized. Sensitization, laser damage, and glass solarization are also covered. A short discussion of materials that can be used in integrated optics is included. A tabulation of the spectroscopic, optical, thermal, and mechanical properties of selected laser glasses comprises the last section. Glasses listed are currently available commercially, plus a fluoroberyllate glass. This table covers only a small fraction of the thousands of glasses that have been melted with rare-earth impurities. However, the glasses selected have technologically important properties and are typical examples of the different laser glass types.

SPECIFIC LASERS

Observed Laser Transitions

The only transition metal ions producing stimulated emission in glass are members of the lanthanide group (rare earths). These ions are listed in Table 2.3.1 along with their laser wavelengths, initial and final energy states, excitation sources, operating temperatures, and sensitizing ions. The listed references are those in which laser action was first reported in the different glass types. The absorption spectra, laser transitions, and energy levels of the seven rare earth lasing ions are shown in Figure 2.3.1a through f.

Only Nd-, Er-, and Yb- glass lasers have been studied extensively. The Nd-glass laser is the only one employed commercially because of its high efficiency, room temperature operation, and conveniently detectable wavelength. These desirable characteristics result from the four-level nature of the Nd ion, the good match between the spectra of the Nd absorption bands and the xenon flashlamps commonly used for pumping, the high radiative efficiency of the $^4F_{3/2}$ state ($\sim$0.9), and its long radiative lifetime (150 to >800 μsec). The fluorescence spectrum Nd and the fluorescence and absorption spectra of the Er and Yb lasing transitions are shown in Figures 2.3.2 to 2.3.4.

Operating Modes, Techniques, and Characteristics
Oscillators

Glass laser oscillators are operated either cw, long-pulsed (0.100 to 10 msec), Q-switched (10 to 100 nsec), or mode-locked ($\sim$1 to 100 psec). The operational characteristics of selected oscillators are listed in Table 2.3.2. CW operation can be obtained, but with low overall efficiency. In long-pulsed operation, however, up to 7.6% efficiency has been seen in a 19 $\phi \times$ 940 mm rod.[37] The highest observed pulse energy from a single element was 5000 J in a 30 mm diameter by $\approx$ 1 m silicate rod.[38] Q-switched, Nd-glass oscillators with apertures up to 40 mm diameter can produce pulsed power up to 30 GW, long pulse energies up to 100 J, and pulse widths of 5 to 35 nsec.[39] However, most Q-switched oscillators used at the present time contain Nd-doped YAG or LiYF$_4$ crystals as the active material because of the higher repetition rate available. On the other hand, laser-pumped glass oscillators can provide wavelength versatility because of their wide fluorescence bandwidths. An ED-2 6$\phi \times$ 75 mm glass rod has been tuned from 1.0532 to 1.0893 μm using two Fabry-Perot etalons.[49]

Picosecond pulses from passively mode-locked Nd glass lasers vary from 2.5 to 20 psec in total duration with subpicosecond structure as short as 0.25 psec.[50-57] The energy of individual pulses varies from 0.2 to 10 mJ, depending on their position in the pulse train. The spectral width of the pulses is initially 0.3 nm and broadens to 8 nm later in the train.[58] The spectral frequency is swept or "chirped" within the pulse.[59] However, the pulse characteristics are very sensitive to the detailed behavior of the saturable dye cell; therefore, significant variations in the above numbers are observed.

Er and Yb-glass lasers are not as efficient as Nd:glass. However when Er-glass is pumped by a Nd laser, good conversion efficiencies ($\sim$40%) from 1.06 μm to 1.54 μm can be obtained.[48]

Amplifiers

Laser glass makes an excellent amplifier of short laser pulses because it can store high energy densities ($\sim$0.5 J cm^{-3}). The premier examples of this capability are the large laser systems used for fusion research.[60] In 1977-78 the Nd:glass Shiva Laser at Lawrence Livermore Laboratory produced > 10 KJ in a 0.8 nsec pulse and 20 TW in a 0.06 nsec pulse at a 20 cm aperture. This system and others like it consist of a crystalline laser oscillator providing 0.1- to 1.1-nsec 100 μJ pulses, which are then amplified approximately a million times by rod and disk amplifiers.

Rod

A typical rod amplifier schematic is shown as Figure 2.3.5. The rod ends are coated

Table 2.3.1
LASER TRANSITIONS IN GLASS

(Listed by active ion in order of increasing atomic weight and chronologically for each ion)

Active ion	Glass	Laser wavelength (μm)	Transition states	Excitation source	Temperature	Sensitizing ions	Ref.
Nd³⁺	K-Ba silicate	1.061	$^4F_{3/2}-^4I_{11/2}$	Xe flashlamp	R.T.[a]		1
Nd³⁺	Borate	1.06	$^4F_{3/2}-^4I_{11/2}$	Hg arc lamp (cw)	R.T.		18
Nd³⁺	Zn-Li phosphate	1.054	$^4F_{3/2}-^4I_{11/2}$	Xe flashlamp (100 Hz)	R.T.		19
Nd³⁺	Fluoroberyllate	1.047	$^4F_{3/2}-^4I_{11/2}$	Ar laser ($\lambda = 0.5145$ μm)	R.T.		20
Nd³⁺	Li-germanate	1.06	$^4F_{3/2}-^4I_{11/2}$	Xe flashlamp	R.T.		21
Nd³⁺	Glass-ceramic	1.06	$^4I_{13/2}-^4I_{11/2}$	Ar laser ($\lambda = 0.5145$ μm)	R.T.		22,23
Nd³⁺	Fused silica	1.08	$^4F_{3/2}-^4I_{11/2}$	Krypton arc lamp (cw)	R.T.		24
Nd³⁺	Silicate	1.06	$^4F_{3/2}-^4I_{11/2}$		80 K		25
Nd³⁺	Na-Ca silicate	0.918	$^4F_{3/2}-^4I_{9/2}$	Dye laser ($\lambda = 0.58$ μ)	R.T.		26
Nd³⁺	Silicate	0.92	$^4F_{3/2}-^4I_{9/2}$	Xe flashlamp	R.T.		27
Nd³⁺	La-Ba-Th borate	1.37	$^4F_{3/2}-^4I_{13/2}$	Xe flashlamp	R.T.		28
Gd³⁺	Li-Mg-Al silicate	0.3125	$^6P_{7/2}-^8S_{7/2}$	Xe flashlamp	78 K		3[b]
Tb³⁺	Borate	0.54	$^5D_4-^7F_5$	Xe flashlamp	R.T.		29[c]
Ho³⁺	Li-Mg-Al silicate	2.08	$^5I_7-^5I_8$	Xe flashlamp	77 K		3
Ho³⁺	Mg-Li-Al silicate	2.06-2.10	$^5I_7-^5I_8$	Xe flashlamp	77 K	Er, Yb	30
Er³⁺	Na-K-Ba silicate	1.55	$^4I_{13/2}-^4I_{15/2}$	Xe flashlamp	77 K		31
Er³⁺	Na-K-Ba silicate	1.543	$^4I_{13/2}-^4I_{15/2}$	Xe flashlamp	R.T.	Yb³⁺	4
Er³⁺	Fluorophosphate	1.54	$^4I_{13/2}-^4I_{15/2}$	Xe flashlamp	R.T.	Yb³⁺	32
Er³⁺	Al-Zn phosphate	1.54	$^4I_{13/2}-^4I_{15/2}$	Xe flashlamp	R.T.	Yb³⁺	33
Tm³⁺	Li-Mg-Al silicate	1.85	$^3F_4-^3H_6$	Xe flashlamp	80 K		5
Tm³⁺	Li-Mg-Al silicate	2.015	$^3F_4-^3H_6$	Xe flashlamp	R.T.	Yb³⁺, Er³⁺	5
Yb³⁺	Li-Mg-Al silicate	1.015	$^2F_{5/2}-^2F_{7/2}$	Xe flashlamp	77 K		2
Yb³⁺	Li-Mg-Al silicate	1.015 1.06(Nd)	$^2F_{5/2}-^2F_{7/2}$	Xe flashlamp	R.T.	Nd³⁺	34
Yb³⁺	Ca-Li borate	1.018	$^2F_{5/2}-^2F_{7/2}$	Xe flashlamp	R.T.	Nd³⁺	35
Yb³⁺	K-Ba silicate	1.06	$^2F_{5/2}-^2F_{7/2}$	Xe flashlamp	R.T.	Nd³⁺	36

[a] Room temperature.

[b] Reference 3 contains the only reported observation of stimulated emission in Gd³⁺ - activated glass. Evidence for lasing was the large nonlinear increase in 312.5 nm radiation under strong UV excitation. However, later attempts to reproduce these results were unsuccessful. (From Ginther, R. J., private communication.)

[c] This reference contains the only reported observation of stimulated emission in Tb³⁺ - activated glass. Proof of laser action was based on the appearance of emission spikes.

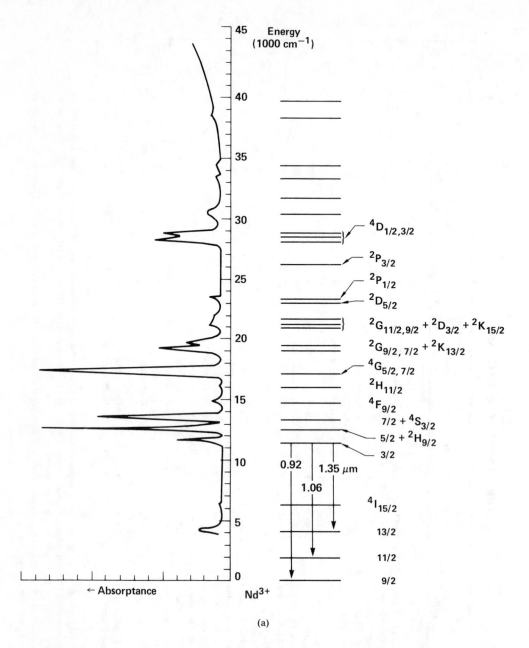

(a)

FIGURE 2.3.1.   Absorption spectra, energy levels, laser transitions, and approximate wavelengths of rare-earth ions in glass: (a), Nd^{3+}, (b) Tb^{3+}, (c) Ho^{3+}, (d) Er^{3+}, (e) Tm^{3+}, (f) Yb^{3+}. (The Ho^{3+} spectrum is from Patek.[9]) Stark splittings of the J states are not shown.

with an antireflective multilayer film or cut at a shallow angle relative to the long axis of the rod in order to prevent laser oscillations. These "parasitic" oscillations decrease the stored energy available for amplification.[64] As additional protection against unwanted oscillations an index-matched absorbing liquid or solid surrounds the rod.[65]

The Nd concentrations commonly used in glass rods are a compromise between the need for high concentrations to absorb the greatest amount of pump light and low concentrations for uniform gain over the beam aperture. Generally lower Nd concentrations are found in rods of larger diameters.

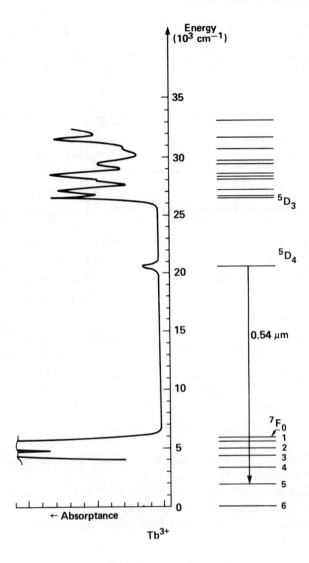

FIGURE 2.3.1(b).

The performance characteristics of typical Nd:glass rod amplifiers are listed in Table 2.3.3.

Disk

The use of several laser glass disks set at Brewster's angle for amplification dates from 1966.[66,67] A disk amplifier schematic appears in Figure 2.3.6. At large apertures disk amplifiers have significant advantages over rod amplifiers, such as, more uniform gain across the aperture, more gain per cm, and easier manufacture of the glass material. Disk amplifiers have been constructed as large as 46 cm in diameter, containing two elliptical disks with major and minor axes of 89 and 49 cm, respectively.[68]

The long optical paths possible in a glass disk make parasitic oscillations likely unless reflections at the disk edge are controlled.[69-72] These edge reflections are eliminated by using an index-matched 1.06 μm-absorbing solid cladding. Ions employed in edge-cladding glass to absorb 1.06 μm light include Sm^{3+}, V^{3+}, and Cu^{2+}, with Cu^{2+} being the ion of choice. The cladding glass is applied to the disk either as a frit, which is sintered

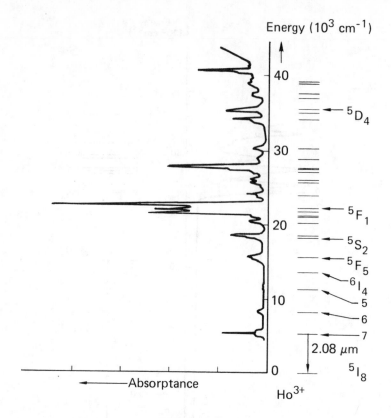

FIGURE 2.3.1(c).

to the laser glass, or poured and cast around the disk.[73] Edge claddings lower the disk edge reflectivity below 0.01%.[74]

The performance characteristics of typical Nd:glass disk amplifiers are found in Table 2.3.4. In addition, Figure 2.3.7 illustrates how the small signal gain in a 4.8 cm disk amplifier containing different laser glasses varies with pump energy.

Other

Disk amplifiers that consist of a rectangular slab of laser glass face-pumped by flashlamps have been constructed. In these amplifiers the laser beam propagates in zigzag fashion down one of the long axes by total internal reflection.[75] In another type of disk amplifier the laser glass is pumped through one face, while the laser beam passes through the other face and is reflected from the back of the disk by a mirror coating.[76] This multilayer coating reflects at the laser wavelength and is transparent to the pump light.

Thin film laser amplifiers, which may find a use in integrated optical systems, have been fabricated and tested.[77] The optical gain coefficient of these film amplifiers is about 1 cm⁻¹ at 1.058 μm when pumped by a dye laser.

GENERAL PROPERTIES OF LASER GLASS

Spectroscopic Properties

The spectroscopic properties of the active ion determine the glass laser characteristics. Knowledge of these properties allows one to predict laser performance.[78] The gain coefficient in a laser material is a product of the ion excited state density and the

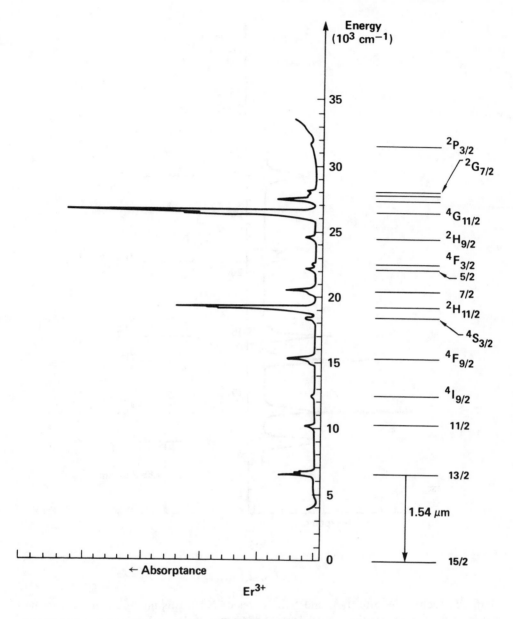

Energy
(10^3 cm^{-1})

35

$^2P_{3/2}$

30

$^2G_{7/2}$

$^4G_{11/2}$

25

$^2H_{9/2}$

$^4F_{3/2}$
5/2

20

7/2
$^2H_{11/2}$

$^4S_{3/2}$

15

$^4F_{9/2}$

$^4I_{9/2}$

10

11/2

13/2

5

1.54 μm

0

15/2

← Absorptance

Er^{3+}

FIGURE 2.3.1(d).

emission cross section of the laser transition. The density of excited ions produced by the pump source depends on the wavelength positions and intensities of the absorption bands, and the lifetime of the emitting state. The thermally activated population of the terminal state determines the operating temperature range. The range of emission wavelengths, effective linewidths, cross sections, terminal state energies, excited state densities for 1% gain per cm, and excited state lifetimes for rare-earth-ion glass lasers are listed in Table 2.3.5.

The most important spectroscopic parameter to know is the emission cross section. If the terminal state of the lasing transition has a significant population at ambient temperature as it does for the Ho^{3+}, Er^{3+}, Tm^{3+}, and Yb^{3+} lasers, the transition strength and thus, the cross section, can be obtained from the absorption and fluorescence spectra of the laser transition. However, if the terminal state energy is much higher

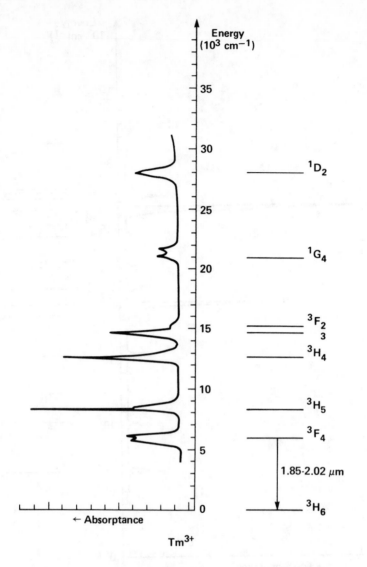

FIGURE 2.3.1(e).

than kT, such as for the $^4I_{11/2}$ and $^4I_{13/2}$ states of Nd^{3+}, this method is inaccurate because of the low absorption coefficients.[79,80] Alternatively, emission cross sections can be calculated through the use of the Judd-Ofelt treatment of spectral intensities.[81-84] This analysis technique provides a successful, convenient, and consistent method of treating spectral intensities of rare-earth f-f transitions.

In the Judd-Ofelt treatment, the line strength S of a transition between two J states is given by:

$$S(aJ:bJ') = \sum_t \Omega_t \, | < aJ \parallel U^{(t)} \parallel bJ' > |^2 ,$$

$$t = 2,4,6 \tag{1}$$

where a and b denote other quantum numbers specifying the eigenstates. The doubly reduced matrix elements of the unit tensor operator $U^{(t)}$ are calculated in an intermediate coupling approximation. The Judd-Ofelt Ω parameters are determined from the integrated absorption band intensities per ion and vary from glass to glass. Once these

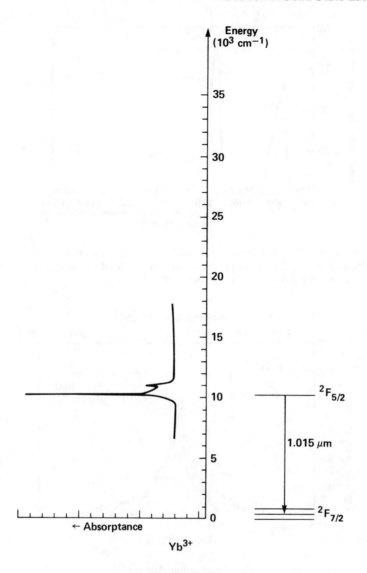

FIGURE 2.3.1(f).

parameters are known, the peak emission cross section can be calculated from the expression

$$\sigma (\lambda_p) = \frac{8\pi^3 e^2}{27hc \, (2J + 1)} \; \frac{\lambda_p}{\Delta\lambda_{eff}} \; \frac{(n^2 + 2)^2}{n} \; S \, (aJ: bJ') \qquad (2)$$

where λ_p is the peak fluorescence wavelength, $\Delta\lambda_{eff}$ is the effective linewidth, and n is the refractive index. The cross section calculated in this manner is accurate to ±10%, assuming that the rare-earth ion concentration is accurately measured.[85]

Other spectroscopic methods used to derive the Nd emission cross sections (σ) are characterized by measuring:

1. $^4F_{3/2}$ fluorescence branching ratios and the $^4I_{9/2} \rightarrow {}^4F_{3/2}$ absorption strength[86]
2. $^4F_{3/2}$ fluorescence branching ratios and the $^4I_{9/2} \rightarrow {}^4F_{7/2}$, $^4S_{3/2}$ absorption strength[87]
3. $^4F_{3/2}$ decay time coupled with quantum efficiency[88]

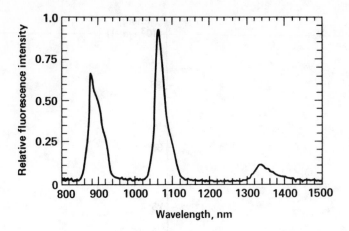

FIGURE 2.3.2.   Normalized $^4F_{3/2}$ fluorescence spectrum of Nd^{3+} in a silicate glass at 295 K. The intensity has been corrected for instrument response and is in terms of relative photon flux.

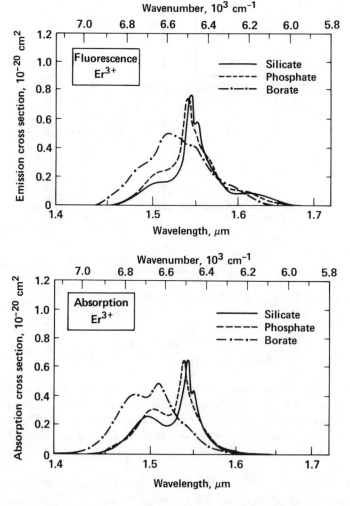

FIGURE 2.3.3.   Fluorescence and absorption spectra of Er^{3+} in different glass types. Data is from Sandoe, Sarkies, and Parke.[133]

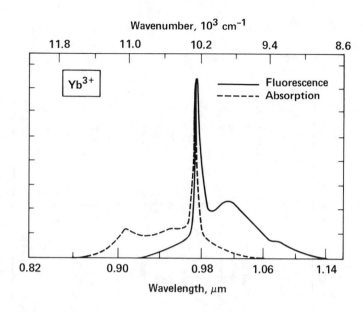

FIGURE 2.3.4.   Fluorescence and absorption spectra of Yb³⁺ in a silicate
glass. Data from Sandoe, Sarkies, and Parke.[133]

Laser measurements that have been employed to determine σ include the following:

1.   Gain reduction after amplification of a high-energy pulse[89,91]
2.   Threshold gain and population inversion[92]
3.   Laser output and fluorescence intensity[93]
4.   Saturation flux in laser oscillators[94]

The decay time (τ_{21}) from the terminal state of the laser transition to the ground state is important in short pulse operation. When the laser pulse width is shorter than τ_{21}, a four–level laser system becomes equivalent to a three-level system. Several attempts have been made to measure τ_{21} for the Nd $^4I_{11/2}$ state. The resulting experimental values span the range between 0.1 and 100 nsec.[95-101] At this time τ_{21} in glass is still considered unknown.

The above discussion of spectroscopic properties assumes a homogeneous system. However, rare-earth ions in glass reside in various local environments and thus, have different homogeneous widths, energy levels, transition strengths, and electron-phonon coupling coefficients.

The measured homogeneous width of the $^4F_{3/2} \rightarrow {}^4I_{11/2}$ transition of Nd is 20 to 25 cm⁻¹ at room temperature.[102] The inhomogeneous width of the same transition is about 130 cm⁻¹.[103] As a result, the calculated Judd-Ofelt parameters and emission cross sections are only averages over the individual ionic properties.

The different transition strengths and radiative decay rates of individual ions produce nonexponential fluorescent decays, even in the absence of energy migration or concentration quenching. The nonexponential decays of the Nd $^4F_{3/2}$ state are illustrated in Figure 2.3.8a,b. Silicate, fluoroberyllate, and fluorophosphate glasses have sizable variations in decay rates, whereas phosphate and tellurite glasses generally have nearly exponential fluorescence decays.

The inhomogeneous nature of ions in glass is also evident by the presence of ''holes'' in fluorescence spectra after passage of an intense resonant laser pulse.[10,104-105] In addition, fluorescence line narrowing (FLN) techniques demonstrate the inhomogeneous nature of glass and are used to probe individual rare-earth ion environments.[106-109]

Table 2.3.2
CHARACTERISTICS OF GLASS LASER OSCILLATORS

(Listed in order of increasing aperture)

Nd³⁺:silicate

Laser wavelength	µm	1.062	1.06	1.062	1.062	1.062	1.062
Glass		Ba-crown silicate	Selfoc	ED-2 silicate	ED-2 silicate	ED-2 silicate	Ba-crown silicate
Rod size: Diameter	mm	0.1 (1.0 with cladding)	1.5	12	20	19.1	30 (38 with cladding)
Length	mm	30	138	150(100 pumped)	250(200 pumped)	940	950
Pump source		Mercury arc	Krypton arc	Xenon flashlamp	Xenon flashlamp	Single xenon lamp	Xenon flashlamp
Pump energy per active volume @ threshold	$J \cdot cm^{-3}$	5.8×10^6 W/cm²		133	72	3.8	
Output mirror reflectivity	%	99.8		45	45	28	
Laser output/pulse	J	~10 µW(cw)	1 mW(cw)	70	205	600	5000
Pump energy density	$J \cdot cm^{-3}$			266	398	29.6	268
Pulse Widths				~600 µsec	~600 µsec	2.2 msec	3 msec
Beam divergence	mr			10			10
Efficiency (output/input)	%	~10^{-6}		1.4	0.83	7.6	
Slope efficiency	%			2.0		9.1	2.8
Operating temperature	K	R.T.ᵃ	R.T.	R.T.	R.T.	R.T.	R.T.
Ref.		18	47	40	40	37	38

Nd³⁺:phosphate

Laser wavelength	µm	1.054	1.06	1.06	1.06	1.053	1.06	1.05
Glass		LHG-8 phosphate	Li-La phosphate	Li-Nd-La phosphate	Li-Nd-La phosphate	Q-98 phosphate	Phosphate	LGS-40 phosphate
Rod size: Diameter	mm	2	5	5	5	6.4	Slab 6x46	10
Length	mm	20	50	50	50	8.3	280	130

Parameter	Units							
Pump source		Ar laser (λ = 0.5145 μm)	Xenon flashlamp	Xenon flashlamp	Xenon flashlamp	Xenon flashlamp	Xenon flashlamp	Xenon flashlamp
Pump energy per active volume @ threshold	J·cm⁻³	0.210 W/cm²(cw)	20.4	6.8	13.6	1.0		6.9
Output mirror reflectivity	%	98	27	79	60	85		4
Laser Output/Pulse	J	0.011 W(cw)	1.6	0.35	0.16	1.08	2	1.2
Pump energy density	J·cm⁻³	0.370 W/cm²(cw)	244	2.94	36.2	12	6.5	9.8
Pulse widths			~200 μsec	60 μsec (1 Hz)	Q-switched	110 μsec	250 μsec (10 Hz)	60 nsec
Beam divergence	mr		0.58				2.5X diffraction limit	
Efficiency (output/input)	%	3	7.5	1.3	0.45	3.4	0.4	1.2
Slope efficiency	%	7.5		1.8	0.76	4.9		
Operating temperature	K	R.T.	R.T.	R.T.	R.T.	R.T.	R.T.	R.T.
Ref.		44	46	43	43	42	41	45

Other ions:glass

Parameter	Units			
Ion		Er³⁺	Er³⁺	Yb³⁺
Laser wavelength	μm	1.536	1.54	1.018
Sensitizers		Yb³⁺, Nd³⁺	Yb³⁺	Nd³⁺
Glass		Phosphate	Phosphate	Ca-Li borate
Rod Size: Diameter	mm	4	5	2.5
Length	mm	76	125	35
Pump Source		Xenon flashlamp	1.06 μm laser	Xenon flashlamp
Pump energy per active volume @ threshold	J·cm⁻³		4 ± 1	17.0
Output mirror reflectivity	%	99.5	30	98
Laser output/pulse	J	0.86, 0.18	27	
At a pump density of	J·cm⁻³	356, 157	31	
Pulse widths		3.3 msec, 25 nsec	20-25 msec	
Beam divergence	mr		2.5	
Efficiency (output/input)	%	0.25, 0.12	35	
Slope efficiency	%		43	
Operating temperature	K	R.T.	R.T.	
Ref.		33	48	35

[a] R.T. = room temperature

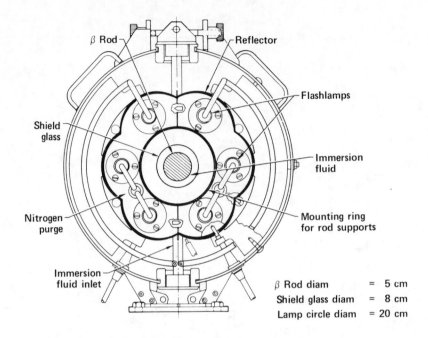

β Rod diam	= 5 cm
Shield glass diam	= 8 cm
Lamp circle diam	= 20 cm

FIGURE 2.3.5. End-on view of the Shiva laser 5-cm rod amplifier. The length of the glass rod is 40 cm. (From Lawrence Livermore Laboratory.[61])

The inhomogeneous nature of the rare-earth-ion transitions in glass has important consequences when the laser oscillator or amplifier is stimulated to emit a sizable fraction of its stored energy. These effects are seen most clearly by comparing the saturated gains of homogeneous and inhomogeneous systems. The gain coefficient per cm (g) is given by

$$g\,(\nu,t) \;=\; \sum_i \sigma_i\,(\nu)\, N_i^*(t), \tag{3}$$

where σ_i and N_i^* are, respectively, the stimulated emission cross section at frequency ν and the excited state population of the different rare-earth sites. For small signal gain conditions Equation 3 reduced to $g = \sigma N^*$, where σ is an effective cross section and N^* is the total number of excited ions. Since the stimulated emission rate is proportional to σ_i, different sites in laser glasses are de-excited at different rates. For large-signal or saturated gain conditions, less energy is extracted from an inhomogeneous system than that from a homogeneous of the same gain coefficient. This difference is sensitive to the distribution of individual ionic cross sections at the lasing wavelength and polarization.

Nonradiative transitions between rare-earth ion energy states are necessary to populate the initial lasing state from the higher energy states excited by the pump radiation.[110,111] However, the radiative quantum efficiency and thus the lifetime of the initial lasing state, will be lowered by nonradiative decay to lower energy states. In the pulsed mode operation typical of glass lasers, low radiative quantum efficiency makes pumping more difficult by requiring short pump pulses.

Empirically, the nonradiative decay rate is given approximately by

$$W_{nr} \;=\; W_o\,(T)\,\exp\,(-\,\alpha\,\Delta E) \tag{4}$$

Table 2.3.3
CHARACTERISTICS OF Nd:GLASS ROD AMPLIFIERS

Clear aperture	mm	23	46	62	87	25	50
Small signal gain		190	19	18	6.5	59	13.4
Pump energy	kJ	32	40	67	67	22.5	22.5
Rod dimensions	mm	250×400	500×400	640×360	900×360	250×380	500×380
Glass		ED-2	ED-2	LHG-8	LHG-8	LHG-7	LHG-7
Glass type		Silicate	Silicate	Phosphate	Phosphate	Phosphate	Phosphate
Nd concentration	wt% Nd_2O_3	1.2	0.6	0.55	0.4	1.2	0.6
Immersion fluid		90% $ZnCl_2$ + 10% $SmCl_2$	90% $ZnCl_2$ + 10% $SmCl_2$	Ethylene glycol	Ethylene glycol	De-ionized water	De-ionized water
No. of lamps		6	6	12	12	6	6
Lamp dimensions:							
bore	mm	13	15	19	19	15	15
length	mm	370	370	300	300	300	300
Lamp envelope material		Ce-doped silica	Ce-doped silica	Fused silica	Fused silica	Ce-doped silica	Ce-doped silica
Pump pulse length	μsec	600	750	400	400	567	567
Reflector		Silver-plated	Silver-plated	Alzac	Alzac	Silver plated	Silver plated
Shield glass		Fused silica	Fused silica	Pyrex	Pyrex	Fused silica	Fused silica
Ref.		61	61	62	62	63	63

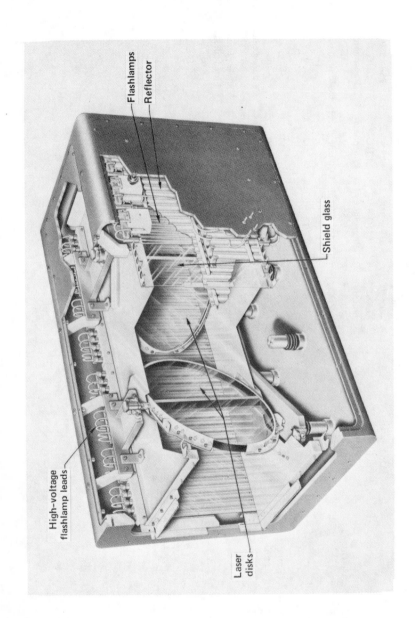

FIGURE 2.3.6. Schematic view of a disk amplifier. This particular amplifier has a 46 cm clear aperture and contains two disks. These disks are split along their minor axes for ease of fabrication and suppression of parasitic oscillations. (From Lawrence Livermore Laboratory.)

Table 2.3.4

CHARACTERISTICS OF Nd:GLASS DISK AMPLIFIERS

Clear aperture	mm	94	150	208	67	104	100	200
Small signal gain		4.3	2.9	2.1	8	17	8.5	3.2
Pump energy	kJ	170	300	400	80	170	220	440
No. of disks		6	4	3	6	6	6	3
Disk size	mm	108 × 202	164 × 309	223 × 418	70 × 140	110 × 216	114 × 214	214 × 400
Disk thickness	mm	24	30	32	20	30	24	32
Glass		ED-2	ED-2	LSG-91H	EV-2	Q-88	LHG-7	LHG-7
Glass type		Silicate	Silicate	Silicate	Phosphate	Phosphate	Phosphate	Phosphate
Nd concentration	wt% Nd$_2$O$_3$	2.2	2.2	1.9	2.7	3	2.9	1.9
Edge cladding		EI-4(Cu-doped)	EI-6(Cu-doped)	HBL-306(Cu-doped)	Frit-type (Cu²⁺)	Poured-type (Cu²⁺)	LCG-7(Cu²⁺)	LCG-7(Cu²⁺)
No. of lamps		16	24	32	20	16	16	32
Lamp dimensions:								
bore	mm	15	15	15	10	15	15	15
length	mm	1120	1120	1120	780	1200	1270	1270
Pump pulse length	μsec	600	600	600	350	500	556	556
Reflector		Silver-plated	Silver-plated	Silver-plated	Aluminum	Aluminum	Silver-plated	Silver-plated
Shield glass		Fused silica	Fused silica	Fused silica	Fused silica	Fused silica	Fused silica	Fused silica
Lamp envelope material		Ce-doped silica	Ce-doped silica	Ce-doped silica	Fused silica	Fused silica	Ce-doped silica	Ce-doped silica
Ref.		61	61	61	NRL[a] (1978)	NRL[a] (1978)	63	63

[a]   J. M. McMahon, Naval Research Laboratory, Washington, D.C.

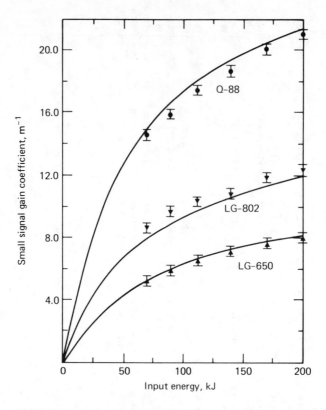

FIGURE 2.3.7.   Small-signal gain coefficient in a disk amplifier as a function of the flash lamp input energy. The edges of the 48 × 84 mm, 15 mm-thick disks are surrounded by a 90% $ZnCl_2$ + 10% $SmCl_2$ liquid solution. Q-88 is a phosphate glass: LG-802, a fluorophosphate; and LG-650, a silicate. Data from Linford et al.[78]

where W_o (T) depends on the host glass and temperature, α depends on host glass, and ΔE is the energy gap between neighboring states.[112,113] By measuring decay times of different states of rare earths in a variety of glasses, Layne et al.[114,115] obtained the multiphonon decay rates shown in Figure 2.3.9. The rates are proportional to the frequency of the highest-energy phonons in the different glass types.

When OH^- groups are present in a glass, the nonradiative rates are increased because of coupling between the rare earth states and the high energy ($\sim$3200 cm^{-1}) vibrations of OH^-.[116] Phosphate glasses can retain significant amounts of OH^- and consequently, the lifetimes of the rate earth states are shortened. For example, the lifetime (τ) of the $^4F_{3/2}$ state of Nd^{3+} in a phosphate glass depends on the OH^- concentration, as measured by the absorption coefficient at 2.85 μm (α in cm^{-1}), according to the empirical expression[117]

$$\frac{1}{\tau} = 3.3 \times 10^3 + 2.2 \times 10^2\ \alpha \qquad (5)$$

where τ is measured in seconds.

Linear and Nonlinear Indices of Refraction

At the high laser powers present in a laser oscillator or amplifier, intensity-dependent nonlinearities in the refractive index can produce self-focusing of the laser beam.[118] In

Table 2.3.5
SPECTROSCOPIC PROPERTY RANGES OF RARE-EARTH IONS IN LASER GLASSES

		Nd³⁺	Nd³⁺	Nd³⁺	Ho³⁺
Ion					
Transition		$^4F_{3/2} \to {}^4I_{11/2}$	$^4F_{3/2} \to {}^4I_{13/2}$	$^4F_{3/2} \to {}^4I_{9/2}$	$^5I_7 \to {}^5I_8$
Wavelength	μm	1.047—1.08	1.314—1.37	0.918—0.921	2.08
Emission cross section	10^{-20}cm²	1.0—5.1	0.3—1.3	0.3	0.45
Effective linewidth	nm	19—43	45—67		
Excited state lifetime	μsec	140—1000	140—1000	140—1000	800
Inversion density for 1% cm⁻¹ gain	10^{18}cm⁻³	0.2—1.0	0.7—3.3	3.5	2.2
Terminal state energy	cm⁻¹	≈1950	≈4070	470	290

		Er³⁺	Tm³⁺	Yb³⁺	Yb³⁺
Ion					
Transition		$^4I_{13/2} \to {}^4I_{15/2}$	$^3F_4 \to {}^3H_6$	$^2F_{5/2} \to {}^2F_{7/2}$	$^2F_{5/2} \to {}^2F_{7/2}$
Wavelength	μm	1.536—1.55	1.85, 2.02	1.015	1.06
Emission cross section	10^{-20}cm²	0.4—1.2		0.36—0.8	0.09—0.2
Effective linewidth	nm				
Excited state lifetime	μsec	6,000—13,000	550	1300—2200	1300—2200
Inversion density for 1% cm⁻¹ gain	10^{18}cm⁻³	0.8—2.5		1.3—2.8	5 —11
Terminal state energy	cm⁻¹	40—110	380, 840	400	830

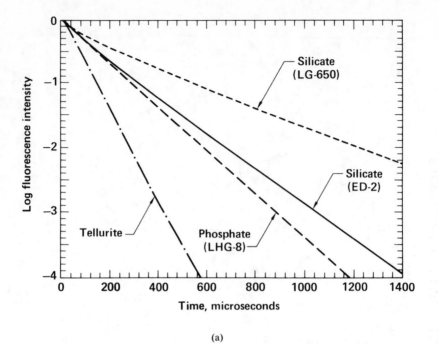

(a)

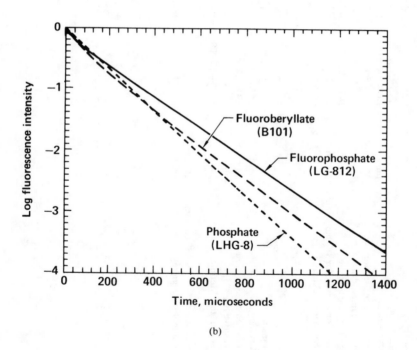

(b)

FIGURE 2.3.8. Nd^{3+}: $^4F_{3/2}$ fluorescence decay curves for various glasses. The LHG-8 exponential decay is included for comparison in both (a) and (b).

addition, small-scale spatial irregularities in the beam profile will grow exponentially as e^B, where

$$B = \frac{2\pi}{\lambda} \int \gamma I \, dz \tag{6}$$

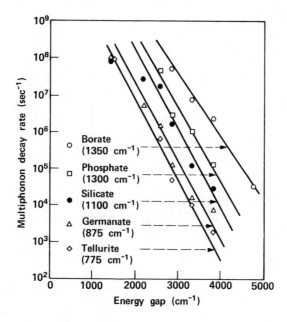

FIGURE 2.3.9. Multiphonon decay rates of rare-earth ions in various glass types. The highest frequency optical phonon in each glass type is listed. Data from Layne, Lowdermilk, and Weber.[114]

λ is the wavelength, γ is the nonlinear refractive index coefficient, and I is the optical intensity.[119,120] The nonlinear refractive index coefficient, γ, is defined by $\Delta n = \gamma I$ where Δn is the change in refractive index and I is the intensity. Alternatively $\Delta n = n_2 \langle E^2 \rangle$ where E is the amplitude of the optical electric field; thus, $n_2(\text{esu}) = (cn/40\pi) \gamma (m^2/W)$.

Studies of γ in glasses and crystals demonstrate that, in the long wavelength limit, γ can be estimated from the linear index and dispersion.[121] The dispersion is characterized by the Abbe number, ν, defined by,

$$\nu = (n_D - 1) / (n_F - n_C) \tag{7}$$

where n_F, n_D, and n_C are the linear refractive indices at 486.1, 589.3, and 656.3 nm, respectively. In terms of n_D and ν:

$$\gamma = \frac{K (n_D - 1) (n_D^2 + 2)^2}{n_D \nu [1.52 + (n_D^2 + 2)(n_D + 1) \nu / 6 n_D)]^{1/2}} \tag{8}$$

where the constant K, which is equal to 2.8×10^{-18} m^2/W, is obtained from fitting experimental results at 1064 nm.[122-127] Measured γ values are compared with γ values calculated from Equation 8, for several different glasses, in Figure 2.3.10. Limitations of this approach to determine γ, and a discussion of empirical relationships for predicting nonlinear refractive-index changes in optical solids, are given in References 121, 128, and 129. Equation 8 fails to predict γ accurately at high values of n_D and ν.

Lines of constant γ, calculated from Equation 8, are shown on a plot of n_D vs. ν in Figure 2.3.11. Regions of known oxide and fluoride optical glasses are indicated in the figure.

Effects of Glass Composition

The spectroscopic properties of rate earth ions in glass are sensitive to the local

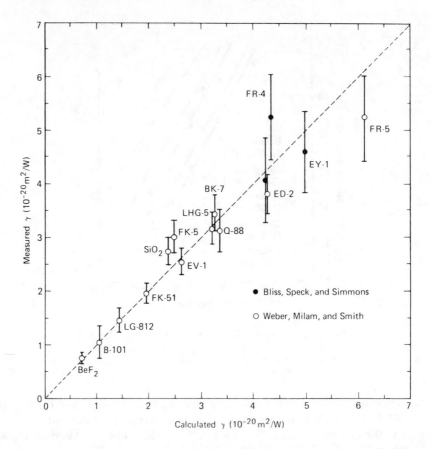

FIGURE 2.3.10.   Comparison of measured and calculated nonlinear index coefficients in different glasses. Data from Bliss, Speck, and Simmons[122] and Weber, Milam, and Smith.[127]

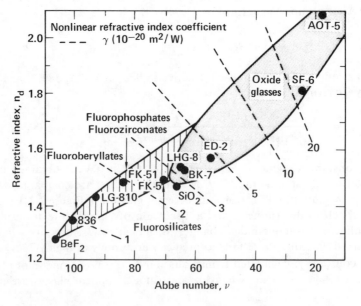

FIGURE 2.3.11.   Refractive index n_D and Abbe number v of optical glasses. Dashed lines of constant n_2 (intensity-dependent index change) are calculated from Equation 8. Figure from Weber (to be published).

Table 2.3.6

OBSERVED VARIATIONS IN THE SPECTROSCOPIC PROPERTIES OF
THE Nd^{3+} $^4F_{3/2} \rightarrow ^4I_{11/2}$ TRANSITION[a]

Glass	Nonlinear index $n_2(10^{-13}$ esu.)	Cross section $\sigma(10^{-20}$ cm^2)	Effective linewidth[b] $\Delta\lambda_{eff}$(nm)	Lifetime $\tau_R(\mu sec)$	Peak wavelength λ_p(nm)
Silicate	≥ 1.2	1.0—3.6	34—43	170—950	1057—1088
Germanate	≥ 1.0	1.6—3.5	22—36	200—500	1057—1065
Phosphate	≥ 1.0	1.8—4.7	23—34	320—560	1060—1063
Borate	≥ 0.9	1.8—4.8	23—43	100—500	1052—1057
Tellurite	≥ 10	3.0—5.1	26—31	140—240	1054—1063
Fluorophosphate	≥ 0.5	2.2—4.3	27—34	350—600	1050—1056
Fluorozirconate	≥ 1.2	2.0—3.4	31—43	300—500	1049
Fluoroberyllate	≥ 0.3	1.7—4.0	19—28	550—1000	1046—1050

[a]   The data listed here were obtained primarily from a study of Nd-doped laser glasses at the Lawrence Livermore Laboratory, Livermore, CA.

[b]   Defined as the ratio of fluorescence intensity integrated over wavelength divided by the peak intensity.

environment, which, in turn, is determined by the overall glass structure. By varying glass formers and glass modifiers, substantial changes can be made in the spectroscopic properties of rare-earth ions.

Laser glasses melted and studied over the past two decades are based on the following glass formers, which are listed along with their common concentration range (in cationic %):

SiO_2	silicates	35 to 65
GeO_2	germanates	50 to 60
TeO_2	tellurites	65 to 80
B_2O_3	borates	50 to 80
P_2O_5	phosphates	45 to 70
BeF_2	fluoroberyllates	45 to 60
ZrF_4	fluorozirconates	50 to 60

Mixed-anion (flourine-oxygen) glasses that have been investigated include fluorosilicates and fluorophosphates with flourine-to-oxygen ratios up to 0.1 and 30, respectively.

The observed variation in properties of the Nd^{3+} $^4F_{3/2} \rightarrow ^4I_{11/2}$ transition with glass former are given in Table 2.3.6. Nd^{3+} absorption and fluorescence spectra in the various glass types are shown in Figures 2.3.12 and 2.3.13. Refer to Figure 2.3.3 for the effects of glass former on the Er^{3+} $^4I_{13/2} \rightarrow ^4I_{15/2}$ transitions.

Within a given glass type systematic changes in laser properties can be made by varying the amount and species of the modifying ions in the glass composition.[130-143] The alkali ions (Li, Na, K), alkaline earth ions (Mg, Ca, Sr, Ba), aluminum, yttrium, and zinc are the most commonly employed glass modifiers. The modifier ions produce the general changes in spectroscopic properties summarized in Table 2.3.6.

Current commercial laser glasses are of the silicate, phosphate, or fluorophosphate glass types. A typical silicate composition (in mole %) with cross section $\simeq 2.7 \times 10^{-20}$ cm^2 is the following:

SiO_2	—	60
Li_2O	—	27.5
CaO	—	10.0
Al_2O_3	—	2.5

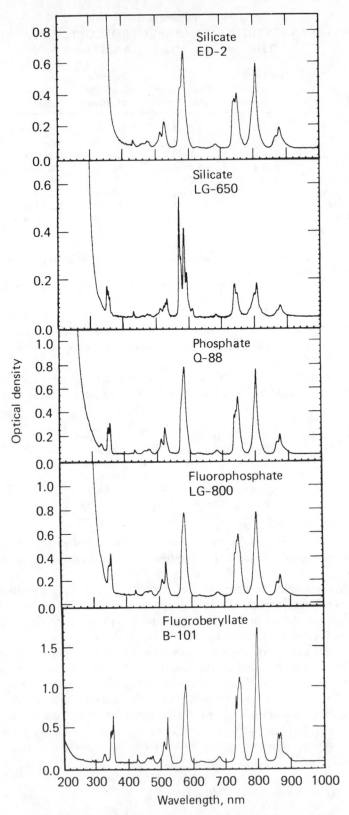

FIGURE 2.3.12. Nd^{3+} absorption spectra for various glasses at 295 K. Data from Linford et al.[78]

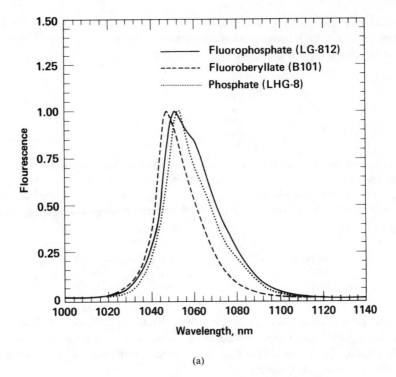

(a)

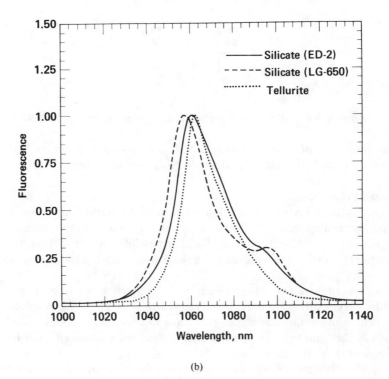

(b)

FIGURE 2.3.13.   Relative Nd^{3+}: $^4F_{3/2}$ fluorescence spectra in various glasses: (a) fluorophosphate, fluoroberyllate, and phosphate; (b) high-cross-section silicate (ED-2), low-cross-section silicate, and tellurite glasses.

By replacing Li and Ca by K and Ba, the cross section is lowered to $\sim 1 \times 10^{-20}$ cm². On the other hand, high cross section phosphates ($\sigma \sim 4.5 \times 10^{-20}$ cm²) contain K and Ba in the following typical composition:

P_2O_5	—	59
BaO	—	8
K_2O	—	25
Al_2O_3	—	5
SiO_2	—	3

Fluorophosphate glasses have low nonlinear indexes because of the use of fluorine as the primary anion. The glass former is nominally P_2O_5, but its content in the glass can be very small (≈ 2 mole %). Using AlF_3 along with three or four alkaline earth fluorides makes the glass stable enough to be melted in large sizes. A typical composition (in mole %) is

$Al(PO_3)_3$	—	4	CaF_2	—	30
AlF_3	—	36	SrF_2	—	10
MgF_2	—	10	BaF_2	—	10

The only pure fluoride laser glasses are those based on BeF_2 or ZrF_4. Although pure BeF_2 forms a glass, its strongly hygroscopic nature confines it to the research laboratory. By adding alkali and alkaline earth modifiers, BeF_2 glass is made more durable. One such composition (in mole%) is[144]

BeF_2	—	47
KF	—	27
CaF_2	—	14
AlF_3	—	10
NdF_3	—	2

A typical composition for a fluorozirconate laser glass is 60% ZrF_4, 34% BaF_2, and 6% NdF_3.[145]

Borate glasses are not usually used for Nd^{3+} laser glasses because the high nonradiative transition rates in this glass[106] reduce the $^4F_{3/2}$ state lifetime to < 100 μsec.

Concentration Quenching

Interactions between rate earth ion pairs lead to nonradiative deexcitation, or quenching, of excited states.[146] In the case of Nd^{3+}, a nonradiative transition between the pair states ($^4F_{3/2}$, $^4I_{9/2}$) and ($^4I_{15/2}$, $^4I_{15/2}$) is possible with or without the assistance of phonons. This process is believed to be the primary mechanism for Nd self-quenching.

In addition, the excited state population is reduced when quenching centers are reached by energy diffusion among the rare-earth ions.[147,148] The possible quenching centers include other rare-earth ions that have nearby energy states, transition metal ions, such as Cu^{2+} and Fe^{2+}, and color centers. Thus, the amount of quenching depends on the impurity concentration.

The Nd $^4F_{3/2}$ lifetime in selected glasses as a function of the Nd concentration is shown in Figure 2.3.14. The severity of concentration quenching depends on the glass type and composition, being least in phosphate glasses and in silicates with low oscillator strengths, such as LG-650. The theoretical aspects of concentration quenching in glasses, along with some experimental results can be found in Reference 157, and references therein.

Table 2.3.7

COMPOSITIONAL VARIATION OF OPTICAL AND SPECTROSCOPIC PROPERTIES WITH ALKALI AND ALKALINE EARTH SPECIES

(Cation % of the alkali/alkaline earth is given in parentheses)
(A) Nd:Glass

Property	ALKALI SPECIES (Li→Rb)				ALKALINE EARTH SPECIES (Mg→Ba)		
	Silicate (50%)	Fluoroberyllate (27%)	Phosphate (38%)	Germanate (20%)	Silicate (17%)	Fluoroberyllate (14%)	Phosphate (33%)
Refractive index @ 589.2 nm	1.54→1.50 (K)	1.344→1.354			1.510→1.569	1.337→1.355 (Sr)	
Nonlinear refractive index, 10^{-13} esu.	0.30→0.33				1.2→1.6	0.31→0.34 (Sr)	
Intensity parameters, 10^{-20} cm²							
Ω_2	3.4→5.7	~0→0.4		Little effect			
Ω_4	4.5→2.2	3.6→3.9		~2× decrease			
Ω_6	4.6→1.9	3.7→4.8					
$^4F_{3/2}\to{}^4I_{11/2}$ Transition							
Effective linewidth, nm	33.8→37.7	25.6→23.0	23.1→21.2	Decreases	41.2→36.0	Less than 10% variation	33.0→26.7
Radiative lifetime, μsec	400→950	700→590		Increases	550→660		
Cross section, 10^{-20} cm²	2.6→1.0	2.3→3.3			1.7→1.4	1.8→3.0	
Ref.	130, 131	132	19	141	130, 131	132	19, 143

Table 2.3.7 (continued)

COMPOSITIONAL VARIATION OF OPTICAL AND SPECTROSCOPIC PROPERTIES WITH ALKALI AND ALKALINE EARTH SPECIES

(Cation % of the alkali/alkaline earth is given in parentheses)

(B) Er:Glass

	ALKALI SPECIES (Li→Rb)	ALKALINE EARTH SPECIES (Mg→Ba)	
Absorption strength	Borate (14%)	Silicate (11%)	Phosphate (33%)
$^4I_{15/2} \rightarrow {}^4I_{13/2}$ transition 10^{-8} cm^2 sec^{-1}	6.2→6.0	4.0→3.0	3.1→4.3
Ref.	133	133	133

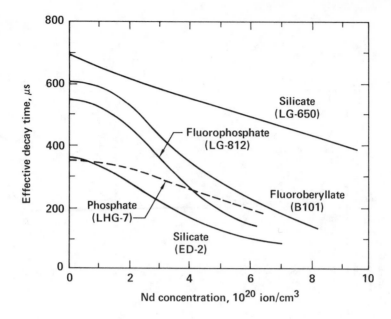

FIGURE 2.3.14.   Time required for the Nd^{3+}: $^4F_{3/2}$ fluorescence to decay to 1/e of its initial value plotted against the Nd^{3+} ion density.

Sensitization

Co-doping laser glass with a sensitizing ion, i.e., an ion that absorbs flashlamp energy not absorbed by a lasing ion and transfers it to the lasing ion, can raise amplifier or oscillator efficiency. An ion must have the following characteristics to be a good sensitizer:

1.   Absorption bands in spectral regions where the lasing ion does not absorb
2.   Efficient energy transfer to the active ion
3.   No significant quenching of the initial lasing state

Sensitizing schemes that have been investigated are listed in Table 2.3.8, which includes transfer rates and efficiencies. The most useful of these schemes are the Nd-sensitized Yb laser and the Yb-sensitized Er laser.   In addition, broad-band pump radiation from flashlamps has been converted to a narrower spectrum matching the active-ion absorption by surrounding them with fluorescent dye solutions. These results are listed in Table 2.3.9.

Laser Damage

The intrinsic bulk damage thresholds of laser glasses are in the range of 40 to 100 J/cm^2 and 12 to 21 J/cm^2 for 40-nsec and 30-psec 1064-nm pulses, respectively.[185-187] These threshold flux levels are not of concern in most practical applications. On the other hand, bulk damage in the glass due to extrinsic material inclusions can be produced at flux levels below 2 J/cm^2.[187] Bulk damage in laser glasses has, in most cases, been due to small platinum particle inclusions. These particles are generated by chemical attack on the platinum crucible used for melting the glass. When irradiated by high laser fluxes, the particles are heated to very high temperatures, fracturing the glass through the resulting thermal stresses.[192] These platinum inclusions have been eliminated from most laser glasses.

Table 2.3.8
SENSITIZER IONS STUDIED IN LASER GLASS

Acceptor ion	Sensitizer ion	Glass	Rate of nonradiative transfer (sec⁻¹)	Acceptor and sensitizer concentrations (10^{20} cm⁻³)	Nonradiative transfer efficiency	Effect on laser performance	Ref.
Nd^{3+}	*Ce^{3+}	Li-Mg-Al silicate	2.6×10^7	1.0, 1.0 (cation %)	0.65		158
Nd^{3+}	*Ce^{3+}	Li-Ca silicate(ED-2)	0.9×10^7	4.0, 0.017	0.44		159
Nd^{3+}	Mn^{2+}	Ca-phosphate	3.7×10^2	2.0, 6.0 (cation %)	0.55		160
Nd^{3+}	Mn^{2+}	Mg-Ca phosphate		1.5, 8.3 (cation %)	>0.9		161
Nd^{3+}	UO_2^{2+}	Ba-silicate	3.2×10^5	3.5, 0.37	0.86	Self-Q-switching	162
Nd^{3+}	UO_2^{2+}	Ba-silicate	1.8×10^4	1.0, 0.1 (cation %)	0.86		163
Nd^{3+}	UO_2^{2+}	Ba-borate	1.2×10^4	3.0, 0.25 (wt. % of oxide)	0.70		164
Nd^{3+}	Tb^{3+}	Borosilicate	6.0×10^2	3.0, 3.0 (wt. % of oxide)	0.60	Nd fluorescence up by 10%	165
Nd^{3+}	Bi^{3+}	Germanate					166
Nd^{3+}	Eu^{3+}	Borosilicate	1.2×10^6	1.0, 1.0 (wt. % of oxide)	0.31	Nd fluorescence up by 10%	165
Nd^{3+}	Eu^{3+}	Zn-Ba silicate				No increase in Nd fluorescence due to quenching by Eu	167
Nd^{3+}	Cr^{3+}	Silicate		3.0, 0.5 (wt. % of oxide)	>0.9		168
Nd^{3+}	Cr^{3+}	Silicate		4.0, 0.2 (wt. % of oxide)	>0.9	Little change in threshold	169
Nd^{3+}	Cr^{3+}	Li-Sr silicate	1.6×10^5	3.0, 0.08	0.8	Cr absorption at 1060 nm decreases efficiency	170
Nd^{3+}	Cr^{3+}	Li-La phosphate	$>10^6$	8.1, 2.2	>0.9		171
Tb^{3+}	Ce^{3+}	Borate	1.4×10^7	3.0, 0.25 (wt. % of oxide)	0.32		172
Tb^{3+}	Cu^+	Ca-phosphate					160
Ho^{3+}	Yb^{3+}	Li-Mg-Al silicate	1.2×10^4	3.0, 3.0	0.79		163
Ho^{3+}	Tb^{3+}	Ca-Li borate	1.3×10^3	3.0, 2.0 (wt. % of oxide)	0.76		173
Ho^{3+}	*Er^{3+}, Yb^{3+}	Li-Mg-Al silicate		4.0(Ho), 2.0(Er), 5.0(Yb) (wt. % of oxide)		Threshold is a minimum with 2% Er_2O_3	30
Er^{3+}	*Yb^{3+}	Silicate	4.6×10^3	0.47, 18.4	0.92		48
Er^{3+}	*Yb^{3+}	Phosphate	10.7×10^3	0.3, 22.0	0.86	35% conversion 1.06 to 1.54 μm	48

						Ref.
Er^{3+}	*Yb^{3+}	Fluorophosphate	0.68, 11.0	4.1×10^3	0.82		48
Er^{3+}	*Yb^{3+}	Ba-Al phosphate	0.25, 15.0	$>9.0 \times 10^3$	0.93		174
Er^{3+}	*Yb^{3+}	Fluorophosphate	1.56, 10.5			Room temp. laser operation	32
Er^{3+}	*Yb^{3+}	Li-Mg-Al silicate	3.3, 3.3	2.4×10^4	>0.95	12-fold decrease in threshold	4, 34
Er^{3+}	*Yb^{3+}	Mg-Li silicate	1.5, 8.3	10^4	>0.95		175
Tm^{3+}	Er^{3+}	Li-Mg-Al silicate	3.3, 3.3	5.7×10^3	>0.95		34
Tm^{3+}	Yb^{3+}	Li-Mg-Al silicate	3.3, 3.3	10^4	0.90		34
Tm^{3+}	*Yb^{3+}, Er^{3+}	Li-Mg-Al silicate	3.3, 6.6 (Er^{3+}), 6.6 (Yb^{3+})	8.3×10^4	>0.95	10-fold decrease in threshold	34
Tm^{3+}	Ce^{3+}	Borate	1.0, 0.13 (wt. % of oxide)	1.5×10^7	0.52		176
Yb^{3+}	Ce^{3+}	Li-Mg-Al silicate	3.2, 3.2	5.2×10^7	0.78		163
Yb^{3+}	*Nd^{3+}	Na-K-Ba silicate	2.5, 2.5	8.0×10^3	0.95		163
Yb^{3+}	*Nd^{3+}	Li-Mg-Al silicate	3.2, 3.2	4.6×10^4	0.92		163
Yb^{3+}	*Nd^{3+}	Ca-Li borate	1.8, 0.9	3.0×10^4			177
Yb^{3+}	*Nd^{3+}	Silicate	2.1, 2.1	4.6×10^3			178
Yb^{3+}	*Nd^{3+}	Na-K-Ba silicate	1.8, 4.5	3.5×10^3			179
Yb^{3+}	*Nd^{3+}	Mg-Li silicate	8.3, 2.0	3.5×10^4			175
Yb^{3+}	Cr^{3+}	Silicate	2.0, 0.1 (wt. % of oxide)	$\sim 10^6$	>0.95	Yb fluorescence is quenched by Cr	180

Note: Those sensitizer ions actually used for a glass laser are indicated by a star.

Table 2.3.9

LUMINESCENT LIQUIDS USED FOR PUMP
CONVERSION IN Nd: GLASS LASERS

Dye	Concentration (M)	Solute	Effect on laser performance	Ref.
Rhodamine 6G	3×10^{-4}	Ethanol	38% increase in output	181
	2×10^{-4}	Ethanol	32% reduction in threshold; 33% increase in slope efficiency	182
	10^{-3}	Ethanol	30% reduction in threshold; 50 to 125% increase in efficiency	183
Dye 1351	10^{-2}	Ethylene gly-colethers	20% increase in output	184

Table 2.3.10

SURFACE DAMAGE THRESHOLDS OF LASER
GLASSES

Glass	Type	Threshold (J/cm^2)	Wavelength (nm) Pulse width (nsec)	Ref.
ED-2	Silicate	13.8 ± 2.0^a	1064, 0.125	189
		100	1064, 30	190
Q-88	Phosphate	13.7 ± 2.0	1064, 0.125	189
LSG-91H	Silicate	13.1 ± 2.0	1064, 0.125	189
FK-51	Fluorophosphate	10.0 ± 1.5	1064, 0.125	189
		7.2 ± 1.1	1064, 0.125	189
		20 ± 3	1064, 1.0	188
LG-802	Fluorophosphate	13	1064, 1.0	188
		40	1064, 40	185
E-309	Fluorophosphate	16 ± 2	1064, 1.0	188
B-101	Fluoroberyllate	$12 - 15$	1064, 1.0	188

[a]   D. Milam, in a private communication, states that a more sensitive technique of detecting surface damage using Nomarski microscopy indicates that the values in Reference 189 may be high by ∼50%.

Surface damage thresholds are generally lower, 10 to 20 J/cm^2 for 1-nsec pulses, than bulk thresholds. The measured surface damage thresholds of several laser glasses are listed in Table 2.3.10. The measured thresholds with >30 nsec pulses are found to vary with surface finishing techniques.[191]

Solarization

Changes in the optical absorption of a glass resulting from its exposure to UV or blue light is defined as solarization.[193] The induced background absorption in solarized glass, which is due to color centers or to impurity ion valence changes, is detrimental to laser pump efficiency.[194,195] Transient solarization can also result in self-Q-switching in laser glass.[196] Glass solarization can be avoided by preventing exposure to UV light

Table 2.3.11

HIGHLY CONCENTRATED Nd LASER GLASSES

Glass	NdP$_5$O$_{14}$	Li-phosphate	Li-tellurite	Al-phosphate	Li-phosphate
Nd ion concentration (10^{20}cm^{-3})	39	27	9.0	27	27
Fluorescence decay time (μsec)		38	50	50	
Quantum efficiency			0.28	~0.1	0.23
Laser wavelength (μm)	1.05	1.055	1.061	1.05	1.054, 1.065, 1.323
Excitation source	laser	Raman laser (λ = 0.74, 0.8 μm)	Argon laser (λ = 0.5145 μm)	Dye laser (λ = 0.583 μ)	Lamp
Temperature	R.T.[a]	R.T.	R.T.	R.T.	300 K, 400 K, 300 K
Emission cross section (10^{-20} cm^2)		3.8	4.7	1.96[b], 2.7	3.8
Comments	Threshold ~ 5 times higher that of NdP$_5$O$_{14}$ crystal				
Ref.	201	202	203	204	205

[a]   Ambient or room temperature.
[b]   Obtained by two different methods.

or by stabilizing the impurity oxidation states.[197] An example of the first technique is the use of Ce^{3+} in the flashlamp envelopes or in the glass to absorb UV light.[194,198,199] Ions such as Sb^{3+}, Mo^{6+}, Nb^{5+}, and Ti^{4+}, are used in laser glasses to stabilize the oxidation potential, thereby preventing long-term solarization.[200]

Glass for Miniature Lasers

Very small glass laser oscillators or amplifiers can be used in integrated or fiber optic applications. However, because of their small size, these devices require glasses with high Nd concentrations to obtain sufficient pump absorption. Highly concentrated Nd laser glasses are listed in Table 2.3.11.

SPECIFIC LASER GLASS PROPERTIES

Table 2.3.12 contains the spectroscopic, optical, thermal, and mechanical properties of selected Nd laser glasses. These glasses are representative of laser glasses that can be melted in large sizes with high optical quality. Most of these glasses have been offered commercially. Glass types include silicate, phosphate, fluorophosphate, and fluoroberyllate. Below is a short description of the properties listed in Table 2.3.12.

Spectroscopic Properties

Peak fluorescence wavelength — Value measured at 295 K. (When the value for the $^4F_{3/2} \rightarrow {}^4I_{13/2}$ transition is available, it is given in parentheses.)

Linewidth (FWHM) — Full width at 50% of peak intensity.

Linewidth (effective) — Defined as the ratio of fluorescence intensity integrated over wavelength divided by the peak intensity. (When the value for the $^4F_{3/2} \rightarrow {}^4I_{13/2}$ transition is available, it is given in parentheses.)

Peak emission cross sections — Value determined from the calculated Judd-Ofelt line strength and the measured fluorescence linewidth. (When the value for the $^4F_{3/2} \rightarrow {}^4I_{13/2}$ transition is available, it is given in parentheses.)

Calculated radiative lifetime — Calculated from the Judd-Ofelt intensity parameters.

Zero-concentration lifetime — Time for the fluorescence intensity to decay to e^{-1} of its initial value (first e-folding time) as measured in a very dilute sample, or by extrapolation to zero concentration.

e-Folding times — Times for the fluorescence intensity to decay to e^{-1}, from e^{-1} to e^{-2}, and from e^{-2} to e^{-3} of its initial value, respectively.

Absorption loss coefficient — Measured at 1064 nm.

Judd-Ofelt parameters — Optical intensity parameters determined from a least-squares fit to the measured absorption band strengths.

Optical Properties

Refractive index — Measured at 295 K.

Abbe number — Reciprocal dispersion calculated from the measured values of refractive index at 486.1 nm (n_F), 589.3 nm (n_D), and 656.3 nm (n_C).

Index temperature coefficient — Change in the refractive index with temperature, measured at the given wavelength and standard pressure (760 mmHg) for the temperature range indicated.

Thermal coefficient of optical path length — Change in the optical path length with temperature, $\partial n / \partial T + (n - 1)\alpha$, for the temperature range and wavelength indicated.

Nonlinear refractive index — Measured value if available; calculated value is obtained from Equation 8.

Table 2.3.12
PROPERTIES OF SELECTED LASER GLASSES*

Spectroscopic properties

Glass / Glass type		LG-650[b] Silicate	LG-660[b] Silicate	ED-2[d] Silicate	LSG-91H[g] Silicate
Peak-fluorescence wavelengths	nm	1057 (1325[a])	1057	1061 (1335[a])	1061.5 (1335[a])
Linewidths (FWHM)	nm	23.5	24.9	27.8	27.4
Linewidths (effective)	nm	34.3 (52[a])	33.3	34.4 (67[a])	34.4 (64[a])
Peak emission cross sections	10^{-20} cm^2	1.1 (0.31[a])	2.0	2.7[a] (0.72[a])	2.4[a] (0.65[a])
Calculated radiative lifetime	μsec	926	540	359	412
Zero-concentration lifetime	μsec	840	530	410	440
e-Folding times	μsec	430, 599, 637	474, 553, 631	309, 360, 358	287, 314, 362
@ Nd concentration	10^{20} cm^{-3}	4.70	1.32	1.83	3.01
Absorption loss coefficient	cm^{-1}	<0.002	<0.002	<0.002	<0.002
Judd-Ofelt parameters: Ω_2	10^{-20} cm^2	3.7 ± 0.2	4.1 ± 0.2	3.2 ± 0.2	3.5 ± 0.2
Ω_4	10^{-20} cm^2	2.2 ± 0.3	3.3 ± 0.3	4.6 ± 0.3	4.0 ± 0.3
Ω_6	10^{-20} cm^2	1.8 ± 0.1	3.6 ± 0.1	4.8 ± 0.1	4.4 ± 0.1

Optical properties

Glass / Glass type		LG-650[b] Silicate	LG-660[b] Silicate	ED-2[d] Silicate	LSG-91H[g] Silicate
Refractive index: 589.3 nm		1.5214	1.520[b]	1.5672[d]	1.5612[g]
@ peak λ		1.5118	1.511[b]	1.5554[d]	1.5487[g]
Abbe number		56.2	58.5[b]	58.2[d]	56.6[g]
Index temperature coefficient	10^{-7} °C^{-1}	−19[b]		29[n], 38 ± 5[p,r]	27 ± 5[h]
Thermal coefficient of optical path length	10^{-7} °C^{-1}	30[b]		80[n], 81[p,r]	81
Nonlinear refractive index	10^{-13} esu	1.4[c]	1.33[c]	1.41 ± 0.14[l]	1.5[c]
Stress optical coefficients: ΔB	nm·cm·kg^{-1}	2.7[b]		2.06[d], 2.02 ± 0.15[p,q]	2.16, 2.12 ± 0.15[p,q]
B⊥	nm·cm·kg^{-1}			2.60 ± 0.52[p,s]	
B∥	nm·cm·kg^{-1}			−0.20 ± 0.7[p,x]	

* Manufacturers may make compositional changes or change designations; therefore, the manufacturer's literature should be consulted for the most recent property data. Also, availability of these glasses is subject to change.

Table 2.3.12 (continued)
PROPERTIES OF SELECTED LASER GLASSES

Glass / Glass type		LG-650[b] Silicate	LG-660[b] Silicate	ED-2[d] Silicate	LSG-91H[a] Silicate
Thermal properties					
Thermal expansion coefficient	10^{-7} °C^{-1}	95[b] (−30 to 70°C)	121[b] (20-300°C)	92.6[d] (25-100°C)	105[g] (100-300°C)
Specific heat at constant pressure	J·g^{-1}·°C^{-1}	0.97[b] (25°C)		0.92[d] (25°C)	0.63[g] (50°C)
Thermal heat conductivity	W·m^{-1}·°C^{-1}			1.35[d](100°C), 1.25[p,u]	1.03[g], 1.05[p,u]
Strain point ($\eta = 10^{14.5}$)	°C			454[d]	
Transformation point	°C	485[b]	399[b]		465[g]
Anneal point ($\eta = 10^{13.0}$)	°C			468[d]	
Softening point ($\eta = 10^{7.6}$)	°C	682[b]	573[b]	582[d]	505[g]
Mechanical properties					
Density	g·cm^{-3}	2.625	2.602[b]	2.539	2.814
Knoop hardness	kg·mm^{-2}	418[b]		599[d], 612 ± 18[p,t]	590[g], 478 ± 14[p,t]
Young's modulus	kg·mm^{-2}	6320[b]		9189[d], 9620 ± 130[p]	8890[g], 8607 ± 130[p]
Shear modulus	kg·mm^{-2}			3698[d], 3633 ± 130[p]	3590[g], 3389 ± 130[p]
Poisson ratio				0.242[d]	0.237

Glass / Glass type		EV-4[a] Phosphate	Q-88[i] Phosphate	LHG-5[s] Phosphate
Spectroscopic properties				
Peak-fluorescence wavelengths	nm	1053.5	1054	1053 (1323[q])
Linewidths (FWHM)	nm	19.6	21.9	22.0
Linewidths (effective)	nm	23.6	26.3	26.1 (49[q])
Peak emission cross sections	10^{-20} cm^2	4.2	4.0	4.1 (0.94[q])

Property	Units			
Calculated radiative lifetime	μsec	362	326	322
Zero-concentration lifetime	μsec	340[d]	380[p]	350[p]
e-Folding times	μsec	257, 271, 284	273, 289, 341	269, 297, 342
@ Nd concentration	10^{20} cm^{-3}	2.83	2.91	0.96
Absorption loss coefficient	cm^{-1}	<0.003		
Judd-Ofelt parameters: Ω_2	10^{-20} cm^2	3.6 ± 0.3	3.3 ± 0.2	4.6 ± 0.3
Ω_4	10^{-20} cm^2	4.7 ± 0.4	5.1 ± 0.3	5.1 ± 0.5
Ω_6	10^{-20} cm^2	5.5 ± 0.2	5.6 ± 0.1	6.0 ± 0.2

Optical properties

Property	Units			
Refractive index:				
589.3 nm		1.5138[d]	1.5449[i]	1.5409[g]
@ peak λ		1.5034[d]	1.5362	1.5308[g]
Abbe number		68.5[d]	64.8[i]	63.5[g]
Index temperature coefficient	10^{-7} °C^{-1}	−68[d]	−6 ± 5[p,q]	0.0[g], −2 ± 5[p,q]
Thermal coefficient of optical path length	10^{-7} °C^{-1}	0 ± 5[d]	42[p,q]	46[g]
Nonlinear refractive index	10^{-13} esu	1.04[c]	1.14 ± 0.15[k]	1.16 ± 0.12[i]
Stress optical coefficient: ΔB	nm·cm·kg^{-1}	1.9[d]	2.07 ± 0.1[p,q]	2.26[g], 1.97 ± 0.1[p,q]
B⊽	nm·cm·kg^{-1}		2.9 ± 0.15[p,s]	3.5 ± 0.15[p,s]
B∥	nm·cm·kg^{-1}		0.0 ± 0.7[p,s]	2.2 ± 0.15[p,s]

Thermal properties

Property	Units			
Thermal expansion coefficient	10^{-7} °C^{-1}	138[d] (0-250°C)	92[i] (100-300°C)	98[g] (100-300°C)
Specific heat at constant pressure	J·g^{-1}·°C^{-1}	0.570(0°C)[d]	0.88[i]	0.71[g] (20°C)
Thermal heat conductivity	W·m^{-1}·°C^{-1}	0.67(25°C)[d]	0.72[i], 0.74[h] (25°C)	0.99[i], 0.73[h] (25°C)
Strain point ($\eta = 10^{14.5}$)	°C	347[d]	367[i]	455[g]
Transformation point	°C			
Anneal point ($\eta = 10^{13.0}$)	°C	375[d]	384[i]	486[g]
Softening point ($\eta = 10^{7.6}$)	°C	467[d]		

Table 2.3.12 (continued)
PROPERTIES OF SELECTED LASER GLASSES

Glass Glass type		LG-650[b] Silicate	LG-660[b] Silicate	ED-2[d] Silicate	LSG-91H[a] Silicate
Mechanical properties					
Density	$g \cdot cm^{-3}$	2.165		2.713	2.674
Knoop hardness	$kg \cdot mm^{-2}$			$418 \pm 13^{p,t}$	$497^t, 423 \pm 13^{p,t}$
Young's modulus	$kg \cdot mm^{-2}$	5473^d		7123 ± 130^p	$6429^g, 7258 \pm 130^p$
Shear modulus	$kg \cdot mm^{-2}$	2158^d		2680 ± 130^p	$2631^g, 2707 \pm 130^p$
Poisson ratio		0.267^d		0.24	0.184^g

Glass Glass type		LG-700[b] Phosphate	LHG-8[a] Phosphate	LHG-10[a] Fluorophosphate
Spectroscopic properties				
Peak-fluorescence wavelengths	nm	1053.5	1053	1050.5
Linewidths (FWHM)	nm	22.3	21.8	26.5
Linewidths (effective)	nm	26.6	25.9	31.3
Peak emission cross sections	cm^2 10^{-20}	3.7	4.0	2.6
Calculated radiative lifetime	μsec	363	338	475
Zero-concentration lifetime	μsec		380^a	550
e-Folding times	μsec		290, 293, 278	319, 385, 414
@ Nd concentration	cm^{-3} 10^{-20}		2.98	3.09
Absorption loss coefficient	cm^{-1}	<0.001	<0.002	<0.002
Judd-Ofelt parameters: Ω_2	cm^2 10^{-20}	4.2 ± 0.2	4.4 ± 0.3	1.6 ± 0.2
Ω_4	cm^2 10^{-20}	4.6 ± 0.3	5.1 ± 0.4	4.0 ± 0.4
Ω_6	cm^2 10^{-20}	5.4 ± 0.1	5.6 ± 0.2	4.7 ± 0.2

Optical properties

Refractive index:					
589.3 nm			1.5253	1.5312	1.4592
@ peak λ			1.5173	1.5234	1.4539
Abbe number			68	70.2	89.9
Index temperature coefficient	$°C^{-1}$	10^{-7}	-27^b, $-31 \pm 5^{p,g}$	-53^g, -46 ± 5^h	$-49 \pm 5^{h,q,r}$
Thermal coefficient of optical path length	$°C^{-1}$	10^{-7}	26^b, 18^p	6^g	$16^{h,q,r}$
Nonlinear refractive index	esu	10^{-13}	1.08^c	1.05^c	0.58^c
Stress optical coefficient: ΔB	$nm \cdot cm \cdot kg^{-1}$		2.1, $2.07 \pm 0.15^{p,q}$	1.93^g, $1.75 \pm 0.1^{p,q}$	$0.61 \pm 0.02^{i,q}$, $0.54 \pm 0.1^{p,q}$

Thermal properties

Thermal expansion coefficient	$°C^{-1}$	10^{-7}	103^b $(-30$ to $70°C)$	112^g $(100\text{-}300°C)$	158^g $(100\text{-}300°C)$
Specific heat at constant pressure	$J \cdot g^{-1} \cdot °C^{-1}$		0.79^b	$0.75^g (25°C)$	$0.84^{h,u}$
Thermal heat conductivity	$W \cdot m^{-1} \cdot °C^{-1}$		0.79^b, $0.69^{p,u}$	$0.52^h (25°C)$	
Strain point $(\eta = 10^{14.5})$	$°C$		430^b	485^g	435^g
Transformation point	$°C$		460^b		
Anneal point $(\eta = 10^{13.0})$	$°C$		472^b	520^g	459^g
Softening point $(\eta = 10^{7.9})$	$°C$		543^b		

Mechanical properties

Density	$g \cdot cm^{-3}$		2.73	2.849	3.641
Knoop hardness	$kg \cdot mm^{-2}$		320^b, $372 \pm 11^{p,t}$	321^g, $344 \pm 10^{p,t}$	355^g
Young's modulus	$kg \cdot mm^{-2}$		6180^b, 6427 ± 130^p	5109^g, 5619 ± 130^p	7389 ± 130^h
Shear modulus	$kg \cdot mm^{-2}$		2480^b, 2377 ± 130^p	2030^g, 2113 ± 130^p	2630 ± 130^h
Poisson ratio			0.243^b	0.26^g	

Table 2.3.12 (continued)
PROPERTIES OF SELECTED LASER GLASSES

Glass Glass type		LG-812[b] Fluorophosphate	E-309[d] Fluorophosphate	B101ℓ Fluoroberyllate
Spectroscopic properties				
Peak-fluorescence wavelengths	nm	1051	1050.5	1047.5
Linewidths (FWHM)	nm	26.1	26.6	19.4
Linewidths (effective)	nm	31.0	31.3	23.2
Peak emission cross sections	10^{-20} cm^2	2.6	2.5	3.2
Calculated radiative lifetime	μsec	495	510	611
Zero-concentration lifetime	μsec	550		613
e-Folding times	μsec	335, 404, 419	443, 483, 507	271, 345, 367
@ Nd concentration	10^{20} cm^{-3}	2.95	2.08	4.86
Absorption loss coefficient	cm^{-1}	<0.002	<0.002	
Judd-Ofelt parameters: Ω_2	10^{-20} cm^2	1.7 ± 0.3	1.6 ± 0.2	0.2 ± 0.3
Ω_4	10^{-20} cm^2	4.1 ± 0.4	3.9 ± 0.3	3.9 ± 0.4
Ω_6	10^{-20} cm^2	4.8 ± 0.2	4.6 ± 0.2	4.6 ± 0.2
Optical properties				
Refractive index:				
589.3 nm		1.435[b]	1.4414	1.3459
@ peak λ		1.429[b]	1.436	1.3422
Abbe number		90.8[b]	90.4	96
Index temperature coefficient	10^{-7} °C^{-1}	−75[b]		−98[v]
Thermal coefficient of optical path length	10^{-7} °C^{-1}	−12[b]		−31[v]
Nonlinear refractive index	10^{-13} esu	0.49 ± 0.08[k]	0.54[c]	0.32 ± 0.1[m]
Stress optical coefficient: ΔB	nm·cm·kg^{-1}	0.91[b], 0.91 ± 0.02[j,q]		1.02[h,q]

Thermal properties

Thermal expansion coefficient	10^{-7}	$°C^{-1}$	165^b (20-300°C)		200^v
Specific heat at constant pressure		$J \cdot g^{-1} \cdot °C^{-1}$	0.71^b		
Thermal heat conductivity		$W \cdot m^{-1} \cdot °C^{-1}$	1.06^b		0.77^v
Strain point ($\eta = 10^{14.5}$)		°C			
Transformation point		°C	437^b		
Anneal point ($\eta = 10^{13.0}$)		°C			
Softening point ($\eta = 10^{7.6}$)		°C	566^b		

Mechanical properties

Density		$g \cdot cm^{-3}$	3.194	3.391	2.6213
Knoop hardness		$kg \cdot mm^{-2}$	330^b		215^v
Young's modulus		$kg \cdot mm^{-2}$	7680^b		4595^v
Shear modulus		$kg \cdot mm^{-2}$	3010^b		1817^v
Poisson ratio			0.275		0.264

[a] **Jacobs, R. R. and Weber, M. J.**, Induced-emission cross sections for the $^4F_{3/2} \rightarrow {}^4I_{13/2}$ transition in neodymium laser glasses, *IEEE J. Quantum Electron.*, QE-11, 846, 1975.

[b] Schott Optical Glass, Duryea, Pa.

[c] Calculated.

[d] Owens-Illinois, Toledo, Ohio. (This glass is now manufactured by Scholt Optical Glass under the name LG-670.)

[e] For other values, see **Krupke, W. F.**, Induced-emission cross section in neodymium laser glasses, *IEEE J. Quantum Electron.*, QE-10, 450, 1974.

[f] **Milam, D. and Weber, M. J.**, Measurement of nonlinear refractive index coefficients using time-resolved interferometry: Application to optical materials for high-power neodymium lasers, *J. Appl. Phys.*, 47, 2497, 1976.

[g] Hoya Corp., Tokyo.

[h] **Jacobs, S.**, *Laboratory for Laser Energetics*, University of Rochester.

[i] Kigre, Toledo, Ohio.

[j] **Waxler, R.**, National Bureau of Standards.

[k] **Weber, M. J., Milam, D., and Smith, W. L.**, Nonlinear refractive index of glasses and crystals, *Opt. Eng.*, 17, 463, 1978.

[l] Composition (mol. %): 47 BeF_2, 27 KF, 14 CaF_2, 10 AlF_3, and 2 NdF_3.

[m] **Weber, M. J., Cline, C. F., Smith, W. L., Milam, D., Heiman, D., and Hellwarth, R. W.**, Measurements of the electronic and nuclear contributions to the nonlinear refractive index of beryllium fluoride glasses, *Appl. Phys. Lett.*, 32, 403, 1978.

Table 2.3.12 (continued)
PROPERTIES OF SELECTED LASER GLASSES

[n]   25 to 100°C, 633 nm.
[o]   University of Rochester, Laboratory for Laser Energetics Annual Report 1977, Rochester, New York, 1978, 36.
[q]   @ 1064 nm.
[r]   10 to 40°C.
[s]   @ 632.8 nm.
[t]   200 g load for 20 sec.
[u]   @ 20°C.
[v]   Reference 132.

Stress optical coefficient — Measured difference in the refractive indexes for light polarized parallel and perpendicular to an applied uniaxial stress. Measurement wavelength is indicated.

Thermal Properties

Thermal expansion coefficient — Linear coefficient of expansion over the indicated temperature range.

Specific heat at constant pressure — Measured at the indicated temperature.

Thermal heat conductivity — Measured at the indicated temperature.

Strain point — Temperature at which the glass viscosity is 10^{14}Poise.

Transformation point — Temperature at which the thermal expansion coefficient undergoes a significant change in value (glass viscosity is $\sim 10^{13}$P).

Anneal point — Temperature at which the glass viscosity is 10^{13}P.

Softening point — Temperature at which the glass viscosity is $10^{7.6}$P, also known as the yield point.

Mechanical Properties

Density — Measured at 295 K.

Knoop hardness — Hardness measured for the indicated load.

Young's modulus — Ratio of tensile stress to linear strain.

Shear modulus — Radio of shear stress to shear strain.

Poisson's ratio — Ratio of lateral contractional strain to the longitudinal extensional strain. Calculated from the relation $\sigma = (E/2G) - 1$.

REFERENCES

1. Snitzer, E., Optical maser action of Nd^{+3} in a barium crown glass, *Phys. Rev. Lett.*, 7, 444, 1961.
2. Etzel, H. W., Gandy, H. W., and Ginther, R. J., Stimulated emission of infrared radiation from ytterbium-activated silicate glass, *Appl. Opt.*, 1, 534, 1962.
3. Gandy, H. W. and Ginther, R. J., Stimulated emission from holmium activated silicate glass, *Proc. IRE*, 50, 2113, 1962.
4. Snitzer, E. and Woodcock, R., Yb^{3+} - Er^{3+} glass laser, *Appl. Phys. Lett.*, 6, 45, 1965.
5. Gandy, H. W., Ginther, R. J., and Weller, J. F., Stimulated emission of Tm^{3+} radiation in silicate glass, *J. Appl. Phys.*, 38, 3030, 1967.
6. Andreev, S. I., Bedilov, M. R., Karapetyan, G. O., and Likhachev, V. M., Stimulated radiation of glass activated by terbium, *Sov. J. Opt. Tech.*, 34, 819, 1967; *Opt.-Mekh. Promst.*, 34, 60, 1967.
7. Simmons, W. W., Speck, D. R., and Hunt, J. T., Argus laser system: performance summary, *Appl. Opt.*, 17, 999, 1978; Lawrence Livermore Laboratory, *Laser Program Annual Report*, 1977, UCRL-50021-77, Lawrence Livermore Laboratory, Livermore, Calif., 1978.
8. Koechner, W., *Solid-State Laser Engineering*, Springer Series in Optical Sciences Vol. 1, Springer-Verlag, New York, 1976.
9. Patek, K., *Glass Lasers*, Butterworth, London, 1970.
10. Snitzer, E. and Young, C. G., Glass lasers, in *Lasers*, Vol. 2, Levine, A. K., Ed., Marcel Dekker, New York, 1966, chap. 2.
11. Weber, M. J., Solid state lasers, in *Methods of Experimental Physics*, Vol. 15A, Academic Press, New York, 1979, 167.
12. Reisfeld, R. and Jorgensen, C. K., *Lasers and Excited States of Rare Earths*, Springer-Verlag, New York, 1977.
13. Wybourne, B. G., *Spectroscopic Properties of Rare Earths*, Interscience, New York, 1965.
14. Wong, J. and Angell, A., *Structure by Spectroscopy*, Marcel Dekker, New York, 1976.
15. Weyl, W. A., *Coloured Glasses*, Society of Glass Technology, Sheffield, 1951 (republished, 1976).
16. Hüfner, S., *Optical Spectra of Transparent Rare Earth Compounds*, Academic Press, New York, 1978.
17. Dieke, G. H., *Spectra and Energy Levels of Rare-Earth Ions in Crystals*, John Wiley & Sons, New York, 1968.

18. Young, C. G., Continuous glass laser, *Appl. Phys. Lett.,* 2, 151, 1963.

19. Deutschbein, O., Pautrat, C., and Svirchevsky, I. M., Phosphate glasses, new laser materials, *Rev. Phys. Appl.,* 1, 29, 1967.

20. Petrovksii, G. T., Tolstoi, M. N., Feofilov, P. P., Tsurikova, G. A., and Shapovalov, V. N., Luminescence and stimulated emission of neodymium in beryllium fluoride glass, *Opt. Spectrosc. USSR,* 21, 72, 1966; *Opt. Spektrosk.,* 21, 126, 1966.

21. Birnbaum, M., Fincher, C. L., Dugger, C. O., Goodrum, J., and Lipson, H., Laser characteristics of neodymium-doped lithium germinate glass, *J. Appl. Phys.,* 41, 2470, 1970.

22. Rapp, C. F. and Chrysochoos, J., Neodymium-doped glass-ceramic laser material, *J. Mater. Sci.,* 7, 1090, 1972.

23. Muller, G. and Neuroth, N., Glass ceramic — a new host material, *J. Appl. Phys.,* 44, 2315, 1973.

24. Stone, J. and Burrus, C. A., Neodymium-doped silica lasers in end-pumped fiber geometry, *Appl. Phys. Lett.,* 23, 388, 1973.

25. Galaktionova, N. M., Garkavi, G. A., Zubkova, V. S., Mak, A. A., Soms, L. N., and Khaleev, M. M., Continuous Nd-glass laser, *Opt. Spectrosc. USSR,* 37, 90, 1974; *Opt. Spektrosk.,* 37, 162, 1974.

26. Maurer, R. D., Operation of a Nd^{3+} glass optical maser at 9180 A, *Appl. Opt.,* 2, 87, 1963.

27. Artem'ev, E. P., Murzin, A. G., and Fromzel, V. A., Room-temperature laser action at 0.92 μm in neodymium glasses, *Sov. Phys. Tech. Phys.,* 22, 274, 1977; *Zh, Tekh. Fiz.,* 47, 456, 1977.

28. Mauer, P. B., Laser action in neodymium-doped glass at 1.37 microns, *Appl. Opt.,* 3, 153, 1964.

29. Andreev, S. I., Bedilov, M. R., Karapetyan, G. O., and Likhachev, V. M., Stimulated radiation of glass activated by terbium, *Sov. J. Opt. Tech.,* 34, 819, 1967; *Opt.-Mekh. Promst.,* 34, 60, 1967.

30. Veinberg, T. I., Zhmyreva, I. A., Kolobkov, V. P., and Kudryashov, P. I., Laser action of Ho^{3+} ions in silicate glasses coactivated by holmium, erbium, and ytterbium, *Opt. Spectrosc. USSR,* 24, 441, 1968; *Opt. Specktrosk,* 24, 823, 1968.

31. Gandy, H. W., Ginther, R. J., and Weller, J. F., Laser oscillations in erbium activated silicate glass, *Phys. Lett.,* 16, 266, 1965.

32. Auzel, F., Stimulated emission of Er^{3+} in a fluorophosphate glass, *C. R. Acad. Sci. B,* 263, 765, 1966.

33. Snitzer, E., Woodcock, R. F., and Segre, J., Phosphate glass Er^{3+} laser, *IEEE J. Quantum Electron.,* QE-4, 360, 1968.

34. Gandy, H. W. and Ginther, R. J., Simultaneous laser action of neodymium and ytterbium ions in silicate glass, *Proc. IRE,* 50, 2114, 1962.

35. Pearson, A. D. and Porto, S. P. S., Non-radiative energy exchange and laser oscillation in Yb^{3+} - Nd^{3+} - doped borate glass, *Appl. Phys. Lett.,* 4, 202, 1964.

36. Snitzer, E., Laser emission at 1.06 μ from Nd^{3+} - Yb^{3+} glass, *IEEE J. Quantum Electron.,* QE-2, 562, 1966.

37. Owens-Illinois Technical Data.

38. Young, C. G., Glass laser delivers 5000-joule output, *Laser Focus,* 3, 36, 1967.

39. Current information on laser oscillators available from publications such as the *Laser Focus Buyers' Guide,* Advanced Technology Publications, Newton, Mass., and *The Optical Industry and Systems Directory,* Optical Publishing, Pittsfield, Mass.

40. Koechner, W., *Solid-State Laser Engineering,* Vol. 1, Springer-Verlag, New York, 1976, 108.

41. Jones, W. B. and Hulme, G. J., Nd: slab face-pumped lasers in *Proc. Electro-Optics/Laser 78 Conference,* Ind. Sci. Conf. Manage., Chicago, 1978, 515.

42. Myers, Jr., J. D., *Q-98, A New Laser Glass,* Kigre, Inc., Toledo, 1980.

43. Avanesov, A. G., Vasil'ev, I. V., Voron'ko, Yu. K., Denker, B. I., Zinov'ev, S. V., Kuznetsov, A. S., Osiko, V. V., Pashinin, P. P., Prokhorov, A. M., and Semenov, A. A., Investigation of the laser characteristics of Li-Nd-La phosphate glass active elements, *Sov. J. Quantum Electron.,* 9, 937, 1979; *Kvantovaya Elektron.,* 6, 1586, 1979.

44. Kishida, S., Washio, K., Yoshikawa, S., and Kato, Y., CW oscillation in a Nd-phosphate glass laser, *Appl. Phys. Lett.,* 34, 273, 1979.

45. Kravchenko, V. B. and Rudnitskii, Yu. P., Phosphate laser glasses (review), *Sov. J. Quantum Electron.,* 9, 399, 1979; *Kvantovaya Elektron.,* 6, 661, 1979.

46. Vodop'yanov, K. L., Denker, B. I., Maksimova, G. V., Malyutin, A. A., Osiko, V. V., Pashinin, P. P., and Prokhorov, A. M., Characteristics of stimulated emission from Li-Nd-La phosphate glass, *Sov. J. Quantum Electron.,* 8, 403, 1978; *Kvantovaya Elektron.,* 5, 686, 1978.

47. Washio, K., Koizumi, K., and Ideda, Y., Actively initiated, quasi-passive mode-locking in a CW selfoc Nd-glass laser, *Opt. Commun.,* 18, 86, 1976.

48. Galant, E. I., Kalinin, V. N., Lunter, S. G., Mak, A. A., Przhevuskii, A. K., Prilezhaev, D. S., Tolstoi, M. N., and Fromzel, V. A., Stimulated emission from laser-pumped ytterbium-and erbium-activated glasses, *Sov. J. Quantum Electron.,* 6, 1190, 1977; *Kvantovaya Elektron.,* 76, 2187, 1976.

49. Booth, D. J., Kobayashi, T., and Inaba, H., A widely-tunable narrow-linewidth Nd-glass laser, *Opt. Quantum Electron.,* 11, 370, 1979.

50. DeMaria, A. J., Stetser, D. A., and Heynau, H., Self-mode-locking of laser with saturable absorbers, *Appl. Phys. Lett.*, 8, 174, 1966; DeMaria, A. J., Statser, D. A., and Glenn, W. H., Jr., Ultra-short light pulses, *Science*, 156, 1557, 1967.

51. Armstrong, J. A., Measurement of picosecond laser pulse widths, *Appl. Phys. Lett.*, 10, 16, 1967.

52. Giordmaine J. A., Rentzepis, P. M., Shapiro, S. L., and Wecht, K. W., Two-photon excitation of fluorescence by picosecond light pulses, *Appl. Phys. Lett.*, 11, 216, 1967.

53. Glenn, W. H. and Brienza, M. J., Time evolution of picosecond optical pulses, *Appl. Phys. Lett.*, 10, 221, 1967.

54. Eckardt, R. C. and Lee, C. H., Optical third harmonic measurements of subpicosecond light pulses, *Appl. Phys. Lett.*, 15, 425, 1969.

55. Duguay, M. A., Hansen, J. W., and Shapiro, S. L., Nd:glass laser radiation, *IEEE J. Quantum Electron.*, QE-6, 725, 1970.

56. Bradley, D. J., New, G. H. C., Sutherland, B., and Caughey, S. J., Subpiscosecond structure in Nd:glass laser relaxation oscillations, *Phys. Lett.*, A, 28, 532, 1969; Bradley, D. J., New, G. H. C., and Caughey, S. J., Subpicosecond structure in mode-locked Nd:glass lasers, *Phys. Lett.*, A, 30, 78, 1969.

57. Bradley, D. J. and Sibbett, W., Streak-camera studies of picosecond pulses from a mode-locked Nd-glass laser, *Opt. Commun.*, 9, 17, 1973.

58. Eckardt, R. C., Lee, C. H., and Bradford, J. N., Temporal and spectral development of mode-locking in a ring-cavity Nd:glass laser, *Appl. Phys. Lett.*, 19, 420, 1971.

59. Treacy, E. B., Direct demonstration of picosecond-pulse frequency sweep, *Appl. Phys. Lett.*, 17, 14, 1970.

60. Lawrence Livermore Laboratory, *Laser Program Annual Reports —* 1975, 1976, 1977, 1978, UCRL — 50021-75, 76, 77, 78.

61. Lawrence Livermore Laboratory, *Laser Program Annual Report —* 1976, UCRL — 50021-76, 1977, 2—66 to 2—84.

62. Seka, W., Soures, J., Lewis, O., Bunkenburg, J., Brown, D., Jacobs, S., Mourou, G., and Zimmerman, J., High-power phosphate-glass laser system: design and performance characteristics, *Appl. Opt.*, 19, 409, 1980.

63. Institute of Laser Engineering, Osaka University, *Annual Progress Report on Laser Fusion Program*, University of Osaka, Japan, 1978, 1—26.

64. Sooy, W. R., Congleton, R. S., Dobratz, B. E., and Ng, W. K., Dynamic limitations on the attainable inversion in ruby lasers, *Quantum Electronics 3*, Columbia University Press, New York, 1964, 1203.

65. Linford, G. J. and Hill, L. W., Nd:YAG long lasers, *Appl. Opt.*, 13, 1387, 1974.

66. Almasi, J. C., Chernoch, J. P., Martin, W. S., and Tomiyasu, K., Face pumped laser, G. E. Report to ONR AD 467468/5ST, Office of Naval Research, Washington, D.C., 1966.

67. Swain, J. E., Kidder, R. E., Pettipiece, K., Rainer, F., Baird, E. D., and Loth, B., Large-aperture glass disk laser system, *J. Appl. Phys.*, 40, 3973, 1969.

68. Lawrence Livermore Laboratory, *Laser Program Annual Report —* 1978, UCRL-50021-78, 1979, 7-1.

69. Trenholme, J. B., Fluorescence amplication and parasitic oscillation limitations in disc lasers, *Naval Research Lab. Memorandum Rep.*, 2480, 1972.

70. McMahon, J. M., Emmett, J. L., Holzrichter, J. F., and Trenholme, J. B., A glass-disk-laser amplifier, *IEEE J. Quantum Electron.*, QE-9, 992, 1973.

71. Brown, D. C., Parasitic oscillations in large aperture Nd^{3+}:glass amplifiers revisited, *Appl. Opt.*, 12, 2215, 1973.

72. Glaze, J. A., Guch, S., and Trenholme, J. B., Parasitic suppression in large aperture Nd:glass disk laser amplifiers, *Appl. Opt.*, 13, 2808, 1974.

73. Bennett, R. B., Shillito, K. R., and Linford, G. J., Progress in claddings for laser glasses, in *Laser Induced Damage In Optical Materials:1977*, NBS Spec. Publ. 509, Washington, 1978, 434.

74. Hirota, S. and Izumitani, T., Reflection measurements at the interface between disk laser glass and edge cladding glass, *Appl. Opt.*, 18, 97, 1979.

75. Jones, W. B., Goldman, L. M., Chernoch, J. P., and Martin, W. S., The mini-FPL-a face-pumped laser: concept and implementation, *IEEE J. Quantum Electron.*, QE-8, 534, 1972.

76. Brown, D. C., Jacobs, S. D., and Nee, N., Parasitic oscillations, absorption, stored energy density and heat density in active-mirror and disk amplifiers, *Appl. Opt.*, 17, 211, 1978.

77. Chen, Bor-Uei and Tang, C. L., Nd-Glass thin-film waveguide: An active medium for Nd thin-film laser, *Appl. Phys. Lett.*, 28, 435, 1976.

78. Linford, G. J., Saroyan, R. A., Trenholme, J. B., and Weber, M. J., Measurements and modeling of gain coefficients for neodymium laser glasses, *IEEE J. Quantum Electron.*, QE-15, 510, 1979.

79. Dube, G., Calorimetric measurements of stimulated emission cross sections and absorption coefficients at 1.06 μm, *Appl. Opt.*, 16, 1484, 1977.

80. Mauer, P., Amplification coefficient of neodymium-doped glass at 1.06 microns, *Appl. Opt.*, 3, 433, 1964.

81. **Judd, B. R.**, Optical absorption intensities of rare-earth ions, *Phys. Rev.*, 127, 750, 1962.

82. **Ofelt, G. S.**, Intensities of crystal spectra of rare-earth ions, *J. Chem. Phys.*, 37, 511, 1962.

83. **Peacock, R. D.**, The intensities of lanthanide f↔f transitions, *Struct. Bonding*, 22, 83, 1975.

84. **Krupke, W. F.**, Induced-emission cross sections in neodymium laser glasses, *IEEE J. Quantum Electron.*, QE-10, 450, 1974.

85. **Haugen, G. R. and Rigdon, L. P.**, *Spectrophotometric Determination of Neodymium in Laser Glass*, UCRL-51652, Lawrence Livermore Laboratory, Livermore, Calif., 1974.

86. **Dianov, E. M., Karasik, A. Ya., Kornienko, L. S., Prokhorov, A. M., and Shcherbakov, I. A.**, Measurement of the laser transition cross section of neodymium glasses, *Sov. J. Quantum Electron.*, 5, 901, 1975; *Kvantovaya Elektron.*, 2, 1965, 1975.

87. **Deutschbein, O. K.**, A simplified method for determination of the induced-emission cross section of neodymium-doped glasses, *IEEE J. Quantum Electron.*, QE-12, 551, 1976.

88. **Singh, S., Smith, R. G., and Van Uitert, L. G.**, Stimulated-emission cross section and fluorescent quantum efficiency of Nd^{3+} in yttrium aluminum garnet at room temperature, *Phys. Rev. B*, 10, 2566, 1974.

89. **Guba, B. S., Prilezhaev, D. S., Raba, O. B., and Sedov, B. M.**, Measurement of cross section for induced transitions in neodymium glasses, *Opt. Spectrosc. USSR*, 47, 67, 1979; *Opt. Spektrosk.*, 47, 121, 1979.

90. **Belan, V. R., Grigoryants, V. V., and Zhabotinski, M. E.**, Use of lasers in measurements of the stimulated emission cross section, *IEEE J. Quantum Electron.*, QE-3, 425, 1967.

91. **Belan, V. R., Grigoryants, V. V., and Zhabotinski, M. E.**, Use of laser to measure the cross section of stimulated emission of matter, *JETP Lett.*, 6, 200, 1967; *ZETF Pisma*, 6, 721, 1967.

92. **Birnbaum, M. and Gelbwachs, J. A.**, Stimulated-emission cross section of Nd^{3+} at 1.06 μm in POC_3, YAG, $CaWO_4$, ED-2 Glass and LG55 Glass, *J. Appl. Phys.*, 43, 2335, 1972.

93. **Dumachin, R., Farcy, J., Michon, M., and Vincent, P.**, Analysis of giant pulse amplification in Nd^{3+} doped glass, *IEEE J. Quantum Electron.*, QE-7, 53, 1971.

94. **Lomheim, T. S. and DeShazer, L. G.**, Determination of optical cross sections by the measurement of saturation flux using laser-pumped laser oscillators, *J. Opt. Soc. Am.*, 68, 1576, 1978.

95. **Rudnitskii, Yu, P., Smirnov, R. V., and Chernyak, V. M.**, Behavior of the population of the $^4I_{11/2}$ level of Nd^{3+} in glasses interacting with high-power coherent radiation, *Sov. J. Quantum Electron.*, 6, 1107, 1976; *Kvantovaya Elektron.*, 3, 2035, 1976.

96. **Grigoryants, V. V., Zhabotinski, M. E., and Markushev, V. M.**, Determining the relaxation period τ_{21} of the $^4I_{11/2}$ level of the Nd ions in glass, *J. Appl. Spectrosc.*, 14, 58, 1971; *Zh. Prikl. Spektrosk.*, 14, 73, 1973.

97. **Magnante, P. C.**, Influence of the lifetime and degeneracy of the $^4I_{11/2}$ level on Nd-glass amplifiers, *IEEE J. Quantum Electron.*, QE-8, 440, 1972.

98. **Mak, A., Prilezhaev, D., Serebryakov, V., and Straikov, A.**, Measurement of relaxation rates in glasses activated with Nd^{3+} ions, *Opt. Spectrosc. USSR*, 33, 381, 1971; *Opt. Spektrosk.*, 33, 698, 1971.

99. **Duston, D.**, Measurement of terminal level lifetime in Nd-doped laser glass, *IEEE J. Quantum Electron.*, QE-6, 3, 1970.

100. **Michon, M.**, The influence of Nd^{3+} ion properties in a glass matrix on the dynamics of a Q-spoiled laser, *IEEE J. Quantum Electron.*, QE-2, 612, 1966.

101. **Michon, M., Ernest, J., Hanus, J., and Auffret, R.**, Influence of the $^4I_{11/2}$ level lifetime on the effective use of the population in a Q-spoiled neodymium doped glass laser, *Phys. Lett.*, 19, 219, 1965.

102. **Nikitin, V. I., Soskin, M. S., and Khizhnyak, A. I.**, Influence of uncorrelated inhomogeneous broadening of the 1.06 μm band of the Nd^{3+} ions on laser properties of neodymium glasses, *Sov. J. Quantum Electron.*, 8, 788, 1978; *Kvantovaya Elektron.*, 5, 1375, 1978.

103. **Mann, M. and DeShazer, L.**, Energy levels and spectral broadening of neodymium ions in laser glass, *J. Appl. Phys.*, 41, 2951, 1970.

104. **Nikitin, V. I., Soskin, M. S., and Khizhnyak, A. I.**, New data about internal 1.06 μm luminescence band structure of Nd^{3+} in silicate glass, *Sov. Tech. Phys. Lett.*, 2, 64, 1976; *Pis'ma Zh, Tekh. Fiz.*, 2, 172, 1976.

105. **Nikitin, V. I., Soskin, M. S., and Khizhnyak, A. I.**, Uncorrelated nonuniform spreading-A basic reason for narrow-band generation in phosphate glass with Nd^{3+}, *Sov. Tech. Phys. Lett.*, 3, 5, 1977; *Pis'ma Zh. Tekh. Fiz.*, 3, 14, 1977.

106. **Brecher, C., Riseberg, L. A., and Weber, M. J.**, Variations in the transition probabilities and quantum efficiency of Nd^{3+}ions in ED-2 laser glass, *Appl. Phys. Lett.*, 30, 475, 1977.

107. **Brecher, C., Riseberg, L. A., and Weber, M. J.**, Site-dependent variation of spectroscopic relaxation parameters in Nd glasses, *Luminescence*, 18/19, 651, 1979.

108. **Brawer, S. A. and Weber, M. J.**, Observation of fluorescence line narrowing, hole burning, and ion-ion energy transfer in neodymium laser glasses, *Appl. Phys. Lett.*, 35, 31, 1979.

109. **Alimov, O. K., Basiev, T. T., Voron'ko, Yu, K., Gribkov, Yu. V., Karasik, Ya., Osiko, V. V., Prokhorov, A. M., and Shcherbakov, I. A.,** Selection-laser-excitation study of the structure of inhomogeneously broadened spectra of Nd^{3+} ions in glass, *Sov. Phys. JETP,* 47, 29, 1978; *Zh, Eksp. Teor. Fiz.,* 74, 57, 1978.

110. **DeShazer, L. G. and Komai, L. G.,** Fluorescence conversion efficiency of neodymium glass, *J. Opt. Soc. Am.,* 55, 940, 1965.

111. **Brandewie, R. A. and Telk, C. L.,** Quantum efficiency of Nd^{3+} in glass, calcium tungstate, and yttrium aluminum garnet, *J. Opt. Soc. Am.,* 57, 1221, 1967.

112. **Riseberg, L. A. and Weber, M. J.,** Relaxation phenomena in rare-earth luminescence, in *Progress in Optics.,* Vol. 14, Wolf, E., Ed., North-Holland, Amsterdam, 1976, 89.

113. **Reisfeld, R.,** Radiative and nonradiative transitions of rare earth ions in glasses, *Struct. Bonding,* 22 129, 1975.

114. **Layne, C. B., Lowdermilk, W. H., and Weber, M. J,.** Multiphonon relaxation of rare-earth ions in oxide glasses, *Phys. Rev. B,* 16, 10, 1977.

115. **Layne, C. B. and Weber, M. J.,** Multiphonon relaxation of rare-earth ions in beryllium-fluoride glass, *Phys. Rev. B,* 16, 3259, 1977.

116. **Zhmyreva, I. A., Kovaleva, I. V., Kolobkov, V. P., and Tatarintsev, B. V.,** Effect of deuteration on the emissive power of rare-earth elements in tellurite glasses, *J. Appl. Spectrosc.,* 29, 1119, 1978; *Zh. Prikl. Spektrosk.,* 29, 515, 1978.

117. **Bondarenko, E. G., Galant, E. I., Lunter, S. G., Przhevuskii, A. K., and Tolstoi, M. N.,** The effect of water in glass on the quenching of the luminescence of the rare-earth activator, *Sov. J. Opt. Technol.,* 42, 333, 1976; *Opt.-Mekh. Prom.,* 42, 42, 1975.

118. **Suydam, B. R.,** Self-focusing of very powerful laser beams II, *IEEE J. Quantum Electron.,* QE-10, 837, 1974.

119. **Trenholme, J. B.,** Small-scale instability growth: Review of small signal theory, in the *1974 Lawrence Livermore Laboratory Laser Program Ann. Rep.* (UCRL-50021-74), Lawrence Livermore Laboratory, Livermore, Calif., 1975, 179.

120. **Bliss, E. S., Hunt, J. T., Renard, P. A., Sommargren, G. E.,and Weaver, H. J.,** Effects of nonlinear propagation on laser focusing properties, *IEEE J. Quantum Electron.,* QE-12, 402, 1976.

121. **Boling, N. L., Glass, A. J., and Owyoung, A.,** Empirical relationships for predicting nonlinear refractive-index changes in optical solids, *IEEE J. Quantum Electron.,* QE-14, 601, 1978.

122. **Bliss, E. S., Speck, D. R., and Simmons, W. W.,** Direct interferometric measurements of the nonlinear refractive index coefficient n_2 in laser materials, *Appl. Phys. Lett.,* 25, 718, 1974.

123. **Milam, D. and Weber, M. J.,** Measurement of nonlinear refractive index coefficients using time-resolved interferometry: application to optical materials for high-power neodymium lasers, *J. Appl. Phys.,* 47, 2497, 1976.

124. **Milam, D. and Weber, M. J.,** Nonlinear refractive index coefficients for Nd phosphate laser glasses, *IEEE J. Quantum Electron.,* QE-13, 512, 1976.

125. **Milam, D., Weber, M. J., and Glass, A. J.,** Nonlinear refractive index of fluoride crystals, *Appl. Phys. Lett.,* 31, 822, 1977.

126. **Weber, M. J., Cline, C. F., Smith, W. L., Milam, D., Heiman, D., and Hellwarth, R. W.,** Measurements of the electronic and nuclear contributions to the nonlinear refractive index of beryllium fluoride glasses, *Appl. Phys. Lett.,* 32, 403, 1978.

127. **Weber, M. J., Milam, D., and Smith, W. L.,** Nonlinear refractive index of glasses and crystals, *Opt. Eng.,* 17, 463, 1978.

128. **Hellwarth, R. W., Cherlow, J., and Yang, T-T.,** Origin and frequency dependence of nonlinear optical susceptibilities of glasses, *Phys. Rev. B,* 11, 964, 1975.

129. **Owyoung, A., Hellwarth, R. W., and George, N.,** Intensity-induced changes in optical polarizations in glasses, *Phys. Rev. B,* 5, 628, 1972.

130. **Jacobs, R. R. and Weber, M. J.,** Dependence of the $^4F_{3/2} \rightarrow {}^4I_{11/2}$ induced-emission cross section for Nd^{3+} on glass composition, *IEEE J. Quantum Electron.,* QE-12, 102, 1976.

131. **Sarkies, P. H., Sandoe, J. N., and Parke, S.,** Variation of Nd^{3+} cross section for stimulated emission with glass composition, *J. Phys. D.: Appl. Phys.,* 4, 1642, 1971.

132. **Cline, C. F. and Weber, M. J.,** Beryllium fluoride optical glasses: preparation and properties, *Wiss. Z. Friedrich-Schiller-Univ. Jena, Math.-Naturwiss. Reihe,* 28, Jg., 1979, H. 2/3, p. 351, 1979; also available from the Lawrence Livermore Laboratory, Livermore, Calif. as UCRL-81168.

133. **Sandoe, J. N., Sarkies, P. H., and Parke, S.,** Variation of Er^{3+} cross section for stimulated emission with glass composition, *J. Phys. D,* 5, 1788, 1972.

134. **Neilson, G. F. and Weinberg, M. C.,** The spectroscopic behavior of high CaO content, laser-type glasses, *J. Non-Cryst. Solids,* 28, 209, 1978.

135. **Egorova, V. F., Zubkova, V. S., Karapetyan, G. O., Mak, A. A., Prilezhaev, D. S., and Reichakhrit, A. L.,** Influence of glass composition on the luminescence characteristics of Nd^{3+} ions, *Opt. Spectrosc. USSR,* 23, 148, 1967; *Opt. Spektrosk.,* 23, 275, 1967.

136. **Lipson, H. G., Buckmelter, J. R., and Dugger, C. O.,** Neodymium ion environment in germanate crystals and glasses, *J. Non-Cryst. Solids,* 17, 27, 1975.

137. **Brachkovskaya, N. B., Grubin, A. A., Lunter, S. G., Przhevuskii A. K., Raaben, E. L., and Tolstoi, M. N.,** Intensities of optical transitions in absorption and luminescence spectra of neodymium in glasses, *Sov. J. Quantum Electron.,* 6, 534, 1976.

138. **Brachkovskaya, G. O., Karapetyan, G. O., Reinshakhrit, A. L., and Tolstoi, M. N.,** Luminescence of neodymium in alkali silicate glasses, *Opt. Spectrosc.,* 29, 173, 1970.

139. **Hauptmanova, K., Pantoflicek, J., and Patek, K.,** Absorption and fluorescence of Nd^{3+} ion in silicate glass, *Phys. Status Solidi,* 9, 525, 1965.

140. **Hirayama, C.,** Nd fluorescence in alkali borate glasses, *Phys. Chem. Glasses,* 7, 52, 1966.

141. **Hirayama, C., Camp, F. E., Melamed, N. T., and Steinbruegge, K. B.,** Nd^{3+} in germanate glasses: spectral and laser properties, *J. Non-Cryst. Solids,* 6, 342, 1971.

142. **Alekseev, N. E., Izyneev, A. A., Kopylov, Yu. L., Kravchenko, V. B., Rudnitskii, Yu. P., and Udovenko, N. F.,** Activated Nd^{3+} laser glasses based on the metaphosphates of divalent metals, *J. Appl. Spec.,* 24, 691, 1976; *Zh. Prikl. Spektros.,* 24, 976, 1976.

143. **Alekseev, N. E., Izyneev, A. A., Kopylov, Yu. L., Kravchenko, V. B., and Rudnitskii, Yu. P.,** A study of neodymium glasses based on alkali metal metaphosphates, *J. Appl. Spec.,* 26, 87, 1977; *Zh. Prikl. Spektros.,* 26, 116, 1977.

144. **Deutschbein, O.,** Glass for Lasers, French Patent 1,502,709, 1967.

145. **Lucas, J., Chanthanasinh, M., Poulain, M., Poulain, M., Brun, P., and Weber, M. J.,** Preparation and optical properties of neodymium fluorozirconate glasses, *J. Non-Cryst. Solids,* 27, 273, 1978.

146. **Dexter, D. L.,** A theory of sensitized luminescence in solids, *J. Chem. Phys.,* 21, 836, 1953.

147. **Dexter, D. L. and Schulman, J. H.,** Theory of concentration quenching in inorganic phosphors, *J. Chem. Phys.,* 22, 1063, 1954.

148. **Forster, T.,** *Ann. Phys.,* 2, 55, 1948.

149. **Weber, M. J.,** Optical properties of Yb^{3+} and Nd^{3+} energy transfer in $YAlO_3$ *Phys. Rev.,* B4, 3153—3159, 1971.

150. **Danielmeyer, H. G.,** Efficiency and fluorescence quenching of rare earth laser materials, *J. Luminescence,* 12/13, 179, 1976.

151. **Sakun, V. P.,** Kinetics of the hopping mechanism of luminescence quenching, *Sov. Phys. Solid State,* 21, 390, 1979; *Fiz. Tverd. Tela.,* 21, 662, 1979.

152. **Yokota, M. and Tanimoto, O.,** Effects of diffusion on energy transfer by resonance, *J. Phys. Soc. Jpn.,* 22, 779, 1967.

153. **Motegi, N. and Shionoya, S.,** Excitation migration among inhomogeneously broadened levels of Eu^{3+} ions, *J. Luminescence,* 8, 1, 1973.

154. **Yamada, N. S., Shionoya, S., and Kushida, T.,** Phonon-assisted energy transfer between trivalent rare earth ions, *J. Phys. Soc. Jpn.,* 32, 1577, 1972.

155. **Holstein, T., Lyo, S. K., and Orbach, R.,** Energy transfer in random systems, *J. Luminescence,* 18, 634, 1979.

156. **Eisenthal, K. B. and Siegel, S.,** Influence of resonance transfer on luminescence decay, *J. Chem. Phys.,* 41, 652, 1964.

157. **Denker, B. I., Osiko, V. V., Prokhorov, A. M., and Shcherbakov, I. A.,** Concentration dependencies of the quantum efficiency of the luminescence emitted by neodymium-activated laser matrices and microscopic determination of these dependencies., *Sov. J. Quantum Electron.,* 8, 485, 1978; *Kvantovaya Elektron.,* 5, 847, 1978.

158. **Gandy, H. W., Ginther, R. J., and Weller, J. F.,** Energy transfer in silicate glass coactivated with cerium and neodymium, *Phys. Lett.,* 11, 213, 1964.

159. **Jacobs, R. R., Layne, C. B., Weber, M. J., and Rapp, C.,** $Ce^{3+} \rightarrow Nd^{3+}$ energy transfer in silicate glass, *J. Appl. Phys.,* 47, 2020, 1976.

160. **Shionoya, S. and Nakazawa, E.,** Sensitization of Nd^{3+} luminescence by Mn^{2+} and Ce^{3+} in glasses, *Appl. Phys. Lett.,* 6, 117, 1965.

161. **Melamed, N. T., Hirayama, C., and Davis, E. K.,** Laser action in neodymium-doped glass produced through energy transfer , *Appl. Phys. Lett.,* 7, 170, 1965.

162. **Melamed, N. T., Hirayama, C., and French, P. W.,** Laser action in uranyl-sensitized Nd-doped glass, *Appl. Phys. Lett.,* 6, 43, 1965.

163. **Gandy, H. W., Ginther, R. J., and Weller, J. F.,** Energy transfer in triply activated glasses., *Appl. Phys. Lett.,* 6, 46, 1965.

164. **Joshi, J. C., Pandey, N. C., Joshi, B. C., and Joshi,J.,** Energy transfer from $UO_2 \rightarrow Nd^{3+}$ in barium borate glass , *J. Luminescence,* 16, 435, 1978.

165. **Cabezas, A. Y. and DeShazer, L. G.,** Radiative transfer of energy between rare-earth ions in glass, *Appl. Phys. Lett.,* 4, 37, 1964.

166. **Reisfeld, R. and Kalisky, Y.,** Energy transfer between Bi^{3+} and Nd^{3+} in germanate glass, *Chem. Phys. Lett.,* 50, 199, 1977.

167. **Sharp, E. J., Weber, M. J., and Cleek, G.,** Energy transfer and fluorescence quenching in Eu- and Nd-doped silicate glasses, *J. Appl. Phys.,* 47, 364, 1976.

168. **Karapetyan, G. O., Kovalyov, V. P., and Lunter, S. G.,** Chromium sensitization of the neodymium luminescence in glass, *Opt. Spectrosc. USSR,* 19, 529, 1965; *Opt. Specktrosk.,* 19, 951, 1965.

169. **Dauge, G.,** Nonradiative energy transfer in silicate glass, *IEEE J. Quantum Electron.,* QE-2, lviii, 1966.

170. **Edwards, J. G. and Gomulka, S.,** Enhanced performance of Nd laser glass by double doping with Cr, *J. Phys. D.,* 12, 187, 1979.

171. **Avanesov, A. G., Voron'kov, Yu. K., Denker, B. I., Maosimova, G. V., Osiko, V. V., Prokhorov, A. M., and Shcherbakov, I. A.,** Nonradiative energy transfer from Cr^{3+} to Nd^{3+} ions in glasses with high neodymium concentrations, *Sov. J. Quantum Electron.,* 9, 935, 1979; *Kvantovaya Elektron.,* 6, 1583, 1979.

172. **Reisfeld, R. and Hormadaly, J.,** Quantum yield of cerium (3 +) ion and energy transfer between cerium (3 +) and terbium(3 +) ions in borax glasses, *J. Solid State Chem.,* 13, 283, 1975.

173. **Joshi, J. C., Pandey, N. C., Joshi, B. C., Belwal, R., and Joshi, J.,** Diffusion limited energy transfer from $Tb^{3+} \rightarrow Ho^{3+}$ in calibo glass, *J. Non-Cryst. Solids,* 27, 173, 1978.

174. **Gapontsev, V. P., Zhabotinskii, M. E., Izyneev, A. A., Kravchenko, V. B., and Rudnitskii, Yu. P.,** Effective $1.054 \rightarrow 1.54 \mu m$ stimulated emission, *JETP Lett.,* 18, 251, 1973.

175. **Edwards, J. G. and Sandoe, J. N.,** A theoretical study of the Nd-Yb-Er glass laser, *J. Phys. D,* 7, 1078, 1974.

176. **Reisfield, R. and Eckstein, U.,** Energy transfer from Ce^{3+} to Tm^{3+} in borate and phosphate glasses, *Appl. Phys. Lett.,* 26, 253, 1975.

177. **Peterson, G. E., Pearson, A.D., and Bridenbaugh, P. M.,** Energy exchange from Nd^{3+} to Yb^{3+} in calibo glass, *J. Appl. Phys.,* 36, 1962, 1965.

178. **Kovalev, V. P. and Karapetyan, G. O.,** Sensitizing of the luminescence of trivalent ytterbium by neodymium in silicate glasses, *Opt. Spectrosc. USSR,* 43, 102, 1965.

179. **Snitzer, E.,** Glass lasers, *Appl. Opt.,* 5, 1487, 1966.

180. **Sharp, E. J., Miller, J. E., and Weber, M. J.,** Chromium-ytterbium energy transfer in silicate glass, *J. Appl. Phys.,* 44, 4098, 1973.

181. **Morey, W. W.,** Active filtering for neodymium lasers, *IEEE J. Quantum Electron.,* QE-8, 818, 1972.

182. **Bhawalkar, D. D. and Pandit, L.,** Improving the pumping efficiency of a Nd^{3+} glass laser using dyes, *IEEE J. Quantum Electron.,* QE-9, 43, 1972.

183. **Vodop'yanov, K. L., Il'ichev, N. N., Malyutin, A. A., Matyushin, G. A., and Podgaetskii, V. M.,** Enhancement of the efficiency of neodymium lasers by conversion of the pump radiation in a luminescent liquid, *Sov. J. Quantum Electron.,* 9, 1059, 1979; *Kvantovaya Elektron.,* 6, 1795, 1979.

184. **Levin, M. B., Leschiner, M. E., Matyushin, G. A., Podgaetskii, V. M., Slivka, L. K., and Cherkasov, A. S.,** Calculation and experimental verification of the efficiency of luminescence filters in neodymium glass lasers, *Opt. Spectrosc. USSR,* 46, 301, 1979; *Opt. Spektrosk.,* 46, 543, 1979.

185. **Deutschbein, O., Faulstich, M., Jahn, W., Krolla, G., and Neuroth, N.,** Glasses with a large laser effect: Nd-phosphate and Nd-fluorophosphate, *Appl. Opt.,* 17, 228, 1978.

186. **Smith, W. L., Bechtel, J. H., and Bloembergen, N.,** Dielectric-breakdown threshold and nonlinear-refractive-index measurements with picosecond laser pulses, *Phys. Rev. B,* 12, 706, 1975.

187. **Smith, W. L.,** Laser-induced breakdown in optical materials, *Opt. Eng.,* 17, 489, 1978.

188. **Stokowski, S. E., Milam, D., and Weber, M. J.,** Laser-induced damage in fluoride glasses: a status report, in *Laser Induced Damage in Optical Materials: 1978,* NBS Spec. Publ. 541, Washington, D.C., 1978, 99.

189. **Milam, D.,** Laser-induced damage at 1064 nm, 125 psec, *Appl. Opt.,* 16, 1204, 1977.

190. **Boling, N. L., Crisp, M. D., and Dube, G.,** Laser induced surface damage, *Appl. Opt.,* 12, 650, 1973.

191. **Milam, D., Smith, W. L., Weber, M. J., Guenther, A. H., House, R. A., and Bettis, J. R.,** The effects of surface roughness on 1064-nm, 150-ps laser damage, in *Laser Induced Damage in Optical Materials: 1977,* NBS Spec. Publ. 509, Washington, D.C., 1977, 166.

192. **Hopper, R. W. and Uhlmann D. R.,** Mechanism of inclusion damage in laser glass, *J. Appl. Phys.,* 41, 4023, 1970.

193. **Weyl, W.,** *Coloured Glasses,* Society of Glass Technology, Sheffield, England, 1976, 497.

194. **Buzhinskii, I. M., Koryagina, E. I., and Mamonov, S. K.,** Photostability of laser-active elements from neodymium-activated glass, *Radio Eng. Electron Phys.,* 14, 1743, 1969; *Radio Tekh. Elektron.,* 14, 2017, 1969.

195. **Buzhinskii, I. M. and Mamanov, S. K.,** Deterioration of the parameters of neodymium glasses during use, *J. Appl. Spectrosc.,* 8, 438, 1968; *Zh. Prikl. Spektrosk.,* 8, 731, 1968.

196. **Landry, R. J., Snitzer, E., and Bartrum, R. H.,** Ultraviolet-induced transient and stable color centers in self-q-switching laser glass, *J. Appl. Phys.,* 42, 3827, 1971.

197. **Bamford, C. R.,** *Colour Generation and Control in Glass,* Elsevier, New York, 1977.

198. **Buzhinskii, I. M., Avakyants, L. I., and Surkova, V. F.,** Effect of cerium and iron on the gamma-stability of glasses activated by neodymium, *Zh, Prikl. Spektroski.*, 28, 238, 1975.

199. **Buzhinskii, I. M., Koryagina, E. I., and Mamonov, S. K.,** Effect of cerium on the photostability of neodymium silicate glasses under radiation generation conditions, *J. Appl Spectrosc.*, 22, 250, 1975; *Zh. Prikl. Specktrosk.*, 22, 326, 1975.

200. **Buzhinskii, I. M., Momonov, S. K., and Surkova, V. F.,** Role of titanium in stabilization of the lasing parameters of neodymium glasses, *J. Appl. Spectrosc.*, 27, 657, 1977; *Zh. Prikl. Spektrosk.*, 27, 657, 1977.

201. **Weber, H. P., Damen, T. C., Danielmeyer, H. G., and Topfield, B. C.,** Nd-ultra-phosphate laser, *Appl. Phys. Lett.*, 22, 534, 1973.

202. **Voronko, Yu. K., Denker, B. I., Zlenko, A. A., Karasik, A. Ya., Kuz'minov, Yu. S., Maksimova, G. V., Neustruyeve, V. B., Osiko, V. V., Prokhorov, A. M., Sychugov, V. A., Shipulo, G. P., and Shcherbakov, I. A.,** Spectral and lasing properties of Li-Nd phosphate glass, *Opt. Commun.*, 18, 88, 1976.

203. **Michel, J. C., Morin, D., and Auzel, F.,** Spectroscopic properties and cw laser action of two Nd^{3+} highly doped, lithium tellurite and cadmium phosphate glasses, *Phys. Appl.*, 13, 859, 1978.

204. **Lempicki, A., Klein, R. M., and Chinn, S. R.,** Spectroscopy and lasing of a high concentration Al-phosphate glass, *IEEE J. Quantum Electron.*, QE-14, 283, 1978.

205. **Kaminskii, A. A., Sarkisov, S. E., Ngoc, T., Denker, B. I., Osiko, V. V., and Prokhorov, A. M.,** Stimulated emission spectroscopy of concentrated lithium-neodymium phosphate glasses in $^4F_{3/2} \rightarrow {}^4I_{11/2}$ and $^4F_{3/2} \rightarrow {}^4I_{13/2}$ transitions, *Phys. Status Solidi A*, 50, 745, 1978.

Note: I have attempted to reference only published works. However, I do reference special reports that are obtainable as follows:

Lawrence Livermore Laboratory reports (UCRL's) and reports with AD numbers from National Technical Information Service, U.S. Dept. of Commerce, 5285 Port Royal Road, Springfield, VA, 22161; NBS Special Publications from Superintendent of Documents, U. S. Government Printing Office, Washington, D. C., 20434.

2.4 FIBER RAMAN LASERS

R. H. Stolen and Chinlon Lin

INTRODUCTION

The fiber Raman laser[1] is a new type of laser-pumped tunable laser which utilizes both the long interaction lengths of low-loss optical fibers and the broad Raman bandwidths in glasses. These sources have so far found their primary application in the study of loss and pulse dispersion in optical fibers. Fiber Raman lasers take two general forms; pulsed single-pass lasers or oscillators pumped by either cw or repetitively pulsed lasers. Operation has been obtained in both the visible and near infrared with oscillator thresholds as low as 200 mW. The intrinsic bandwidths as determined by the Raman spectrum of fused silica is about 500 cm^{-1}, but this range can be extended many times with cascaded Stokes emission.

SINGLE-PASS FIBER RAMAN LASERS

The Stimulated Raman Effect

In its simplest form, the fiber Raman laser is an optical fiber with pump laser light focused into one end and stimulated Raman light appearing at the output[2] as pictured in Figure 2.4.1. Most of the light is produced on the low frequency or "Stokes" side although some anti-Stokes light is produced.[3] Stokes light is amplified with energy supplied by the pump laser. The amplification is proportional to exp(G) where G depends on the pump intensity, the interaction length L, and the Raman gain coefficient g.[4]

$$G = \frac{gPL}{A} \tag{1}$$

Here P is the pump power and A is the spot size.

The important difference between fiber Raman lasers and Raman lasers using conventional geometries lies in the factor L. In low-loss fibers this length can be several kilometers which can lead to enhancements as large as 10^8 over the focal region of a gaussian beam with comparable spot size.[5]

If the factor gPL/A is large enough, the Stokes power will build up from noise until most of the pump energy is Raman-shifted. There is no definite threshold but a critical power is defined[6] as the level where the Stokes power equals the pump power which occurs at G $\approx$ 16. When G $\gg$ 16, Stokes conversion and pump depletion take place in a small fraction of the total fiber length. The process will then cascade until many orders of Stokes radiation appear at the output. Note that total conversion takes place at the center of the pulse.

Fiber Characteristics

Materials

The Raman gain curve of fused silica as derived from the spontaneous Raman spectrum[4] is shown in Figure 2.4.2. The stimulated Stokes wavelength will be close to the peak of the curve where the gain is highest. In general, dopants such as Boron, Germanium, or Phosphorus do not appreciably modify the silica gain spectrum[7] although a new stimulated line appears at 1330 cm^{-1} in heavily phosphorus-doped fibers.[8] It is, of course, possible to make nonsilicate glass[9] fibers as well as hollow fibers filled with liquids.[10] A comparison of peak Raman gains as compared to fused silica along with

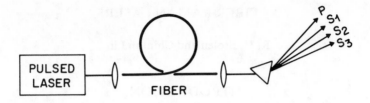

FIGURE 2.4.1.   Single-pass fiber Raman laser. The stimulated Raman process cascades at high power to produce many orders of Stokes radiation at the output.

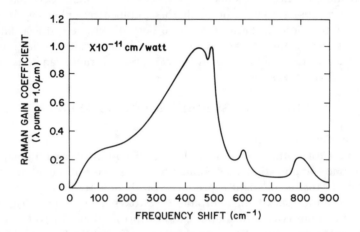

FIGURE 2.4.2.   Raman gain curve for fused silica for a pump at 1.0 μm. The gain varies with pump wavelength as $1/\lambda$.

Table 2.4.1
RAMAN GAINS OF GERMANIA GLASS AND LIQUID BENZENE RELATIVE TO FUSED SILICA AND GAIN BANDWIDTHS (FWHM) FOR ALL THREE MATERIALS

	Silica	Germania	Benzene
Peak Raman Gain	1.0	9.2	323
Bandwidth (FWHM)	240 cm^{-1}	90 cm^{-1}	2.15 cm^{-1}

the bandwidths for two glasses and one liquid are given in Table 2.4.1. It is clear that silica has a relatively low Raman gain but a bandwidth broad enough to be useful in a tunable laser. Although there are many applications where higher gain is desirable, silica based fibers are preferred because of their lower loss and higher damage resistance. The damage threshold of pure fused silica[11] is around 10^{10} W/cm^2.

Other Factors Affecting Gain
   If we neglect the frequency dependence of the refractive index and take the pump and Stokes frequencies to be approximately the same, the Raman gain coefficient will depend linearly on pump frequency.[4]
   Raman gains are usually quoted assuming that linear polarization is maintained.[4] This is in fact possible in specially designed fibers,[12] but in most single and multimode fibers depolarization occurs. The effect of this depolarization is to reduce the Raman gain by a factor of two.[13]

Effective Length

Raman gain will decrease along the fiber because of linear pump absorption (α_p). The effect of pump absorption is included in Equation 1 by taking P as the value of pump power at the fiber input end and replacing L with an effective length[5] defined as:

$$L_{eff} = \frac{1-e^{-\alpha l}}{\alpha} \tag{2}$$

If the fiber is short, L_{eff} is the actual length (l) while for very long fibers L_{eff} becomes $1/\alpha_p$ which is the absorption length. The actual Stokes amplification must also include linear loss at the Stokes wavelength (α_s) so:

$$\text{Stokes amplification} \sim e^{gPL_{eff}/A}e^{-\alpha_s l} \tag{3}$$

Usually $\alpha_s \approx \alpha_p$ except near an absorption peak in the fiber loss spectrum.

The gain can be reduced from pulse spreading or physical separation of pump and Stokes pulses due to group velocity dispersion. These effects can be quite complicated when a pulse contains sharp microstructure.

Effective Core Area

In the present review, the spot size in Equation 1 is considered to be the core area of the fiber. This is not a bad approximation in large-core multimode fibers and in single-mode fibers close to the edge of the single-mode regime. At longer wavelengths or small V numbers[14] the fundamental mode becomes large and the area can be a factor of two larger. More detailed treatments of the effective core area entail computation of mode overlap integrals.[1,4]

Generation of Discrete Stokes Lines

Most of the single-pass stimulated Raman work in optical-fibers has been done with Q-switched Nd:YAG lasers at 1.06 μm, or their second harmonic radiation at 0.53 μm. Using a 450 m single-mode fiber and a pump laser at 0.53 μm, up to 26 orders of stimulated Raman emission have been observed.[15] In the near infrared region similar Raman laser sources provide the basis of techniques for measuring properties of optical fibers. For example, subnanosecond pulses from a Q-switched and mode-locked Nd:YAG laser at 1.06 μm have been used to pump single-mode fibers and generate discrete Stokes lines covering the 1.1-1.65 μm spectral range. These wavelength-selectable pulses have been used in wavelength-dependent pulse dispersion measurements in the spectral region near the zero material dispersion wavelengths of doped silica fibers.[16,17] The same near-IR fiber Raman laser source has also been used for study of modal dispersion and loss measurements.[17]

Another convenient pump source is a flashlamp-pumped pulsed dye laser or a nitrogen-laser pumped dye laser. The tunability of the dye lasers can be translated to the tunability of the Stokes components. Care is needed to obtain good mode quality from these pulsed dye lasers because poor mode quality would result in inefficient coupling into the fiber and damage to the fiber input end.

Continuum Generation

The series of Stokes lines produced by cascaded stimulated Raman scattering becomes progressively broader in higher orders. The broadening appears to come from self-phase modulation[18] due to the intensity dependent refractive index, but the details are not well understood. The broadening becomes more pronounced at higher laser powers.

If the pump has a bandwidth comparable to the Raman shift, the result is a continuum. By using a nitrogen-laser-pumped broadband blue dye laser a white light pulse of 10 ns duration has been produced.[19] These nanosecond continuum sources can be useful for time resolved studies of excited-state absorption.

In the near infrared, a continuum extending to beyond 2 μm has been generated in a low-loss, low-water-content multimode silica fiber by using high power (50 kW) Q-switched Nd:YAG laser pulses at 1.06 μm as the pump.[20] The wavelength limit to the cascaded Stokes generation is determined by the fiber loss.

FIBER RAMAN OSCILLATORS

Feedback and Tuning

A Raman amplifier will oscillate when feedback is provided and can be tuned over the broad gain curve if a frequency selective element is inserted in the resonator. Figure 2.4.3 illustrates some of the feedback and tuning schemes for cw oscillators. Figures 2.4.3a and 3b show the most common form of oscillator.[21-23] Here pump light is coupled through mirror M1 which is highly reflecting at the Stokes wavelength and M2 is the output mirror. Usually, M1 is the output mirror of the pump laser and AR coated microscope objectives are used for coupling. Feedback with a high reflectivity mirror can be better than 90%. At threshold, Raman amplification equals the losses from feedback and from the fiber itself. Various parameters such as length and the pump power can be chosen to optimize threshold or Stokes power output.[24] With two prisms in the cavity, tuning can be achieved over about half the gain curve with linewidths of about 1 Å.[22] Fabry-Perot etalons can also be used in the cavity either by themselves or along with prisms.

At high pump powers the Stokes power in the simple cavity of Figure 2.4.3a is high enough to pump second Stokes and several cascaded Stokes orders have been observed in this way. The same thing happens in the tunable oscillator of Figure 2.4.3b, but now a mirror must be added for each Stokes order as shown in Figure 2.4.3c.[25]

Fiber Raman oscillators can also be pumped with repetitive pulses. The requirement is that the pumping be synchronous — that is, the pump pulse spacing must be an integral multiple of the cavity round trip time. There are long or short pulse regimes depending on whether there is significant spatial separation of pump and Stokes pulses due to group-velocity dispersion. The long-pulse regime is similar to the cw case and tuning can be achieved using prisms or Fabry-Perot etalons. In the short pulse regime, walk-off occurs unless corrected by group velocity matching using waveguide modes.[26] Group velocity differences can sometimes be used to advantage in "time-dispersion tuning" in which either the cavity length or the pumping rate is varied to synchronize the desired Stokes wavelength pulses with the pump pulse at the fiber input.[27]

Pulsed Raman oscillators can take either the form of the simple oscillator shown in Figure 2.4.3a or the ring laser of Figure 2.4.3d where coupling is by a dichroic mirror for efficient pump coupling and Stokes feedback.[28] Combination arrangements combine time-dispersion and prism tuning.[27,29] In this case, the moving mirror must be simultaneously tilted. Raman lasers have also been made in which the fiber is entirely within the pump cavity as in Figure 2.4.3e.[30]

Actual Oscillators

cw Raman oscillation has been observed pumped by the laser lines listed in Table 2.4.2. As a general rule, fiber lengths in the visible are about 100 m and are around 1 km in the infrared[31] where the loss is lower. A convenient rule of thumb seems to be that the length should be about $\frac{1}{2}\alpha$ where α is the linear absorption coefficient. Threshold powers are lowest using small-core fibers in which linear polarization is

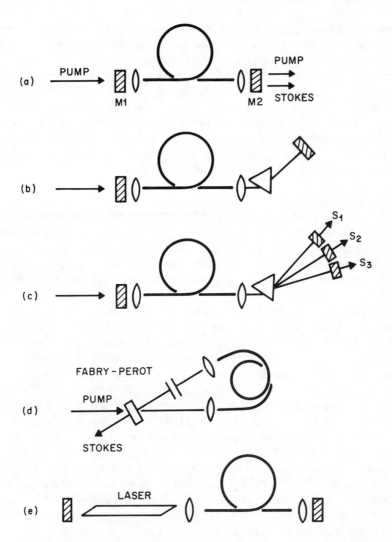

FIGURE 2.4.3.   Several types of fiber Raman oscillator. (a) Simple form in which light is fed back directly into the fiber. The first mirror is usually the output mirror of the pump laser. (b) Prism tuned oscillator. (c) Multiple cavity oscillator for producing tunable cascaded Stokes output. (d) Ring laser with coupling through a dichroic mirror and with an intracavity Fabry-Perot interferometer for tuning. (e) Fiber Raman laser within the pump laser cavity.

Table 2.4.2

LIST OF PUMP WAVELENGTHS AND LASERS USED IN FIBER RAMAN LASERS

Pump Source	Wavelength	CW	Pulsed	Mode-locked	Cavity-dumped
Nd:YAG Laser	$1.064\mu m$	x	x	x	
Nd:YAG Laser	$0.5032\mu m$		x		
Argon Laser	5145Å	x		x	x
Argon Laser	4965Å	x			
Argon Laser	4880Å	x		x	
Argon Laser	4765Å	x			
Dye lasers	4000Å-6500Å		x		

maintained[13] and are typically around half a watt although thresholds as low as 200 mW have been obtained. Cascaded Stokes operation has been obtained in both the visible[25] and near IR with pump powers of a few watts.[32]

Synchronous operation has been obtained using mode-locked[27] and cavity-dumped[28] Agron lasers at 514.5 nm and mode-locked[29] Nd:YAG lasers at 1.06 μm. In the case of the mode-locked pumping, the cavity round trip time which is more than 1 μsec is much larger than the time between pump pulses (about 10 nsec) so the Stokes pulse is synchronized not with the next pump pulse but one much later. The best operation with mode-locked pumps is obtained with a prism in the cavity which acts to decouple fed-back pump light from the mode-locked laser.[27,29]

Limits to Operation

All lasers have their idiosyncracies and fiber Raman lasers are no exception. Instabilities caused by feedback of pump light are particularly serious because the condition for good coupling to the fiber is also the condition for good feedback of pump light. Mode-locked pumps are affected more than cw lasers while cavity-dumped lasers are almost immune to such interference. There is hope that fiber Faraday isolators[33] may remedy this problem.

Care must be taken to avoid heating of the fiber input end. This is usually caused by pump light which is not coupled into the fiber core and is dumped by some part of the fiber holder. The effect is to move the end of the fiber slightly which results in a power dependent input coupling. It is important to note that position and tilt of the fiber end are more critical for single-mode fibers than for large-core multimode fibers.

There are some more fundamental problems, in particular stimulated Brillouin scattering[34] and four-photon mixing.[35] Stimulated Brillouin scattering is a competing nonlinear process which can occur at lower threshold powers than stimulated Raman scattering. There are various ways to avoid stimulated Brillouin scattering. These include utilizing the fact that the Brillouin threshold is much higher for pump linewidths greater than 1 GHz and the use of frequency selective elements in the cavity. Problems will, however, arise when the Stokes linewidth becomes narrow enough to cause stimulated Brillouin scattering itself. Four-photon mixing acts as a power dependent broadening mechanism which acts to keep the linewidth greater than about 10 GHz. One interesting aspect of this mechanism is that it may also provide intensity stabilization of fiber Raman oscillators.[36]

CALCULATION OF THRESHOLD POWER

The basic results on Raman gain and threshold power are summarized in Table 2.4.3. Example A at 1.0 μm provides a convenient reference. For this example, gL/A is 5.36 W^{-1} so the Stokes amplification, Exp(G), is 1.71 for a power of 100 mW before correction for linear absorption at the Stokes wavelength. Example B illustrates a different fiber at a different wavelength, again in the limit where the fiber is much longer than the absorption length. The results are translated from example A using the fact that g~1/α, the approximation that the core area is the spot size, and the limiting value for the effective length of 1/α. Example C illustrates the use of a fiber where 1 = ½α which is a convenient length for both oscillator and single-pass applications. These examples all assume that linear polarization is maintained in the fiber; for polarization-scrambling fibers g is reduced and critical power is increased by a factor of two.

Threshold in an oscillator occurs when the round-trip Stokes amplification equals the sum of fiber and feedback loss. As an example, we use the prism tuned oscillator of Figure 2.4.3b, the fiber and wavelength from Table 2.4.1c and assume the feedback

Table 2.4.3

RAMAN GAINS, gL/A, AND CRITICAL POWERS, P_c, FOR FIBERS WHICH
ARE MUCH LONGER THAN THE ABSORPTION LENGTH at 1.0 μm (A) and
0.5 μm (B)[a]

A	B	C	Notes
$\lambda = 1.0\,\mu$m	$\lambda = 0.5\,\mu$m	$\lambda = 0.5\,\mu$m	$g \sim 1/\lambda$
d = 10.0 μm	d = 3.0 μm	d = 3.0 μm	A = $\pi d^2/4$
1 dB/km	20 dB/km	20 dB/km	
$\ell \gg 1/\alpha$	$\ell \gg 1/\alpha$	$\ell = 1/2\alpha = 108.6$ m	
L = $1/\alpha$ = 4.34 km	L = $1/\alpha$ = 217.1 m	L = $(1-e^{-\alpha\ell})/\alpha$ = 85.5 m	
gL/A = 5.36 W^{-1}	gL/A = 5.96 W^{-1}	gL/A = 2.35 W^{-1}	
P_c = 3.0 W	P_c = 2.7 W	P_c = 6.8 W	P_c = 16A/gL

[a] Example C is a case where the length is $\frac{1}{2}\alpha$ which is a convenient length for both oscillator and single-pass applications. It is assumed that linear polarization is maintained in the fiber. For polarization scrambling fibers the gain is reduced and P_c is increased by a factor of two. The table also uses the approximation that A is the fiber core area which is useful for both single and multimode fibers.

at the input and output ends to be 0.9 and 0.8. The round-trip Stokes transmission is 0.72 exp($-2\alpha\ell$) which is 0.265. Raman gain is the same in the backward and forward directions so threshold exp($2gLP_t/A$) = 1/0.265 or P_t = 283 mW. If the configuration of Figure 2.4.3a were used, pump light would also be fed back into the fiber which would lower the input power required for threshold. The power in the fiber for strong Stokes conversion is typically twice the threshold power. Note that input coupling efficiencies range between 0.5 and 0.9 so actual laser powers will be somewhat higher.

In general, measurements of Raman gain[4] and oscillator thresholds[13,22,24] in single-mode fibers are in excellent agreement with calculated values although accurate comparisons require more careful treatment of the effective core area than the core-size approximation used here. Stimulated thresholds in single-mode fibers have been consistant with the rule[6] that G = 16. Some care is required in these measurements since pulsed lasers often have strong sharp spikes within a broader envelope which can contribute to an apparent low threshold.

Threshold in large-core multimode fibers is less certain than for single-mode fibers for two reasons. First is the question of the applicability of the threshold value of G = 16 since this number was calculated for a single-mode fiber. An accurate calculation includes the number of modes, which decreases the required value of G,[6] and overlap integrals between the modes, which increases it again. Second is the problem that the gain will depend on the distribution of energy between the modes[37] and the core-size approximation is expected to apply only to the case of uniform excitation of all modes. Published threshold powers for multimode fibers[38,39] are in the range of a few kilowatts and fall about a factor of two below predictions based on Table 2.4.1.

SUMMARY

The single pass configuration is the simplest form of fiber Raman laser and has already found application in fiber loss and dispersion measurements and time resolved spectroscopy while the various oscillator forms are still under development. It is interesting to note that while a single-pass system is the simplest to construct it is actually the more complex in terms of the nonlinear processes involved. Operation is usually at high enough powers so that very high levels of pump depletion are involved and numerous other nonlinear processes play a role. As a result, the output is extremely dependent on the characteristics of the pump laser. Ultimately, the greater opportunities for precise control should favor the oscillators.

REFERENCES

1. For recent reviews on fiber Raman lasers and nonlinear optics in fibers see: **Stolen, R. H.**, Fiber Raman lasers, in *Fiber and Integrated Optics*, Ostrowsky, D. B., Ed., Plenum Press, N.Y., 1979; **Hill, K. O., Kawasaki, B. S., Johnson, D. C., and Fujii, Y.**, Nonlinear effects in optical fibers, in *Fiber Optics — Advances in Research and Development*, Bendow, B. and Mitra, S. S., Eds., Plenum Press, N.Y., 1979; **Stolen, R. H.**, Nonlinear properties of optical fibers, in *Optical Fiber Telecommunications*, Miller, S. E. and Chynoweth, A. G., Eds., Academic Press, New York, 1979.
2. **Stolen, R. H., Ippen, E. P., and Tynes, A. R.**, Raman oscillation in glass optical waveguide, *Appl. Phys. Lett.*, 20, 62, 1972.
3. **Stolen, R. H., Bjorkholm, J. E., and Askin, A.**, Phase-matched three-wave mixing in silica fiber optical waveguides, *Appl. Phys. Lett.*, 24, 308, 1974; **Stolen, R. H.**, Phase-matched stimulated four-photon mixing in silica-fiber waveguides, *IEEE J. Quantum Electron.*, QE-11, 100, 1975.
4. **Stolen, R. H. and Ippen, E. P.**, Raman gain in glass optical waveguides, *Appl. Phys. Lett.*, 22, 276, 1973.
5. **Ippen, E. P.**, Nonlinear effects in optical fibers, in *Laser Applications to Optics and Spectroscopy*, Jacobs, S. F., Scully, M. O., and Sargent, M., Eds., Addison-Wesley, Reading, Mass., 1975.
6. **Smith, R. G.**, Optical power handling capacity of low loss optical fibers as determined by stimulated Raman and Brillouin scattering, *Appl. Opt.*, 11, 2489, 1972.
7. **Walrafen, G. E. and Stone, J.**, Raman characterization of pure and doped fused silica optical fibers, *Appl. Spectrosc.*, 29, 337, 1975.
8. **Galeener, F. L., Mikkelsen, Jr., J. C., Geils, R. H., and Mosby, W. J.**, The relative Raman cross-sections of vitreous SiO_2, GeO_2, B_2O_3 and P_2O_5, *Appl. Phys. Lett.*, 32, 34, 1978; **Grigoryants, V. V., Davydov, B. L., Zhabotinski, M. E., Zolin, V. F., Ivanov, G. A., Smirnov, V. I., and Chamorouski, Y. K.**, Spectra of stimulated Raman scattering in silica fiber waveguides, *Opt. Quantum Electron.*, (Short communication), 9, 351, 1977.
9. **Lin, Chinlon, Cohen, L. G., Stolen, R. H., Tasker, G. W., and French, W. G.**, Near-infrared sources in the 1-1.3 μm region by efficient stimulated Raman emission in glass fibers, *Opt. Commun.*, 20, 426, 1977.
10. **Ippen, E. P.**, Low power quasi-cw Raman ocillator, *Appl. Phys. Lett.*, 16, 303, 1970; **Stone, J.**, "cw Raman fiber amplifier," *Appl. Phys. Lett.*, 26, 163, 1975.
11. **Smith, W. L., Bechtel, J. H., and Bloembergen, N.**, Dielectric — breakdown threshold and nonlinear-refractive-index measurements with picosecond laser pulses, *Phys. Rev. B*, 12, 706, 1975.
12. **Stolen, R. H., Ramaswamy, V., Kaiser, P., and Pleibel, W.**, Linear polarization in birefringent single-mode fibers, *Appl. Phys. Lett.*, 33, 699, 1979.
13. **Stolen, R. H.**, Polarization effects in fiber Raman and Brillouin lasers, *IEEE J. Quantum Electron.*, QE-15, 1137, 1979.
14. **Gloge, D.**, Weakly guiding fibers, *Appl. Opt.*, 10, 2252, 1971.
15. **Stolen, R. H. and Lin, Chinlon**, unpublished results.
16. **Cohen, L. G. and Lin, Chinlon**, Pulse delay measurements in the zero material dispersion wavelength region for optical fibers, *Appl. Opt.*, 16, 3136, 1977.
17. **Lin, Chinlon, Cohen, L. G., French, W. G., and Foertmeyer, V. A.**, Pulse-delay measurements in the zero-material-dispersion region for germanium- and phosphorus-doped silica fibers, *Electron. Lett.*, 14, 170, 1978; **Cohen, L. G. and Lin, Chinlon**, A universal fiber optic (UFO) measurement system based on a near IR fiber Raman laser, *IEEE J. Quantum Electron.*, QE-14, 855, 1978.
18. **Stolen, R. H. and Lin, Chinlon**, Self-phase-modulation in silica optical fibers, *Phys. Rev.*, 417, 1448, 1978.
19. **Lin, Chinlon, and Stolen, R. H.**, New nanosecond continuum for excited-state spectroscopy, *Appl. Phys. Lett.*, 29, 216, 1976.
20. **Lin, Chinlon, Nguyen, V. T., and French, W. G.**, Wideband near-IR continuum (0.7-2.1 μm) generated in low-loss optical fibers, *Electron. Lett.*, 14, 822, 1978.
21. **Hill, K. O., Kawasaki, B. S., and Johnson, D. C.**, Low-threshold cw Raman laser, *Appl. Phys. Lett.*, 29, 181, 1976.
22. **Jain, R. K., Lin, Chinlon, Stolen, R. H., Pleibel, W., and Kaiser, P.**, A high-efficiency tunable cw Raman oscillator, *Appl. Phys. Lett.*, 30, 162, 1977.

23. Johnson, D. C., Hill, K. O., Kawasaki, B. S., and Kato, D., Tunable Raman fiber-optic laser, *Electron. Lett.*, 13, 53, 1977.
24. AuYeung, J. and Yariv, A., Theory of cw Raman oscillation in optical fibers, *J. Opt. Soc. Am.*, 69, 803, 1979.
25. Jain, R. K., Lin, Chinlon, Stolen, R. H., and Ashkin, A., A tunable multiple stokes cw fiber Raman oscillator, *Appl. Phys. Lett.*, 31, 89, 1977.
26. Lin, Chinlon, Stolen, R. H., and Jain, R. K., Group velocity matching in optical fibers, *Opt. Lett.*, 1, 205, 1977.
27. Stolen, R. H., Lin, Chinlon, and Jain, R. K., A time-dispersion-tuned fiber Raman oscillator, *Appl. Phys. Lett.*, 30, 340, 1977.
28. Stolen, R. H., Lin, Chinlon, Shah, J., and Leheny, R. F ., A fiber Raman ring laser, *IEEE J. Quantum Electron.*, QE-14, 860, 1978.
29. Lin, Chinlon, Stolen, R. H., and Cohen, L. G., A tunable 1.1 μm fiber Raman oscillator, *Appl. Phys. Lett.*, 31, 97, 1977.
30. Dianov, E. M., Isaev, S. K., Kornienko, L. S., Kravtsor, N. V., and Firsor, V. V., Raman Laser with optical fiber resonator, *Kvantovaya Elektron. (Moscow)*, 5, 1305, 1978 [*Sov. J. Quantum Electron*, 8, 744, 1978]; Johnson, D. C., Hill, K. O., and Kawasaki, B. S., Continuous-wave optical-fiber Raman oscillator employing a two-mirror resonator configuration, *Appl. Opt.*, 17, 3032, 1978.
31. Lin, Chinlon, Stolen, R. H., French, W. G., and Melone, T. G., A cw tunable near infrared (1.085-1.175 μm) Raman oscillator, *Opt. Lett.*, 1, 96, 1977.
32. Lin, Chinlon and French, W. G., A near-infrared fiber Raman oscillator tunable from 1.07 to 1.32 μm, *Appl. Phys. Lett.*, 34, 666, 1979.
33. Turner, E. H. and Stolen, R. H., Fiber Faraday circulator, *J. Opt. Soc. Am.*, 69, 1483, 1979.
34. Ippen, E. P. and Stolen, R. H., Stimulated Brillouin scattering in optical fibers, *Appl. Phys. Lett.*, 21, 539, 1972; Hill, K. O., Kawasaki, B. S., and Johnson, D. C., cw Brillouin laser, *Appl. Phys. Lett.*, 28, 608, 1976.
35. Hill, K. O., Johnson, D. C., and Kawasaki, B. S., cw Three-wave-mixing in single-mode optical fibers, *J. Appl. Phys.*, 49, 5098, 1978.
36. Labudde, P., Weber, H. P., and Stolen, R. H., Bandwidth reduction in cw fiber Raman lasers, *IEEE J. Quantum Electron.*, QE-16, 115, 1980.
37. Capasso, F. and DiPorto, P., Coupled-mode theory of Raman amplification in lossless optical fibers, *J. Appl. Phys.*, 47, 1472, 1976.
38. Jensen, S. M. and Barnoski, M. K., Stimulated Raman scattering in multimode fibers, in Technical Digest of the OSA/IEEE Topical Meeting on Optical Fiber Transmission II, Williamsburg, Va., 1977, paper TuD7.
39. Lin, C. H., Marshall, B. R., Nelson, M. A., and Theobald, J. K., Backward stimulated Raman scattering in multimode fiber, *Appl. Opt.*, 17, 2486, 1978.

2.5 TABLE OF WAVELENGTHS OF SOLID STATE LASERS

In Table 2.5.1 solid-state lasers are arranged by increasing laser wavelength (in air). Phonon-terminated lasers, color-center lasers, and semiconductor lasers involve broadband transitions and therefore are tunable over a limited range. When a range of lasing wavelengths is reported, the laser is listed by the shortest wavelength. Since these ranges are dependent on the experimental conditions, they are not necessarily the extreme wavelengths possible.

The emission wavelengths in solids are generally temperature dependent, therefore small changes in lasing wavelengths can be made by using operating temperatures different from those given. Several examples of measured temperature tuning ranges for Nd^{3+} and other lasers are included in the table.

Glass lasers involve inhomogeneously broadened transitions and can be tuned over this linewidth. For rare-earth lasers operating on 4f - 4f transitions, the linewidths arising from the combined Stark splitting and inhomogeneous broadening are typically in the order of several hundred cm^{-1}.

Tables where additional operating data and references to the original reports of laser action can be found are listed in the final column of the table.

Table 2.5.1

KEY TO TABLE REFERENCES

a	= 2.1.1.2	e	=	2.2.2
b	= 2.1.2.2	f	=	2.3.1
c	= 2.1.3.1	g	=	2.3.11
d	= 2.2.1	h	=	2.3.12

Wavelength (μm)	Activator ion/center	Host	Temperature (K)	Table number
0.286	Ce^{3+}	LaF_3	300	a
0.306—0.315	Ce^{3+}	$LiYF_4$	300	a
0.3125	Gd^{3+}	Silicate glass	78	f
0.3146	Gd^{3+}	$Y_3Al_5Of_{12}$	300	a
0.323—0.328	Ce^{3+}	$LiYF_4$	300	a
0.33		ZnS		d
0.33—0.49		ZnCdS		d
0.36		GaN		d
0.37		ZnO		d
0.4526	Pr^{3+}	$LiYF_4$	300	a
0.46		ZnSe		d
0.479	Pr^{3+}	$LiYF_4$	300	a
0.4892	Pr^{3+}	$LaCl_3$	5.5—14	a
0.4892	Pr^{3+}	$PrCl_3$	14	b
0.49		CdS		d
0.5298	Pr^{3+}	$LaCl_3$	35	a
0.53		ZnTe		d
0.532	Pr^{3+}	$LaBr_3$	≤300	a
0.54	Tb^{3+}	Borate glass	300	f
0.5445	Tb^{3+}	$LiYF_4$	300	a
0.5512	Ho^{3+}	CaF_2	77	a
0.5515	Ho^{3+}	$B(Y,Yb)_2F_8$	77	a
0.5540	Er^{3+}	BaY_2F_8	77	a
0.5617	Er^{3+}	BaY_2F_8	77	a

Table 2.5.1 (continued)

KEY TO TABLE REFERENCES

a	=	2.1.1.2	e	=	2.2.2
b	=	2.1.2.2	f	=	2.3.1
c	=	2.1.3.1	g	=	2.3.11
d	=	2.2.1	h	=	2.3.12

Wavelength (μm)	Activator ion/center	Host	Temperature (K)	Table number
0.59—0.60		GaSe		d
0.59—0.69		CdSSe		d
0.59—0.90		InGaP		d
0.5932	Sm^{3+}	TbF_3	116	a
0.5985	Pr^{3+}	LaF_3	77	a
0.6113	Eu^{3+}	Y_2O_3	220	a
0.6164	Pr^{3+}	$LaCl_3$	65	a
0.6164	Pr^{3+}	$(Pr, La)Cl_3$	65	b
0.6193	Eu^{3+}	YVO_4	90	a
0.62—0.90		AlGaAs		d
0.62—1.60		AlGaAsSb		d
0.62—3.20		InGaPAs		d
0.621	Pr^{3+}	$LaBr_3$	≤300	a
0.63		InGaAsP-GaAsP	77	e
0.63—0.90		GaAsP		d
0.632	Pr^{3+}	$LaBr_3$	<70	a
0.6374	Pr^{3+}	PrP_5O_{14}	300	b
0.638		$Al_{0.68}Ga_{0.32}As$	77	e
0.6452	Pr^{3+}	$LaCl_3$	300	a
0.647	Pr^{3+}	$LaBr_3$	≤300	a
0.6700	Er^{3+}	$Ba(Y,Yb)_2F8$	77	a
0.6709	Er^{3+}	BaY_2F_8	77	a
		$Ba(Y,Yb)_2F_8$	77	
0.675		GaAsP-InGaP	273	e
~0.680	Cr^{3+}	$BeAl_2O_4$	77	a
0.6804	Cr^{3+}	$BeAl_2O_4$	300	a
0.6874	Cr^{3+}	$Y_3Al_5O_{12}$	~77	a
0.69		CdSe		d
0.6929	Cr^{3+}	Al_2O_3	300	a
0.6934	Cr^{3+}	Al_2O_3	77	a
0.6943—0.6952	Cr^{3+}	Al_2O_3	300—500	a
0.6969	Sm^{2+}	SrF_2	412	a
0.7009	Cr^{3+}-Cr^{3+}	Al_2O_3	77	a
0.701—0.818	Cr^{3+}	$BeAl_2O_4$	300	a
0.703		GaAsP-InGaP	300	e
0.7037	Er^{3+}	BaY_2F_8	77	a
0.7041	Cr^{3+}-Cr^{3+}	Al_2O_3	77	a
0.7083	Sm^{2+}	CaF_2	65—90	a
0.7085	Sm^{2+}	CaF_2	20	a
0.7207	Sm^{2+}	CaF_2	85—90	a
0.7287	Sm^{2+}	CaF_2	110—130	a
0.7310	Sm^{2+}	CaF_2	155	a
0.745	Sm^{2+}	CaF_2	210	a
0.7498	Ho^{3+}	$LiYF_4$	90	a
0.7505	Ho^{3+}	$LiYF_4$	300	a
0.77		$CdSiAs_2$		d
0.79		CdTe		d

Table 2.5.1 (continued)

KEY TO TABLE REFERENCES

a	= 2.1.1.2	e	=	2.2.2
b	= 2.1.2.2	f	=	2.3.1
c	= 2.1.3.1	g	=	2.3.11
d	= 2.2.1	h	=	2.3.12

Wavelength (μm)	Activator ion/center	Host	Temperature (K)	Table number
$\sim$0.81		AlGaAsP-GaAsP	300	e
0.82—1.05	F_2^+	LiF		c
0.83—0.92		GaAs		d
0.8430	Er^{3+}	CaF_2-YF_3	77	a
0.845		AlGaAsP-GaAsP	300	e
0.8456	Er^{3+}	CaF_2	77	a
		CaF_2-YF_3	77	a
0.8468	Er^{3+}	$KGd(WO_4)_2$	300	a
0.84965	Er^{3+}	$YAlO_3$	77	a
0.84975	Er^{3+}	$YAlO_3$	300	a
0.85	Er^{3+}	$LiYF_4$	300	a
		$KLu(WO_4)_2$	300	a
0.85—3.20		InGaAs		d
0.8503	Er^{3+}	$LiYF_4$	110, 300	a
0.85165	Er^{3+}	$YAlO_3$	77	a
0.8535	Er^{3+}	$LiYF_4$	110	a
0.8546	Er^{3+}	CaF_2	77	a
0.8594	Er^{3+}	$YAlO_3$	300	a
0.8621	Er^{3+}	$KY(WO_4)_2$	300	a
0.8627	Er^{3+}	$Y_3Al_5O_{12}$	77, 300	a
0.8632	Er^{3+}	$Lu_3Al_5O_{12}$	300	a
0.853	Er^{3+}	$Bi_4G_3O_{12}$	77	
0.86325	Er^{3+}	$Lu_3Al_5O_{12}$	77	a
$\sim$0.89		$InGa_{0.5}P_{0.5}$-GaAs		d
0.89-0.91		InP		d
0.8910	Nd^{3+}	$Y_3Al_5O_{12}$	300	a
0.8999	Nd^{3+}	$Y_3Al_3O_{12}$	300	a
0.9		GaAsP-GaAs	300	e
0.90—3.20		InAsP		d
0.9137	Nd^{3+}	$KY(WO_4)_2$	77	a
0.9145	Nd^{3+}	$CaWO_4$	77	a
0.918	Nd^{3+}	Silicate glass	80	f
0.92	Nd^{3+}	Silicate glass	300	f
0.930	Nd^{3+}	$YAlO_3$	300	a
0.9385	Nd^{3+}	$Y_3Al_5O_{12}$	300	a
0.941	Nd^{3+}	$CaY_2Mg_2Ge_3O_{12}$	300	a
0.9460	Nd^{3+}	$Y_3Al_5O_{12}$	300	a
0.9473	Nd^{3+}	$Lu_3Al_5O_{12}$	77	a
0.95—1.60		GaAsSb		d
0.97		InSe		d
0.979	Ho^{3+}	$LiHoF_4$	90	b
0.9794	Ho^{3+}	$LiYF_4$	90	a
0.98		GaAsSb-AlGaAsSb	300	e
0.99—1.22	F_2^+	NaF		c
1.0143	Ho^{3+}	$LiYF_4$	90	a
1.015	Yb^{3+}	Silicate glass	77,300	f
1.018	Yb^{3+}	Borate glass	300	f

Table 2.5.1 (continued)

KEY TO TABLE REFERENCES

a	= 2.1.1.2	e	=	2.2.2
b	= 2.1.2.2	f	=	2.3.1
c	= 2.1.3.1	g	=	2.3.11
d	= 2.2.1	h	=	2.3.12

Wavelength (μm)	Activator ion/center	Host	Temperature (K)	Table number
1.0230		$Yb^{3+}Lu_3Ga_5O_{12}$	77	a
1.0232	Yb^{3+}	$Gd_3Ga_5O_{12}$	77	a
1.0233	Yb^{3+}	$Y_3Ga_5O_{12}$	77	a
1.0293	Yb^{3+}	$Y_3Al_5O_{12}$	77	a
		$(Y,Yb)_3Al_5O_{12}$	77	a
1.0294	Yb^{3+}	$Lu_3Al_5O_{12}$	77	a
1.0296	Yb^{3+}	$Y_3Al_5O_{12}$	77	a
1.0297	Yb^{3+}	$Lu_3Al_5O_{12}$	77	a
1.0297	Yb^{3+}	$Y_3Al_5O_{12}$	200	a
1.0298	Yb^{3+}	$Y_3Al_5O_{12}$	210	a
1.0299	Yb^{3+}	$Gd_3Sc_2Al_3O_{12}$	77	a
		$Lu_3Sc_2Al_3O_{12}$	77	a
1.0336	Yb^{3+}	CaF_2	120	a
1.0369	Nd^{3+}	$CaF_2\text{-}SrF_2$	300	a
1.0370	Nd^{3+}	CaF_2	300	a
		SrF_2	460	a
1.0370-1.0395	Nd^{3+}	SrF_2	300—530	a
1.04	Pr^{3+}	$SrMoO_4$		
1.0400	Nd^{3+}	LaF_3	77	a
1.0404	Nd^{3+}	CeF_3	77	a
1.0406—1.0410	Nd^{3+}	LaF_3	300—430	a
1.04065	Nd^{3+}	LaF_3	300	a
1.0410	Nd^{3+}	CeF_3	300	a
1.0437	Nd^{3+}	SrF_2	77	a
1.0445	Nd^{3+}	SrF_2	300	a
1.0446	Nd^{3+}	SrF_2	500	a
1.0448	Nd^{3+}	CaF_2	50	a
1.0451	Nd^{3+}	LaF_3	77	a
1.0456	Nd^{3+}	CaF_2	50	a
1.0457	Nd^{3+}	CaF_2	77	a
1.0461	Nd^{3+}	CaF_2	300	a
		$CaF_2\text{-}YF_3$	300	a
1.0461—1.0468	Nd^{3+}	CaF_2	300—530	a
1.0466	Nd^{3+}	CaF_2	50	a
1.0467	Nd^{3+}	CaF_2	77	a
1.0468	Pr^{3+}	$CaWO_4$	77	a
1.047	Nd^{3+}	Fluoroberyllate glass	300	f, h
1.0471	Nd^{3+}	$LiYF_4$	300	a
1.048	Nd^{3+}	$LiNdP_4O_{12}$	300	b
		$Li(Nd,La)P_4O_{12}$	300	b
		$Li(Nd,Gd)P_4O_{12}$	300	b
1.0480	Nd^{3+}	CaF_2	50	a
1.048	Nd^{3+}	$K_5NdLi_2F_{10}$	300	b
1.0486	Nd^{3+}	$LaF_3\text{-}SrF_2$	300	a
1.0491	Nd^{3+}	$SrAl_{12}O_{19}$	300	a
1.0493	Nd^{3+}	$Sr_2Y_5F_{19}$	300	a
1.0497	Nd^{3+}	$CaAl_{12}O_{19}$	300	a

Table 2.5.1 (continued)

KEY TO TABLE REFERENCES

a	= 2.1.1.2	e	=	2.2.2
b	= 2.1.2.2	f	=	2.3.1
c	= 2.1.3.1	g	=	2.3.11
d	= 2.2.1	h	=	2.3.12

Wavelength (μm)	Activator ion/center	Host	Temperature (K)	Table number
1.0498	Nd^{3+}	$Ca_2Y_5F_{19}$	300	a
1.05	Nd^{3+}	$Na_3Nd(PO_4)_2$	300	b
		Fluorophosphate glass	300	f
		Phosphate glass	300	g
1.0505	Nd^{3+}	Fluorophosphate glass	300	h
1.0506	Nd^{3+}	$5NaF \cdot 9YF_3$	300	a
1.0507	Nd^{3+}	CaF_2	50	a
1.051	Nd^{3+}	Fluorophosphate glass	300	
1.051	Nd^{3+}	$NaNdP_4O_{12}$	300	b
		YP_5O_{14}	300	a
		LaP_5O_{14}	300	a
		CeP_5O_{14}	300	a
		GdP_5O_{14}	300	a
		$(Nd,Sc)P_5O_{14}$	300	b
		$(Nd,In)P_5O_{14}$	300	b
1.0511	Nd^{3+}	$(Nd,La)P_5O_{14}$	300	b
1.0512	Nd^{3+}	NdP_5O_{14}	300	b
1.0513	Nd^{3+}	NdP_5O_{14}	300	b
1.0515	Nd^{3+}	YP_5O_{14}	300	a
1.052	Nd^{3+}	$KNdP_4O_{12}$	300	b
		$K_5NdLi_2F_{10}$	300	b
1.0521	Nd^{3+}	BaF_2-YF_3	300	a
		$Y_3Al_5O_{12}$	300	a
		NdP_5O_{14}	300	b
1.0523	Nd^{3+}	BaF_2-LuF_3	300	a
		LaF_3	77	a
1.0525	Nd^{3+}	YP_5O_{14}	300	a
1.0526	Nd^{3+}	BaF_2-GdF_3	300	a
1.0528	Nd^{3+}	SrF_2-GdF_3	300	a
1.0529	Nd^{3+}	NdP_5O_{14}	300	b
1.053	Nd^{3+}	Phosphate glass	300	h
1.0530	Nd^{3+}	$LiYF_4$	300	a
1.0534—1.0563	Nd^{3+}	BaF_2-LaF_3	300—920	a
1.0535	Nd^{3+}	Phosphate glass	300	h
1.0535—1.0547	Nd^{3+}	$CaF_2-SrF_2-BaF_2-YF_3-LaF_3$	300—550	a
1.0535	Nd^{3+}	$Lu_3Al_5O_{12}$	300	a
1.0538	Nd^{3+}	BaF_2-LaF_3	77	a
1.0539	Nd^{3+}	α-$NaCaYF_6$	300	a
1.0539—1.0549	Nd^{3+}	α-$NaCaYF_6$	300—550	a
1.054	Nd^{3+}	Phosphate glass	300	f
		Fluorophosphate glass	300	f
1.0540	Nd^{3+}	CaF_2-YF_3	300	a
1.0543	Nd^{3+}	BaF_2-CeF_3	300	a
1.05436	Nd^{3+}	$Ba_2MgGe_2O_7$	300	a
1.05437	Nd^{3+}	$Ba_2ZnGe_2O_7$	300	a
1.0551	Nd^{3+}	$Pb_5(PO_4)_3F$	300	a
1.0554	Nd^{3+}	$LiNdP_4O_{12}$	300	b

Table 2.5.1 (continued)

KEY TO TABLE REFERENCES

a	= 2.1.1.2	e	=	2.2.2
b	= 2.1.2.2	f	=	2.3.1
c	= 2.1.3.1	g	=	2.3.11
d	= 2.2.1	h	=	2.3.12

Wavelength (μm)	Activator ion/center	Host	Temperature (K)	Table number
1.0556	Nd^{3+}	SrF_2-LuF_3	300	a
1.0566	Nd^{3+}	$SrAl_4O_7$	77	a
1.0567	Nd^{3+}	SrF_2-YF_3	300	a
1.0568	Nd^{3+}	$SrAl_4O_7$	77	a
1.057	Nd^{3+}	Silicate glass	300	h
1.0573	Nd^{3+}	$CaMoO_4$	295	a
1.0574	Nd^{3+}	$SrWO_4$	77	a
1.0575	Nd^{3+}	$Y_3Sc_2Ga_3O_{12}$	77	a
1.05755	Nd^{3+}	$Gd_3Sc_2Ga_3O_{12}$	77	a
1.0576	Nd^{3+}	$SrAl_4O_7$	300	a
		$SrMoO_4$	295	a
1.0580	Nd^{3+}	BaF_2-LaF_3	77	a
		$Gd_3Sc_2Ga_3O_{12}$	77	a
1.0582—1.0597	Nd^{3+}	$CaWO_4$	300—700	a
1.0583	Nd^{3+}	LaF_3	77	a
		$Y_3Sc_2Ga_3O_{12}$	300	a
		$Y_3Ga_5O_{12}$	77	a
1.0584	Nd^{3+}	$Gd_3Ga_5O_{12}$	77	a
		$CaY_2Mg_2Ge_3O_{12}$	300	a
1.0585	Nd^{3+}	$YAlO_3$	300	a
		$LiLa(MoO_4)_2$	300	a
		$Na_2Nd_2Pb_6(PO_4)_6Cl_2$	300	b
		$Sr_5(PO_4)_3F$	300	a
		$KLa(MoO_4)_2$	77	a
1.0585—1.0623	Nd^{3+}	$CaF_2-SrF_2-BaF_2-YF_3-LaF_3$	700—300	a
1.0586	Nd^{3+}	$PbMoO_4$	300	a
		$SrLa_4(SiO_4)_3O$	300	a
1.0587	Nd^{3+}	$CaWO_4$	300	a
		$KLa(MoO_4)_2$	300	a
		$Y_3Sc_2Al_3O_{12}$	77	a
		$Lu_3Ga_5O_{12}$	77	a
1.0588	Nd^{3+}	$Ca(NbO_3)_2$	300	a
1.0589	Nd^{3+}	$SrF_2-CeF_3-GdF_3$	300	a
		$Y_3Ga_5O_{12}$	300	a
1.05895	Nd^{3+}	$CaAl_4O_7$	77	a
1.05896	Nd^{3+}	$CaMg_2Y_2Ge_3O_{12}$	300	a
1.0590	Nd^{3+}	SrF_2-CeF_3	300	a
		$Ca_2MgSi_2O_7$	300	a
1.059	Nd^{3+}	$SrMoO_4$	77	a
		$CaMg_2Y_2Ge_3O_{12}$	300	a
1.0591	Nd^{3+}	$Gd_3Ga_5O_{12}$	300	a
		$Lu_3Sc_2Al_3O_{12}$	77	a
1.05915	Nd^{3+}	$Gd_3Sc_2Al_3O_{12}$	77	a
1.0594	Nd^{3+}	$Lu_3Ga_5O_{12}$	300	a
1.0595	Nd^{3+}	$5NaF \cdot 9YF_3$	300	a
		$Y_3Sc_2Al_3O_{12}$	300	a
		$NaLa(MoO_4)_2$	300	a

Table 2.5.1 (continued)

KEY TO TABLE REFERENCES

a	= 2.1.1.2	e	=	2.2.2
b	= 2.1.2.2	f	=	2.3.1
c	= 2.1.3.1	g	=	2.3.11
d	= 2.2.1	h	=	2.3.12

Wavelength (μm)	Activator ion/center	Host	Temperature (K)	Table number
1 0595—1.0613	Nd^{3+}	LaF_3	380—820	a
1.0596	Nd^{3+}	$CaAl_4O_7$	300	a
1.0597	Nd^{3+}	SrF_2-LaF_3	300	a
		$Ca_3Ga_2Ge_3O_{12}$	300	a
1.0597—1.0583	Nd^{3+}	SrF_2-LaF_3	300—800	a
1.0597—1.0629	Nd^{3+}	α-$NaCaYF_6$	1000—300	a
1.05975	Nd^{3+}	$Y_3Ga_5O_{12}$	77	a
1.0599	Nd^{3+}	$Gd_3Ga_5O_{12}$	77	a
		$Lu_3Sc_2Al_3O_{12}$	300	a
		$LiGd(MoO_4)_2$	300	a
1.05995	Nd^{3+}	$Gd_3Sc_2Al_3O_{12}$	300	a
1.06	Nd^{3+}	BaF_2	77	a
		$NaGd(WO_4)_2$	300	a
		$Na(Nd,Gd)(WO_4)_2$	77	b
		$K_3Nd(PO_4)_2$	300	b
		$K_3(Nd,La)(PO_4)_2$	300	b
		Borate glass	300	f
		Silicate glass	300	f
		Glass-ceramic	300	f
		Germanate glass	300	f
	Yb^{3+}	Silicate glass	300	f
1.0601	Nd^{3+}	$CaWO_4$	77	a
1.06025	Nd^{3+}	$Y_3Ga_5O_{12}$	77	a
		$Lu_3Ga_5O_{12}$	77	a
1.0603	Nd^{3+}	$Y_3Ga_5O_{12}$	300	a
1.0603—1.0632	Nd^{3+}	CaF_2-YF_3	950—300	a
1.0604	Nd^{3+}	HfO_2-Y_2O_3	300	a
1.06045	Nd^{3+}	$Gd_3Sc_2Ga_3O_{12}$	77	a
1.0605	Nd^{3+}	$Lu_3Al_5O_{12}$	77	a
1.0606	Nd^{3+}	$Gd_3Ga_5O_{12}$	300	a
		$Gd_2(MoO_4)_3$	300	a
1.0607	Nd^{3+}	$SrWO_4$	77	a
1.0608	Nd^{3+}	$(Y,Lu)_3Al_5O_{12}$	77	a
		ZrO_2-Y_2O_3	300	a
1.0609	Nd^{3+}	$Lu_3Ga_5O_{12}$	300	a
1.061	Nd^{3+}	$CaMoO_4$	300	a
1.061	Nd^{3+}	Silicate glass	300	f, h
		Tellurite glass	300	h
1.0610	Nd^{3+}	$Y_3Al_5O_{12}$	77	a
		$Lu_3Al_5O_{12}$	300	a
1.0610—1.0627	Nd^{3+}	$Y_3Al_5O_{12}$	77—600	a
1.0611	Nd^{3+}	$SrMoO_4$	77	a
1.0612	Nd^{3+}	$Ca(NbO_3)_2$	77	a
		$Gd_3Sc_2Ga_3O_{12}$	300	a
		$CaLa_4(SiO_4)_3O$	300	a
1.0613	Nd^{3+}	$Ca_4La(PO_4)_3O$	300	a
		$Ba_2NaNb_5O_{15}$	300	a

Table 2.5.1 (continued)

KEY TO TABLE REFERENCES

a	= 2.1.1.2	e	=	2.2.2
b	= 2.1.2.2	f	=	2.3.1
c	= 2.1.3.1	g	=	2.3.11
d	= 2.2.1	h	=	2.3.12

Wavelength (μm)	Activator ion/center	Host	Temperature (K)	Table number
1.0614	Nd^{3+}	$Ca(NbO_3)_2$	77	a
		$Y_3Ga_5O_{12}$	77	a
1.0615	Nd^{3+}	$Ca(NbO_3)_2$	300	a
		$Y_3Al_5O_{12}$	300	a
		$Y_3Sc_2Ga_3O_{12}$	300	a
		$Ba_{0.25}Mg_{2.75}Y_2Ge_3O_{12}$	300	a
1.0615	Nd^{3+}	Silicate glass	300	h
1.0615—1.0625	Nd^{3+}	$Ca(NbO_3)_2$	300—650	a
1.0616	Nd^{3+}	$Lu_3Ga_5O_{12}$	77	a
1.062	Nd^{3+}	$Ca_{0.25}Ba_{0.75}(NbO_3)_2$	295	a
1.0620	Nd^{3+}	$Gd_3Sc_2Al_3O_{12}$	300	a
		$Lu_3Sc_2Al_3O_{12}$	300	a
1.06205	Nd^{3+}	$Y_3Ga_5O_{12}$	300	a
1.0621	Nd^{3+}	$SrAl_{12}O_{19}$	300	a
		$Gd_3Ga_5O_{12}$	300	a
1.0622	Nd^{3+}	$Y_3Sc_2Al_3O_{12}$	300	a
1.0623	Nd^{3+}	CaF_2-SrF_2-BaF_2-YF_3-LaF_3	300	a
		$Lu_3Ga_5O_{12}$	300	a
1.0623—1.0628	Nd^{3+}	CaF_2	560—300	a
1.0624	Nd^{3+}	$LaNbO_4$	300	a
1.0625	Nd^{3+}	YVO_4	300	a
1.0626	Nd^{3+}	$Ca(NbO_3)_2$	77	a
1.0627	Nd^{3+}	$SrAl_4O_7$	77	a
		$SrMoO_4$	77	a
		$SrWO_4$	77	a
1.0629	Nd^{3+}	α-$NaCaYF_6$	300	a
		$Bi_4Si_3O_{12}$	77, 300	a
1.0629—1.0656	Nd^{3+}	CdF_2-YF_3	600—300	a
1.063	Nd^{3+}	$SrWO_4$	295	a
		$Na_5Nd(WO_4)_4$	300	b
		$NdAl_3(BO_3)_4$	300	b
1.0630	Nd^{3+}	$Ca_5(PO_4)_3F$	300	a
1.06305	Nd^{3+}	LaF_3	77	a
1.0632	Nd^{3+}	CaF_2-YF_3	300	a
		CaF_2-YF_3-NdF_3	300	a
1.0632—1.0642	Nd^{3+}	LaF_3	400—700	a
1.0633—1.0653	Nd^{3+}	α-$NaCaCeF_6$	920—300	a
1.06335	Nd^{3+}	LaF_3	300	a
1.0634	Nd^{3+}	$CaWO_4$	77	a
1.0635	Nd^{3+}	LaF_3-SrF_2	300	a
		$NdAl_3(BO_3)_4$	300	b
		$NaLa(WO_4)_2$	300	a
		$(Nd,Gd)Al_3(BO_4)_4$	300	b
		$Bi_4(Si,Ge)_3O_{12}$	300	a
1.0636	Nd^{3+}	$(Y,Lu)_3Al_5O_{12}$	77	a
1.0637	Nd^{3+}	$Y_3Al_5O_{12}$	170	a
1.0637—1.0670	Nd^{3+}	$Y_3Al_5O_{12}$	170—900	a

Table 2.5.1 (continued)

KEY TO TABLE REFERENCES

a	= 2.1.1.2	e	=	2.2.2
b	= 2.1.2.2	f	=	2.3.1
c	= 2.1.3.1	g	=	2.3.11
d	= 2.2.1	h	=	2.3.12

Wavelength (μm)	Activator ion/center	Host	Temperature (K)	Table number
1.06375	Nd^{3+}	$Lu_3Al_5O_{12}$	120	a
10638	Nd^{3+}	CeF_3	300	a
		$CaAl_4O_7$	300	a
		NdP_5O_{14}	300	b
1.0638—1.0672	Nd^{3+}	$Lu_3Al_5O_{12}$	120—900	a
1.0639	Nd^{3+}	CeF_3	77	a
		$Ca_3Ga_2Ge_3O_{12}$	300	a
1.0640	Nd^{3+}	$SrMoO_4$	77	a
1.0640—1.0657	Nd^{3+}	CaF_2-CeF_5	700—300	a
1.06405—1.0654	Nd^{3+}	$YAlO_3$	77—500	a
1.06405	Nd^{3+}	$YAlO_3$	77	a
1.0641	Nd^{3+}	YVO_4	300	a
1.06415	Nd^{3+}	$Y_3Al_5O_{12}$	300	a
1.0642	Nd^{3+}	$(Y,Lu)_3Al_5O_{12}$	295	a
1.06425	Nd^{3+}	$Lu_3Al_5O_{12}$	300	a
		$Bi_4Ge_3O_{12}$	77	a
1.0644	Nd^{3+}	$YAlO_3$	300	a
		Y_2SiO_5	77	a
		$Bi_4Ge_3O_{12}$	300	a
1.0645	Nd^{3+}	CaF_2-LaF_3	300	a
		$SrMoO_4$	300	a
1.0647	Nd^{3+}	$CeCl_3$	300	a
1.0648	Nd^{3+}	CaF_2	50	a
		YVO_4	300	a
1.0649	Nd^{3+}	$CaWO_4$	77	a
		$CaY_2Mg_2Ge_3O_{12}$	300	a
1.065	Nd^{3+}	CdF_2-YF_3-LaF_3	300	a
1.0652	Nd^{3+}	$CaWO_4$	300	a
		$SrMoO_4$	77	a
1.0652—1.0659	Nd^{3+}	$YAlO_3$	310—500	a
1.0653	Nd^{3+}	α-$NaCaCeF_6$	300	a
		$NaLa(MoO_4)_2$	300	a
1.0653—1.0665	Nd^{3+}	$NaLa(MoO_4)_2$	300—750	a
1.0654	Nd^{3+}	CaF_2-GdF_3	300	a
1.0656	Nd^{3+}	CdF_2-YF_3	300	a
1.0657	Nd^{3+}	CaF_2-CeF_3	300	a
1.0658	Nd^{3+}	$LiLa(MoO_4)_2$	300	a
1.06585	Nd^{3+}	$CaAl_4O_7$	77	a
1.0660	Nd^{3+}	$K_5Nd(MoO_4)_4$	300	b
		$K_5Bi(MoO_4)_4$	300	a
1.0661	Nd^{3+}	CaF_2	120	a
1.0664	Nd^{3+}	YVO_4	300	a
1.0664—1.0672	Nd^{3+}	YVO_4	300—690	a
1.0665	Nd^{3+}	CdF_2-LaF_3	300	a
1.0666	Nd^{3+}	HfO_2-Y_2O_3	300	a
1.0669	Nd^{3+}	$KY(MoO_4)_2$	300	a
1.067	Nd^{3+}	$Ca_3(VO_4)_2$	300	a

Table 2.5.1 (continued)

KEY TO TABLE REFERENCES

a	= 2.1.1.2	e	=	2.2.2
b	= 2.1.2.2	f	=	2.3.1
c	= 2.1.3.1	g	=	2.3.11
d	= 2.2.1	h	=	2.3.12

Wavelength (μm)	Activator ion/center	Host	Temperature (K)	Table number
1.0671	Nd^{3+}	LuAlO$_3$	77	a
1.0672	Nd^{3+}	CaY$_4$(SiO$_4$)$_3$O	300	a
		KGd(WO$_4$)$_2$	300	a
1.0673	Nd^{3+}	CaMoO$_4$	77, 300	a
		ZrO$_2$-Y$_2$O$_3$	300	a
1.0675	Nd^{3+}	LuAlO$_3$	300	a
		Na$_2$Nd$_2$Pb$_6$(PO$_4$)$_6$Cl$_2$	300	b
1.0682	Nd^{3+}	Y$_3$Al$_5$O$_{12}$	300	a
1.0687—1.0690	Nd^{3+}	KY(WO$_4$)$_2$	77—600	a
1.0688	Nd^{3+}	KY(WO$_4$)$_2$	300	a
1.0689	Nd^{3+}	GdAlO$_3$	77	a
1.0690	Nd^{3+}	GdAlO$_3$	300	a
1.0698	Nd^{3+}	La$_2$Be$_2$O$_5$	300	a
1.0701	Nd^{3+}	Gd$_2$(MoO$_4$)$_3$	300	a
1.0710	Nd^{3+}	Y$_2$SiO$_5$	77	a
1.0715	Nd^{3+}	Y$_2$SiO$_5$	300	a
1.0720	Nd^{3+}	CaSc$_2$O$_4$	300	a
1.0721	Nd^{3+}	KLu(WO$_4$)$_2$	300	a
1.07255	Nd^{3+}	YAlO$_3$	77	a
1.0726	Nd^{3+}	YAlO$_3$	300	a
		(Y,Lu)$_3$Al$_5$O$_{12}$	77	a
1.0726—1.0730	Nd^{3+}	YAlO$_3$	77—490	a
1.073	Nd^{3+}	Y$_2$O$_3$	77, 300	a
1.0730	Nd^{3+}	CaSc$_2$O$_4$	77	a
1.0737	Nd^{3+}	Y$_3$Al$_5$O$_{12}$	300	a
1.0740	Nd^{3+}	Y$_2$SiO$_5$	77	a
1.0741	Nd^{3+}	Gd$_2$O$_3$	300	a
1.0742	Nd^{3+}	Y$_2$SiO$_5$	300	a
1.075		InGaP-InGaAs	300	e
1.075	Nd^{3+}	La$_2$O$_2$S	300	a
1.0755	Nd^{3+}	CaSc$_2$O$_4$	77	a
1.0759	Nd^{3+}	GdAlO$_3$	77	a
		LuAlO$_3$	300	a
1.0760	Nd^{3+}	GdAlO$_3$	300	a
1.07655	Nd^{3+}	CaAl$_4$O$_7$	77	a
1.0770	Nd^{3+}	YScO$_3$	77	a
1.0772	Nd^{3+}	CaAl$_4$O$_7$	77	a
1.0776	Nd^{3+}	Gd$_2$O$_3$	77	a
1.078	Nd^{3+}	Y$_2$O$_3$	77	a
1.0781	Nd^{3+}	Y$_2$SiO$_5$	77	a
1.0782	Nd^{3+}	Y$_2$SiO$_5$	300	a
1.0782—1.0787	Nd^{3+}	LiNbO$_3$	590—450	a
1.0785	Nd^{3+}	LuScO$_3$	300	
1.0786	Nd^{3+}	CaAl$_4$O$_7$	300	a
1.0789	Nd^{3+}	Gd$_2$O$_3$	77, 300	a
1.079	Nd^{3+}	La$_2$O$_3$	77	a
		La$_2$Be$_2$O$_5$	300	a

Table 2.5.1 (continued)

KEY TO TABLE REFERENCES

a	= 2.1.1.2	e	=	2.2.2
b	= 2.1.2.2	f	=	2.3.1
c	= 2.1.3.1	g	=	2.3.11
d	= 2.2.1	h	=	2.3.12

Wavelength (μm)	Activator ion/center	Host	Temperature (K)	Table number
1.07925	Nd^{3+}	Lu_2SiO_5	300	a
1.0795—1.0803	Nd^{3+}	$YAlO_3$	77—700	a
1.08	Nd^{3+}	Fused silica		f
1.0804	Nd^{3+}	$LaAlO_3$	300	a
1.08145	Nd^{3+}	Sc_2SiO_5	300	a
1.0828	Nd^{3+}	$SrAl_4O_7$	300	a
1.0831	Nd^{3+}	$LuAlO_3$	120	a
1.0832	Nd^{3+}	$LuAlO_3$	300	a
1.0840	Nd^{3+}	$LiNbO_3$	77	a
		$GdScO_3$	200	a
1.0842	Nd^{3+}	$YAlO_3$	300	a
1.0843	Nd^{3+}	$YScO_3$	300	a
1.0846	Nd^{3+}	$LiNbO_3$	300	a
		$LiNbO_3$	300	a
1.0847	Nd^{3+}	$YAlO_3$	530	a
1.08515	Nd^{3+}	$GdScO_3$	300	a
1.0867	Nd^{3+}	$CaSc_2O_4$	77	a
1.0868	Nd^{3+}	$CaSc_2O_4$	300	a
1.0885	Nd^{3+}	CaF_2	300	a
		CaF_2-CeO_2	300	a
1.0885—1.0889	Nd^{3+}	CaF_2	300—420	a
1.0913	Nd^{3+}	$YAlO_3$	530	
1.0922—1.0933	Nd^{3+}	$LiNbO_3$	620—300	a
1.0933	Nd^{3+}	$LiNbO_3$	300	a
		$LiNbO_3$	300	a
1.0990	Nd^{3+}	$YAlO_3$	300	a
1.0991	Nd^{3+}	$YAlO_3$	500	a
1.1		$In_{0.88}Ga_{0.12}As_{0.23}P_{0.77}$-InP	300	e
1.10		$CdSnP_2$		d
1.10—1.60		AlGaSb		d
1.1119	Nd^{3+}	$Y_3Al_5O_{12}$	300	a
1.1158	Nd^{3+}	$Y_3Al_5O_{12}$	300	a
1.1160	Tm^{2+}	CaF_2	4.2	a
1.1217	V^{2+}	MgF_2	77	a
1.1225	Nd^{3+}	$Y_3Al_5O_{12}$	300	a
1.190	Ho^{3+}	$BaYb_2F_8$	300	a
1.2085	Ho^{3+}	$Gd_3Ga_5O_{12}$	≈110	a
1.2155	Ho^{3+}	$Y_3Al_5O_{12}$	≈110	a
		$Y_3Al_5O_{12}$	≈110	a
1.2160	Ho^{3+}	$Lu_3Al_5O_{12}$	≈110	a
1.22—1.50	F_2^+	KF		c
1.2308	Er^{3+}	$LiYF_4$	≈110, 300	a
1.26	Er^{3+}	CaF_2	77	a
1.3065	Nd^{3+}	$SrAl_{12}O_{19}$	300	a
1.3070	Nd^{3+}	$5NaF \cdot 9YF_3$	300	a
1.3125	Nd^{3+}	LaF_3	77	a
1.3130	Nd^{3+}	CeF_3	77	a

Table 2.5.1 (continued)

KEY TO TABLE REFERENCES

a	= 2.1.1.2	e	=	2.2.2
b	= 2.1.2.2	f	=	2.3.1
c	= 2.1.3.1	g	=	2.3.11
d	= 2.2.1	h	=	2.3.12

Wavelength (μm)	Activator ion/center	Host	Temperature (K)	Table number
1.3144	Ni^{2+}	MgO	77	a
1.3160	Nd^{3+}	SrF_2-LaF_3	77	a
1.3165	Nd^{3+}	CdF_2-YF_3	77	a
		$\alpha-NaCaCeF_6$	77	a
1.317	Nd^{3+}	$LiNdP_4O_{12}$	300	b
1.3170	Nd^{3+}	CeF_3	300	a
		LaF_3-SrF_2	77	a
1.3185	Nd^{3+}	CaF_2-GdF_3	300	a
		BaF_2-LaF_3	300	a
1.3188	Nd^{3+}	$Y_3Al_5O_{12}$	300	a
1.3190	Nd^{3+}	$Ca_2Y_5F_{19}$	300	a
		CaF_2-LaF_3	300	a
		CaF_2-CeF_3	300	a
		$Sr_2Y_5F_{19}$	300	a
		$\alpha-NaCaCeF_6$	300	a
1.32	Nd^{3+}	$NaNdP_4O_{12}$	300	b
		$KNdP_4O_{12}$	300	b
1.3200	Nd^{3+}	$Ca_2Y_5F_{19}$	77	a
		SrF_2-LuF_3	300	a
		BaF_2-YF_3	300	a
		YP_5O_{14}	300	a
1.3209	Nd^{3+}	$Lu_3Al_5O_{12}$	300	a
1.3225	Nd^{3+}	SrF_2-YF_3	77	a
1.323	Nd^{3+}	NdP_5O_{14}	300	b
1.323	Nd^{3+}	Phosphate glass	300	h
1.3235	Nd^{3+}	LaF_3	77	a
		SrF_2-LaF_3	77	a
1.3240	Nd^{3+}	CeF_3	77	a
1.3245	Nd^{3+}	CdF_2-YF_3	300	a
1.325	Nd^{3+}	Silicate glass	300	h
1.3250	Nd^{3+}	SrF_2-LaF_3	300	a
		SrF_2-GdF_3	77	a
1.3255	Nd^{3+}	CaF_2-YF_3	77	a
		SrF_2-CeF_3	300	a
1.3260	Nd^{3+}	SrF_2-GdF_3	300	a
		$\alpha-NaCaYF_6$	77	a
1.3270	Nd^{3+}	CaF_2-YF_3	300	a
1.3275	Nd^{3+}	LaF_3-SrF_2	77	a
1.3280	Nd^{3+}	BaF_2-LaF_3	300	a
1.3285	Nd^{3+}	$\alpha-NaCaYF_6$	300	a
1.3290	Nd^{3+}	BaF_2-LaF_3	77	a
1.3300	Nd^{3+}	SrF_2-YF_3	77	a
		$SrMoO_4$	77	a
1.3305	Nd^{3+}	LaF_3	77	a
		$Y_3Ga_5O_{12}$	300	a
		$HfO_2-Y_2O_3$	300	a
1.3307	Nd^{3+}	$Gd_3Ga_5O_{12}$	77	a

Table 2.5.1 (continued)

KEY TO TABLE REFERENCES

a	= 2.1.1.2	e	=	2.2.2
b	= 2.1.2.2	f	=	2.3.1
c	= 2.1.3.1	g	=	2.3.11
d	= 2.2.1	h	=	2.3.12

Wavelength (μm)	Activator ion/center	Host	Temperature (K)	Table number
1.3310	Nd^{3+}	LaF_3	300	a
		CeF_3	77	a
		$CaWO_4$	77	a
		$Y_3Sc_2Ga_3O_{12}$	300	a
1.3315	Nd^{3+}	LaF_3-SrF_2	300	a
		$Lu_3Ga_5O_{12}$	300	a
		$Gd_3Ga_5O_{12}$	300	a
1.3317	Nd^{3+}	$Ca_3Ga_2Ge_3O_{12}$	300	a
1.3319	Nd^{3+}	$Lu_3Al_5O_{12}$	77	a
1.3320	Nd^{3+}	CeF_3	300	a
		SrF_2-YF_3	77	a
		$SrAl_4O_7$	77	a
		$PbMoO_4$	77	a
		ZrO_2-Y_2O_3	300	a
1.3325	Nd^{3+}	LaF_3-SrF_2	77	a
		$SrMoO_4$	300	a
1.3326	Nd^{3+}	$Lu_3Al_5O_{12}$	300	a
1.3333	Nd^{3+}	$Lu_3Al_5O_{12}$	77	a
1.3338	Nd^{3+}	$Y_3Al_5O_{12}$	300	a
1.3340	Nd^{3+}	$CaWO_4$	300	a
		$PbMoO_4$	300	a
1.3342	Nd^{3+}	$Lu_3Al_5O_{12}$	300	a
1.3345	Nd^{3+}	$CaWO_4$	77	a
		$SrAl_4O_7$	300	a
		$Ca_5(PO_4)_3F$	77	a
1.3347	Nd^{3+}	$Ca_5(PO_4)_3F$	300	a
1.335	Nd^{3+}	Silicate glass	300	h
1.3350	Nd^{3+}	$KLa(MoO_4)_2$	77, 300	a
1.3351	Nd^{3+}	$Y_3Al_5O_{12}$	300	a
1.3354	Nd^{3+}	$CaLa_4(SiO_4)_3O$	300	a
1.3355	Nd^{3+}	SrF_2-LaF_2	77	a
		$NaLa(WO_4)_2$	300	a
1.3360	Nd^{3+}	$Y_3Sc_2Al_3O_{12}$	300	a
		$Gd_3Sc_2Al_3O_{12}$	300	a
		$Lu_3Sc_2Al_3O_{12}$	300	a
1.3370	Nd^{3+}	CaF_2-YF_3	300	a
		$Ca(NbO_3)_2$	77	a
		$CaWO_4$	300	a
		$LiLa(MoO_4)_2$	300	a
1.3372	Nd^{3+}	$CaWO_4$	77	a
1.3375	Nd^{3+}	α-$NaCaYF_6$	300	a
		$PbMoO_4$	77	a
		$LiLa(MoO_4)_2$	77	a
1.3376	Nd^{3+}	$Lu_3Al_5O_{12}$	77	a
1.3380	Nd^{3+}	CaF_2-YF_3	77	a
		$Ca(NbO_3)_2$	300	a
		$NaLa(MoO_4)_2$	77, 300	a

Table 2.5.1 (continued)

KEY TO TABLE REFERENCES

a	= 2.1.1.2	e	= 2.2.2
b	= 2.1.2.2	f	= 2.3.1
c	= 2.1.3.1	g	= 2.3.11
d	= 2.2.1	h	= 2.3.12

Wavelength (μm)	Activator ion/center	Host	Temperature (K)	Table number
1.3381	Nd^{3+}	$Y_3Al_5O_{12}$	300	a
1.3387	Nd^{3+}	$Lu_3Al_5O_{12}$	300	a
1.3390	Nd^{3+}	α-NaCaYF$_6$	77	a
		$CaWO_4$	300	a
1.3391	Nd^{3+}	$YAlO_3$	77	a
1.3400	Nd^{3+}	$CaAl_4O_7$	77	a
		$LiGd(MoO_4)_2$	77, 300	a
1.3407	Nd^{3+}	$Bi_4Si_3O_{12}$	300	
1.3410	Nd^{3+}	$Lu_3Al_5O_{12}$	300	a
1.3413	Nd^{3+}	$YAlO_3$	300	a
1.3415	Nd^{3+}	$Ca(NbO_3)_2$	77	a
		YVO_4	77	a
1.3418	Nd^{3+}	$Bi_4Ge_3O_{12}$	300	
1.3420	Nd^{3+}	$CaAl_4O_7$	300	a
1.3425	Nd^{3+}	$Ca(NbO_3)_2$	300	a
		YVO_4	300	a
		$PbMoO_4$	300	a
1.3430	Nd^{3+}	$NaLa(MoO_4)_2$	77	a
1.3437	Nd^{3+}	$LuAlO_3$	300	a
1.3440	Nd^{3+}	$SrMoO_4$	77	a
		$LiLa(MoO_4)_2$	77	a
		$NaLa(MoO_4)_2$	300	a
1.345	Nd^{3+}	$NdAl_3(BO_3)_4$	300	a
1.3450	Nd^{3+}	$PbMoO_4$	77	a
1.3455	Nd^{3+}	$LiGd(MoO_4)_2$	77	a
1.3459	Nd^{3+}	$CaWO_4$	77	a
1.3475	Nd^{3+}	$CaWO_4$	300	a
1.3482	Nd^{3+}	$KLu(WO_4)_2$	300	
1.3485	Nd^{3+}	$KY(MoO_4)_2$	300	a
1.3499	Nd^{3+}	$Lu_3Al_5O_{12}$	77	a
1.3510	Nd^{3+}	$KGd(WO_4)_2$	300	a
		$La_2Be_2O_5$	300	a
1.3512	Nd^{3+}	$YAlO_3$	300	a
1.3514	Nd^{3+}	$YAlO_3$	300	a
1.3515	Nd^{3+}	$KY(WO_4)_2$	77	a
1.3525	Nd^{3+}	$Ca_2Y_5F_{19}$	300	a
		$KY(WO_4)_2$	300	a
		$Lu_3Al_5O_{12}$	77	a
1.3530	Nd^{3+}	$SrAl_4O_7$	77	a
1.3532	Nd^{3+}	$Lu_3Al_5O_{12}$	300	a
1.3533	Nd^{3+}	$Y_3Al_5O_{12}$	300	a
1.3545	Nd^{3+}	$KY(WO_4)_2$	77, 300	a
1.3550	Nd^{3+}	$LiNbO_3$	300	a
1.3560	Nd^{3+}	$Y_3Al_5O_{12}$	77	a
1.3565	Nd^{3+}	$CaSc_2O_4$	300	a
1.3572	Nd^{3+}	$Y_3Al_5O_{12}$	300	a
1.3575	Nd^{3+}	Lu_2SiO_5	300	

Table 2.5.1 (continued)

KEY TO TABLE REFERENCES

a	= 2.1.1.2	e	=	2.2.2
b	= 2.1.2.2	f	=	2.3.1
c	= 2.1.3.1	g	=	2.3.11
d	= 2.2.1	h	=	2.3.12

Wavelength (μm)	Activator ion/center	Host	Temperature (K)	Table number
1.3580	Nd^{3+}	Y_2SiO_5	77	a
1.3585	Nd^{3+}	CaF_2-YF_3	300	a
		Y_2SiO_5	300	a
1.3595	Nd^{3+}	LaF_3	300	a
1.3600	Nd^{3+}	CaF_2-YF_3	77	a
		α-$NaCaYF_6$	300	a
1.3632	Nd^{3+}	Sc_2SiO_3	300	
1.3644	Nd^{3+}	$YAlO_3$	77	a
1.3665	Nd^{3+}	$SrAl_4O_7$	300	a
1.3670	Nd^{3+}	LaF_3	77	a
1.3675	Nd^{3+}	LaF_3	300	a
		CeF_3	77	a
		$CaAl_4O_7$	77	a
1.3680	Nd^{3+}	$SrAl_4O_7$	300	a
1.3690	Nd^{3+}	CeF_3	300	a
1.37	Nd^{3+}	Borate glass	300	f
1.3710	Nd^{3+}	$CaAl_4O_7$	300	a
1.3745	Nd^{3+}	$LiNbO_3$	300	a
1.3755	Nd^{3+}	$NaLa(MoO_4)_2$	77	a
1.3780	Nd^{3+}	$PbMoO_4$	77	a
1.3790	Nd^{3+}	$SrMoO_4$	77	a
1.3840	Nd^{3+}	$NaLa(MoO_4)_2$	77	a
1.3849	Nd^{3+}	$YAlO_3$	77	a
1.3870	Nd^{3+}	$LiNbO_3$	300	a
		$LiNbO_3$	300	a
		$CaSc_2O_4$	77	a
1.3880	Nd^{3+}	$CaWO_4$	77	a
1.3885	Nd^{3+}	$CaWO_4$	300	a
1.3908	Ho^{3+}	$KY(WO_4)_2$	≈ 110	
1.392	Ho^{3+}	$LiYF_4$	300	a
1.3960	Ho^{3+}	$LiYF_4$	116, 300	a
1.3982	Ho^{3+}	$KGd(WO_4)_2$	≈ 110	
1.40—1.75	F_2^+	NaCl		c
1.4026	Nd^{3+}	$YAlO_3$	77	
1.4028	Ho^{3+}	$YAlO_3$	≈ 110	
1.4040	Ho^{3+}	$Gd_3Ga_5O_{12}$	≈ 110	a
1.4072	Ho^{3+}	$Y_3Al_5O_{12}$	≈ 110	
1.4085	Ho^{3+}	$Lu_3Al_5O_{12}$	≈ 110	
1.4862	Ho^{3+}	$LiYF_4$	90, 116	a
1.50—1.60		GaSb		d
1.5298	Er^{3+}	CaF_2	77	a
1.5308	Er^{3+}	CaF_2	77	a
1.54	Er^{3+}	Phosphate glass	300	f
		Fluorophosphate glass	300	f
1.543	Er^{3+}	Silicate glass	300	f
1.547	Er^{3+}	CaF_2-YF_3	77	a
1.55	Er^{3+}	Silicate glass	77	f

Table 2.5.1 (continued)

KEY TO TABLE REFERENCES

a	= 2.1.1.2	e	=	2.2.2
b	= 2.1.2.2	f	=	2.3.1
c	= 2.1.3.1	g	=	2.3.11
d	= 2.2.1	h	=	2.3.12

Wavelength (μm)	Activator ion/center	Host	Temperature (K)	Table number
1.5500	Er^{3+}	$CaAl_4O_7$	77	a
1.558	Er^{3+}	$Bi_4Ge_3O_{12}$	77	
1.5815	Er^{3+}	$CaAl_4O_7$	77	a
1.6	Ni^{2+}	MnF_2	77	a
1.60		In_2Se		d
1.61	Er^{3+}	$Ca(NbO_3)_2$	77	a
1.61—1.74	Ni^{2+}	MgF_2	80	a
1.6113	Er^{3+}	LaF_3	77	a
1.612	Er^{3+}	$CaWO_4$	77	a
1.617	Er^{3+}	CaF_2	77	a
1.620	Er^{3+}	ZrO_2-Er_2O_3	77	a
		ZrO_2-Er_2O_3	77	a
1.62—1.91	$(F_2^+)_A$	KCl:Na		c
1.623	Ni^{3+}	MgF_2	77	a
1.63—2.11	Co^{2+}	MgF_2	80	a
1.632	Er^{3+}	$Y_3Al_5O_{12}$	300	a
1.636	Ni^{2+}	MgF_2	77	a
1.6449	Er^{3+}	$Y_3Al_5O_{12}$	295	a
1.6452	Er^{3+}	$Y_3Al_5O_{12}$	77	a
1.6459	Er_3^+	$Y_3Al_5O_{13}$	295	a
1.6470	Er^{3+}	$LiYF_4$	$\approx$110	a
1.6525	Er^{3+}	$Lu_3Al_5O_{12}$	77	a
1.6602	Er^{3+}	$Y_3Al_5O_{12}$	77	a
1.6615	Er^{3+}	$Yb_3Al_5O_{12}$	77	a
1.6630	Er_3^+	$Lu_3Al_5O_{12}$	77	a
1.6632	Er^{3+}	$YAlO_3$	300	a
1.664	Er^{3+}	$Bi_4Ge_3O_{12}$	77	a
1.6675	Er^{3+}	$LuAlO_3$	$\approx$90	a
1.673	Ho^{3+}	$LiYF_4$	300	a
1.6734	Ho^{3+}	$LiYF_4$	116	a
1.674—1.676	Ni^{2+}	MgF_2	82—100	a
1.696	Er^{3+}	CaF_2	77	a
1.715	Er^{3+}	CaF_2	77	a
1.7155	Er^{3+}	$KGd(WO_4)_2$	300	a
1.7178	Er^{3+}	$KY(WO_4)_2$	300	a
1.726	Er^{3+}	CaF_2	77	a
1.731—1.756	Ni^{3+}	MgF_2	100—192	a
1.732	Er^{3+}	$LiErF_4$	$\approx$90	a
1.7320	Er^{3+}	$LiYF_4$	110, 300	a
1.7325	Er^{3+}	$KGd(WO_4)_2$	300	a
1.7355	Er^{3+}	$KY(WO_4)_2$	300	
1.7383	Er^{3+}	$KLu(WO_4)_2$	300	
1.750	Co^{2+}	MgF_2	77	a
1.7757	Er^{3+}	$Y_3Al_5O_{12}$	300	a
1.7762	Er^{3+}	$Lu_3Al_5O_{12}$	300	a
1.785—1.797	Ni^{2+}	MgF_2	198—240	a
1.8035	Co^{2+}	MgF_2	77	a
1.821	Co^{2+}	$KMgF_3$	77	a

Table 2.5.1 (continued)

KEY TO TABLE REFERENCES

a	= 2.1.1.2	e	= 2.2.2
b	= 2.1.2.2	f	= 2.3.1
c	= 2.1.3.1	g	= 2.3.11
d	= 2.2.1	h	= 2.3.12

Wavelength (μm)	Activator ion/center	Host	Temperature (K)	Table number
1.833	Nd^{3+}	$Y_3Al_5O_{12}$	293	a
1.85	Tm^{3+}	Silicate glass	80	f
1.8532	Tm^{3+}	$LiNbO_3$	77	a
1.856	Tm^{3+}	$YAlO_3$	$\approx$90	a
1.8580	Tm^{3+}	α-$NaCaErF_6$	150	a
1.860	Tm^{3+}	CaF_2-ErF_3	77	a
1.861	Tm^{3+}	$(Y,Er)AlO_3$	77	a
1.865	Ni^{2+}	MnF_2	20	a
1.872	Tm^{3+}	$ErAlO_3$	77	a
1.880	Tm^{3+}	$(Y,Er)_3Al_5O_{12}$	77	a
1.883	Tm^{3+}	$YAlO_3$	$\approx$90	a
1.8834	Tm^{3+}	$Y_3Al_5O_{12}$	77	a
1.884	Tm^{3+}	$(Y,Er)_3Al_5O_{12}$	77	a
1.8845	Tm^{3+}	$(Er,Lu)AlO_3$	77	a
1.8850	Tm^{3+}	$(Er,Yb)_3Al_5O_{12}$	77	a
1.8855	Tm^{3+}	$Lu_3Al_5O_{12}$	77	a
1.8885	Tm^{3+}	α-$NaCaErF_6$	150	a
$\sim$1.896	Tm^{3+}	ZrO_2-Er_2O_3	77	a
1.9—2.0	Tm^{3+}	$Er_3Al_5O_{12}$	77	a
1.9060	Tm^{3+}	$CaMoO_4$	77	a
1.91	Tm^{3+}	$Ca(NbO_3)_2$	77	a
1.911	Tm^{3+}	$CaWO_4$	77	a
1.9115	Tm^{3+}	$CaMoO_4$	77	a
1.915	Ni^{2+}	MnF_2	77	a
1.916	Tm^{3+}	$CaWO_4$	77	a
1.922	Ni^{2+}	MnF_2	77	a
1.929	Ni^{2+}	MnF_2	85	a
1.9335	Tm^{3+}	$YAlO_3$	$\approx$90	a
1.934	Tm^{3+}	Er_2O_3	77	a
1.939	Ni^{2+}	MnF_2	85	a
1.972	Tm^{3+}	SrF_2	77	a
1.99	Co^{2+}	MgF_2	77	a
$\sim$2.0	Ho^{3+}	$SrY_4(SiO_4)_3O$	77	a
	Tm^{3+}	YVO_4	77	a
2.0—2.5	$(F_2^+)_A$	KCl:Li		c
2.0010	Ho^{3+}	$(Er,Lu)AlO_3$	77	a
2.0132	Tm^{3+}	$Y_3Al_5O_{12}$	77	a
2.014	Tm^{3+}	$(Y,Er)_3Al_5O_{12}$	85	a
2.015	Tm^{3+}	Silicate glass	300	f
2.019	Tm^{3+}	$Y_3Al_5O_{12}$	295	a
2.0195	Tm^{3+}	$(Er,Yb)_3Al_5O_{12}$	77	a
2.0240	Tm^{3+}	$Lu_3Al_5O_{12}$	77	a
2.030	Ho^{3+}	CaF_2-ErF_3	77	a
2.0312	Ho^{3+}	α-$NaCaErF_6$	77	a
2.0318	Ho^{3+}	CaF_2—YF_3	77	a
2.0345	Ho^{3+}	α-$NaCaErF_6$	150	a
2.0377	Ho^{3+}	α-$NaCaErF_6$	77	a

Table 2.5.1 (continued)

KEY TO TABLE REFERENCES

a	= 2.1.1.2	e	=	2.2.2
b	= 2.1.2.2	f	=	2.3.1
c	= 2.1.3.1	g	=	2.3.11
d	= 2.2.1	h	=	2.3.12

Wavelength (µm)	Activator ion/center	Host	Temperature (K)	Table number
2.0412	Ho³⁺	YVO₄	77	a
2.0416	Ho³⁺	ErVO₄	77	a
2.046	Ho³⁺	CaWO₄	77	a
2.047	Ho³⁺	Ca(NbO₃)₂	77	a
2.05	Ho³⁺	CaF₂-ErF₃-TmF₃—YbF₃	100	a
	Co²⁺	MgF₂	77	a
2.050	Ho³⁺	NaLa(MoO₄)₂	90	a
2.0555	Ho³⁺	BaY₂E₈	20	a
2.0556	Ho³⁺	CaMoO₄	77	a
2.0563	Ho³⁺	BaYb₂F₈	300	
2.059	Ho³⁺	CaWO₄	77	a
2.06—2.10	Ho³⁺	Silicate glass	77	f
2.060	Ho³⁺	CaF₂-ErF₃-TmF₃—YbF₃	298	a
		CaY₄(SiO₄)₃O	77	a
2.065	Ho³⁺	BaY₂E₈	77	a
2.0654	Ho³⁺	Li(Y,Er)F₄	300	a
2.066	Ho³⁺	LiYF₄	77	a
2.0672	Ho³⁺	LiYF₄	≈90	a
2.07	Ho³⁺	LaNbO₄	≈90	a
2.0707	Ho³⁺	CaMoO₄	77	a
2.0720	Ho³⁺	K(Y,Er)(WO₄)₂	110—220	a
2.074	Ho³⁺	BaY₂F₈	20	a
		CaMoO₄	77	a
2.0746	Ho³⁺	BaY₂F₈	77	a
2.0786	Ho³⁺	LiNbO₃	77	a
2.079	Ho³⁺	Ca₅(PO₄)₃F	77	a
2.08	Ho³⁺	Silicate glass	77	f
2.085	Ho³⁺	Y₂SiO₅	≈110	
2.086	Ho³⁺	Y₃Fe₅O₁₂	77	a
		Y₃Ga₅O₁₂	77	a
		Ho₃Ga₅O₁₂	77	b
2.0866	Ho³⁺	BaY₂F₈	77	a
2.087	Ho₃⁺	Bi₄Ge₃O₁₂	77	
2.0885	Ho³⁺	Gd₃Ga₅O₁₂	≈110	a
2.089	Ho³⁺	Y₃Fe₅O₁₂	77	a
2.090	Ho³⁺	HoF₃	77	b
2.0914	Ho³⁺	Y₃Al₅O₁₂	77	a
2.0917	Ho³⁺	(Y,Er)₃Al₅O₁₂	77	a
2.092	Ho³⁺	CaF₂	77	a
		Y₂SiO₂	≈110	
2.0960	Ho³⁺	Yb₃Al₅O₁₂	77	a
2.097	Ho³⁺	Ho₃Al₅O₁₂	77	b
2.0975	Ho³⁺	Y₃Al₅O₁₂	77	a
2.0979	Ho³⁺	(Y,Er)₃Al₅O₁₂	77	a
2.0982	Ho³⁺	(Y,Er)₃Al₅O₁₂	77	a
2.0985	Ho³⁺	(Er₃Sc₂Al₃O₁₂	77	a
2.0990	Ho³⁺	(Y,Er)₃Al₅O₁₂	77	a

Table 2.5.1 (continued)

KEY TO TABLE REFERENCES

a	= 2.1.1.2	e	=	2.2.2
b	= 2.1.2.2	f	=	2.3.1
c	= 2.1.3.1	g	=	2.3.11
d	= 2.2.1	h	=	2.3.12

Wavelength (μm)	Activator ion/center	Host	Temperature (K)	Table number
2.1	Ho^{3+}	CaF_2-ErF_3-TmF_3-YbF_3	77	a
		$Y_3Fe_5O_{12}$	77	a
2.10		Cd_3P_2		d
2.1010	Ho^{3+}	$(Er,Tm,Yb)_3Al_5O_{12}$	77	a
2.1020	Ho^{3+}	$Lu_3Al_5O_{12}$	77	a
		$Lu_3Al_5O_{12}$	77	a
2.105	Ho^{3+}	Y_2SiO_5	110—220	
2.1070	Ho^{3+}	$Y_3Fe_5O_{12}$	77	a
2.1135	Ho^{3+}	$Ho_3Ga_5O_{12}$	77	
2.114	Ho^{3+}	$Y_3Ga_5O_{12}$	77	a
2.115	Ho^{3+}	ZrO_2-Er_2O_3	77	a
2.1170	Ho^{3+}	$Ho_3Sc_2Al_3O_{12}$	77	b
2.119	Ho^{3+}	$(Y,Er)AlO_3$	300	a
2.1205	Ho^{3+}	$ErAlO_3$	77	a
		$(Er,Lu)AlO_3$	77	a
2.121	Ho^{3+}	Er_2O_3	145	a
2.122	Ho^{3+}	$Ho_3Al_5O_{12}$	$\approx$90	b
2.1223	Ho^{3+}	$Y_3Al_5O_{12}$	77	a
2.1227	Ho^{3+}	$(Y,Er)_3Al_5O_{12}$	85	a
		$(Y,Er)_3Al_5O_{12}$	85	a
		$Ho_3Al_5O_{12}$	77	b
2.123	Ho^{3+}	$(Y,Er)AlO_3$	300	a
		$(Y,Er)_3Al_5O_{12}$	77	a
2.1285	Ho^{3+}	$(Y,Er)_3Al_5O_{12}$	77	a
		$Ho_3Sc_2Al_3O_{12}$	77	b
2.1288	Ho^{3+}	$(Y,Er)_3Al_5O_{12}$	295	a
2.129	Ho^{3+}	$Ho_3Al_5O_{12}$	77	b
2.13	Ho^{3+}	$(Y,Er)_3Al_3O_{12}$	300	a
2.1348	Ho^{3+}	$LuAlO_3$	$\approx$90	a
2.165	Co^{2+}	ZnF_2	77	a
2.171	Ho^{3+}	BaY_2F_8	295	a
2.234	U^{3+}	CaF_2	77	a
2.25—2.65	$F_B(II)$	KCl:Na		c
2.274	Tm^{3+}	$YAlO_3$	300	a
2.318	Tm^{3+}	$YAlO_3$	300	a
2.324	Tm^{3+}	$Y_3Al_5O_{12}$	300	a
2.34	Tm^{3+}	$YAlO_3$	90, 300	a
2.348	Tm^{3+}	$YAlO_3$	300	
2.349	Tm^{3+}	$YAlO_3$	300	a
2.352	Ho^{3+}	$LiHoF_4$	$\approx$90	a
2.3524	Ho^{3+}	$LiYF_4$	116	a
2.35867	Dy^{2+}	CaF_2	4.2—120	a
2.3659	Dy^{2+}	SrF_2	20	a
2.375	Ho^{3+}	BaY_2F_8	77	
2.377	Ho^{3+}	BaY_2F_8	20	a

Table 2.5.1 (continued)

KEY TO TABLE REFERENCES

a	= 2.1.1.2	e	=	2.2.2
b	= 2.1.2.2	f	=	2.3.1
c	= 2.1.3.1	g	=	2.3.11
d	= 2.2.1	h	=	2.3.12

Wavelength (μm)	Activator ion/center	Host	Temperature (K)	Table number
2.407	U^{3+}	SrF_2	20—90	a
2.439	U^{3+}	CaF_2	77	a
2.5—2.9	$F_A(II)$	KCl:Li		c
2.511	U^{3+}	CaF_2	77	a
2.556	U^{3+}	BaF_2	20	a
2.571	U^{3+}	CaF_2	77	a
2.6	U^{3+}	CaF_2	4.2	a
2.6—3.3	$F_A(II)$	RbCl:Li		c
2.613	U^{3+}	CaF_2	77—300	a
2.6887	Er^{3+}	$K(Y,Er)(WO_4)_2$	300—350	a
2.69	Er^{3+}	CaF_2-ErF_3-TmF_3	298	a
2.6990	Er^{3+}	$(Er,Lu)_3Al_5O_{12}$	300	a
		$Lu_3Al_5O_{12}$	300	a
		$Lu_3Al_5O_{12}$	300	a
2.7222	Er^{3+}	$KGd(WO_4)_2$	300	a
2.7307	Er^{3+}	CaF_2—ErF_3	300	a
2.747	Er^{3+}	$LiYF_4$		
2.7953	Er^{3+}	$Y_3Al_5O_{12}$	300	a
2.7987	Er^{3+}	$Lu_3Al_5O_{12}$	300	a
2.7990	Er^{3+}	$YAlO_3$	300	a
2.8070	Er^{3+}	$KY(WO_4)_2$	300	a
		$K(Y,Er)(WO_4)_2$	300	a
		$KEr(WO_4)_2$	300	a
2.8092	Er^{3+}	$KLu(WO_4)_2$	300	a
2.81	Er^{3+}	$LiYF_4$	300	a
2.8218	Er^{3+}	$Gd_3Ga_5O_{12}$	300	a
2.8298	Er^{3+}	$(Er,Lu)_3Al_5O_{12}$	300	a
		$Lu_3Al_5O_{12}$	300	a
2.8302	Er^{3+}	$Y_3Al_5O_{12}$	300	a
		$(Y,Er)_3Al_5O_{12}$	300	a
2.850	Ho^{3+}	$LiYF_4$	300	
2.8510	Ho^{3+}	$LaNbO_4$	300	
2.870	Er^{3+}	$LiYF_4$	110, 300	
2.9	Ho^{3+}	$Gd_3Ga_5O_{12}$	300	a
2.9073	Ho^{3+}	$BaYb_2F_8$	≈300	a
2.9155	Ho^{3+}	$YAlO_3$	300	a
2.9342	Ho^{3+}	$KGd(WO_4)_2$	300	a
2.9365	Er^{3+}	$Y_3Al_5O_{12}$	300	a
		$(Y,Er)_3Al_5O_{12}$	300	a
2.9367	Er^{3+}	$Er_3Al_5O_{12}$	300	b
2.9395	Er^{3+}	$(Er,Lu)_3Al_5O_{12}$	300	a
	Ho^{3+}	$KY(WO_4)_2$	300	a
2.9403	Ho^{3+}	$Y_3Al_5O_{12}$	300	a
2.9406	Er^{3+}	$Lu_3Al_5O_{12}$	300	a
2.9445	Ho^{3+}	$KLu(WO_4)_2$	300	a
2.9460	Ho^{3+}	$Lu_3Al_5O_{12}$	300	a
2.952	Ho^{3+}	$LiYF_4$	300	a
3.00—3.20		InAs		d

Table 2.5.1 (continued)

KEY TO TABLE REFERENCES

a	= 2.1.1.2	e	=	2.2.2
b	= 2.1.2.2	f	=	2.3.1
c	= 2.1.3.1	g	=	2.3.11
d	= 2.2.1	h	=	2.3.12

Wavelength (μm)	Activator ion/center	Host	Temperature (K)	Table number
3.0132	Ho^{3+}	$YAlO_3$	300	a
3.022	Dy^{3+}	$Ba(Y,Er)_2F_8$	77	a
3.50		PbCdS		d
3.70		Te		d
3.80—4.10		CdHgTe		d
3.914	Ho^{3+}	$LiYF_4$	300	a
4.30		PbS		d
4.40—6.50		PbGeTe		d
4.70—5.50		PbSSe		d
4.78		$Pb_{0.78}Se_{0.22}$	12	e
4.80—5.30		InSb		d
6.50		PbTe		d
6.50—32.0		PbSnTe		d
~8.35—8.9		PbSnTe-PbTe	77	e
8.50		PbSe		d
8.50—32.0		PbSnSe		d

Section 3
Liquid Lasers

3.1 ORGANIC DYE LASERS

R. Steppel

INTRODUCTION

After the discovery of the dye laser by Sorokin and Lankard,[1] numerous reports followed which detailed the study of various classes of fluorescent organic materials. These included various oligoaromatics such as *p*-terphenyl and *p*-quaterphenyl; stilbenes, including various alkoxy-, amino-, and halo-substituted styryl benzenes and aryl-substituted stilbenes; furans; aryl-substituted oxazoles, benzoxazoles, oxadiazoles, and other heterocyclic materials; condensed aromatic hydrocarbons consisting of aryl-, alkyl-, and halo-substituted anthracenes, sulfonated pyrenes and others; quinolones, or the carbostyrils; hydroxy and amino coumarins; pyrylium salts; phthalimides including brilliant sulfaflavine; xanthenes which include the well-known rhodamines, fluoresceins, pyronins, and rosamines; acridines; oxazones; and oxazines, including cresyl violet and nile blue; and numerous examples of the cyanine dyes including the familiar DOTC, DTTC, and HITC dyes.

By 1973 a careful tabulation of these and other dyes had been presented by Drexhage[2] as part of an important contribution outlining many of the important relationships between molecular structure and the lasing properties of these dyes.

EXCITATION AND RELAXATION PROCESSES

The excitation and relaxation processes associated with dye lasers are shown in Figure 3.1. In short, we have the sequence:

1. excitation (a), absorption of light, $S_0 + h\nu \rightarrow S_x$
2. nonradiative relaxation (b) or internal conversion, $S_x \rightarrow S_1$
3. processes from S_1 which include (c_1) internal conversion from $S_1 \rightarrow S_0$, or (c_2) nonradiative intersystem crossing, $S_1 \rightarrow T_1$, or (c_3) fluorescence or light emission, $S_1 \rightarrow S_0 + h\nu$, or ($c_4$) excited state absorption, $S_1 + h\nu \rightarrow S_x$ and relaxation to S_1
4. from T_1, (d_1) internal conversion $T_1 \rightarrow S_0$, or (d_2) phosphorescence, $T_1 \rightarrow S_0 + h\nu$, or ($d_3$) absorption, $T_1 + h\nu \rightarrow T_x$ and relaxation to T_1.

In the first step the dye molecule interacts with electromagnetic radiation which may result in the absorption of light and thus electronic excitation. Assuming a very short burst of intense radiation, the entire population of the ground state, S_0, could be converted to higher singlet electronic states such that the entire population of S_0 is transformed to S_x. Following the nonradiative relaxation of $S_x \rightarrow S_1$, the fluorescence emission or stimulated emission would have the largest possible bandwidth since there would be no absorption of the emitted radiation due to wavelength coincidence. This region can be observed in the absorption and fluorescence spectra where the two overlap, that is, the long wavelength side of the absorption spectrum and the short wavelength side of the fluorescence spectrum (see Figure 3.1.2). However, since the nonradiative relaxation of S_x to S_1 is very fast ($10^{-11} - 10^{-14}$ sec), the normal emission ($S_1 \rightarrow S_0$) may suffer losses due to reabsorption when the pump pulse is long in relation to $S_x \rightarrow S_1$, the severity depending on the degrees of overlap of the spectral regions (see Figure 3.1.2, Case 1 and Case 2) and losses due to excited state absorption, $S_1 \rightarrow S_x$, of the pump radiation. In hypothetical Case 1, there is a large overlap between the absorption spectrum and the fluorescence spectrum whereas in Case 2 the overlap is minimal. The letters represent the wavelength for maximum absorption (A), fluores-

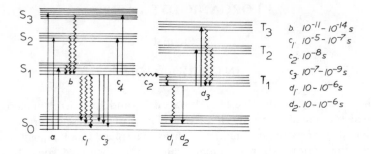

FIGURE 3.1.1.   Schematic energy level diagram showing electronic states and
some of the associated vibrational levels. Typical lifetimes of different relaxation
processes are given at the right.

cence emission (F), and laser emission (L). In Case 1 a significant shift of L from F
may occur due to self absorption whereas in Case 2 the differential is minimal. Thus,
there may be potential efficiency losses due to reabsorption.

Losses in efficiency may also occur due to an energy mismatch between the pump
source and the energy differential of the S_0 - S_1 states. Subsequent to excitation to S_x,
internal conversion occurs. The larger the degree of internal conversion ($S_x \rightarrow S_1$) the
greater the loss of conversion efficiency in the total system. The most efficient situation
exists when the pump energy closely matches the difference in the $S_0 \rightarrow S_1$ transition
or vice versa. Examples of this are found for rhodamine 6G(absorption maximum at
530 nm) with the 514.5-nm line of an argon-ion laser or the 532-nm line of a frequency
doubled Nd:YAG laser which produces conversion efficiencies of about 50% while the
337.1-nm nitrogen laser may give 20% and the 249-nm KF (excimer) laser may give
only 6%. On the other hand, KF pumping of *p*-terphenyl may give efficiencies in the
high 20% even though not pumping the $S_0 \rightarrow S_1$ transition directly but still pumping
with a good spectral match. None of these numbers represent total systems efficiencies.

In the case where the excitation energy is less than that of the S_0 - S_x differential, it
may not be utilized unless the power density is high enough to allow for the possibility
of multi-photon excitation.

The processes of internal conversion (c_1 and d_1), intersystem crossing (c_2), and radiative decay (fluorescence — c_3 and phosphorescence — d_2) may be influenced by numerous factors most important of which include molecular structure, the chemical and
physical nature of the solvents and chemical additives, and certain physical parameters.
Many times the influences are very subtle and hard to characterize and sometimes
reverse themselves.

Drexhage[2] has delineated many of these influences and the reader is encouraged to
refer to this reference and the references therein. Some of these influences are reviewed
in the following paragraphs.

CORRELATION OF STRUCTURAL FEATURES WITH PHYSICAL PROCESSES

The internal conversion from $S_1 \rightarrow S_0$ is promoted by any mechanism which allows
the loss or transfer of the excited state energy from the dye molecule to the surrounding
environment or to another molecular species of lower electronic energy. Since the electronic states are composed of vibrational levels and rotational sublevels, deactivation
through molecular mobility may be severe.

Rigidization

Although unnecessary in many cases, rigidization of the molecular structure may

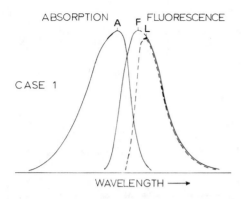
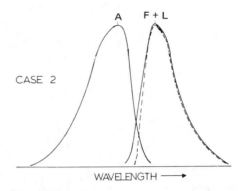

FIGURE 3.1.2.   Hypothetical absorption and emission spectra with large overlap (Case 1) and small over-lap (Case 2). The letters represent the wavelength for maximum absorption (A), fluorescence emission (F), and laser emission (L).

preclude several undesirable dependencies involving temperature, solvent viscosity, and end group substitution. In the Sequence 1 through 4 of Figure 3.1.3 we illustrate the influence of increasing rigidization. In 1, we have a triphenylmethane-type dye characterized by little or no fluorescence ability. Rigidization by introduction of a bridge may be sufficient to provide a molecule of high fluorescence ability, such as rhodamine 6G, where A and B are N-ethylamino (-NHC$_2$H$_5$) substituents, C is an ethylcarboxylate (-COOC$_2$H$_5$) group in the ortho position of the 9-phenyl ring (as shown in 3), and the bridging atom X is oxygen. Indeed, the fluorescence quantum yield is very high, about 95%. The above case, also illustrates the fact that the carboxy group on the ortho position of the 9-phenyl ring limits the otherwise free rotation of this group and thus helps to avoid fluorescence losses.

In the case where the terminal groups contain hydrogen substituents, that is, where A and B are amino (-NH$_2$) or alkylamino (-NHR), internal conversion may result through loss of energy from hydrogen stretching vibrations. Thus, conversion of the lowest vibrational level of the S$_1$ state to a high vibrational level of the S$_0$ state may provide a mechanism for internal conversion. Drexhage[2] has shown that by substitution of deuterium for hydrogen on these terminal groups, the fluorescence efficiency may be enhanced. Where A and B are amino groups and C is a carboxylic acid, the fluorescence efficiency increased from 85% to 92% by exchanging the hydrogens for deuteriums using methanol-Od. For the oxazine dye, cresyl violet, similarly containing unsubstituted amino groups, the fluorescence efficiency dramatically increased from 70% to 90%.

For rhodamine B, the substituents A and B are diethylamino[-N(C$_2$H$_5$)$_2$] and C is a carboxylic acid. In this case the fluorescence efficiency is sensitive to both solvent viscosity and temperature. In ethanol, lowering the temperature increases the fluorescence efficiency and raising the temperature decreases it. Changing the solvent from ethanol, where the quantum efficiency of fluorescence is about 40% at 25°C, to a viscous solvent, glycerol, enhances the efficiency such that the fluorescence quantum yield approaches 100%.

By pursuing even greater structural rigidity in the sequence and progressing to Stage 4, we have a near completely rigidized dye with a fluorescence efficiency of 100%, independent of dialkylamino temperature-dependent rotation problems, solvent viscosity, and free of losses due to substituent hydrogen vibrations.

Loop Rule

Radiationless transfer between the first singlet and the lowest excited triplet, the

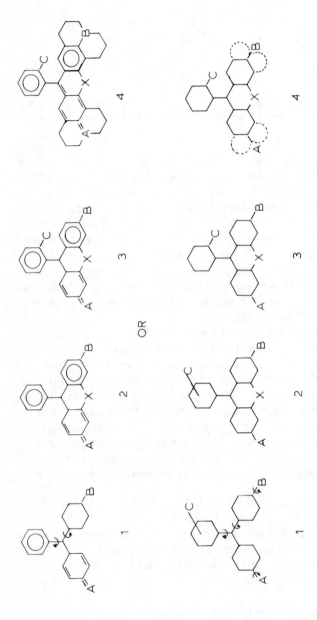

FIGURE 3.1.3.   Structures with varying degrees of mobility; rigidization increases in the sequence 1 through 4.

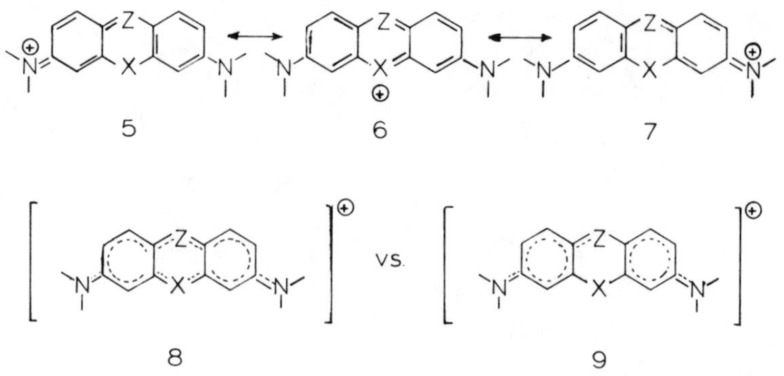

FIGURE 3.1.4.   Some of the individual resonance forms (5-7) which contribute to either the composite forms 8 or 9, depending upon whether or not form 6 contributes to the energy stabilization of the molecule.

process known as intersystem crossing (c_i) may be of significant importance in the consideration of potential fluorescent material for use in a dye laser. Uninfluenced by outside factors, a molecule in the T_1 state may have a lifetime (10 to 10^{-6} sec), many orders of magnitude longer than the lifetime of a molecule in the S_1 state (10^{-11} to 10^{-14} sec). Radiative or nonradiative decay from T_1 to S_0 is forbidden in the quantum mechanical sense due to the difference in multiplicities of the states. Thus, without quenching, the T_1 state could serve as a trap for the excited state population. If the energy differences between the T_1 and higher T_x states is in resonance with some other emission or source of excitation, absorption of $T_1 \to T_x$ and internal conversion, $T_x \to T_1$, could be an important factor in loss of efficiency.

To avoid triplet state problems one should exclude factors which catalyze intersystem crossing by spin-orbit coupling. Spin-orbit coupling is enhanced by heavy atom effects and the loop rule.[2] This rule states that there is a greater propensity for triplet formation when the π electrons of a chromophore can form a loop when oscillating between end groups. This rule has been useful in the prediction of whether or not a particular dye would be a good laser candidate. The important feature of this determination is to analyze whether or not the resonance structure containing the positive charge on the atom, X, contributes to the overall stability of the system relative to the other forms such as depicted in Figure 3.1.4, that is, does form 6 with the positive charge located on X contribute as well as forms 5, 7, or others (not shown) to the energy stabilization of the molecule.

Where resonance structure 6 contributes, electrons may flow through atom X and the loop rule applies. The charge distribution may then be depicted in the composite resonance form 8. Where the resonance structure 6 does not contribute substantially, the composite resonance form may be more accurately depicted by form 9.

Figure 3.1.5 shows the skeletal features of a number of classes of dyes considered as laser dye candidates by Drexhage.[2] Of these, the dyes depicted in the left hand column of the figure tend to follow the loop rule since the resonance forms with the positively charged heteroatoms, sulfur or nitrogen, are important, whereas the dyes in the right hand column do not follow the loop rule, either because the positively charged oxygen species is energetically too high or the development of the positive charge on the central atom X is completely blocked. The latter group of dye families include the pyronines, rhodamines, and oxazines where oxygen is the central atom, and the carbazines and carbopyronins where the central atom is a fully substituted carbon atom.

Monocyclic Precursors

For numerous reasons, exact descriptions of the properties of large molecules are

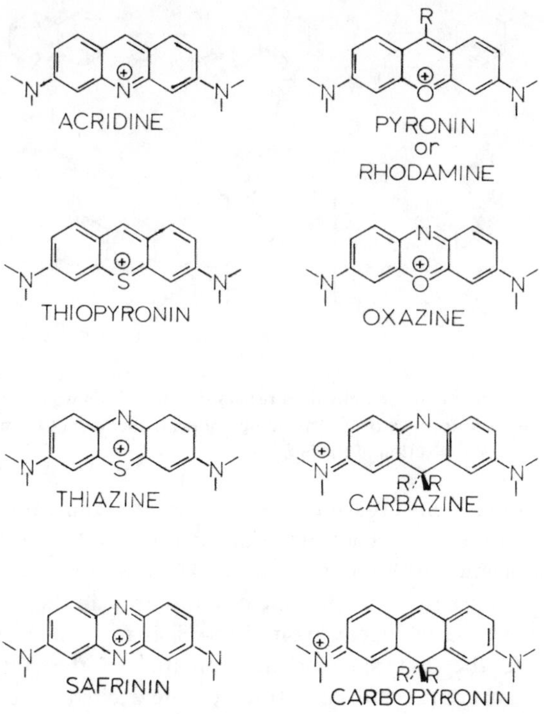

FIGURE 3.1.5. Dyes which tend to follow the loop rule (left hand column) as opposed to dyes which do not (right hand column).

still beyond calculation. The science is thus based on numerous data points and extrapolations. This results in the formulation of empirical rules and relationships, some of which have already been discussed. These correlations have formed the basis for the prediction of numerous factors pertinent to the lasing properties of organic molecules: the general region for absorption wavelength, fluorescence emission, laser emission, factors influencing certain excited state properties, and photochemical stability. However, the proof or lack thereof for any predicted property still lies in the synthesis and study of the chemical and physical properties of such molecules.

A series of six-membered ring monocyclic dyes (A-I) has been presented by Hammond et al.[3] as precursors for many of the well-known laser dyes used today, that is, those dyes with fused six-membered ring extended chromophores of the bicyclic and tricyclic type. Expansion of the grid to include the latter types (A2 through I3) reveal the basic type of structure for some of the well-known dye families, including the bicyclic umbelliferones, quinolones, and coumarins, and the tricyclic acridones, fluoresceins, acridines, pyronins or rhodamines, and carbopyronins as shown in Figure 3.1.6. The numbers in parenthesis for A through I are for the predicted potential lasing wavelengths in alcohol.[3] But, perhaps of even greater interest are the suggested structures for some potentially good, yet untested laser dyes.

Even without the addition of more peripheral substituents, such a grid could be extended. It may be worth noting that by the replacement of the carbon atom in the four position of the monocyclic ring with a heteroatom such as nitrogen (see Figure 3.1.7), one generates an additional 27 laser dye candidates which could be listed in a grid like Figure 3.1.6. The well-known oxazines and carbazines[2] would be included in this list and are included in Figure 3.1.7. Included in both grids are dyes found to have undesireable characteristics for use as laser dyes, such as the acridines and safranines.

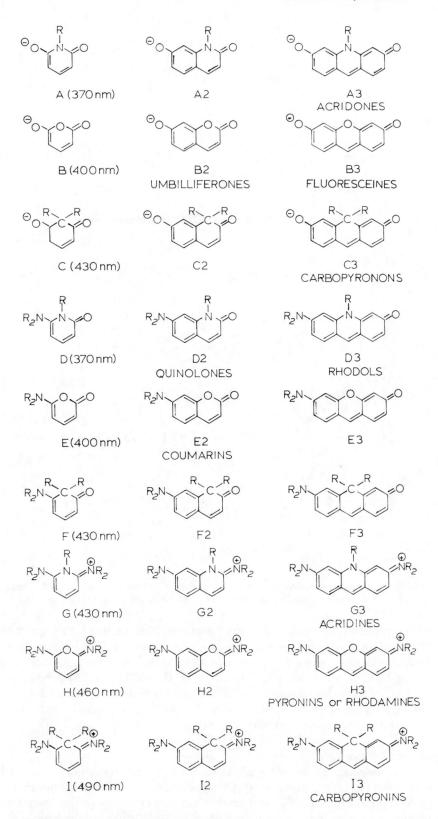

FIGURE 3.1.6.   Basic structures for some monocyclic dyes expanded to include some related bicyclic and tricyclic dyes.

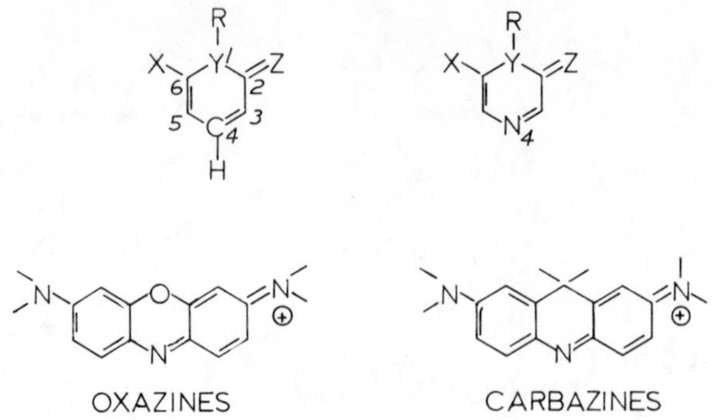

OXAZINES   CARBAZINES

FIGURE 3.1.7.   Additional laser dye candidates.

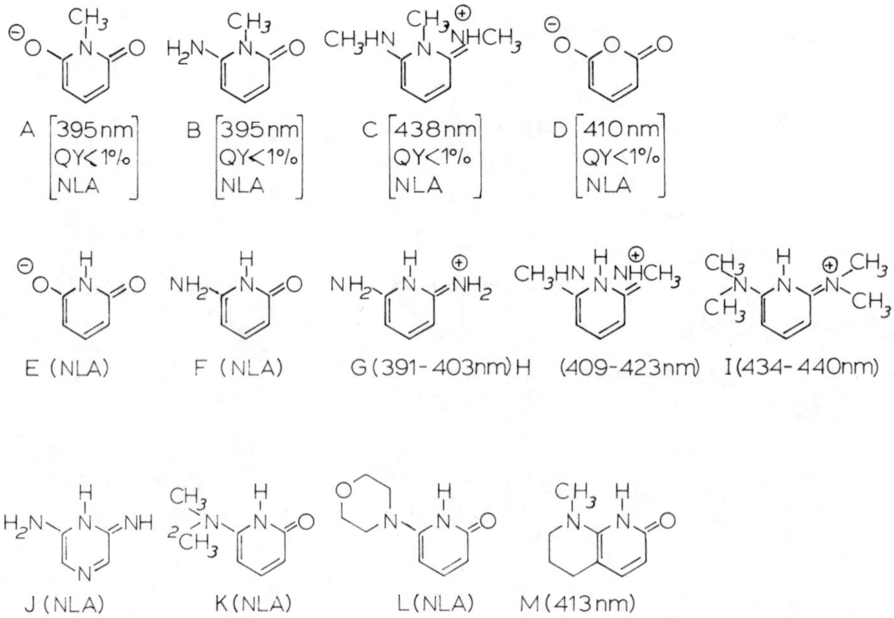

A [395 nm / QY<1% / NLA]   B [395 nm / QY<1% / NLA]   C [438 nm / QY<1% / NLA]   D [410 nm / QY<1% / NLA]

E (NLA)   F (NLA)   G (391-403 nm)   H (409-423 nm)   I (434-440 nm)

J (NLA)   K (NLA)   L (NLA)   M (413 nm)

FIGURE 3.1.8.   Examples of monocyclic dyes investigated for laser action.[3] Laser action under short pulse flashlamp excitation was noted for G, H, and I. NLA denotes no laser action.

Some examples of the monocyclic dyes which have been studied are shown in Figure 3.1.8.[3] The fluorescence maxima in aqueous solution are indicated for A, B, C, D, and M and in some cases the quantum yield (QY) of fluorescence. Laser action under short-pulse flashlamp excitation was noted for G, H, and I when dissolved in ethanol containing 0.01 N perchloric acid but with low efficiencies and low stabilities compared to their bicyclic counterparts. The rigidized dye M lased marginally under nitrogen pumping while others showed no laser action (NLA).

Bicyclic and Related Dyes

As the rhodamines in the red spectral region had been recognized as good laser materials from the early reports, so also had the coumarins. Significant advances have been made for the rhodamines and coumarins; however, unlike the rhodamines a greater amount of research has been applied to the modification of the basic coumarin nucleus. Early contributions were made by Drexhage,[2,4,5] Schimitschek et al.,[6] and

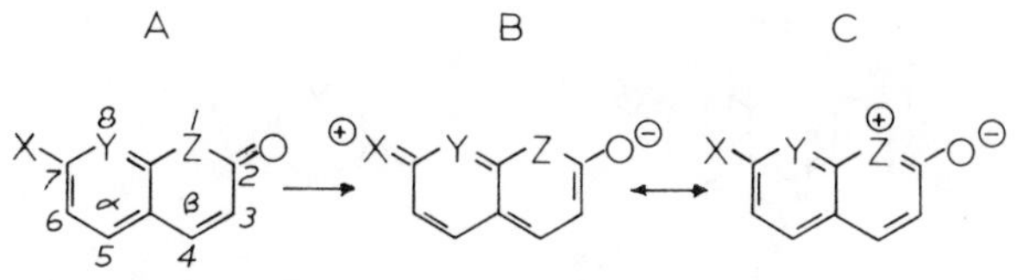

Y	Z	FAMILY
CH	O	COUMARIN
CH	NR	QUINOLONE
N	NR	AZAQUINOLONE
N	O	AZACOUMARIN

FIGURE 3.1.9.   Designation of skeletal position (numbers) and ring (alpha or beta) for the coumarins, quinolones, azaquinolones, and azacoumarins. B and C show resonance structures for the photoexcited state of A.

Hammond et al.[7] Improvements of the range of spectral coverage, efficiency, and stability were made by the application of the concepts of rigidization to the 7-amino substituent of the coumarin ring,[2,7] by attachment of heterocyclic moieties at the 3 position,[2] and by the addition of trifluoromethyl groups onto the coumarin nucleus, mainly at the 4 position.[6,7]

Subsequent reports dealt with water soluble coumarins (Drexhage et al.,[8] Henry[9]), further substitution (Reynolds[10]), and replacement of the elements of the coumarin nucleus to produce azacoumarins,[12,13] quinolones,[11,13] and azaquinolones,[12,13] most invariably involving efforts to restrain the movement of peripheral appendages attached to the nucleus. Considerable time and effort has also been expended in the investigation of the lasing properties of many of these materials in an effort to elucidate the important parameters contributing to their efficiency and stability.[14-20]

In Figure 3.1.9, A depicts the major sites of substitution both nuclear and peripheral and distinguishes between the rings alpha and beta. The nuclear substitution of Y and Z with oxygen, nitrogen, or carbon determines the dye family. The elements of nitrogen and carbon may be substituted with either hydrogen or alkyl groups such as methyl. The 7 position will in most cases involve an amino function, which when incorporated into a five- or six-membered ring with the terminus at the 6 and/or 8 position is said to be rigidized. Positions 3 and 4 provide the other sites where substitutions are made.

Assume the structure as depicted in A, the major contributor to the ground electronic state S_0, is transformed into the charge-separated species B, the major contributor to the excited electronic state S_1. The factors which stabilize S_1 relative to S_0 will cause a red shift in wavelength. Those factors which cause $S_1 - S_0$ to become larger or result in destabilization of the excited electronic state relative to the ground state will cause a blue shift.

Thus, in the excited state B, electron-donating substituents attached directly to the site of positive charge formation, X, will provide direct stabilization to X, and indirectly to ring alpha, while electron-withdrawing substituents on ring beta will help to stabilize a negative charge. Electronegative elements at Z will help to stabilize the negative charge in ring beta. Electron-donating groups on ring beta would have a tendency to destabilize negative charge formation in this ring.

Relative to hydrogen, methyl and ethyl(alkyl) groups and methoxy(alkoxy) are electron donating while groups such as trifluoromethyl (CF_3), acetyl(CH_3CO), carboxyl(COOR), and cyano(CN) are electron withdrawing. For the elements utilized in these dyes, the order of electronegativity is as follows: oxygen > nitrogen > carbon.

Further examination of the structures (Figure 3.1.10, A through D) indicates the

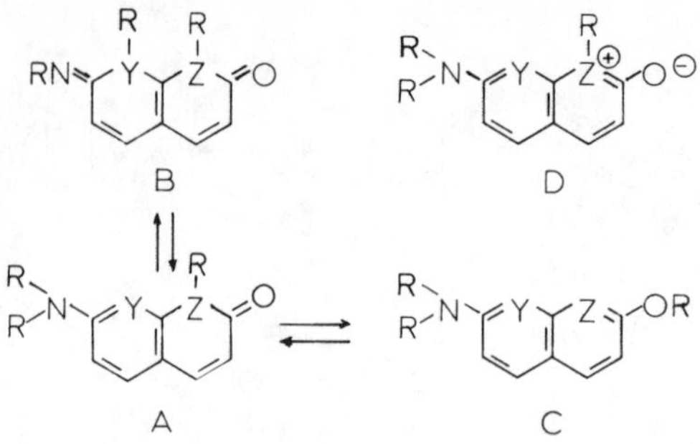

FIGURE 3.1.10. Isomers (B and C) and resonance forms (D) of A in Figure 3.1.9.

possibility of molecular rearrangement where either Z and/or Y are nitrogen atoms and R equals hydrogen. A prototropic shift of the hydrogen from the 7-amino substituent in A to the nitrogen at Y would allow the transformation of A→B and vice versa, B→A. Similarly a prototropic shift from nitrogen to oxygen in ring beta would convert A to C in an equilibrated manner, but this type of tautomeric shift is precluded when R is an alkyl group such as methyl. The lower energy form of each pair would predominate. The former type of isomerization has been suggested as a possible mechanism for lack of or marginal lasing ability of some 7-amino azaquinolones.[12]

On the other hand, D could result when R is either hydrogen or alkyl. Assuming, as in our prior discussion, that the uncharged species A is more important in S_0 and the charged species D is more important in S_1, the latter form of structure may be important in excited state processes. Thus, intersystem crossing may result from spin-orbit coupling where electron flow thru Z is not prohibited (loop rule).[2]

Heavy Atom Effects

As with other dyes, it is important to avoid the incorporation of heavy atoms such as iodine or bromine into the dye nucleus or to use heavy atom solvents which promote spin-orbit coupling resulting in intersystem crossing to the triplet state. A test has been devised to check a molecule's propensity to undergo intersystem crossing.[2] By testing a 25% alcoholic methyl iodide solution of azaquinolone A, quinolone B, azacoumarin C, and coumarin D (see Figure 3.1.11 for structures), it was found that the fluorescence was attenuated by 39%, 74%, 76%, and 0%, respectively.[12] This test suggested that intersystem crossing to the triplet state may be an inherent problem with the aza-substituted derivatives since only the coumarin was unaffected.

Structural Features Influencing Wavelengths

The tabulation of lasing wavelengths correlated to molecular structures may be found in Tables 3.1.1-4. Selected data from these tables has been combined in Figure 3.1.12 to illustrate the extent to which some of the various substitutions influence the wavelength. Extrapolated values appear in parentheses.

Group 1 involves methylation of the 7-amino and 1-aza nitrogen atoms of four quinolones with all other variables held constant. As expected, methylation of the 7-amino group produces a red shift, that is, A1 to C1 ($\Delta\lambda = 11$ nm) and B1 to D1 ($\Delta\lambda = 11$ nm). Methylation of the 1-aza position produces a blue shift, that is, D1 to C1($\Delta\lambda = 3$ nm) and B1 to A1 ($\Delta\lambda = 3$ nm).

Group 2 involves variation of the 4-substituent for three families of dyes for which all other variables within each family have been held constant. In each family the electron-donating 4-methoxy causes a blue shift relative to the 4-methyl, the 4-hydrogen (considered electronically neutral), and the electron-withdrawing 4-trifluorome-

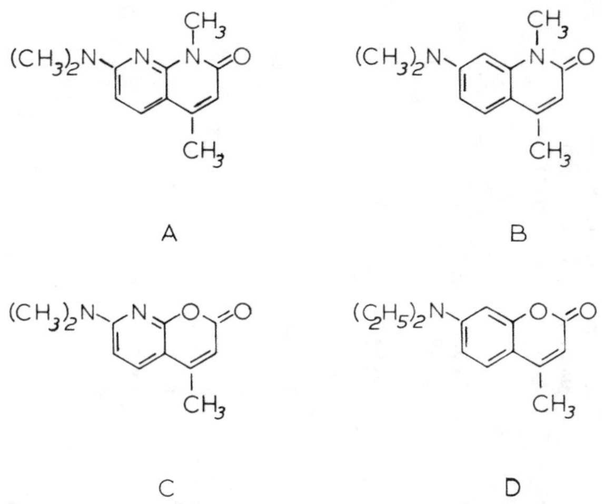

A

B

C

D

FIGURE 3.1.11.   Fluorescence attention of 39%, 74%, 76%, and 0% resulted for A,B,C, and D, respectively, when treated with an alcoholic solution of methyl iodide.[12]

thyl. Of the three families considered in Figure 3.1.13, Group 2 (the azaquinolones (AQ), the quinolones (Q), and the coumarins (C)), only the data for the coumarins are complete. The data for AQ-4H (structure C) and Q-4H (structure G) were extrapolated from similiar cases in Table 3.1.1. However, in each case the overall value for the wavelength increment from methoxy to trifluoromethyl was complete and in good agreement, that is, 62 nm (ΔAD), 68 nm (ΔEH), and 67 nm (ΔIL), respectively, for the three families. Average increments are as follows: OCH_3 ← 16 nm → CH_3 ← 7 nm → H ← 43 → CF_3; overall, OCH_3 ← 66 nm → CF_3 (these values are estimates meant only as a guide for further extrapolation).

Group 3 involves variation of the degree of alkylation at the 7-amino substituent for three series of coumarins where the 4-substituent is varied within a series from CH_3 to H to CF_3. In Group 3, we have six structures which show an incremental red wavelength shift as both the degree of alkylation and rigidization of the 7-amino substituent is enhanced. Although the data is sparse for the 4-methyl series, in going from the nonrigidized 7-amino structure M to the fully rigidized ring structure R, the overall differential increments of 44 nm, 40 nm, and 54 nm for the three respective series, 4-CH_3, 4-H, and 4CF_3 are fairly constant and average 46 nm. Rigidization appears to play as significant a role in red shifting the wavelength as alkylation. Some average values are shown: H_2N ← 25 nm → $(H_5C_2)_2N$ ← 20 nm → R_2N as in structure R. Relative to alkylation, monocyclization affords the largest wavelength shift while formation of the second ring still contributes quite significantly relative to alkylation of the monocyclized dye (compare M to P, P to R, and P to Q, respectively).

Group 4 involves variation of the nuclear elements at the 1 and 8 positions to give one of four possible combinations of azaquinolone (8N-1NR), quinolone (8CH-1NR), azacoumarin (8N-1O), or coumarin (8CH-1O) where the number refers to the position and the letter to the element at that position and the other substituents are held constant within a series. In Group 4, we have four series of dyes, one with a 7-amino substituent with either a 4-CH_3 or a 4-CF_3, and the other with a dimethylamino substituent with a 4-CH_3 or 4-CF_3. The total incremental red wavelength shifts in going from the azaquinolones with structures S and W to the coumarins with structures V and Z are 50 nm, 47 nm, 46 nm, and 67 nm which average 52 nm. The high value of 67 nm for Δ WZ-CF_3 may suggest a nonlinearity.

Table 3.1.1
LASING PROPERTIES OF VARIOUS AZAQUINOLONES (AQ), QUINOLONES (Q), AZACOUMARINS (AC), AND COUMARINS(C)

Item no.	Literature designation[a,7,11-13]	X	Y	Z	4	Other	Lasing wavelength λl (nm)	Stability[b]	Fluorescence quantum yield[18]	Fluorescence wavelength λ_f (nm)[c]
1	7A-4M-AQ	NH$_2$	N	NH	CH$_3$		386	L		
2	7DMA-1M-4MO-AQ	N(CH$_3$)$_2$	N	NCH$_3$	OCH$_3$		390	H	0.52	383
3	7OH-4M-AQ	OH	N	NH	CH$_3$		395[c]	M	0.48	
4	7OH-3,4DM-AQ	OH	N	NH	CH$_3$	3-CH$_3$	405[c]	L		
5	7DMA-1,4DM-AQ	N(CH$_3$)$_2$	N	NCH$_3$	CH$_3$		407	H	0.65	400
6	7A-4TFM-AQ	NH$_2$	N	NH	CF$_3$		437	L		
7	AQ1F	N(CH$_3$)$_2$	N	NCH$_3$	CF$_3$		452			430
8	7DMA-1M-4MO-Q	N(CH$_3$)$_2$	CH	NCH$_3$	OCH$_3$		409	H		396
9	7A-1,4-DM-Q	NH$_2$	CH	NCH$_3$	CH$_3$		409	L		413[a]
10	7A-1M-Q	NH$_2$	CH	NCH$_3$	H		412	M		417[a]
11	7A-4M-Q	NH$_2$	CH	NH	CH$_3$		413	L	0.63	415[a]
12	7A-Q	NH$_2$	CH	NH	H		418	M		425[a]
13	7DMA-1,4DM-Q	N(CH$_3$)$_2$	CH	NCH$_3$	CH$_3$		420	H	0.69	415
14	7A-4,8DM-Q	NH$_2$	CH	NH	CH$_3$	8-CH$_3$	420	L		429[a]
15	Cpd 857-11-c5	HN(CH$_2$)$_2$-6	CH	NH	CH$_3$		422	L		
16	Cpd 857-109	NHCH(CH$_3$) C(CH$_3$)$_2$-6	CH	NH	CH$_3$		423	M	0.68	
17	7DEA-4M-Q	N(C$_2$H$_5$)$_2$	CH	NH	CH$_3$		425	H		444[a]
18	7DMA-4M-Q	N(CH$_3$)$_2$	CH	NH	CH$_3$		426	H	0.63	429[b]
19	7OH-4M-Q	OH	CH	NH	CH$_3$		441[c]	L		437[b]
20	7OH-3,4-DM-Q	OH	CH	NH	CH$_3$	3-CH$_3$	447[c]	H		445[a]
21	7DMA-1M-Q	N(CH$_3$)$_2$	CH	NCH$_3$	H		—	M		
22	Q1F	NH$_2$	CH	NH	CF$_3$		463			
23	Q6F	NHCH(CH$_3$) C(CH$_3$)$_2$-6	CH	NCH$_3$	CF$_3$		473	H	0.8	
24	Q3F	N(CH$_3$)$_2$	CH	NCH$_3$	CF$_3$		477	H		472

No.	Compound								
25	Q4F	$NHCH(CH_3)_2$-6, $C(CH_3)_3$-6	CH	NH	CF_3	477			
26	7MOR-4M-AC	$O(CH_2CH_2)_2N$-	N	O	CH_3	431		0.54	415
27	7OH-4M-AC	OH	N	O	CH_3	431			
28	7DMA-4M-AC	$N(CH_3)_2$	N	O	CH_3	434			420
29	AC2F	$CH_3N(CH_2)_3$, CH_3-6	N	O	CF_3	490		0.86	
30	7A-4MO-C	NH_2	CH	O	OCH_3	417	M		405
31	7A-4M-C	NH_2	CH	O	CH_3	436	H	0.72	429
32	C3H	NH_2	CH	O	H	439			439
33	7DMA-4M-C	$N(CH_3)_2$	CH	O	CH_3	453	H	0.75	452
34	C3F	NH_2	CH	O	CF_3	484		0.74	484
35	C2F	$N(CH_3)_2$	CH	O	CF_3	519		0.21	

a Chemical nomenclature: The chemical names have been abbreviated according to the nomenclature system previously used by Fletcher. The shorthand notation for the parent compounds and substituents are as follows with numbers referring to the skeletal position (see numbered structure): AQ, azaquinolone; Q, quinolone, AC, azacoumarin; C, coumarin; A, amino; M, methyl; DM, dimethyl; DMA, dimethylamino; DEA, diethylamino; MO, methoxy; MOR, morpholino; TFM, trifluoromethyl; and OH, hydroxy. Thus 7A-4M-AQ designates 7-amino-4-methylazaquinolone. Literature description such as AQ1F, Q1F and C2H are not necessarily systematic. The first letter(s) refer to the family whereas the H and F refer to hydrogen and fluorine substitution.

b Stability: Relative stability - low (L), medium (M), and high (H).

c Solvent: Ethanol unless otherwise noted; (a) water, (b) 0.1% sodium carbonate solution-water, (c) 10^{-3} N sodium hydroxide in ethanol.

Table 3.1.2

LASING PROPERTIES OF COUMARINS WITH NONRIGIDIZED AMINO GROUPS

Item no.	Literature designation	Substituent position				Lasing wavelength λl (nm)[a]	Ref.
		R_1	R_2	4	Other		
1	DA-4MO-C	H	H	OCH_3		417	11
2	C120	H	H	CH_3		436	11
	C120	H	H	CH_3		442	10
3	C3H	H	H	H		450	13
4	C3F	H	H	CF_3		484	7
	C151	H	H	CF_3		490	10
5	C311	CH_3	CH_3	CH_3		453	12
	C311	CH_3	CH_3	CH_3		457	18
6	C2H	CH_3	CH_3	H		465	13
7	C2F	CH_3	CH_3	CF_3		519	7
	C152	CH_3	CH_3	CF_3		520	10
8	C1	Et	Et	CH_3		460	13
	Cl	Et	Et	CH_3		460	10
9	C1H	Et	Et	H		466	13
10	C1F	Et	Et	CF_3		—	7
	C35	Et	Et	CF_3		—	10
11	C2	Et	H	CH_3	$6CH_3$	458	13
	C2	Et	H	CH_3	$6CH_3$	454	10
12	—	Et	H	H		—	—
13	C307	Et	H	CF_3	$6CH_3$	502	10

[a] Solvent, ethanol.

Since the lasing wavelength values are known for only one of the four members of the group U and Y, that is, for Y-CH_3, it becomes important for purposes of extrapolation to recognize possible nonlinearities. The extrapolated values again appear in parentheses. After consideration of the differential value within each CH_3-CF_3 series, ΔCH_3-CF_3, the average differential value between the dyes with the four combinations of nuclear elements are: 8N-1NR ← 24 nm → 8CH-1NR ← 7 nm → 8N-1O ← 22 nm → 8CH-1O. The last average value of 22 nm appears high again due to the magnitude of the extrapolated value of Δ YZ-CF_3 of 33 nm.

Although based mostly on extrapleted data the average incremental value for the azacoumarin (8N-1O)-coumarin (8CH-1O) of 22 nm (see above) may be high due to the magnitude of the YZ-CF_3 value (33 nm). Further support may be suggested by noting once again the larger differential of 67 nm for the ΔWZ-CF_3 series as compared to the other series and by comparison of the ΔCH_3-CF_3 value of Z (66 nm) as compared to all the other Δ CH_3-CF_3 values which average about 50 nm. However, the wavelength value of 519 nm for Z-CF_3 appears to be valid and is corroborated by the lasing maximum for this dye in ethanol under nitrogen pumping and Nd:YAG pumping. Another nonlinearity, is suggested by the lasing value of W-CH_3 (407 nm) which appears to be high relative to S-CH_3 by about 10 nm when compared to the differential value of B1 and C1 of Group 1 and the differential between T-CH_3 and X-CH_3.

Figure 3.1.13 shows the basic dye structures with various 4-substituents for many of the coumarins, and a few quinolones, azaquinolones, and azacoumarins. An arbitrary stability classification has been assigned to each dye from 0 to 7 with the lower

Table 3.1.3

LASING PROPERTIES OF COUMARINS WITH MONOCYCLIC RESTRICTION OF THE AMINO GROUP

Item no.	Literature designation	Substituent position				Lasing wavelength λl (nm)	Solvent	Ref.	Absorption wavelength (nm)	Fluorescence wavelength (nm)	Quantum yield
		R	3	4						
1	C4H	H	H	H	477	Ethanol	13	—	—	—
2	C8H	CH$_3$	H	H	475	Ethanol	13	—	—	—
3	C4F	H	H	CF$_3$	513	Ethanol	10	406[10]	500[10]	0.47[10]
	C340	H	H	CF$_3$	522	Ethanol	7	—	—	0.86[18]
4	C8F	CH$_3$	H	CF$_3$	522	Ethanol	7	410[18]	515	0.78[18]
5	C355	C$_2$H$_5$	H	CF$_3$	522	Ethanol	10	412[13]	515[13]	0.23[13]
6	C386	(CH$_2$)$_3$SO$_3$Na	H	CH$_3$	486	Water	8	388[8]	—	—
7	C388	(CH$_2$)$_4$SO$_3$Na	H	CH$_3$	489	Water	8	390[8]	—	—

Table 3.1.4

LASING PROPERTIES OF COUMARINS WITH BICYCLIC RESTRICTION OF THE AMINO GROUP

Item no.	Literature designation	Substituents		Lasing wavelength λl (nm)[a]	Ref.	Absorption wavelength (nm)	Fluorescence wavelength (nm)	Quantum yield
		4	3					
1	C106	-CH$_2$-CH$_2$-CH$_2$-		478	10	386	465	0.57[10]
	C106							0.80[18]
2	C102	CH$_3$	H	480	7	390	465	0.58[10]
	C102			480	10			0.93[18]
3	C6H	H	H	490	13	396	—	0.72[18]
4	C314	H	COOC$_2$H$_5$	504	10	436	480	0.68[10]
								0.83[18]
5	C343	H	COOH	519	10	446	490	0.63[10]
6	C334		COCH$_3$	521	10	452	495	0.69[10]
								0.75[18]
7	C337	H	CN	522	10	443	485	0.64[10]
								0.63[18]
8	C6F	CF$_3$	H	538	7	423	530	0.38[10]
	C153			540	10			0.53[18]
9	C217	CH$_2$COOH	H	514[b]	8	402[b]	—	—
10	C343	H	COOH	518[b]	8	425[b]	—	—

[a] Solvent: ethanol unless otherwise noted.

[b] Water.

number corresponding to those dyes that give the best results. The arbitrary scale is correlated with various ranges of relative half-life as is indicated in Table 3.1.5 and depends on the study and laser system. It should be recognized that there are many factors influencing the relative half-lives of the dyes not the least to which is mirror refectivity and whether or not the uv radiation is filtered or unfiltered. In addition, significant lifetime differences may be noted in broadband flashlamp excitation versus narrow-line laser excitation. The results in Table 3.1.5 and Figure 3.1.13 were taken from studies by Schimitschek et al.,[7,13] Hammond et al.,[11,12] and Fletcher.[16] Some of the dyes were studied in both laser systems as indicated in Table 3.1.5.

Stability

Although the data from Tables 3.1.1 and 3.1.5 came from different flashlamp-pumped dye laser systems, numerous comparisons may be made concerning the apparent dye stability. Attenuation of the dye laser output may result not only from dye degradation but by interference of the resultant photoproducts. All comparisons were done with ethanol as the solvent unless otherwise indicated. Throughout the data the replacement of hydrogen bonded to nitrogen with alkyl groups appears to be of universal importance. This may be especially noted in Table 3.1.1 for the quinolones and azaquinolones. In this discussion and in Figure 3.1.13 and Table 3.1.5, the A will always refer to a dye with a methyl group, B to a hydrogen, and C to a trifluoromethyl substituent in the four position of the basic skeletal structure. A five- to six-fold stability enhancement may be noted in the comparison of the completely alkylated quinolone 8C with nonalkylated 7C and comparison of the partially alkylated 14C to the fully alkylated 13C. Comparison of the nonalkylated 7-amino coumarins 1A, 1B, and 1C to 6A, 6B, and 6C indicates a 2-, 25-, and 11-fold stability increase. Also, comparison of the monoalkylated, monocyclized 4B and 4C to the monocyclized, dialkylated 5B and 5C and to the dicyclized, dialkylated 6B and 6C indicates a two- to eight-fold increase in stability for the latter dyes.

In a comparison of the 4-methyl (A), 4-hydrogen (B), and 4-trifluoromethyl (C) series of coumarins, the 4-methyl series is generally the least stable. Perhaps the greater stability of the coumarins in the C and B series can be accounted for by the elimination of the oxidizable methyl group that is found in the A series. The importance of removing or replacing the methyl has been demonstrated by the report of Winters et al.[21] concerning the photodegradation of 7-diethylamino-4-methylcoumarin (Figure 3.1.14, 3A). Figure 3.1.15 indicates the five photochemical products produced upon photolysis of 3A by flashlamp excitation in 95% ethanol. Two pathways were identified, sequential dealkylation of 3A to form R and then 1A and oxidation of 3A to the carboxylic acid, Q, which has an absorption maximum at 470 nm which essentially overlaps the wavelength of laser emission. Thus, in this case the formation of Q is primarily responsible for the attenuation of laser emission.

Determination of the influence of rigidization on stability in the coumarin series is most impressive for the C series. Comparison of the relatively unstable 2C with 5C and 6C, indicates that rigidization provides some of the best stability. However, the unrigidized dyes with structure 2B and 3B were found to be more stable than their rigidized counterparts 5B and 6B, although the arbitrary stability categories are close, that is, 2, 3, and 4 (see Table 3.1.5).

Comparison of the fully rigidized structures with substituents in the 3 and 4 position would indicate a relative stability as shown: 4-CF_3 (6C) > 3-CN(15I) > $3\text{-COOC}_2\text{H}_5$ (15J) > 3-CO-CH_3(15K) > H(6B) > 4-CH_3(6A). It may be noted that dye 6A with an oxidizable methyl group in the 4-position is the least stable in this series.

Considering the data in Table 3.1.5 as a whole, the dyes providing the best stability would appear to be the fluorinated coumarins, 5C and 6C, and the two azacoumarins of essentially identical structure, 12M and 12N.

Q

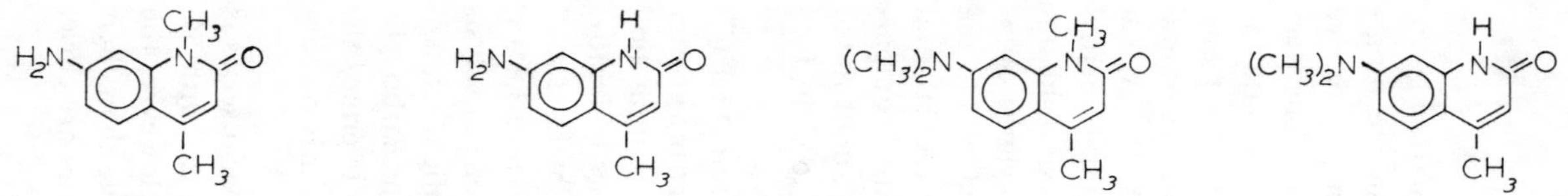

A1 409 nm | B1 412 nm | C1 420 nm | D1 423 nm

GROUP 1

FIGURE 3.1.12. Lasing wavelength for related molecules of four groups: Group 1, quinolones (Q); Group 2, azaquinolones (AQ), quinolones (Q), and coumarins (C); Group 3, coumarins; Group 4, variation of family with constant substitution. Δ values indicate wavelength differentials between dyes within each group; the letters indicate the exact case. Group 4 also shows wavelength differentials within the exact case. Group 4 also shows wavelength differentials within each series, that is, Δ CH_3-CF_3 values.

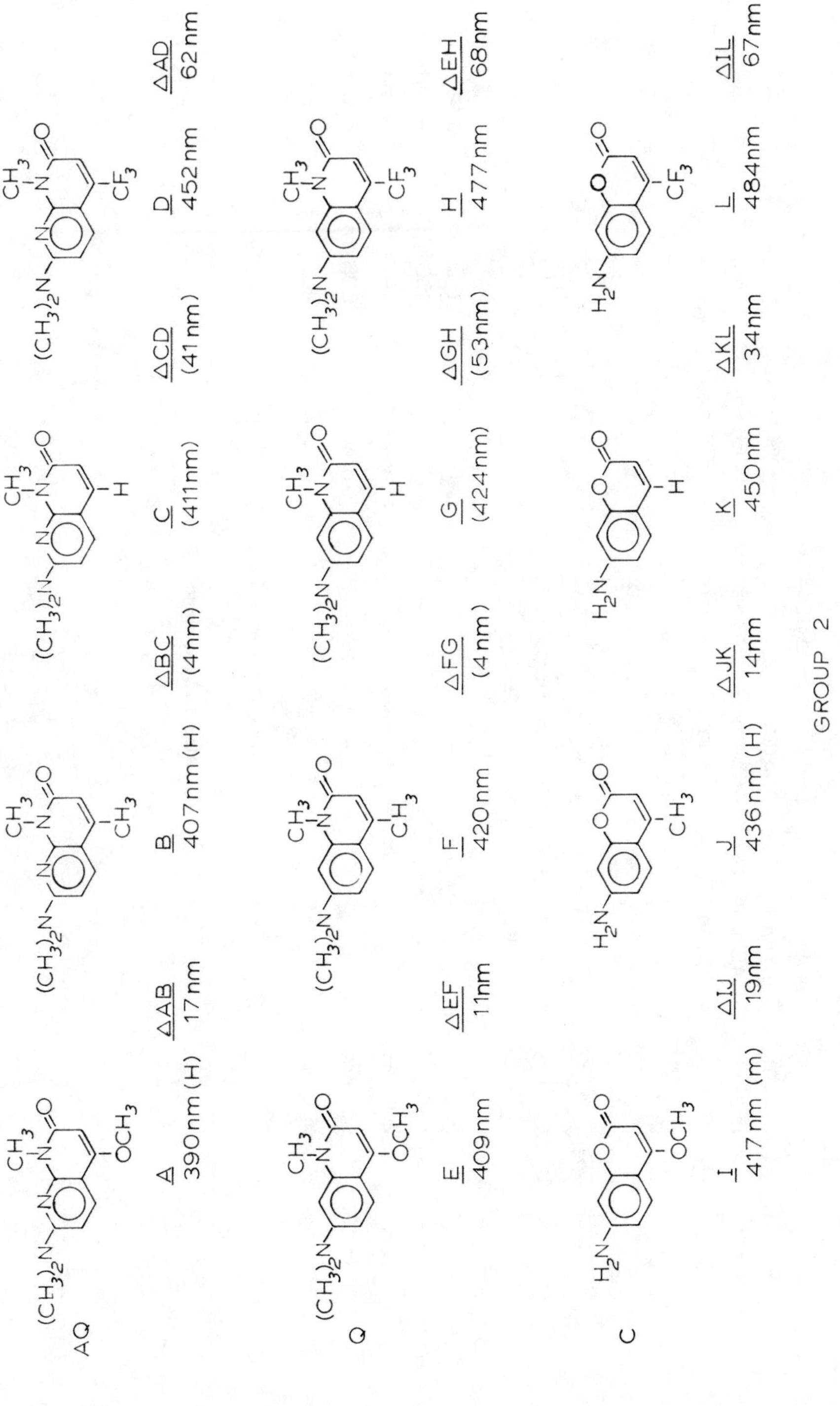

GROUP 2

A 390nm(H)   ΔAB 17nm   B 407nm(H)   ΔBC (4nm)   C (411nm)   ΔCD (41nm)   D 452nm   ΔAD 62nm

E 409nm   ΔEF 11nm   F 420nm   ΔFG (4nm)   G (424nm)   ΔGH (53nm)   H 477nm   ΔEH 68nm

I 417nm (m)   ΔIJ 19nm   J 436nm (H)   ΔJK 14nm   K 450nm   ΔKL 34nm   L 484nm   ΔIL 67nm

FIGURE 3.1.12 (continued).

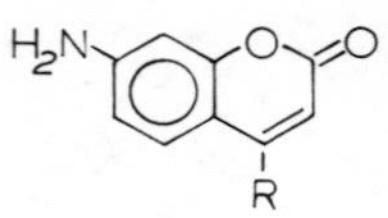

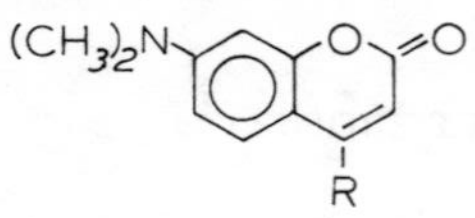

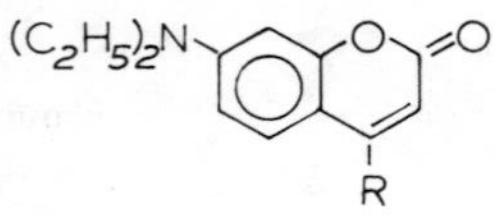

	M	ΔMN	N	ΔNO	O	ΔOP
$R=CH_3$	436nm	17nm	453nm	7nm	460nm	—
R=H	450nm	15nm	465nm	1nm	466nm	11nm
$R=CF_3$	484nm	35nm	519 nm	(1)nm	(520nm)	(2nm)

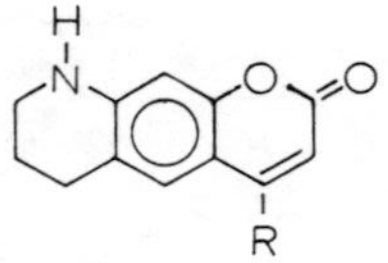

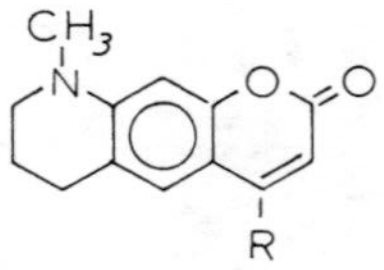

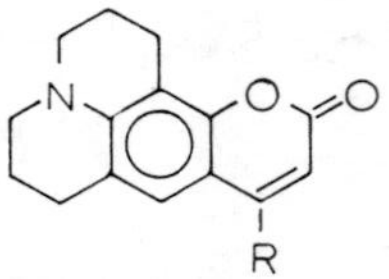

	P	ΔPQ	Q	ΔQR	R	ΔMR
$R=CH_3$	(470nm)	—	—	—	480nm	44nm
R=H	477nm	-2nm	475nm	15nm	490nm	40nm
$R=CF_3$	522nm	0	522 nm	16nm	538nm	54nm

GROUP 3

FIGURE 3.1.12 (continued).

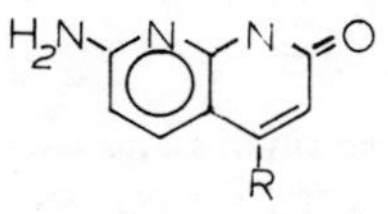

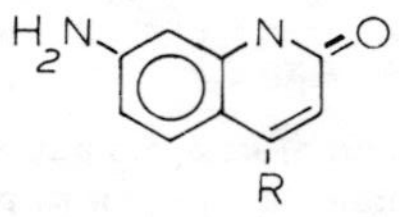

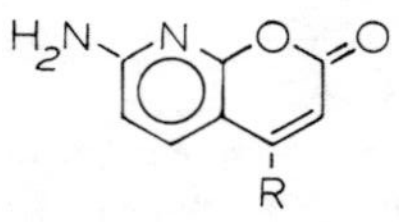

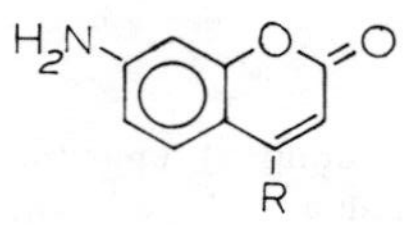

	S	ΔST	T	ΔTU	U	ΔUV	V	ΔSV
CH₃	386nm	27nm	413nm	(4)nm	(417nm)	(19)nm	436nm	50nm
CF₃	437nm	26nm	463nm	4nm	467nm	17nm	484nm	43nm
ΔCH₃−CF₃	51		50		50		48	

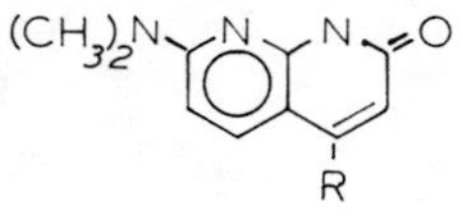

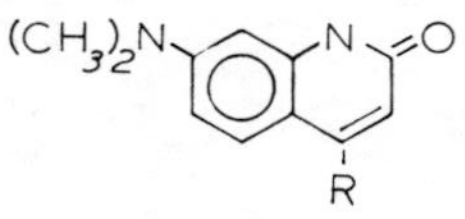

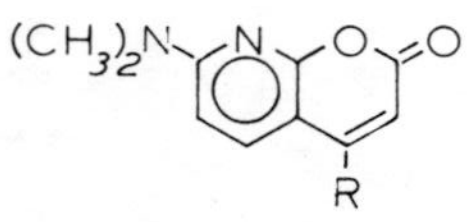

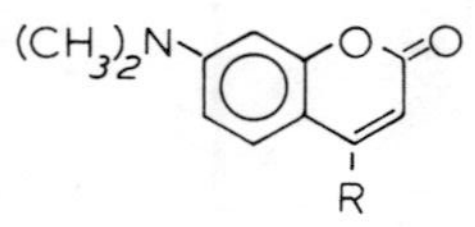

	W	ΔWX	X	ΔXY	Y	ΔYZ	Z	ΔWZ
CH₃	407nm	17nm	420nm	14nm	434nm	19nm	453nm	46nm
CF₃	452nm	25nm	477nm	7nm	(486)nm	(33)nm	519nm	67nm
ΔCH₃−CF₃	45		57		(52)		66	

GROUP 4

FIGURE 3.1.12 (continued).

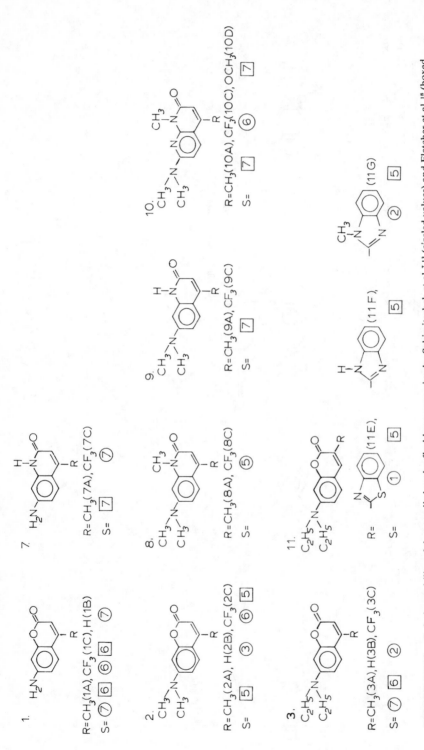

FIGURE 3.1.13.   Relative stability of dyes studied under flashlamp pumping by Schimitschek et al.[7,13] (circled values) and Fletcher et al.[18] (boxed values). See Table 3.1.5. Numbers in O and □ indicates apparent relative stability, the lower the number the higher the stability. The number in parentheses refers to the particular structure and the letter to the particular substituent, that is, A (methyl), B (hydrogen), and C (trifluoromethyl).

4.

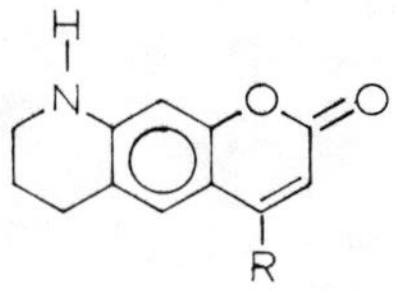

R = CH$_3$(4A), H(4B), CF$_3$(4C)

S = ⑦ ⑤ ⑤

5.

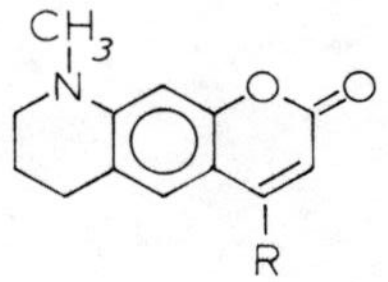

R = CH$_3$(5A), H(5B), CF$_3$(5C)

S = ④ ⓪ ③

6.

R = CH$_3$(6A), H(6B), CF$_3$(6C)

S ⑦ ⑤ ④ ① ②

12.

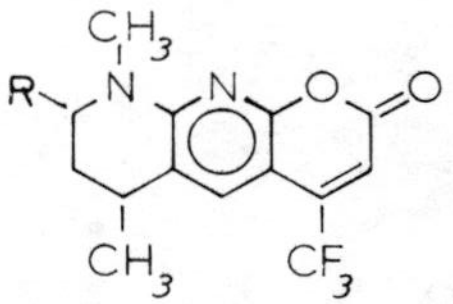

R = CH$_3$(12M), H(12N)

S = ① ① ① ①

13.

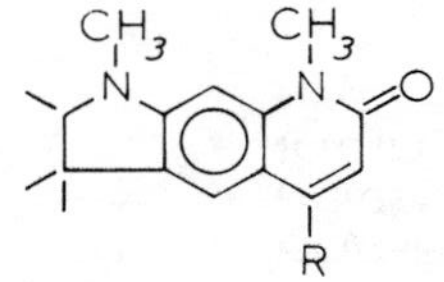

R = CF$_3$(13C)

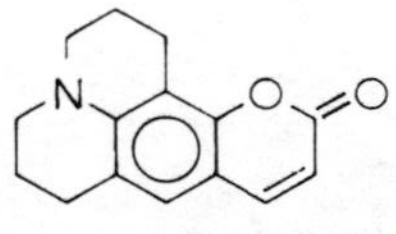

S = ⑤ ⑤

14.

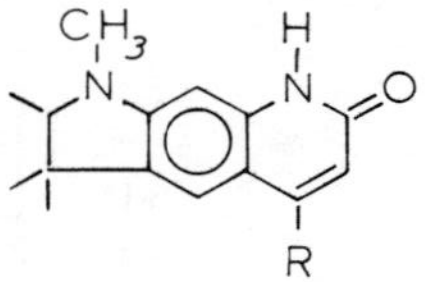

R = CH$_3$(14A), CF$_3$(14C)

S = ⑦ ⑦

15.

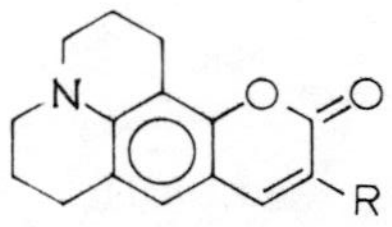

R = CN(15I), COOC$_2$H$_5$(15J), COCH$_3$(15K)

S = ③ ③ ④

FIGURE 3.1.13 (continued).

Table 3.1.5

APPARENT RELATIVE STABILITY OF VARIOUS COUMARINS, AZACOMARINS, QUINOLONES, AND AZAQUINOLONES UNDER FLASHLAMP PUMPING (SEE FIGURE 3.1.13 FOR CORRESPONDING STRUCTURES)

Relative half-life (shots)	1400	650—600	480—450	370—340	300—250	190—140	100—40	30—0
Relative stability category	0	1	2	3	4	5	6	7
Ref. 7, 13	5C(C8F)[a]	12M(AC2F)	3B(C1H)	2B(C2H)	6B(C6H)	13G(Q6F)	10C(AQIF)	4B(C4H)
		6C(C6F)	11G(C30)		5B(C8H)	4C(C4F)	1C(C3F)	14C(Q4F)
		11E(C6)				8C(Q3F)	2C(C2F)	7C(QIF)
								3A(C1)
								6A(C102)
								1B(C3H)
								1A(C120)
Ref. 18		12M(AC2F)	6C(C6F)	5C(C8F)	6B(C6H)	11E(C6)	1C(C3F)	7A
		12N(AC3F)		15I(C337)	15K(C334)	6A(C102)	1A(C120)	10A
				15J(C314)		2C(C2F)	3A(C1)	10D
						13C(Q6F)		9A
						11G(C30)		14A
						4C(C4F)		
						11F(C7)		

[a] Names in parentheses are original literature designations. The number refers to the structure in Figure 3.1.13 in association with a letter A, B, or C which represents the substituent (R), that is methyl, hydrogen, or trifluoromethyl, respectively.

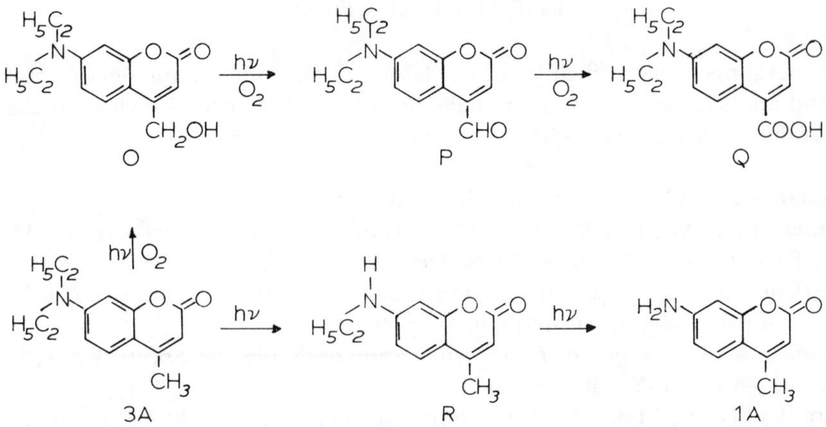

FIGURE 3.1.14. Sequential photooxidation of 7-diethylamino-4-methylcoumarin, 3A, gives the carboxylic acid, Q, and photo-dealkylation leads to the 7-amino-4-methylcoumarin, 1A.

Fletcher and Bliss[18] have related the effects of chemical constitution of coumarin and related dyes to the attenuation of dye laser output with various parameters. An equation was derived relating A, a photodegration factor at the laser wavelength which is inversely proportional to the total input energy per unit volume T, to β, a bleaching or photodegradation constant for a given dye and solvent, and to F, a factor relating the percent absorption at the laser wavelength for a change in dye concentration. c_o is the initial dye concentration.

$$A = \frac{\% \text{ absorption at } \lambda \text{ las}}{T} \cong c_o \beta F$$

The values for the bleaching coefficient β indicated:

1. On an average and with exceptions, the quinolines bleached five to six times faster than the coumarins.
2. Within the coumarins replacement of the 4-CH$_3$ with 4-CF$_3$ reduces β on an average of 20%.
3. Replacement of the 4-CH$_3$ with 4-H and replacement of 3-H with various 3-substituents (see Figure 3.1.14), reduces β by only 6%. Small values of β are the most desirable as is the case with A and F.

Average values for the F factor for the quinolines are only slightly higher than for the coumarins where the 4-substituent is methyl. Within both series, the F factor for the dyes with nonalkylated 7-amino groups is high. However, in the coumarin series with 4-CF$_3$ substituents the value for the dye with the nonalkylated 7-amino group is three times greater than the average of the alkylated 7-amino dyes. Also, in the coumarin series, replacement of the CH$_3$ by CF$_3$ reduces F by a factor of three on the average.

Similar relationships exist for the A term as for the F factor. However, when A was combined with τ, the total input energy required for the output energy to be attenuated to half the original value, it was experimentally determined that Aτ had the same value as the absorption (1.2%) of the dye solution per centimeter at the lasing wavelength when the laser output had been reduced by half.

Thus, having determined τ for a particular dye and laser system one may estimate τ for a new dye, having measured the photodegradation factors for the two dyes by the relation $\tau_1 A_1 = \tau_2 A_2$.[18]

FURTHER READING

The general operating principles of dye lasers, pump sources, and device configurations, and applications of dye lasers have been treated in several review articles. For details of these subjects the reader is referred to the following:

- **Schafer, F. P., Ed.,** *Dye Lasers,* Springer-Verlag, Berlin, 1973.
- **Peterson, O. G.,** Dye lasers, in *Methods of Experimental Physics,* Vol. 15, Part A, Tang, C., Ed., Academic Press, New York, 1979, 251.
- **Wallenstein, R.,** Pulsed dye lasers, in *Laser Handbook,* Vol. 3, Stitch, M. L., Ed., North-Holland, Amsterdam, 1979, 289.
- **Dienes, A.,** Dye lasers, in *Laser Application to Optics and Spectroscopy,* Jacobs, S., Ed., Addison-Wesley, Reading, 1975.
- **Bradley, D. J.,** Methods of generation, in *Ultrashort Light Pulses,* Shapiro, S. L., Ed., Springer-Verlag, Berlin, 1977, 17, and other chapters.
- **Chan, C. K.,** *Synchronously Pumped Dye Lasers,* Spectra Physics Laser Technical Bulletin, No. 8, 1, 1978.

TABLE OF DYE LASERS

The data in Table 3.1.6 supplements that reported earlier by Drexhage[2] and covers the literature through mid-1980.

Laser dyes are tabulated in the following order:

Polyphenyls
    Terphenyls
    Quaterphenyls
    4-Phenylpyridines
Oxadiazoles
Oxazoles
Pteridines
Styryl type compounds (including stilbenes)
    Stilbenes
    Styrylbenzenes
    Styrylbiphenyls
    Triazinylstilbenes
    Triazolstilbenes
    Other fluorescent brightening agents
Pyrylium salts
    Azapyrylium salts
    Pyrylium salts
    Thiapyrylium salts
Quinoxalinones
Phosphorines
Pyrazolines

Quinolones
    Quinolones (see Table 3.1.1)
    Azaquinolones (see Table 3.1.1)
Coumarins*
    Azacoumarins (see Table 3.1.1)
    4,7-Substituted
    3,7-Substituted
    3,7- and 3,4,7-Substituted
    Rigidized, monocyclic
    Rigidized, bicyclic
Naphthalimides
Xanthenes
Oxazines
Other red dyes
    DCM
    LD 690
    LD 700
    Carbazine 720
Cyanines
    Thiacarbocyanines
    Indocarbocyanines
    Oxacarbocyanines

* See also Tables 3.1.2 — 3.1.4.

Table 3.1.6
ORGANIC COMPOUNDS FOR DYE LASERS

POLYPHENYLS
Terphenyls

Laser Dye, Literature Designation:

Substituents position				Lasing wavelength		Pump source[a] (nm)	Solvent[b]	Concentration (molar)	Molecular weight	Absorption λ-maximum (nm)	Fluorescence λ-maximum (nm)	Intensity
R_1	R_2	R_3	R_4	Max. (nm)	Range (nm)							
2,2'-Dimethyl-p-terphenyl												
H-	CH_3-	CH_3-	-H	332	311-353	KrF(248)[38]	Cyclohexane	9.6×10^{-4}		251c	336c	
p-Terphenyl												
H-	H-	H-	-H	337		KrF(248)[81]	Cyclohexane	1.25×10^{-3}	230	276c	339c	
				338	326-358	KrF(248)[80]						
				338		KrF(248)[81]	Ethanol	1×10^{-3}			335c	
				339	322-366	KrF(248)[38]	Cyclohexane	1.1×10^{-3}				
				340	323-364	KrF(249)[46]	Cyclohexane	5×10^{-3}				
				340		KrF(248)[82,83]	p-Dioxane					
				341		FL[47]	DMF	8×10^{-4}				
				415	335-355	FL[48]	DMF	1×10^{-4}				
p,p'-N,N,N',N'-tetraethyldiaminoterphenyl												
$(H_5C_2)_2N$-	H-	H-	-N(C_2H_5)$_2$		417-427	N_2[34]	Ethanol, saturated					

[a] FL (flashlamp), Ar (argon ion laser), p (pulsed), cw (continuous wave), N_2 (nitrogen laser), Kr (krypton ion laser), SF (single frequency), KrF (krypton fluoride), Nd:YAG (neodymium:YAG), Cu (copper vapor).

[b] DPA (N,N-dipropylacetamide), EG (ethylene glycol), HFIP (hexafluoroisopropanol), MeOH/H_2O (methanol/water), DMSO (dimethylsulfoxide), TEA (triethylamine), COT (cyclooctatetraene), LO (Ammonyx LO), TF (trifluoroethanol), DMF (dimethylformamide), THF (tetrahydrofuran), a (acidic), b (basic), bz (benzene), c (cyclohexane), d (DMF), e (ethanol), f (chloroform), m (methanol), p (p-dioxane), s (DMSO), t (toluene).

Table 3.1.6 (continued)
ORGANIC COMPOUNDS FOR DYE LASERS

Quaterphenyls

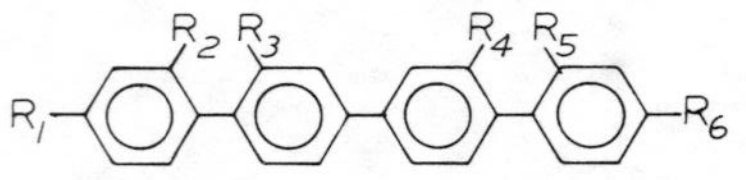

Laser Dye, Literature Designation:

Substituents Position

	R_1	R_2	R_3	R_4	R_5	R_6
3,3',2'',3'''-Tetramethyl-p-quaterphenyl	H-	CH_3,	CH_3,	CH_3,	CH_3,	H-
p-Quaterphenyl	H	H	H	H	H	H
4,4'''-Bis(2-butyloctyloxy)-p-quaterphenyl; BBQ	$H_{13}C_6CHCH_2O-$ C_4H_9	H	H	H	H	$-O-CH_2CHC_6H_{13}$ C_4H_9

Lasing wavelength Max (nm)	Range (nm)	Pump source[a] (nm)	Solvent[b]	Concentration (molar)	Molecular weight	Absorption λ-maximum	Fluorescence λ-maximum	Intensity
362	354-388	Nd:YAG(266)[24]	Cyclohexane	2×10^{-3}				
374	362-390	FL[47] N$_2$[51]	DMF Toluene	7×10^{-3}	306	294[c] 297[f]	365[c]	
380	—	KrF(248)[52]	Cyclohexane		675	306[c] 307[f]	381[c] 391[d]	
382	373-391 381-389 389-395	Nd:YAG(266)[24] FL[22] FL[22]	Cyclohexane Butanol DMF	2.5×10^{-3}				
386	KrF(248)[83]							
386	373-399	N$_2$[50]	Toluene/ethanol	25×10^{-3}				
390	370-410	FL[48]	DMF	8×10^{-3}				
391	380-410	Nd:YAG(355)[114]	EtOH/toluene, 1/1	1×10^{-3}				

4-Phenylpyridine

Laser Dye Literature Designation:

Substituents position			Lasing wavelength		Pump source[a] (nm)	Solvent[a]	Concentration (molar)	Molecular weight	Absorption λ-maximum	Fluorescence λ-maximum	Intensity
R₁	R₂	R₃	Max. (nm)	Range (nm)							
CSA-35											
H-	CH₃-	H-	418		N₂[60]	Ethanol	1×10^{-3}		330	410	Strong
CSA-36											
H-	C₂H₅-	H-	418		N₂[60]	Ethanol	1×10^{-3}		330	408	Strong
CSA-37											
H-	CH₃-	F-	418		N₂[60]	Ethanol	1×10^{-3}		325	406	Strong

N₂[24] Ethanol Saturated

Table 3.1.6 (continued)
ORGANIC COMPOUNDS FOR DYE LASERS

OXADIAZOLES

Laser Dye, Literature Designation:

Substituents position		Lasing wavelength		Pump source[a] (nm)	Solvent[a]	Concentration (molar)	Molecular weight	Absorption λ-maximum (nm)	Fluorescence λ-maximum (nm)	Intensity
R_1	R_2	Max. (nm)	Range (nm)							
2-(2-Fluorophenyl)-5-phenyl-1,3,4-oxadiazole										
		347		KrF(249)[23]	Ethanol	5×10^{-3}				
2-(3-Fluorophenyl)-5-phenyl-1,3,4-oxadiazole										
		347		KrF(249)[23]	Cyclohexane	5×10^{-3}				

Laser Dye, Literature Designation:

| Substituents position | | Lasing wavelength | | Pump source[a] (nm) | Solvent[b] | Concentration (molar) | Molecular weight | Absorption λ-maximum (nm) | Fluorescence λ-maximum (nm) | Intensity |
R₁	R₂	Max. (nm)	Range (nm)							
2-(4-Fluorophenyl)-5-phenyl-1,3,4-oxadiazole		347		KrF(249)[23]	Ethanol	5 × 10⁻³				
Di(2-Fluorophenyl)-1,3,4-oxadiazole		347		KrF(249)[23]	Ethanol	5 × 10⁻³				
Di(3-Fluorophenyl)-1,3,4-oxadiazole		347		KrF(249)[23]	Ethanol	5 × 10⁻³				

Table 3.1.6 (continued)
ORGANIC COMPOUNDS FOR DYE LASERS

Laser Dye, Literature Designation:

Substituents position		Lasing wavelength		Pump source[a] (nm)	Solvent[b]	Concentration (molar)	Molecular weight	Absorption λ-maximum (nm)	Fluorescence λ-maximum (nm)	Intensity
R₁	R₂	Max. (nm)	Range (nm)							

2-(4-Bromophenyl)-5-phenyl-1,3,4-oxadiazole

| | | 353 | | KrF(249)[23] | Ethanol | 5×10^{-3} | | | | |

Di(4-Chlorophenyl)-1,3,4-oxadiazole

| | | 357 | | KrF(249)[23] | Ethanol | 5×10^{-4} | | | | |

2-(3,4-Dichlorophenyl)-5-phenyl-1,3,4-oxadiazole

| | | 357 | | KrF(249)[23] | Ethanol | 5×10^{-3} | | | | |

Laser Dye, Literature Designation:

| Substituents position | | Lasing wavelength | | Pump source[a] (nm) | Solvent[a] | Concentration (molar) | Molecular weight | Absorption λ-maximum (nm) | Fluorescence λ-maximum (nm) | Intensity |
R$_1$	R$_2$	Max. (nm)	Range (nm)							
2-(2,4-Dichlorophenyl)-5-phenyl-1,3,4-oxadiazole		359		KrF(249)[23]	Ethanol	5×10^{-3}				
3-Methyl-alpha-naphthyloxadiazole		372		N$_2$[25]	Toluene	$10^{-2} - 10^{-3}$				
2-Fluoro-alpha-naphthyloxadiazole		373		N$_2$[25]	Toluene	$10^{-2} - 10^{-3}$				

Table 3.1.6 (continued)
ORGANIC COMPOUNDS FOR DYE LASERS

Laser Dye, Literature Designation:

Substituents position		Lasing wavelength		Pump source[a] (nm)	Solvent[b]	Concentration (molar)	Molecular weight	Absorption λ-maximum (nm)	Fluorescence λ-maximum (nm)	Intensity
R$_1$	R$_2$	Max. (nm)	Range (nm)							
3-Fluoro-alpha-naphthyloxadiazole		375		N$_2$[25]	Toluene	10^{-2}–10^{-3}				
4-Bromo-alpha-naphthyloxadiazole		377		N$_2$[25]	Toluene	10^{-2}–10^{-3}				
4-Methoxy-alpha-naphthyloxadiazole		379		N$_2$[25]	Toluene	10^{-2}–10^{-3}				

Laser Dye, Literature Designation:

| Substituents position | | Lasing wavelength | | Pump source[a] (nm) | Solvent[b] | Concentration (molar) | Molecular weight | Absorption λ-maximum (nm) | Fluorescence λ-maximum (nm) | Intensity |
R₁	R₂	Max. (nm)	Range (nm)							
2,5-Bis(4-biphenyl)-1,3,4-oxadiazole		380	372—406	N_2[49]	Toluene	3.5×10^{-3}				
2,5-Dinaphthyl-1,3,4-oxadiazole		390	385—417	N_2[49]	Toluene	4×10^{-3}				
2,5-Bis(4-diethylaminophenyl)-1,3,4-oxadiazole		425		N_2[34]	Methylene chloride	1×10^{-3}				

Table 3.1.6 (continued)
ORGANIC COMPOUNDS FOR DYE LASERS

Oxazoles

Laser Dye, Literature Designation:

Substituents position		Lasing wavelength		Pump source[a] (nm)	Solvent[b]	Concentration (molar)	Molecular weight	Absorption λ-maximum (nm)	Fluorescence λ-maximum (nm)	Intensity
R_1	R_2	Max. (nm)	Range (nm)							
PPO: 2,5-diphenyloxazole										
		365—380 372 381	359—391	N_2[49] KrF(248)[61] FL[47]	Toluene Cyclohexane p-Dioxane	6×10^{-3} 1×10^{-3} 7×10^{-3}	221	303ᶜ	361ᶜ 360ʸ	
alpha-NPO: 2-(1-naphthyl)-5-phenyloxazole										
		400	385—415	FL[47] N_2[51]	Ethanol Cyclohexane	2.5×10^{-4}	271	329ᶜ 332ᶜ	398ᶜ 391ᶜ	
2-Phenyl-5(4-difluoromethylsulfonylphenyl)oxazole										
	$SO_2 \cdot CHF_2$	416	403—437	N_2[27]	Toluene	$\sim 1 \times 10^{-3}$				
4PyPO: 2-(4-pyridyl)-5-phenyloxazole										
		504	395—402 493—512	N_2[37] N_2[37] FL[37]	Ethanol Water + HCl(pH2) Water + HCl(pH2)	7.5×10^{-4}		307,322ᶜ 244,364ᵃ	360ᶜ 470ᵃ	

Laser Dye, Literature Designation:

Substituents position		Lasing wavelength		Pump source[a] (nm)	Solvent[b]	Concentration (molar)	Molecular weight	Absorption λ-maximum (nm)	Fluorescence λ-maximum (nm)	Intensity
R₁	R₂	Max. (nm)	Range (nm)							
4PyPO-TS; 4[2-(5-phenyloxazolyl)]-1-methylpyridinium p-toluenesulfonate										
			495—514	N₂[37]	Water	7.5×10^{-4}		246,270[b]	470[b]	
		506		FL[37]	Water	7.5×10^{-4}				
POPOP; 1,4-bis[2-(5-phenyloxazolyl)]benzene										
H	H	419	415—430	FL[52]	Toluene	3.4×10^{-3}	364	358[c]	410[c]	
		381		N₂[51]	THF					
				e-beam[110]	Vapor (Ar + N₂)					
		393		N₂[111]	Vapor					
		423	414—442	N₂[27]	Toluene	$\sim 10^{-3}$				
		427	411—465	N₂[28]	p-Dioxane	$< g/l$				
			419—424	FL[3]	Ethanol					
Dimethyl POPOP; 1,4-bis[2-(4-methyl-5-phenyloxazolyl)]benzene										
CH₃-	-CH₃	430	418—465	N₂[49]	p-Dioxane	7×10^{-4}				
TP-C-DMP										
CH₃-		~430		FL[37]	p-Dioxane	1×10^{-4}		286,266[d]	428[d]	

Table 3.1.6 (continued)
ORGANIC COMPOUNDS FOR DYE LASERS

Laser Dye, Literature Designation:

Substituents position	Lasing wavelength		Pump source** (nm)	Solvent †	Concentration (molar)	Molecular weight	Absorption λ-maximum	Fluorescence λ-maximum	Intensity
	Max. (nm)	Range (nm)							
Pteridines									
		375—380	N_2,[34]	Methanol	1×10^{-3}				Weak
	460		N_2,[34]	Methanol, alkaline	1×10^{-3}				Weak
	522		N_2,[34]	Water, alkaline	Saturated				Weak
	522		N_3,[34]	Water alkaline	Saturated				Weak

STYRYL TYPE COMPOUNDS (including stilbenes)
Laser Dye, Literature Designation:

| Substituents position | Lasing wavelength | | Pump source** (nm) | Solvent † | Concentration (molar) | Molecular weight | Absorption λ-maximum | Fluorescence λ-maximum | Intensity |
	Max. (nm)	Range (nm)							
Stilbenes									
DPS; Diphenylstilbene	406 409	396—416	N_2[50] FL[47]	p-Dioxane DMF	<1.2 × 10⁻³ 0.66 of saturation	332	34[b]	408[b]	
Stilbene 1	415	392—432	Kr(uv)[26]	EG	1.35 × 10⁻³			347	
Blankophor R	416	399—445	N_2[28]	Methanol	<g/ℓ			340[m]	
Delft Weiss BSW	434	411—467	N_2[28]	Methanol	<g/ℓ			351[m]	

Table 3.1.6 (continued)
ORGANIC COMPOUNDS FOR DYE LASERS

Styrylbenzenes

R$_2$ / R$_1$ — CH=CH — (benzene) — CH=CH — R$_2$ / Y ; (SO$_3$X)$_n$

Laser Dye, Literature Designation:

Substituents position							Lasing wavelength		Pump source[a] (nm)	Solvent[b]	Concentration (molar)	Molecular weight	Absorption λ-maximum	Fluorescence λ-maximum	Intensity
R$_1$	R$_2$	R$_3$	R$_4$	X	n	y	Max. (nm)	Range (nm)							
Bis-dicyanostyryl-2-chlorobenzene															
CN	CN	—	—	H	O	Cl	414	408—423	N$_2$[29]	p-Dioxane,THF	1.8 × 10^{-3}				
1,4-distyrylbenzene															
H	H	—	—	O	H		419	410—439	N$_2$[27]	Toluene	~10^{-3}				
Bis-sulfostyryl-2-chlorobenzene															
H	—	H	—	I	Cl		420	413—431	N$_2$[29]	Methanol	1.8 × 10^{-3}				
Bis-sulfostyrylbenzene															
H	—	H	—	H	H		423	413—431	N$_2$[28]	Methanol	1.8 × 10^{-3}				

Styrylbiphenyls

R$_2$ / R$_1$ — CH=CH — (biphenyl) — CH=CH — R$_3$ / R$_4$; (SO$_3$X)$_n$

Bis-dicyanostyrylbiphenyl											
CN	CN	H	H	—	O	—	413	408—422	N$_2$[29]	p-Dioxane,THF, ethanol	1.8 × 10^{-3}

Substituents position							Lasing wavelength		Pump source[a] (nm)	Solvent[b]	Concentration (molar)	Molecular weight	Absorption λ-maximum	Fluorescence λ-maximum	Intensity
R_1	R_2	R_3	R_4	X	n	y	Max. (nm)	Range (nm)							
Laser Dye, Literature Designation:															
Bis-dichlorostyrylbiphenyl															
CL	CL	H	H	—	O	—	425	416—437	N_2[29]	p-Dioxane,THF	1.8×10^{-3}				
Bis-sulfostyrylbiphenyl; Stilbene 420; Stilbene 3															
H	H	H	H	H	I	—	424	411—436	Nd:YAG (355)[90]	Methanol		562	349[m]		
							425	400—480	Kr(uv)[103]						
							425	408—453	N_2[29]	Methanol	1.8×10^{-3}				
							429	404—460	Nd:YAG (355)[94]	Methanol	1.8×10^{-3}				
							431	415—458	N_2[29]	H_2O + NP—10	1.8×10^{-3}				
							432	406—448	Ar(uv)[32]	EG/methanol, 9/1	2×10^{-3}				
							445	421—468	N_2[29]	H_2O	1.8×10^{-3}				
							449	436—493	Ar(uv)[32]	EG/methanol, 9/1	2×10^{-3}				
								420—470	Ar(uv)[89]	EG	1.5×10^{-3}				

Table 3.1.6 (continued)
ORGANIC COMPOUNDS FOR DYE LASERS

Triazinylstilbenes

Laser Dye, Literature Designation:

| Substituents position | | | | Lasing wavelength | | | | | | | | |
R₁	R₂	X	n	Max. (nm)	Range (nm)	Pump source[a] (nm)	Solvent[b]	Concentration (molar)	Molecular weight	Absorption λ-maximum	Fluorescence λ-maximum	Intensity
4,4'-di(4-anilino-6-methoxytriazinyl)-2,2'-stilbene disulfonic acid												
-OCH₃	$\bigcirc\!\!-\!NH-$	H	—	425	418—443	N₂[31]	Methanol	≈1.5 × 10⁻³				
-NH₂ NH₂ Uvitex CF		Na	—	429	406—465	N₂[28]	Methanol	<g/ℓ			348[ᵐ]	
-OCH₃		Na	—	430	412—462	N₂[28]	Methanol	<g/ℓ			351[ᵐ]	
4,4'-di(4-anilino-6-diethanolaminotriazinyl)-2,2'-stilbene disulfonic acid												
-N(C₂H₄OH)₂	$\bigcirc\!\!-\!NH-$	H	—	430	420—438	N₂[31]	Methanol	≈1.5 × 10⁻³				
4,4'-di(4-sulfoanilino-6-N-methyl-N-ethanolaminotriazinyl)-2,2'-stilbene disulfonic acid												
N(CH₃)C₂H₄OH	$\bigcirc\!\!-\!NH-$ (SO₃H)ₙ	H	1	430	420—445	N₂[31]	Methanol	≈1.5× 10⁻³				

Laser Dye, Literature Designation:

Substituents position				Lasing wavelength		Pump source[a] (nm)	Solvent[b]	Concentration (molar)	Molecular weight	Absorption λ-maximum	Fluorescence λ-maximum	Intensity
R_1	R_3	X	n	Max. (nm)	Range (nm)							
4,4'-di(4-disulfoanilino-6-N-ethyltriazinyl)-2,2'-stilbene disulfonic acid												
$-NHC_2H_5$	H	2		431	419—448	N_2[31]	Methanol	$\sim 1.5 \times 10^{-3}$				
4,4'-sulfoanilino-6-di-iso-propanolaminotriazinyl)-2,2'-stilbene disulfonic acid												
$-N(CH_3\text{-}CH_2\text{-} CH_2OH)_2$	H	1		433	420—447	N_2[31]	Methanol	$\sim 1.5 \times 10^{-3}$				
4,4'-di(4-sulfoanilino-6-diethanolaminotriazinyl)-2,2'-stilbene disulfonic acid-sodium salt												
$-N(C_2H_4OH)_2$	H	1		433	418—461	N_2[31]	Methanol	$\sim 1.6 \times 10^{-3}$				
				447	423—461	N_2[31]	H_2O, 1.5% NP10	$\sim 1.6 \times 10^{-3}$				
4,4'-di(4-disulfoanilino-6-morpholinotriazinyl)-2,2'-stilbene disulfonic acid												
R_2-N (morpholino)	H	2		433	418—448	N_2[31]	Methanol	$\sim 1.5 \times 10^{-3}$				

Table 3.1.6 (continued)
ORGANIC COMPOUNDS FOR DYE LASERS

Laser Dye, Literature Designation:

Substituents position			Lasing wavelength			Solvent[a]	Concentration (molar)	Molecular weight	Absorption λ-maximum	Fluorescence λ-maximum	Intensity
R$_1$	R$_2$	X	n	Max. (nm)	Range (nm)	Pump source[a] (nm)					

4,4′-di(4-disulfoanilino-6-*N*-methyl-*N*-ethanolaminotriazinyl)-2,2′-stilbene disulfonic acid

CH$_3$NC$_2$H$_4$OH	H		2	433	418—449	N$_2$[31]	Methanol	≈1.5 × 10^{-3}			

Leukophor B

-OH		Na	—	436	410—462	N$_2$[28]	Methanol	<g/l		277,351[m]	

Laser Dye, Literature Designation:

Substituents position	Lasing wavelength Max. (nm)	Range (nm)	Pump source** (nm)	Solvent †	Concentration (molar)	Molecular weight	Absorption λ-maximum	Fluorescence λ-maximum	Intensity
Triazolstilbenes 4,4'-Di(4-phenyl-1,2,3-triazol-2-yl)-2,2'-stilbene disulfonic acid-potassium salt	425	414—438	N₂[31]	Methanol	2 × 10⁻³				
	436	417—455	N₂[31]	H₂O, 2.5% NP10	2 × 10⁻³				
Tinopal PCRP	427	414—451	N₂[28]	p-Dioxane	<g/l			371ᵐ	
	430		N₂[34]	Benzene	1 × 10⁻³				
Tinopal RBS	434	414—472	N₂[28]	Methanol/ p-dioxane	<g/l			360ᵐ	
Tinopal GS	440	424—475	N₂[28]	Methanol	<g/l				

Table 3.1.6 (continued)
ORGANIC COMPOUNDS FOR DYE LASERS

Laser Dye, Literature Designation:

Substituents position	Lasing wavelength		Pump source** (nm)	Solvent †	Concentration (molar) 1×10^{-3}	Molecular weight	Absorption λ-maximum	Fluorescence λ-maximum	Intensity
	Max. (nm)	Range (nm)							
	485		N_2[34]	Methanol					
Other fluorescent brightening agents									
Uvitex RS	422	412—432	N_2[28]	Methanol	<g, l			360=	
Uvitex SFC	427	410—459	N_2[28]	Methanol	<g, l			350=	
Heleofor BDC	432	407—460	N_2[28]	Methanol	<g, l			351=	
Uvitex NSI	432	412—464	N_2[28]	Methanol	<g, l			349=	
Leukophor DC	464	447—510	N_2[28]	Methanol	<g, l			374=	

Pyrylium salts
Azapyrylium salts

Laser Dye, Literature Designation:

Substituents position				Lasing wavelength		Pump source** (nm)	Solvent[a]	Concentration (molar)	Molecular weight	Absorption λ-maximum	Fluorescence λ-maximum	Intensity
R₁	R₂	R₃	X	Max. (nm)	Range (nm)							
			BF₄⁻	442		N₂³⁴	Dichloromethane	2×10^{-3}				Weak
			BF₄⁻	448		N₂³⁴	Dichloromethane	1×10^{-3}				Medium
			BF₄⁻	468		N₂³⁴	Dichloromethane	1×10^{-3}				Medium
			BF₄⁻	475		N₂³⁴	Dichloromethane	2×10^{-3}				Weak

Pyrylium salts

Table 3.1.6 (continued)
ORGANIC COMPOUNDS FOR DYE LASERS

Laser Dye, Literature Designation:

R = *p*-Chlorophenyl

Substituents position				Lasing wavelength		Pump source[a] (nm)	Solvent[a]	Concentration (molar)	Molecular weight	Absorption λ-maximum	Fluorescence λ-maximum	Intensity
R$_1$	R$_2$	R$_3$	X	Max. (nm)	Range (nm)							
			BF$_4^-$	478		N$_2^{34}$	Dichloromethane	3×10^{-3}				Strong
	H-		BF$_4^-$	483		N$_2^{34}$	Dichloromethane	6×10^{-3}				Medium
-COOH			BF$_4^-$	488		N$_2^{34}$	Dichloromethane	3×10^{-3}				Weak
			BF$_4^-$	492		N$_2^{34}$	Dichloromethane	1×10^{-3}				Medium
			BF$_4^-$	497		N$_2^{34}$	Dichloromethane	3×10^{-3}				Very strong
			BF$_4^-$	497		N$_2^{34}$	Dichloromethane	3×10^{-3}				Very strong

Laser Dye, Literature Designation:

| Substituents position | | | X | Lasing wavelength | | Pump source*** (nm) | Solvent^a | Concentration (molar) | Molecular weight | Absorption λ-maximum | Fluorescence λ-maximum | Intensity |
R_1	R_2	R_3		Max. (nm)	Range (nm)							
			BF_4^-	501		N_2^{34}	Dichloromethane	Saturated				Strong
			BF_4^-	505		N_2^{34}	Dichloromethane	Saturated				Strong
			BF_4^-	508		N_2^{34}	Dichloromethane	3×10^{-3}				Strong
			BF_4^-	512		N_2^{34}	Dichloromethane	2×10^{-3}				Weak
			BF_4^-	515		N_2^{34}	Dichloromethane	3×10^{-3}				Strong
			BF_4^-	522		N_2^{34}	Dichloromethane	3×10^{-3}				Very strong

R = Phenyl

R = p-Tolyl

Table 3.1.6 (continued)
ORGANIC COMPOUNDS FOR DYE LASERS

Laser Dye, Literature Designation:

Substituents position R₁, R₂, R₃	X	Lasing wavelength Max. (nm)	Range (nm)	Pump source[a] (nm)	Solvent[a]	Concentration (molar)	Molecular weight	Absorption λ-maximum	Fluorescence λ-maximum	Intensity
R₁ = Phenyl R₂ = t-Butyl	BF₄⁻	524		N₂[34]	Dichloromethane	Saturated				Medium
-COOC₂H₅	BF₄⁻	526		N₂[34]	Dichloromethane	3 × 10⁻³				Very strong
-COOC₃H₇	BF₄⁻	527		N₂[34]	Dichloromethane	3 × 10⁻³				Strong
R = Phenyl	BF₄⁻	532		N₂[34]	Dichloromethane	3 × 10⁻³				Medium
R₁ = t-Butyl R₂ = p-Tolyl	BF₄⁻	532		N₂[34]	Dichloromethane	3 × 10⁻³				Strong
R = p-Tolyl	BF₄⁻	534		N₂[34]	Dichloromethane	3 × 10⁻³				Weak

Laser Dye, Literature Designation:

Substituents position				Lasing wavelength Max. (nm)	Range (nm)	Pump source** (nm)	Solvent[a]	Concentration (molar)	Molecular weight	Absorption λ-maximum	Fluorescence λ-maximum	Intensity
R_1	R_2	R_3	X									
(CH₃O–C₆H₄–)	(CH₂O–C₆H₄–CH₃)	(–C₆H₄–OCH₃)	BF_4^-	545		N_2^{34}	Dichloromethane	3×10^{-3}				Strong
(thienyl)	(C₆H₄–CH₃)	(thienyl)	BF_4^-	550		N_2^{34}	Dichloromethane	3×10^{-3}				Medium
(propenyl)	(C₆H₄–CH₃)	(C₆H₅)	ClO_4^-	560		N_2^{34}	Dichloromethane	3×10^{-3}				Medium
(cyclohexenyl)			BF_4^-	623		N_2^{34}	Dichloromethane	1×10^{-3}				Very strong
(CH₃O–C₆H₄–)	–COOH	(–C₆H₄–OCH₃)	BF_4^-	627		N_2^{34}	Dichloromethane	3×10^{-3}				Very strong
(CH₃O–C₆H₄–)	–COOC₂H₅	(–C₆H₄–OCH₃)	BF_4^-	631		N_2^{34}	Dichloromethane	3×10^{-3}				Very strong

R = Phenyl

Table 3.1.6 (continued)
ORGANIC COMPOUNDS FOR DYE LASERS

Laser Dye, Literature Designation: Substituents position				Lasing wavelength		Pump source[a] (nm)	Solvent[a]	Concentration (molar)	Molecular weight	Absorption λ-maximum	Fluorescence λ-maximum	Intensity
R_1	R_2	R_3	X	Max. (nm)	Range (nm)							
			BF_4^-	670		N_2,[34]	Dichloromethane	3×10^{-3}				Strong
Thiapyrylium salts												
			ClO_4^-	513		N_2,[34]	Dichloromethane	3×10^{-3}				Weak
		H	ClO_4^-	601		N_2,[34]	Dichloromethane	3×10^{-3}				Strong
			BF_4^-	612		N_2,[34]	Dichloromethane	2×10^{-3}				Strong
			ClO_4^-	618		N_2,[34]	Dichloromethane	3×10^{-3}				Strong

QUINOXALINONES

Laser Dye Literature Designation:

Substituents position			Lasing wavelength		Pump source[a] (nm)	Solvent[b]	Concentration (molar)	Molecular weight	Absorption λ-maximum	Fluorescence λ-maximum	Intensity
R₁	R₂	R₃	Max. (nm)	Range (nm)							
Quinoxalinone-TNH₂											
H	H	H	489	460—540	N_2[33]	Ethanol			370	482	
Quinoxalinone-MeTNH₂											
H	H	Me	502	474—556	N_2[33]	Ethanol			388	502	
Quinoxalinone-TNMe₂											
Me	H	H	510	478—570	N_2[33]	Ethanol			370	500	
Quinoxalinone-MeTNH₂											
Me	H	Me	511	483—556	N_2[33]	Ethanol			388	505	

Table 3.1.6 (continued)
ORGANIC COMPOUNDS FOR DYE LASERS

Phosphorines

Laser Dye, Literature Designation:

Substituents position					Lasing wavelength		Pump source[**] (nm)	Solvent[a]	Concentration (molar)	Molecular weight	Absorption λ-maximum (nm)	Fluorescence λ-maximum (nm)	Intensity
R_1	R_2	R_3	R_4	R_5	Max. (nm)	Range (nm)							
R = Phenyl					502		N_2[34]	Benzene	3×10^{-3}				Medium
					522		N_2[34]	Benzene	3×10^{-3}				Medium
R = Phenyl	MeO-	MeO-			529		N_2[34]	Benzene	3×10^{-3}				Medium
	MeO-	MeO-			536		N_2[34]	Benzene	3×10^{-3}				Medium

Pyrazolines

Laser Dye, Literature Designation:

Substituents position	Lasing wavelength		Pump source** (nm)	Solvent †	Concentration (molar)	Molecular weight	Absorption λ-maximum	Fluorescence λ-maximum	Intensity
	Max. (nm)	Range (nm)							
2-Pyrazolines	516		N₂³⁴	Methanol	1×10^{-3}				Strong
	544		N₂³⁴	Methanol alkaline	1×10^{-3}				Weak

Table 3.1.6 (continued)
ORGANIC COMPOUNDS FOR DYE LASERS

COUMARINS, 4,7-subsituted

Laser Dye Literature Designation:

7A-4MO-C; 7-amino-4-methoxycoumarin; 7-amino-4-methoxy-2H-1-benzopyran-2-one

CSA-1; Coumarin 120; Coumarin 440; 7-amino-4-methyl-2H-1-benzopyran-2-one

Substituents position			Lasing wavelength								
7	4	Other	Max. (nm)	Range (nm)	Pump source[a] (nm)	Solvent[a]	Concentration (molar)	Molecular weight	Absorption λ-maximum	Fluorescence λ-maximum	Intensity
-NH₂	OCH₃	—	417		FL[11]	Ethanol					
-NH₂	CH₃	—	436		FL[11,60]	Ethanol	5 × 10⁻³	175	354ᶜ	430ᶜ	
			437	420—457	N₂[80]	Ethanol	3 × 10⁻³		351ᵐ		
			438	419—466	N₂[35]	Ethanol	2 × 10⁻⁴		343ⁿ		
			440	419—469	FL[56]	Methanol	2 × 10⁻⁴				
			440	423—462	FL[48]	Ethanol	3 × 10⁻⁴				
			441		Nd:YAG(355)[96]	Ethanol	7 × 10⁻⁴				
			442	425—443	FL[48]	Methanol	3 × 10⁻⁴				
			442	435—455	FL[57]	MeOH/H₂O, 1/1	2.8 × 10⁻⁴				
			442	426—458	Nd:YAG(355)[90]	Methanol					
			442		FL[10]	Ethanol					
			450	420—470	Ar(351/364)[58]	20% aq. DPA, COT	1 × 10⁻³				
			450	427—477	Ar(cw)[59]	EG					
			453		FL[8]	H₂O	2.8 × 10⁻⁴				

Laser Dye Literature Designation:

Substituents position			Lasing wavelength		Pump source* (nm)	Solvent[b]	Concentration (molar)	Molecular weight	Absorption λ-maximum	Fluorescence λ-maximum	Intensity
7	4	Other	Max. (nm)	Range (nm)							
CSA-8; 7-n-propylamino-4-methylcoumarin; 7-n-propylamino-4-methoxy-2H-1-benzopyran-2-one											
-NH(CH$_2$)$_2$CH$_3$	CH$_3$	—	440		N$_2^{60}$	Ethanol					
			485		N$_2^{60}$	Ethanol + HCl					
CSA-9; 7-n-butylamino-4-methylcoumarin; 7-n-butylamino-4-methoxy-2H-1-benzopyran-2-one											
-NH(CH$_2$)$_3$CH$_3$	CH$_3$	—	440		N$_2^{60}$	Ethanol	1 × 10^{-3}				
			485		N$_2^{60}$	Ethanol + HCl	1 × 10^{-3}				
CSA-10; 7-n-pentylamino-4-methylcoumarin; 7-n-pentylamino-4-methoxy-2H-1-benzopyran-2-one											
-NH(CH$_2$)$_4$CH$_3$	CH$_3$	—	440		N$_2^{60}$	Ethanol	1 × 10^{-3}				
			485		N$_2^{60}$	Ethanol + HCl	1 × 10^{-3}				
CSA-6; Coumarin 445; 7-ethylamino-4-methylcoumarin; 7-(ethylamino)-4-methyl-2H-1-benzopyran-2-one											
-NHEt	CH$_3$	—	445	423—445	FL^{104}	MeOH/H$_2$O, 1/1		203	363^m	430^r	
			445		N$_2^{60}$	Ethanol					
			460	435—484	Ar(cw)^{45}	EG					
			485		N$_2^{60}$	Ethanol + HCl	1 × 10^{-3}				

Table 3.1.6 (continued)
ORGANIC COMPOUNDS FOR DYE LASERS

Laser Dye Literature Designation:

Substituents position			Lasing wavelength		Pump source[a] (nm)	Solvent[b]	Concentration (molar)	Molecular weight	Absorption λ-maximum	Fluorescence λ-maximum	Intensity
7	4	Other	Max. (nm)	Range (nm)							

C3H; 7-aminocoumarin; 7-amino-2H-1-benzopyran-2-one

| -NH$_2$ | H | — | 450 | | FL[13] | Ethanol | | | | | |

CSA-5; 7-pyrrolidino-4-methylcoumarin; 7-pyrrolidino-2H-1-benzopyran-2-one

| CH$_3$CH$_2$CH$_2$C H$_2$- | CH$_3$ | — | 450 | | N$_2$[60] | Ethanol | 1×10^{-3} | | | | |

CSA-4; 7-(2-pyridylamino)-4-methylcoumarin; 7-(2-pyridylamino)-2H-1-benzopyran-2-one

| [structure] -NH- [pyridyl N] | CH$_3$ | — | 450 | | N$_2$[60] | Ethanol | 1×10^{-3} | | | | |

Coumarin 450; Coumarin 2; 7-ethylamino-4,6-dimethylcoumarin; 7-(ethylamino)-4,6-dimethyl-2H-1-benzopyran-2-one

-NHEt	CH$_3$	6CH$_3$	446		FL[61]	Ethanol	1×10^{-2}	217	366[c]	435[c]	
			446	428—465	N$_2$[50]	Ethanol	1×10^{-4}				
			449	435—470	FL[48]	Ethanol/ 1.5% LO					
			450	427—488	FL[56]	Methanol	2×10^{-4}				
			450	435—485	Ar or Kr(uv)[62]	EG	3×10^{-3}				
			452	430—492	Ar(cw)[59]	EG					
			454		FL[10]	Methanol					
			458		FL[13]	Methanol					
			460	430—480	Ar(351/364)[58]	20% aq. DPA	2×10^{-3}				
			460	445—482	FL[57]	MeOH/ H$_2$O, 4/6	1.7×10^{-4}				
			454		Nd:YAG(355)[96]	Methanol	7×10^{-4}				
			455	433—474	Nd:YAG(355)[90]	Methanol					

Laser Dye Literature Designation:

Substituents position			Lasing wavelength		Pump source* (nm)	Solvent[b]	Concentration (molar)	Molecular weight	Absorption λ-maximum	Fluorescence λ-maximum	Intensity
7	4	Other	Max. (nm)	Range (nm)							
Coumarin 311; 7-dimethylamino-4-methyl-2H-1-benzopyran-2-one											
-N(CH₃)₂	CH₃	—	453		FL[12]	Ethanol					
			457		FL[18]	Ethanol					
CSA-11											
-N(Et)₂	CH₂COOH	—	456		N₂[60]	Ethanol	1 × 10⁻³				
Coumarin 175											
-NH (CH₂SO₃Na)	CH₃	—	457		FL[8]	Water			353[w]		
Coumarin 460; Coumarin 1; 7-diethylamino-4-methylcoumarin; 7-(diethylamino)-4-methyl-2H-1-benzopyran-2-one											
-N(Et)₂	CH₃	—	457	450—484	FL[48]	Ethanol	1.5 × 10⁻⁴	231	373[v]	445[c]	
			457	440—478	N₂[50]	Ethanol	1 × 10⁻²		375[m]		
			457		Nd:YAG(355)[96]	Ethanol	7 × 10⁻⁴				
			460	442—490	FL[98]	Ethanol	2 × 10⁻⁴				
			460		Nd:YAG(355)[94]						
			460		FL[13]	Ethanol					
			460		FL[10]	Ethanol					
			461	448—489	FL[48]	Methanol	1.5 × 10⁻⁴				
			470	450—495	Ar or Kr(uv)[62]	EG	3 × 10⁻³				
			471	448—505	Kr(uv)[103]						
			472	446—506	Ar(cw)[59]	EG					

Table 3.1.6 (continued)
ORGANIC COMPOUNDS FOR DYE LASERS

Laser Dye Literature Designation:

Substituents position			Lasing wavelength		Pump source[a] (nm)	Solvent[b]	Concentration (molar)	Molecular weight	Absorption λ-maximum	Fluorescence λ-maximum	Intensity
7	4	Other	Max. (nm)	Range (nm)							
C2H; 7-dimethylaminocoumarin; 7-(dimethylamino)-2H-1-benzopyran-2-one											
-N(CH₃)₂	H	—	465		FL[13]	Ethanol					
ClH; LD 466; 7-diethylaminocoumarin; 7-(diethylamino)-2H-1-benzopyran-2-one											
-N(Et)₂	H	—	465	452—480	N₂[38]	EtOH		217	368ᵐ		
			464	446—492	N₂[54]	EtOH/ p-dioxane, 3/5	7.5 × 10⁻³				
			467	459—477	Nd:YAG(355)[yi]	Ethanol	2.5 × 10⁻⁴				
			474	462—490	FL[48]	Methanol + LO	8 × 10⁻⁵				
Coumarin 378											
NH[(CH₃)₃SO₃ Na]	CH₃	—	468		FL[8]	Water			361ʷ		
Coumarin 360											
-NH(C₂H₄OH)	CH₃	—	470		FL[8]	Water			359ʷ		

Laser Dye Literature Designation:

Substituents position			Lasing wavelength		Pump source* (nm)	Solvent[b]	Concentration (molar)	Molecular weight	Absorption λ-maximum	Fluorescence λ-maximum	Intensity
7	4	Other	Max. (nm)	Range (nm)							
Coumarin 380 -NH[(CH₂)₄ SO₃Na]	CH₃	—	470		FL[8]	Water			362~		
Coumarin 379 -N[(CH₃)₃ SO₃Na]	CH₃	—	473		FL[8]	Water			376~		
Coumarin 381 -N[(CH₃)₄ SO₃Na]₂	CH₃	—	478		FL[8]	Water			382~		

CIF; CSA-27; Coumarin 481; Coumarin 35; 7-diethylamino-4-trifluoromethylcoumarin; 7-(diethylamino)-4-(trifluoromethyl)-2H-1-benzopyran-2-one

7	4	Other	Max. (nm)	Range (nm)	Pump source* (nm)	Solvent[b]	Concentration (molar)	Molecular weight	Absorption λ-maximum	Fluorescence λ-maximum	Intensity
-N(Et)₂	CF₃	—	480		N₂[60]	Ethanol	1 × 10⁻³	285	390[p]	465[c]	
			481		FL[7]	Ethanol			402[m]		
			481		FL[10]	Ethanol					
			481		FL[6,7]	p-Dioxane					
			481	460—518	N₂[55]	p-Dioxane	2 × 10⁻²				
			481	475—490	FL[48]	p-Dioxane	1.5 × 10⁻⁴				
			481		KrF(248)[83]						
			483	460—517	N₂[50]	p-Dioxane	1 × 10⁻²				
			483	463—516	N₂[85]	p-Dioxane	1 × 10⁻²				
			489		Nd:YAG(355)[96]	p-Dioxane	7 × 10⁻⁴				
			495	480—522	N₂	p-Dioxane					
			507	481—540	N₂[85]	p-Dioxane/ EtOH, 2/1	1 × 10⁻²				
			513	488—558	N₂	Ethanol	1 × 10⁻²				
			515	492—545	N₂[85]	Ethanol					
			516	490—566	N₂[55]	Ethanol	1.5 × 10⁻²				

Table 3.1.6 (continued)
ORGANIC COMPOUNDS FOR DYE LASERS

Laser Dye Literature Designation:

Substituents position			Lasing wavelength		Pump source[a] (nm)	Solvent[b]	Concentration (molar)	Molecular weight	Absorption λ-maximum	Fluorescence λ-maximum	Intensity
7	4	Other	Max. (nm)	Range (nm)							

C2F; Coumarin 485; Coumarin 152; 7-dimethylamino-4-trifluoromethylcoumarin;
7-(dimethylamino)-4-((trifluoromethyl))-2H-1-benzopyran-2-one

7	4	Other	Max. (nm)	Range (nm)	Pump source[a] (nm)	Solvent[b]	Concentration (molar)	Molecular weight	Absorption λ-maximum	Fluorescence λ-maximum	Intensity
-N(CH₃)₂	CF₃	—	479		FL[7]	p-Dioxane	1×10^{-3}	257	397[c]	510[c]	
			485		N₂[60]	Ethanol			395[m]		
			520	482—517	FL[10]	Ethanol					
			500		N₂[63]	p-Dioxane					
			519		FL[7,64]	Ethanol					
			520	490—562	N₂[50]	Ethanol	1×10^{-2}				
			523		FL[48]	Methanol	1×10^{-4}				
			525	502—573	Nd:YAG(355)[94]	Methanol					

C3F; CSA-29; Coumarin 490; Coumarin 151; 7-amino-4-trifluoromethylcoumarin;
7-amino-4-((trifluoromethyl))-2H-1-benzopyran-2-one

7	4	Other	Max. (nm)	Range (nm)	Pump source[a] (nm)	Solvent[b]	Concentration (molar)	Molecular weight	Absorption λ-maximum	Fluorescence λ-maximum	Intensity
-NH₂	CF₃	—	455		N₂[60]	Ethanol	1×10^{-3}	229	382[c]	480[c]	
			484		FL[7]	Ethanol			376[m]		
			484		FL[7]	Ethanol					
			488		FL[64]	Ethanol					
			489	467—510	FL[20]	Methanol	2×10^{-4}				
			490		FL[10]	Ethanol					
			495	477—515	FL[48]	Methanol	1×10^{-4}				
			496		FL[7]	Methanol + LO EtOH/H₂O					

Coumarin 316

7	4	Other	Max. (nm)	Range (nm)	Pump source[a] (nm)	Solvent[b]	Concentration (molar)	Molecular weight	Absorption λ-maximum	Fluorescence λ-maximum	Intensity
-NH(CH₂SO₂Na)	—	3 ⬡	494		FL[8]	Water			375[n]		

Laser Dye Literature Designation:

Substituents position			Lasing wavelength		Pump source[a] (nm)	Solvent[a]	Concentration (molar)	Molecular weight	Absorption λ-maximum	Fluorescence λ-maximum	Intensity
7	4	Other	Max. (nm)	Range (nm)							
CSA-28: Coumarin 500; 7-ethylamino-4-trifluoromethylcoumarin; 7-(ethylamino)-4-(trifluoromethyl)-2H-1-benzopyran-2-one											
-NHEt	CF₃		500		N₂[60]	Ethanol	1×10^{-2}	257	392[m]	495[c]	
			500	473—547	N₂[50]	Ethanol	1×10^{-3}				
			500		KrF(248)[83]	Ethanol	1×10^{-3}				
			500	494—504	Nd:YAG(355)[91]	Ethanol					
			508	481—573	Nd:YAG(355)[94]	Methanol					
			514	482—552	Nd:YAG(355)[90]	Methanol					
			522	500—548	FL[104]	MeOH/H₂O, 1/1					
Coumarin 503; Coumarin 307; 7-(ethylamino)-6-methyl-4-(trifluoromethyl)-2H-1-benzopyran-2-one											
-NHEt	CF₃	6CH₃	498	477—531	FL[48]	Ethanol	1×10^{-4}	271	395[c]	490[c]	
			502		FL[10]	Ethanol			393[m]		
			504	481—530	FL[48]	Methanol + LO	1×10^{-4}				

Table 3.1.6 (continued)
ORGANIC COMPOUNDS FOR DYE LASERS

COUMARINS, 3,7-substituted

Laser Dye, Literature Designation:

Substituents position		Lasing wavelength		Pump source[a]	Solvent[b]	Concentration (molar)	Molecular weight	Absorption λ-maximum (nm)	Fluorescence λ-maximum (nm)	Intensity
R_1	3	Max. (nm)	Range (nm)							
		414		N_2[34]	Benzene	1×10^{-3}				Medium
		421		N_2[34]	Benzene	1×10^{-3}				Strong
		423		N_2[34]	Benzene	1×10^{-3}				Medium
		426 430		N_2[34] N_2[34]	Benzene Dichloromethane	1×10^{-3} 1×10^{-3}				Strong Weak
		434		N_2[34]	Benzene	1×10^{-3}				Strong

Laser Dye, Literature Designation:

Substituents position		Lasing wavelength		Pump source[a] (nm)	Solvent[b]	Concentration (molar)	Molecular weight	Absorption λ-maximum (nm)	Fluorescence λ-maximum (nm)	Intensity
R₁	R₂	Max. (nm)	Range (nm)							
		434		N₂[34]	Benzene	1 × 10⁻³				Strong
		437		N₂[4]	Benzene	1 × 10⁻³				Strong
		438		N₂[34]	Benzene	1 × 10⁻³				Strong
		439		N₂[34]	Benzene	1 × 10⁻³				Strong
		446		N₂[34]	Methanol	1 × 10⁻³				Weak
		450 458		N₂[34]	Water	1 × 10⁻³				Strong

Table 3.1.6 (continued)
ORGANIC COMPOUNDS FOR DYE LASERS

Laser Dye, Literature Designation:

| Substituents position | | Lasing wavelength | | Pump source[a] (nm) | Solvent[b] | Concentration (molar) | Molecular weight | Absorption λ-maximum (nm) | Fluorescence λ-maximum (nm) | Intensity |
R₁	R₂	Max. (nm)	Range (nm)							
		457		N_2[34]	Water	1×10^{-3}				Strong
		458		N_2[34]	Methanol	1×10^{-3}				Strong
		467		N_2[34]	Water	1×10^{-3}				Medium
		502		N_2[34]	Methanol	1×10^{-3}				Medium

COUMARINS, 3,7- AND 3,4,7-SUBSTITUTED

Laser Dye, Literature Designation:

Substituents position					Lasing wavelength		Pump source** (nm)	Solvent*	Concentration (molar)	Molecular weight	Absorption λ-maximum (nm)	Fluorescence λ-maximum (nm)	Intensity
R₁	R₂	3	4	Other	Max. (nm)	Range (nm)							
$-C_2H_5$	(benzimidazolium, CH_3 / N^+–CH_3, ClO_4^-)	—	—		508		N_2,[34]	Methanol	Saturated				Strong

Coumarin 515; Coumarin 30; 7-(diethylamino)-3-(1-methyl-1H-benzimidazol-2-yl)-2H-1-benzopyran-2-one

R₁	R₂	3	4	Other	Max. (nm)	Range (nm)	Pump source** (nm)	Solvent*	Concentration (molar)	Molecular weight	Absorption λ-maximum (nm)	Fluorescence λ-maximum (nm)	Intensity
$-C_2H_5$	(1-methylbenzimidazol-2-yl, N/N–CH_3)				505	495—515	Ar(cw,458)[58]	20%aq.DPA,COT	1×10^{-3}	347	413*		
					508	477—548	Kr(violet)[103]						
					510	492—550	Ar(458)[22]	EG	1×10^{-3}				
						482—507	N_2He(428)[86]	Ethanol	1×10^{-2}				
					515	495—545	Kr(400—420)[22]	EG					
$-C_2H_5$	(benzothiazol-2-yl, N/S)			$-CH_3$									

CSA-25; 3-(2-benzothiazolyl)-4-(methyl)-7-(ethylamino)-2H-1-benzopyran-2-one

R₁	R₂	3	4	Other	Max. (nm)	Range (nm)	Pump source** (nm)	Solvent*	Concentration (molar)	Molecular weight	Absorption λ-maximum (nm)	Fluorescence λ-maximum (nm)	Intensity
$-H$	$-C_2H_5$	(benzothiazol-2-yl, N/S)	$-CH_3$	6-CH₃	527		N_2,[60]	Ethanol	1×10^{-3}		417	495	Strong

Table 3.1.6 (continued)
ORGANIC COMPOUNDS FOR DYE LASERS

Laser Dye, Literature Designation:

Substituents position				Lasing wavelength		Pump source** (nm)	Solvent[a]	Concentration (molar)	Molecular weight	Absorption λ-maximum (nm)	Fluorescence λ-maximum (nm)	Intensity
R_1	R_2	3	4	Other	Max. (nm)	Range (nm)							

Coumarin 531, CSA-23; 3-[2-(6'-ethoxy)benzothiazolyl]-4-(methyl)-7-(diethylamino)-2H-1- benzopyran-2-one

-C₂H₅	-C₂H₅		-CH₃		531		N_2[60]	Ethanol	1×10^{-3}		420	504	Strong

CSA-24; 3-[2-(6'-ethoxy)benzothiazolyl]-4-(methyl)-7-(ethylamino)-2H-1-benzopyran-2-one

-H	-C₂H₅		-CH₃		532		N_2[60]	Ethanol	1×10^{-3}		415	505	Strong

Coumarin 535; Coumarin 7; 3-(1H-benzimidazol-2-yl)-7-(diethylamino)-2H-1-benzopyran-2-one

-C₂H₅	-C₂H₅				525	500—575	Ar(cw,477)[58]	20%aq.,DPA + LO,COT	4×10^{-4}	333	436[a]	490[c]	
					535	500—565	Ar(477)[62]	EG	5×10^{-3}				
					535	505—565	Kr(400-420)[62]	EG					

Coumarin 540; Coumarin 6; 3-(2-benzothiazolyl)-7-(diethylamino)-2H-1-benzopyran-2-one

-C₂H₅	-C₂H₅				531	507—529	N,He(428)[86]	ethanol	1×10^{-2}	350	458[c]	505[c]	
					538	510—556	FL[48]	methanol	1×10^{-4}				
					538	516—562	FL[96]	methanol	2×10^{-4}				
					540	521—551	Ar(458,514)[62]	EG	1.25×10^{-3}				
						515—585	Ar(488)[58]	20% aq.,DPA + LO,COT					
					540	515—566	Ar(cw)[59]	EG					
					544	526—570	FL[104]	methanol					
					560	510—570	Ar(488)[88]	EG/benzyl alcohol	1.8×10^{-3}				

COUMARINS, rigidized, monocyclic

Laser Dye Literature Designation:

Substituents position			Lasing wavelength		Pump source* (nm)	Solvent [a]	Concentration (molar)	Molecular weight	Absorption λ-maximum	Fluorescence λ-maximum	Intensity
R	3	4	Max. (nm)	Range (nm)							
C8H; 6,7,8,9-tetrahydro-9-methyl-2H-pyrano[3,2-g]quinolin-2-one											
CH₃	H	H	475		FL [13]	Ethanol					
C4H; 6,7,8,9-tetrahydro-2H-pyrano[3,2-g]quinolin-2-one											
H	H	H	477		FL [13]	Ethanol					
Coumarin 386											
(CH₂)₃SO₃Na	H	CH₃	486		FL [a]	Water			388~		
Coumarin 388											
(CH₂)₃SO₃Na	H	CH₃	489		FL [a]	Water			390~		

Table 3.1.6 (continued)
ORGANIC COMPOUNDS FOR DYE LASERS

Laser Dye Literature Designation:

Substituents position			Lasing wavelength		Pump source[a] (nm)	Solvent[b]	Concentration (molar)	Molecular weight	Absorption λ-maximum	Fluorescence λ-maximum	Intensity
R	3	4	Max. (nm)	Range (nm)							
C4F; Coumarin 340; 6,7,8,9-tetrahydro-4-(trifluoromethyl)-2H-pyrano[3,2-g]quinolin-2-one											
H	H	CF$_3$	513		FL[10]	Ethanol			406[10]	500[10]	
C340			522		FL[7]	Ethanol					
C8F; Coumarin 522; 6,7,8,9-tetrahydro-9-methyl-4-(trifluoromethyl)-2H-pyrano[3,2-g]quinoline-2-one											
CH$_3$	H	CF$_3$	520	498—556	FL[48]	Ethanol	2 × 10^{-4}	283	412[7]	515[7]	
			522		FL[7]	Ethanol					
			522	500—572	FL	Ethanol					
			522		FL[7]	Ethanol					
			526	501—568	FL[48]	Methanol + LO	2 × 10^{-4}				
			533	515—570	FL[57]	0.3L DMF/ 1.85L MeOH/ 1.85L H$_2$O	2.1 × 10^{-4}				
Coumarin 355; 6,7,8,9-tetrahydro-9-ethyl-4-(trifluoromethyl)-2H-pyrano[3,2-g]quinolin-2-one											
C$_2$H$_5$	H	CF$_3$	522		FL[10]	Ethanol			412[13]	515[13]	

COUMARINS, rigidized, bicyclic

Laser Dye, Literature Designation:

Substituents position		Lasing wavelength		Pump source[a] (nm)	Solvent[b]	Concentration (molar)	Molecular weight	Absorption λ-maximum (nm)	Fluorescence λ-maximum (nm)	Intensity
4	3	Max. (nm)	Range (nm)							
Coumarin 478; Coumarin 106; 2,3,6,7,10,11-hexahydro-1H,5H-cyclopental[3,4][1]benzopyrano[6,7,8-ij]quinolizin-12(9H)-one										
-CH₂-CH₂-CH₂-		478		FL[10]	Ethanol			386	465	
Coumarin 480; Coumarin 102; 2,3,6,7-tetrahydro-9-methyl-1H,5H,11H-[1]benzopyran[6,7,8-ij]quinolizin-11-one										
CH₃	H	480		FL[7]	Ethanol			390	465	
		480		FL[10]	Ethanol					
C6H, LD 490; 2,3,6,7-tetrahydro-1H,5H,11H-[1]benzopyrano[6,7,8-ij]quinolizin-11-one										
H	H	490		FL[13]				396	—	
Coumarin 504; Coumarin 314; 2,3,6,7-tetrahydro-11-oxo-1H,5H,11H-[1]benzopyrano[6,7,8-ij]quinolizine-10-carboxylic acid, ethyl ester										
H	COOC₂H₅	499	484—537	FL[48]	Methanol + 1 × 10⁻⁴ + 1% LO LO		313	437ᵐ	480ᵛ	
		504		FL[10]	Ethanol			436ᶜ		
		505	-495—517	FL[104]	MeOH/H₂O, 1/1					
		520	-506-544-	FL[57]	MeOH/H₂O, 1/1	1.6 × 10⁻⁴				

Table 3.1.6 (continued)
ORGANIC COMPOUNDS FOR DYE LASERS

Laser Dye, Literature Designation:

Substituents position		Lasing wavelength		Pump source* (nm)	Solvent[b]	Concentration (molar)	Molecular weight	Absorption λ-maximum (nm)	Fluorescence λ-maximum (nm)	Intensity
4	3	Max. (nm)	Range (nm)							

Coumarin 217

CH_3COOH	H	514		FL[a]	Water			402~		

Coumarin 519: Coumarin 343; 2,3,6,7-tetrahydro-11-oxo-1H,5H,11H-[1]benzopyrano[6,7,8-ij]quinolizine-10-carboxylic acid

H	COOH	501	490—513	FL[48]	Methanol	2×10^{-4}	285	446~	490~	
		518		FL[a]	Water			436~		
		519		FL[16]	Ethanol			425~		

Coumarin 521; Coumarin 334; 10-acetyl-2,3,6,7-tetrahydro-1H,5H,11H-[1]benzopyrano[6,7,8-ij]quinolizin-11-one

H	$COCH_3$	521		FL[10]	Ethanol		283	452~	495~	

Coumarin 523; Coumarin 337; 2,3,6,7-tetrahydro-11-oxo-1H,5H,11H-[1]benzopyrano[6,7,8-ij]quinolizine-10-carbonitrile

H	CN	522		FL[10]	Ethanol		266	443~	443~	485~

C6F: Coumarin 540A; Coumarin 153; 2,3,6,7-tetrahydro-9-(trifluoromethyl)-1H,5H,11H-[1]benzopyrano[6,7,8-ij]quinolizine-11-one

CF_3	H	504	485—530	N_2	p-Dioxane		309	423~	530~	
		507		N_2[85]	p-Dioxane					
		536	515—583	N_2[50]	Ethanol	1×10^{-2}				
		536	517—576	FL[48]	Ethanol	1×10^{-4}				
		538		FL[7]	Ethanol					
		540		FL[10]	Ethanol					
		540	516—590	Nd:YAG(355)[90]	Methanol					
		541	520—586	FL[48]	Methanol	1×10^{-4}				
		543		FL[64]	Ethanol					
		562	520—575-	Ar(476)[88]	EG/benzyl alc., 11/1	4×10^{-3}				

NAPHTHALIMIDES

Laser Dye, Literature Designation:

Fluorol 555; Fluorol 7GA; 6-n-butylamino-2,3-dihydro-2-(4-n-butyl)-1,3-dioxo-1H-benz[de]isoquinoline-5-sulfonic acid, monosodium salt

Substituents position	Lasing wavelength		Pump source** (nm)	Solvent †	Concentration (molar)	Molecular weight	Absorption λ-maximum	Fluorescence λ-maximum	Intensity
	Max. (nm)	Range (nm)							
		550—580	FL[42]	50ccMeOH/ 120cc LO/3.9L H₂O/COT pH = 6(HCl)	F555 (1.8 × 10⁻⁴), C540(1 × 10⁻⁴)	324	442[m]		
	552		FL[66]		2.3 × 10⁻⁴				
	555	530—608	FL[67]	Alcohol or water					
	570	540—590	FL[48]	Ethanol	1 × 10⁻⁴				
	574	542—592	FL[48]	Methanol	1 × 10⁻⁴				
		535—590	FL(m-1)[42]	Methanol	1 × 10⁻⁴				

Brilliant Sulfaflavine; 6-amino-2,3-dihydro-2-(4-methylphenyl)-1,3-dioxo-1H-benz[de]isoquinoline-5-sulfonic acid, monosodium salt

| | 562 | 522—618 | FL[104] | Methanol/ COT | | 404 | 443[r] | | |
| | | 508—573 | FL[71,108] | | | | | | |

Table 3.1.6 (continued)
ORGANIC COMPOUNDS FOR DYE LASERS

XANTHENES

Laser Dye, Literature Designation:

Lasing wavelength		Substituents position	Pump source** (nm)	Solvent †	Concentration (molar)	Molecular weight	Absorption λ-maximum	Fluorescence λ-maximum	Intensity
Max. (nm)	Range (nm)								

Disodium Fluorescein: 3',6'-dihydroxy-spiro[isobenzofuran-1(3H),9'-[9H]xanthen]-3-one disodium salt

Max. (nm)	Range (nm)		Pump source	Solvent	Conc.	MW	Abs.	Fluor.	Int.
552	537—580		Ar[59]	EG,COT	3×10⁻³	412	501[r]	531[r]	
552	538—573		Ar(458,514)[62]	EG	2.7×10⁻³				

Rhodamine 560 Chloride; Rhodamine 110; 2-(6-amino-3-imino-3H-xanthen-9-yl)benzoic Acid, monohydrochloride

552	536—602		Ar(cw)[59]	EG		367	506[m]	532[r]	
560			FL[5]	Ethanol(b)					
563	541—583		FL[104]	Methanol	5×10⁻⁵				
565	544—589		FL[48]	Methanol	5×10⁻⁵				
567	546—587		FL[48]	Ethanol					
569	533—600		Kr(Blue/Green)[103]	EG	1.25×10⁻³				
570	540—600		Ar(458,514)[62]	Ethanol(a)					
570			FL[5]						

Rhodamine 575

575	563—602		FL[5]	Ethanol(b)			518[m]		
577			FL[48]	Ethanol	1×10⁻⁴		528(a)[r]		
585	566—610		FL[5]	Ethanol(a)		414	518(b)[r]		
590			FL[57]	MeOH/H₂O(b), 1/1	1.1×10⁻⁴				

Rhodamine 590 as Chloride, Perchlorate or Tetrafluoroborate; Rhodamine 6G; 2-[6-(ethylamino)-3-(ethylimino)-2,7-dimethyl-3H-xanthen-9-yl]benzoic acid, ethyl ester, as, monohydrochloride, monohydroperchlorate or monohydrotetrafluoroborate

550	548—580		Nd:YAG(532)[91]	Methanol	3×10⁻⁴	479(Cl) 530(BF₄) 543(ClO₄)	530[r]	560[r]	
560			Nd:YAG(532)[94]	Methanol					
280*			Nd:YAG(532)[94]	Methanol					
562	546—592		Nd:YAG(532)[92]	Methanol					
563	550—590		Nd:YAG(532)[95]	Methanol					
564			Nd:YAG(532)[90]	Methanol					
282*	552—607		Nd:YAG(532)[90]	Methanol					
572	-564—600		Cu(511,578)[68]	Ethanol	1×10⁻³				
576	555—618		N₂[107]	Ethanol					
578	565—612		FL[48]	Methanol	5×10⁻⁵				
579	568—605		N₂[50]	Ethanol	5×10⁻³				
580			Kr:F(248)[51]	Ethanol	1×10⁻³				
584	570—618		FL[48]	Ethanol	5×10⁻⁵				
586	563—625		FL[56]	Methanol	5×10⁻⁵				
590	570—650		Ar(458,514)[62]	EG	2×10⁻³				
590			Kr:F(248)[93]	p-Dioxane					
590			FL[98]	Methanol	8×10⁻⁵				
596	577—614		FL[104]	MeOH/H₂O, 1/3					
598	577—625		FL[57]	MeOH/H₂O, 1/1	1.3×10⁻⁴				
600	567—657		Ar(cw)[59]	EG					
600			FL[98]	4% LO/H₂O	1.2×10⁻⁴				
602	560—654		Kr(Blue/Green)[103]						
610	585—633		FL[57]	4·LO/H₂O	1.3×10⁻⁴				

* Frequency doubled.

Laser Dye, Literature Designation:

Substituents position	Lasing wavelength Max. (nm)	Range (nm)	Pump source** (nm)	Solvent †	Concentration (molar)	Molecular weight	Absorption λ-maximum	Fluorescence λ-maximum	Intensity
Rhodamine 610 as Chloride or Perchlorate: Rhodamine B; N-[9-(2-carboxyphenyl)-6-(diethylamino)-3H-xanthen-3-ylidine] N-ethyl-ethanaminium as chloride or perchlorate									
	579	570—596	nD:YAG(532)[92]			479(Cl)	552[c]	588[c]	
	586	570—606	Nd:YAG(532)[94]			543(ClO$_4$)	554(a)[r]	580(a)[r]	
	293*		Nd:YAG(532)[94]				544[m]		
	587	579—601	Nd:YAG(532)[91]	Methanol	3 × 10^{-4}				
	590	578—610	Nd:YAG(532)[95]	TFE(b)	1.3 × 10^{-3}				
	591	-582—618-	Cu(511,578)[88]	Methanol					
	592	-578—629-	Nd:YAG(532)[90]						
	295*		Nd:YAG(532)[90]						
	609	594—643	N$_2$[50]	Ethanol	5 × 10^{-3}				
	613	596—645	FL[48]	Methanol	8 × 10^{-5}				
	617	598—647	FL[69]	Ethanol					
	620	596—647	FL[48]	Ethanol	8 × 10^{-5}				
	630	601—675	Ar(458—514)[62]	EG	2 × 10^{-3}				
	637	608—682	Ar(cw)[59]	EG					
Kiton Red 620; Sulforhodamine B; N-[6-(diethylamino)-9-(2,4-disulfophenyl)-3H-xanthen-3-ylidene]- N-ethyl-ethanaminium hydroxide, inner salt									
	583	570—604	Nd:YAG(532)[94]	Methanol	1.7 × 10^{-3}	581	554[c]	575[c]	
	291*		Nd:YAG(532)[94]	Methanol					
	617	-595—639-	Cu(511,578)[88]	TFE	1.7 × 10^{-3}				
	620	580—630	FL[69]	Ethanol					
	621	608—634	FL[104]	Methanol + COT	2 × 10^{-4} (+ R590)				
	623	598—649	FL[48]	Ethanol + COT	3 × 10^{-5}				
	627	595—629	FL[48]	Methanol + COT	3 × 10^{-5}				
	628	603—647	N$_2$[107]	TFE					
	631	600—660	FL[56,69]	Methanol	1 × 10^{-4}				
	636	603—670	FL[51]	EG					
	637		FL[107]	DMSO					
	638	610—670	Ar(cw)[59]	EG					
	642	622—665	FL[57]	4% LO/H$_2$O	1.1 × 10^{-4}				

* Frequency doubled.

Table 3.1.6 (continued)
ORGANIC COMPOUNDS FOR DYE LASERS

Laser Dye, Literature Designation:

Substituents position	Lasing wavelength		Pump source** (nm)	Solvent †	Concentration (molar)	Molecular weight	Absorption λ-maximum	Fluorescence λ-maximum	Intensity
	Max. (nm)	Range (nm)							
Rhodamine 640 Perchlorate: Rhodamine 101; 9-(2-carboxyphenyl)-2,3,6,7,12,13,16,17-octahydro-1H,5H,11H,15H-xantheno[2,3,4-ij;5,6,7-i'j']diquinolizin-4-ium, perchlorate									
	602	589—623	Nd:YAG(532)[92]	Methanol		591	575(a)[m]		
	602	592—624	Nd:YAG(532)[94]	Methanol					
	301*		Nd:YAG(532)[94]						
	611		Nd:YAG(532)[91]	Methanol	5 × 10⁻⁴				
	612	598—640	Nd:YAG(532)[95]	Methanol					
	613	-602—657-	Nd:YAG(532)[90]	Methanol					
	306*		Nd:YAG(532)[90]						
	630		FL[s]	Ethanol(b)					
	640		FL[s]	Ethanol (a)	5 × 10⁻³				
	640	620—680	N₂[70]	Ethanol					
	642	627—657	FL[104]	Methanol	1 × 10⁻⁴				
	643	623—657	FL[48]	Ethanol	5.7 × 10⁻³				
	644	620—673	N₂[87]	Ethanol	1.5 × 10⁻³ (R640)				
	645	620—690	Ar(458—514)[82]	EG	and 1.5 × 10⁻³ (R590)				
	648	608—710	Kr(568)[103]	Methanol	1.2 × 10⁻⁴				
	650		FL[98]	MeOH/H₂O,	1.1 × 10⁻⁴				
	652	620—687	FL[57]	MeOH/H₂O, 6/4					
	671	634—704	N₂[107]	DMSO + HCl					
Sulforhodamine 640; Sulforhodamine 101; 2',3',6',7',12,13,16,17-octahydro-spiro[3H-2,1-benzoxathiole-3,9'-[1H,5H,9H,11H,15H]xanthenol[2,3,4-ij;5,6,7-i'j']diquinolizine]-6-sulfonic acid, 1,1-dioxide, sodium salt									
	662	648—682	FL[104]	MeOH/H₂O, 1/1		606	576[c]	602[c]	
	656		(cw)[109]	MeOH/H₂O, 1/1	8.3 × 10⁻⁴				
	668	590—640	Nd:YAG(532)[111]	Ethanol (a)	1.5 × 10⁻⁴				
	668	646—680	Ar[41]	EG	3 × 10⁻³ (SR640) and 1 × 10⁻³ (R590)				

* Frequency doubled.

OXAZINES

Laser Dye, Literature Designation:

Cresyl Violet 670 Perchlorate: 5-imino-5H-benzo[a]phenoxazin-9-amine,monoperchlorate

Substituents position	Lasing wavelength Max. (nm)	Range (nm)	Pump source** (nm)	Solvent †	Concentration (molar)	Molecular weight	Absorption λ-maximum	Fluorescence λ-maximum	Intensity
	633	615—655	Nd:YAG(32)[92]			362	601[c]	630[c]	
	637	620—660	Nd:YAG(532)[94]	Methanol			594[m]		
	317*		Nd:YAG(532)[94]						
	639	620—670	Nd:YAG(532)[95]						
		645—705	FL[71]	Methanol					
	640	620—670	Nd:YAG(532)[72]	MeOH/H₂O					
	320*	310—335	Nd:YAG(532)[72]	MeOH/H₂O					
	315*	302—326	Nd:YAG(532)[72]	HFIP/H₂O					
	646	625—660	Nd:YAG(532)[91]	Methanol	2×10^{-4}				
	647		Nd:YAG(532)[73]		4×10^{-4}				
	655	646—697	FL[48]	Methanol	5×10^{-5}				
	659	650—695	FL[48]	Ethanol	5×10^{-5}				
	660	641—687	N₂[50]	Ethanol	$2.5 \times 10^{-3}/3.3 \times 10^{-3}$				
					R590 + CV670				
	664	631—705	FL[84a]	Methanol					
	673	650—696	Ar(cw)[55a]	EG					
	695	675—708	Ar(458—514)[62]	EG	2.4×10^{-3}/R590				

Nile Blue A Perchlorate; Nile Blue 690 Perchlorate; 5-imino-9-(diethylamino)-benzo[a]phenoxazin-7-amine, monoperchlorate

Substituents position	Lasing wavelength Max. (nm)	Range (nm)	Pump source** (nm)	Solvent †	Concentration (molar)	Molecular weight	Absorption λ-maximum	Fluorescence λ-maximum	Intensity
	681	662—710	Nd:YAG(532)[95]	Methanol	2×10^{-4}	418	624[c]		
	683		Nd:YAG → CV670 (647)[73]				628[c]		
			laser[5]	Ethanol					
	690	683—710	N₂[50]	Ethanol	$3.8 \times 10^{-3}/8 \times 10^{-4}$				
	696				R610/NB690				
	705	689—750	FL[48]	Methanol					
	717		FL[84a]	Methanol					
	722		FL[69b]	Methanol					
	730	692—782	Kr(cw)[59]	EG	1×10^{-3}				
	750	710—790	R590(Ar)[62]	EG					

* Frequency doubled.

Table 3.1.6 (continued)
ORGANIC COMPOUNDS FOR DYE LASERS

Laser Dye, Literature Designation:

Substituents position	Lasing wavelength Max. (nm)	Range (nm)	Pump source** (nm)	Solvent †	Concentration (molar)	Molecular weight	Absorption λ-maximum	Fluorescence λ-maximum	Intensity
Oxazine 720 Perchlorate; Oxazine 170 Perchlorate; 5-(ethylimino)-methyl-5H-benzo[a]phenoxazin-9-amino.monoperchlorate									
	668	649—700	Nd:YAG(532)[94]	Methanol		432	627[r]	650[r]	
	671	613—708	Nd:YAG(532)[90]	Methanol			620[m]		
	672		Nd:YAG(532)[73]	Ethanol	4×10^{-4} + R610 or SR640				
	692	676—698	FL[48]	Methanol	5×10^{-3}				
	698	682—720	FL[104]	Methanol					
	699	675—711	FL[48]	Ethanol	5×10^{-3}				
	705	675—730	FL[57]	MeOH/H₂O (a)	7.5×10^{-3} + R590[72]				
	710	690—740	FL[64]	Methanol					
Oxazine 725 Perchlorate; Oxazine 1 Perchlorate; 3,7-bis(diethylamino)phenoxazin-5-ium perchlorate									
	681		FL → R610(622)[100]	CH₂Cl₂	4×10^{-5}	424	645[c]	680[c]	
	690		Nd:YAG(532)[73]	CH₂Cl₂	4×10^{-4}				
	695		FL → R610(622)[100]	DMSO	1×10^{-4}				
	715		FL[5]	Ethanol					
	723	688—800	Kr(Red)[103]	DMSO/EG, 1/3 + COT	1×10^{-3}				
		687—826	Kr(647)[105]	Methanol					
	724	695—761	Nd:YAG(532)[90]	Ethanol	$5 \times 10^{-3}/5 \times 10^{-3}$ R610/O × 725				
	725	705—750	N₂[50]	MeOH/R590					
	725	705—745	FL[56]	CH₂Cl₂,	3.3×10^{-3}				
	740	720—758	FL[48]	DMSO & EG or G	1.1×10^{-3}				
	745	645—810	Kr(647,676)[76b]						
	750	695—801	Kr(647,676)[62]	EG/DMSO, 84/16	6×10^{-4}				
Oxazine 750 Perchlorate and Oxazine 750 Chloride*									
	722	704—786	Nd:YAG(532)[90]	Methanol	2.5×10^{-3}	470	662[m]	705[m]	
	760	760—775	Nd:YAG → R610(585)[102]	DMSO					
	745	700-785	FL → R640(660)[48]	Ethanol	3×10^{-4}				
	760	715—785	R640	EG/DMSO, 84/16					
	770	750—835	Kr(647)[84]	EG/DMSO, 4/1	8.5×10^{-4}				
	775	747—885	Kr(647,676)[62]	EG/DMSO, 84/16	6×10^{-4}				
	776	747—801	Kr(647,676)[39]	EG/DMSO, 2/1	1.2×10^{-3}				
	*780	749—825	Kr(647,676)[39]	EG	1.4×10^{-3}	406			

OTHER RED DYES

Laser Dye, Literature Designation:

Substituents position	Lasing wavelength		Pump source** (nm)	Solvent †	Concentration (molar)	Molecular weight	Absorption λ-maximum	Fluorescence λ-maximum	Intensity
	Max. (nm)	Range (nm)							
DCM: 4-(dicyanomethylene)-2-methyl-6-(p-dimethylaminostyryl)-4H-pyran; [2-[2-[4-(dimethylaminophenyl)ethenyl]-6-methyl-4H-pyran-4-ylidene]propanedinitrile									
	640	605—680	Ar(514,m-1)[a]	BzOH/EG	~3 × 10⁻³	303	482[c]		
	640	610—680	Nd:YAG(532)[94]	Methanol	5.6 × 10⁻⁴/oscillator				
	649	615—688	Cu(511,578)[98]	DMSO	2.6 × 10⁻³				
	649	601—716	FL[107]	DMF					
	654		N₂[107]	DMSO					
	655		FL[107]	DMSO					
LD 690 Perchlorate									
	660	655—705	Nd:YAG(532)[93]	Methanol	2.8 × 10⁻⁴	396	616[c]		
	668		N₂[87]	Ethanol	1.8 × 10⁻³/2.8 × 10⁻³				
	670	660—716	N₂[87]	DMSO/	2.5 × 10⁻³/2.8 × 10⁻³				
				EtOH,2/1	(R610)				
		696—780	Kr(cw)[44]	EG	5 × 10⁻³				
LD 700 Perchlorate									
	690		Nd:YAG → R610(585)[102]	Alcohol	9.3 × 10⁻³ + 5 × 10⁻³ (R640)	538	647[c]		
	706	692—752	N₂[87]	EtOH	1.5 × 10⁻³ + 4.4 ×				
	720	698—758	N₂[87]	EtOH	10⁻³(R640)				
	737	700—810	Kr(SF)[45]	EG					
	740	700—820	Kr(647,676)[45]	EG					

Table 3.1.6 (continued)
ORGANIC COMPOUNDS FOR DYE LASERS

Laser Dye, Literature Designation:

Substituents position	Lasing wavelength		Pump source** (nm)	Solvent †	Concentration (molar)	Molecular weight	Absorption λ-maximum	Fluorescence λ-maximum	Intensity
	Max. (nm)	Range (nm)							
Carbazine 122; Carbazine 720; 7-hydroxy-2',3',5',6'-tetramethyl-spiro [acridine-9(2H),1'-[2,5]cyclohexadiene]-2,4'-dione, ion									
	690	684—736	Nd:YAG(532)[74]	Ethanol (b)		345	651(b)[γ]		
	699	680—740	Nd:YAG(532)[95]	Ethanol (b)					
	700	680—738	Nd:YAG(dbld)[73]	Ethanol (b)	4×10^{-4}				
	700		FL[56]	Methanol, TEA					
	718		Nd:YAG(dbld)[73]	DMSO (b)	4×10^{-4}				
	720		FL[63,2]	Ethanol (b)					
	720	680—760	Nd:YAG(dbld)[72]	Aq. Ammonyx TEA					
	368*	340—380*	Nd:YAG(dbld)[72]	DMSO (b)					
	740		FL[63,2]						
	734	683—776	Kr(SF)[60]	EG(b,ethanol)					
	747	687—811	Kr(647,676)[40]	EG(b,ethanol)	$\sim 2 \times 20^{-3}$				

* Frequency doubled.

Thiacarbocyanines

Dye-counter ion	Max	Range	Pump source (nm)	Solvent	Concentration (molar)	MW	abs	Comments	n	Ref.
DTDC-I		705—735	Ruby(694.3)	DMSO	1×10^{-4}	518	653*	Very stable to photochemical degradation, lases equally well in H₂O-surfactant, alcohol.	2	74
-I	744	(bb)	N₂(337)	DMSO	1×10^{-1}			>50 hr of use to half power point, 100 kw/10 nsec pump		124
-P	746	(bb)	N₂(337)	DMSO	1×10^{-1}	491		Good lifetime		124
-I		720—775	Kr	DMSO						75B
10-Cl-5,6-Dbz-DTDC-I -I	711		Ruby(694.3)	Acetone	1×10^{-4}				2	132
Dmo-DTDC-I -I	714		Ruby(694.3)	Acetone	1×10^{-4}				2	132
10-Cl-4,5 Dbz-DTDC-I -I	785	710—755	Ruby(694.3)	DMSO	1.4×10^{-4}	578	660ᵐ	Good efficiency, poor stability	2	74
-I		(bb)	N₂(337)	DMSO	2×10^{-1}	654	686ᵐ	Good stability	2	124
DTTC-I	774	813—859	Ruby(694.3)	Acetone	1×10^{-4}	544	763*		3	132
-I	828		Nd:YAG(532)	DMSO	2×10^{-2}					90
-I	829		Ruby(694.3)	Acetone	1×10^{-4}					132
-I	834		Nd:YAG→C720(700)	Ethanol(b)	2×10^{-4}					101
-I	863	816—855	Ruby(694.3)	DMSO	0.9×10^{-4}	494	760ᵐ	Very unstable to 694 nm light	3	74
-Br		810—830	Ruby(694.3)	DMSO	6×10^{-5}			Poor stability		74
-I		790—871	Ruby(694.3)	DMSO				DMSO best solvent		115
-I		840—870	Kr	DMSO				Poor efficiency		756
-I	876	(bb)	N₂(337)	DMSO	2×10^{-1}			Good lifetime		124
-I	889	820—900	FL	DMSO	2×10^{-4}					78,79
-I		815—870	FL→KR620(Red)	DMSO						97
-I			FL→R640(660)	EG	1.5×10^{-4}					48
4,5-Dbz-DTTC-I -I		834—900	Ruby(694.3)	DMSO	6×10^{-5}	645	797ᵐ	DMSO most efficient solvent	3	115
-I	928	(bb)	N₂(337)	DMSO	2×10^{-1}			Good stability		124
-I	860		Ruby(694.3)	Acetone	1×10^{-4}					132
IR-123-I (DTTC-derivative) -I	830		FL	DMSO	1×10^{-4}	638	745*		3	128
5,6-Temo-DTTC-I -I		795—815	Ruby(694.3)	DMSO	1.6×10^{-4}	664	793ᵐ	Bleached rapidly	3	74
-I		855—885	Ruby(694.3)	DMSO	3×10^{-4}			Poor stability		74
IR-109-I (DTTC-derivative) -I	853		Ruby(694.3)	Acetone	1×10^{-4}	635			3	132
-I	875		FL	DMSO	1×10^{-7}					128
5-Dmo-DTTC-I -I	883	820—875	Ruby(694.3)	DMSO	1.7×10^{-4}	608		Lases efficiently in EG	3	74
IR-139-P (DTTC-derivative)			FL	DMSO	1×10^{-7}	686			3	128

Table 3.1.6 (continued)
ORGANIC COMPOUNDS FOR DYE LASERS

Dye-counter ion	Max	Range	Pump source (nm)	Solvent[a]	Concentration (molar)	MW	abs	Comments	n	Ref.
IR-116-1 (DTTC-derivative) -P	885		FL	DMSO	1×10^{-4}	644			3	128
IR-134-P (DTTC-derivative)	888		FL	DMSO	1×10^{-4}	798			3	128
IR-141-1 (DTTC-derivative)	946		FL	DMSO	1×10^{-4}	748			3	128
IR-140-P (DTTC-derivative)	893	882—913	Nd:YAG	DMSO		779	823[c]		3	90
-P	898		Nd:YAG→C720-(700)	DMSO	2×10^{-4}		776[c]	Excellent efficiency > 10%		101
-P	950		FL	DMSO	1×10^{-4}					128
	950	862—1013	Kr(752,799)	EG/DMSO, 1/1	0.5g/1					125
-P	927	850—930	Ruby(694.3)	DMSO	0.8g/1			Good stability		74
		858—1030	Kr(752)	EG/DMSO, 3/1				Excellent efficiency, good stability		129
		887—986	Kr(752,799)	EG/DMSO, 3/1	5.1×10^{-4}			30 hr without degradation of output power, COT did not increase output, superior to HDITC		105
-P	910	$\Delta\lambda = 20$(bb)	Nd:YAG(532)	DMSO	5×10^{-4}			Good conversion efficiency 9%		121
-P	908	$\Delta\lambda = 15$(bb)	Nd:YAG(532)	DMSO	4×10^{-4}	835,	550[c]	Peak conversion efficiency 5%, power reduced to 50% after only 10^5 shots per ½ 1 solution		116
-P	961	875—1015	Kr(IR)	Ethanol	1×10^{-4}					103
-P		805—872	FL→Ox720 (690—709)							48
-P		875—916	FL→Ox720 (709)	DMSO	2×10^{-4}					48
IR-137-P (DTTC-derivative)	927	855—1032	Kr(752)	EG/DMSO, 3/1	0.8g/1	672		Excellent efficiency, good stability	3	129
-P	950		FL	DMSO	1×10^{-4}	954	830[c]	Poor efficiency, good stability	4	128
IR-132-P (DTTC-type)	950	875—920	Ruby(694.3)	DMSO	1.2×10^{-4}				3	74
-P	972	863—1048	FL	DMSO	1×10^{-4}			High threshold, poor stability		128
			Kr(752)	EG/DMSO, 3/1	0.8g/1					129
-P		916—984	Kr(752,799)	EG/DMSO, 3/1	0.4g/1			Less efficient than IR-140		105
IR-143-P (DTTC-derivative) -P	910	$\Delta\lambda = 15$(bb)	Nd:YAG(532)	DMSO	5×10^{-4}	810		Low conversion efficiency, 1.3%	3	121
	972		FL	DMSO	1×10^{-4}					128

Dye	Peak	Range (nm)	Pump (nm)	Solvent	Conc. (M)	λ	λ′	Remarks	N	Ref
-P	970	894—1095	Kr(752)	EG/DMSO, 3/1	0.8g/1			Very poor stability, extends useful λ past 1000 nm, best dye from 1006—1020 nm		129
-P	960	913—1020	Kr(752,799)	EG/DMSO, 1/1	0.8g/1					125
12A-DTTeC-P	970	920—950	Ruby(694.3)	DMSO	0.8×10^{-4}	601	872″	Poor shelf life	4	74
-P		915—1058	Kr(752)	EG/DMSO, 3/1	0.8g/1			Poor efficiency, high threshold, poor stability		129
-P		935—1019	Kr(752,799)	EG/DMSO, 1/1	0.5g/1			IR-143 better, 30% output increase by deoxygenation		125
9,11,15,17-Dnp-DTPC-P (DNTPC-P)	1124	1102—1148	Nd:YAG(1064)	DMSO	1×10^{-4}	705	~1060	Conversion efficiency >10%	5	117
9,11,15,17-Dnp-5,6-Temo-DTPC-P (DNXTPC-P)	1140	1107—1187	Nd:YAG(1064)		1×10^{-4}	825	~1060	>10^5 shots for 20% power reduction for ½ ℓ of 10^{-4} M solution	5	118
9,11,15,17-Dnp-6,7-Dbz DTPC-P (DNDTPC-P)	1231	1192—1285	Nd:YAG(1064)	DMSO	2×10^{-3}	805	~1060	Conversion efficiency > 10%	5	118
	1172	1151—1198	Nd:YAG(1064)		1×10^{-4}					117
		1084—1125	Nd:YAG(1064)	DMSO	1.5×10^{-4}			10^6 shots to ½ power output for 50 cc of 1.5×10^{-4} M solution		119,120
Indocarbocyanines										
HIDC-I	675	826—850	Nd:YAG(532)→R610 (585)	Methanol	2×10^{-4}	510	641′		2	102
-I	720	788—832	Nd:YAG(532)→R610 (585)	Methanol	8×10^{-4}					102
HITC-I	839	826—850	N₂	DMSO	1.25×10^{-3}					50
-I	806	788—832	Nd:YAG(532)	DMSO	2×10^{-4}					90
-I	822	(bb)	Nd:YAG→C720 (700)	DMSO	2×10^{-4}	536	751′ 743′		3	101
-I	836	(bb)	Ruby(694.3)	EG						76
-I	862		N₃(337)	DMSO	2×10^{-4}			803—842 nm covered by mixtures of HITC and DOTC		124
-I	865	825—912	Kr(Red)	DMSO	5×10^{-4}			Conversion efficiency ~2%		103
-I	849	(bb)	Nd:YAG(532)	EG	1.5×10^{-3}			Conversion efficiency ~3%		121
-I	869	832—888	Kr(647,676)	DMSO	3.7×10^{-4}			Good efficiency, good lasing in water and Triton X		127
-I		790—840	Ruby(694.3)	DMSO						74
-I		770—830	FL→KR620(Red)	EG						97
-I		800—882	FL→R640(664)	DMSO	1×10^{-4}			DMSO best solvent		48
-I		780—883	Ruby(694.3)	EG/DMSO, 3/1	6×10^{-5}			Good efficiency		115
-I	873	819—937	Kr(752)	EG/DMSO, 3/1	2.8×10^{-4}					129
-I		832—911	Kr(647)	EG/DMSO, 3/1	1×10^{-3}			COT		105
-P	819	812—929	Ruby(694.3)	Acetone	1×10^{-4}	509	751′	Conversion efficiency 12.5%, enhanced 1.5 times by COT		132
-P	870		Kr(752,799)	DMSO + EG	4×10^{-4}			COT		75b
-P	870	828—909	Kr(647,676)	DMSO + EG	1.3×10^{-3}			Conversion efficiency 8%		75b
-P	875	840—940	Kr(647,676)	EG/DMSO, 84/16	7.4×10^{-4}					62

Table 3.1.6 (continued)
ORGANIC COMPOUNDS FOR DYE LASERS

Dye-counterion	Lasing wavelength Max	Range	Pump source (nm)	Solvent	Concentration (molar)	MW	abs	Comments	n	Ref.
6,7-Dbz-HITC-P (HDITC-P)	920	880—965	Kr(752,799)	DMSO + EG	4×10^{-4}	619		Conversion efficiency 6%, COT + N_2 enhanced output 3 2X	3	75b
IR-125-I (HITC-type)	863	846—907	Nd:YAG	DMSO		774	795*		3	90
-I (HITC-type)	903		Nd:YAG→C720 (700)	DMSO	2×10^{-4}					101
-I	940	840—920	FL	DMSO	1×10^{-4}			Excellent photochemical stability		128
-I		(bb)	Ruby(694.3)	DMSO	2.3×10^{-4}					74
-I			Nd:YAG(532)	DMSO	5×10^{-4}			Conversion efficiency ~2.6%		121
IR-144 (HITC-type)	913 (bb)	844—885	Nd:YAG→C720 (700)	DMSO	2×10^{-4}	1008	745*			101—
	863									48
	949,880	800—870	FL→R640(660)	EG	1×10^{-4}		752*		3	128
			FL	DMSO	1×10^{-4}			Good efficiency		74
		835—890	Ruby(694.3)	DMSO	2.2×10^{-4}			Good efficiency		105
		834—892	Kr(752,799)	EG/DMSO, 3/1	3.9×10^{-4}					121
	869	(bb)Δλ = 14	Nd:YAG(532)	DMSO	5×10^{-4}			Conversion efficiency 11%		121
	874	(bb)Δλ = 12	Nd:YAG(532)	DMSO	5×10^{-4}			Conversion efficiency 5.5%, good lifetime, half life>2×10^6 shots		116
Oxacarbocyanines										
DODC-I	633	(bb)	Nd:YAG(532)	Methanol	5×10^{-5}	486	582*	Mixed with 10^{-4}Rhodamine B, conversion efficiency 33% broadband	2	130
DOTC-I (DEOTC)		725—765	Ruby(694.3)	DMSO	5×10^{-5}	512	687*	Good efficiency	3	74
-I		785—800	FL→R610(622)	MeOH/N_2/COT	3×10^{-5}					100
-I	732		Nd:YAG→R610 (585)	DMSO	3.9×10^{-3}					102
-I	740		Nd:YAG R610 (585)	DMSO	1.9×10^{-4}					102
-I	770		Nd:YAG→R610 (585)	DMSO	7.8×10^{-4}					102
-I	745		FL→R610(622)	DMSO	1×10^{-4}					100
-I	756	736—793	Nd:YAG(532)	Methanol	1×10^{-4}					90
-I		750—825	FL→R640(660)	EG	1×10^{-4}					48
-I	808	756—871	Kr(Red)	EG/DMSO, 3/1	1×10^{-3}					103
-I		742—874	Kr(647)		1×10^{-3}			COT		105
-I	800	755—870	Kr(647,676)	DMSO + EG or glycerin	1×10^{-3}			Conversion efficiency 32%, COT enhanced output by 3X using the DOTC-P		75b
-I	783	750—833	Kr(647,676)	EG	7.8×10^{-4}			Conversion efficiency ~3%		127

Compound	λ	(bb)	Pump laser	Solvent	Concentration	λ (emission)		Remarks	Ref
-1	754	(bb)Δλ = 23	Nd:YAG(532)	DMSO	2.6×10^{-4}	680'		Conversion efficiency 8%, 10⁵ shots to 80% power/½ liter	116
-1	782	(bb)	N₂(337)	DMSO	2.5×10^{-1}			Fairly good stability under N₂ pumping	124
-1	742		Ruby(694.3)	Acetone	1×10^{-4}				132
-P	795	765-875	Kr(646,676)	EG/DMSO, 84/16	6×10^{-4}	485			62
DmOTC-1		725—780	Ruby(694.3)	DMSO	1×10^{-4}	484 682[m]	3	Good efficiency, unstable in light, poor shelf life	74
-1	~800	750—810	Ruby(694.3)	DMSO	8×10^{-5}			DMSO most efficient solvent	115
-1		750—864	Kr(647,676)	DMSO + EG or glycerin					75b
-1	788	(bb)	N₂ (337)	DMSO	1×10^{-1}			Fairly good stability under N₂ pumping	124
-1	711		Ruby (694.3)	Acetone	1×10^{-4}				132
Quinocarbocyanines									
Dm-4-QC-1	749		Ruby(694.3)	Glycerin	1×10^{-4}	452 709[m]	1		132
D-4-QC-1	751		Ruby(694.3)	Glycerin		480 708[m]	1		132
-Br	754		Ruby(694.3)	Glycerin		433 710[m]			132
D-2-QDC-1		740—770	Ruby(694.3)	EG	1×10^{-4}	506 710[m]	2	Poor efficiency	74
D-4-QDC-1	745	845—920	Ruby (694.3)	DMSO	5.3×10^{-4}	506 820[m]	2	Poor shelf life	74
11-Br-Dm-2-QDC-1			Ruby (694.3)	Glycerin	1×10^{-4}	557 691[m]	2		132
11-Br-D-2-QDC-1	815		Ruby (694.3)	Glycerin	1×10^{-4}	585 694[m]	2		132
11-Br-D-4-QDC-1	830		Ruby (694.3)	Methanol	1×10^{-4}	585 794[m]	2		132
D-2-QTC-1	898	865—920	Ruby (694.3)	Acetone	1×10^{-4}	532 817[m]	3	Poor efficiency	132
-1			Ruby(694.3)	DMSO	8×10^{-5}				115
D-4-QTC-1	1000	983—1081	Ruby (694.3)	Acetone	1×10^{-1}	532 923[m]	3		132
-1			Ruby (694.3)	DMSO	1×10^{-4}				74

CYANINE DYES
Common Name and/or Structure

The names of various dyes found in the literature have in many instances been reduced to relatively simple pseudo acronyms by reducing the important parts of the name to letters or by the assignment of a number to more complex molecules. The system used here for the cyanine dyes is an extension of that found in the literature with an attempt to further delineate the structure in a simple manner by designating the substituents independently of the basic chromophore. Thus, the parent molecule consists of the chromophore and standard substituents, and is represented by three or four capital letters. Substituents are indicated by letters and their position by numbers which generally preceed the letters of the basic nucleus. For example, DTTC is the shorthand notation for 3,3′-diethylthiatricarbocyanine (considered to be the basic nucleus) and 4,5-Dbz-DTTC is the shortened version of 3,3′-diethyl-2,2′-(4,5,4′,5′-dibenzo) thiatricarbocyanine. Both have the basic DTTC nucleus, but the second dye has the additional benzo substitution. The appropriate letters have been underlined to indicate the derivation of the acronym type abbreviation for the two dyes. (For additional examples see the underlined portions of the first two listings.) Thus, the cyanine dyes have been listed in the table with an abbreviated form for the basic nucleus preceeded by various substituents. For each of these cationic dyes a counter ion may be noted where the literature so designated. The counter-ion usually is iodide(I), bromide(Br), or perchorate(P) and may be noted following the dye name.

The basic cyanine dyes are the thiacarbocyanines(T), indocarbocyanines(I), oxacarbocyanines(O), quino-carbocyanines(Q), and those with mixed chromophores. In most cases the dyes have symmetrical structure about a vinylogous carbon chain. Numbers are used to indicate the position of substitution on the chromophore and primed numbers on the second chromophore. Where there are substituents on the carbon chain, their position is indicated by the extension of the numbering sequence from the first chromophore (See the structure below).

DTTC-1

With the exception of the quinocarbocyanines which have two possible sites of substitution to the carbon chain, the abbreviated names do not show the common sites of substitution, that is, -3,3′-diethyl and 2,2′-carbon chain linkages to the chromophore. They are the same in each dye. For example, compare the names of various thiacarbocyanines. However, the names and numbered positions of other substituents are indicated. In symmetrical cases the primed numbers of the substituents are droped, that is, 5,6,5′,6′-dibenzo becomes 5,6-Dbz.

The letters used in the acronym type abbreviations are indicated below: C(carbocyanine), T(thia), I(indo), O(oxa), Q(quino), D(diethyl), Dm(dimethyl), Dmo(dimethoxy), Dbz(dibenzo), T(thia or tri), Te(tetra), H(hexamethyl), A(acetoxy). The counter-ions, I(iodide), Br(bromide), and P(perchorate) always appear at the end of the name or abbreviation. Thus I may be taken as indo or iodide, however, their relative position in the acronym allows one to discern the correct usage. Thus, the exact sequence of letters is of major importance in understanding the acronym. For example, HITC-I represents 1,3,3,1′,3′3′-hexamethyl-2,2′-indotricarbocyanine iodide. Similarly, T represents both thia and tri, however, only in the thiacarbocyanines does T represent thia while the next letter represents the number of carbo units making up the connecting chain. Thus in DTTC, the first T represents thia while the second represents tri and in DOTC the O represents oxa and again the T represents tri or the number or ethylene linkages between the two chromophores.

Counter-ions associated with each dye where reported are shown in the data listing but not below. Counter-ions such as perchlorate, bromide, and iodide are most frequently associated with the cyanine dye.

Thiacarbocyanines*
General structure

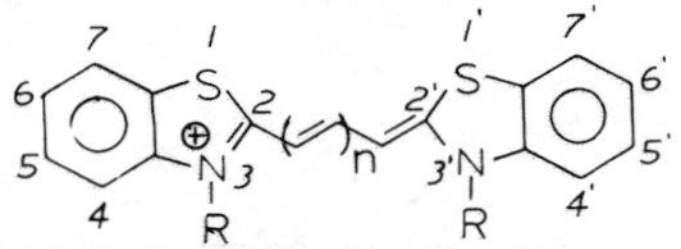

DTDC	3,3′-Diethyl-2,2′-thiadicarbocyanine
10-Cl-5,6-Dbz-DTDC	3,3′-Diethyl-10-chloro-2,2′-(5,6,5,′6′-dibenzo)thiadicarbocyanine
Dmo-DTDC	3,3′-Diethyl-2,2′-(6,6′-dimethoxy)thiadicarbocyanine
10-Cl-4,5-Dbz-DTDC	3,3′-Diethyl-10-chloro-2,2′(4,5,4′,5′-dibenzo)thiadicarbocyanine
DTTC	3,3′-Diethyl-2,2′-thiatricarbocyanine
4,5-Dbz-DTTC	3,3′-Diethyl-2,2′-(4,5,4′,5′-dibenzo)thiatricarbocyanine
IR 123 (DTTC-derivative)	3,3′-Diethyl-9,11-(oxy-o-phenylene)thiatricarbocyanine
Temo-DTTC	3,3′-Diethyl-2,2′-(5,6,5′,6′-tetramethoxy)thiatricarbocyanine
IR-109 (DTTC-Derivative)	3,3′-Diethyl-10,12-ethylene-11-morpholinothiatricarbocyanine

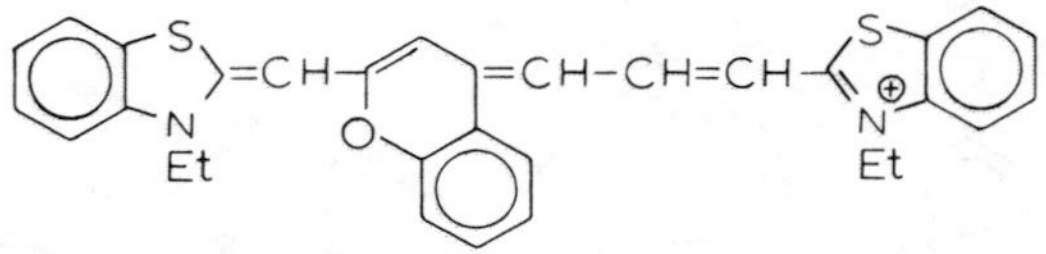

IR-123

IR-109

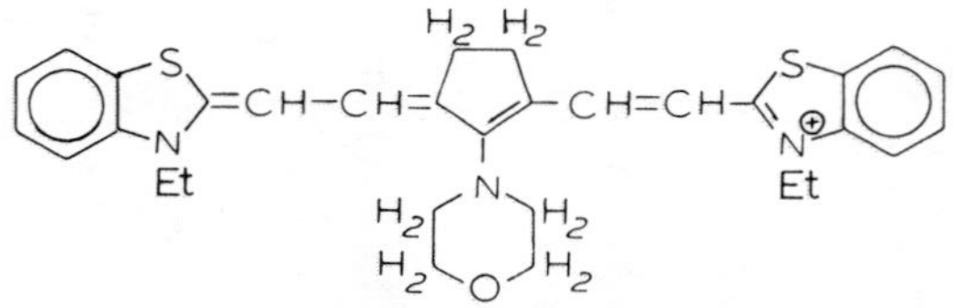

* The corresponding counter-ion has not been included in the name for any of the cyanine dyes - see the table.

CYANINE DYES (continued)

Dmo-DTTC　　3,3′-Diethyl-2,2′-(5,5′-dimethoxy)thiatricarbocyanine

IR-139 (DTTC-derivative)　11-Dimethylamino-3,3′-diethyl-10,12-ethylene-4,5,4′,5′-dibenzothiatricarbocyanine

IR-116 (DTTC-derivative)　3,3′-Diphenylthiatricarbocyanine

IR-134 (DTTC-derivative)　11-(4-Ethoxycarbonylpiperidino)-3,3′-diethyl-10,12-ethylene-4,5,4′,5′ dibenzothiatricarbocyanine

IR-141 (DTTC-derivative)　5,5′-Dichloro-3,3′-diethyl-10,12-ethylene-11-(N-methylanilino)thiatricarbocyanine

IR-140 (DTTC-derivative)　5,5′-Dichloro-11-diphenylamino-3,3′-diethyl-10,12-ethylenethiatricarbocyanine

IR-137 (DTTC-derivative) 3,3'-Diethyl-10,12-ethylene-11-(N-methylanilino)thiatricarbocyanine

IR-132 (DTTC-type) 3,3'-Di(3-acetoxypropyl)-11-diphenylamino-10,12-ethylene-5,6,5',6' dibenzothiatricarbocyanine

IR-143 (Dttc-derivative) 11-Diphenylamino-3,3'-diethyl-10,12-ethylene-4,5,4',5'-dibenzothiatricarbocyanine

12A-DTTeC 3,3'-Diethyl-12-acetoxy-2,2'-thiatetracarbocyanine

9,11,15,17-Dnp-DTPC (DNTPC) 3,3'-Diethyl-9,11,15,17-dineopentylene thiapentacarbocyanine

9,11,15,17,-Dnp-5,6-Temo-DTPC (DNXTPC) 3,3'-Diethyl-9,11,15,17-dineopentylene (5,6,5',6'-tetramethoxy) thiapentacarbocyanine

9,11,15,17-Dnp-6,7-Dbz-DTPC (DNDTPC) 3,3-Diethyl-9,11,15,17-dineopentylene(6,7,6',7'-dibenzo)thiapentacarbocyanine

Indocarbocyanines
General
structure

CYANINE DYES (continued)
Common Name and/or Structure

HIDC   1,3,3,1′,3′,3′-Hexamethyl-2,2′-indodicarbocyanine
HITC   1,3,3,1′,3′,3′-Hexamethyl-2,2′-indotricarbocyanine
6,7-Dbz-HITC   1,3,3,1′,3′,3′-Hexamethyl-2,2′-(6,7,6′,7′-dibenzo)indotricarbocyanine
IR-125 (HITC-type)   3,3,3′,3′-Tetramethyl-1,1′-di(4-sulfobutyl)-4,5,4′,5′-dibenzoindotricarbocyanine Iodide, monosodium salt

IR-125

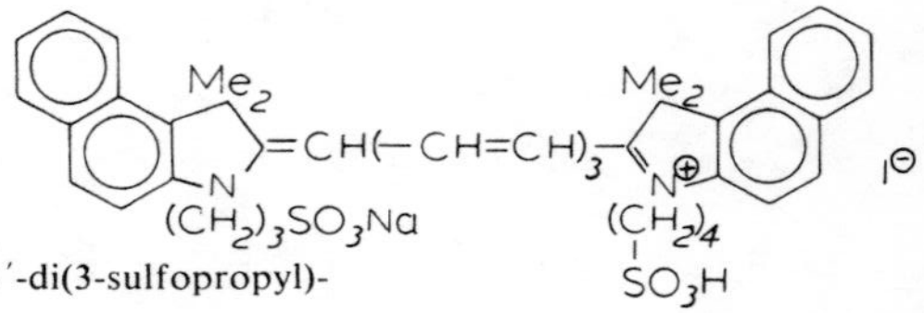

IR-144 (HITC-type)   Anhydro-11-(4-ethoxycarbonyl-1-piperazinyl)-10,12-ethylene-3,3,3′,3′-tetramethyl-1,1′-di(3-sulfopropyl)-4,5,4′,5′-dibenzoindotricarbocyanine hydroxide, triethylamnonium salt

IR-144

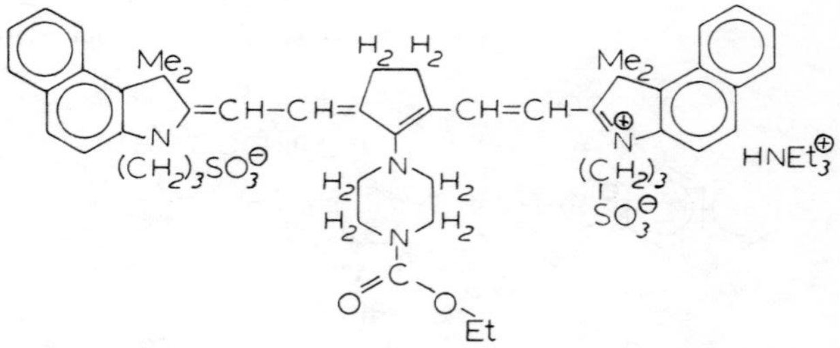

Oxacarbocyanines
   General
   structure

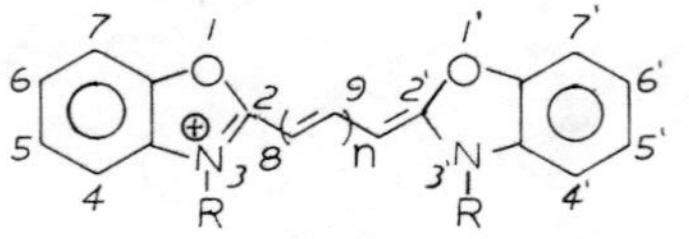

DODC   3,3′-Diethyl-2,2′-oxadicarbocyanine
DOTC   3,3′-Diethyl-2,2′-oxatricarbocyanine
DmOTC   3,3′-Dimethyl-2,2′-oxatricarbocyanine

Quinocarbocyanines
General
structures

Dm-4-QC	1,1'-Dimethyl-4,4'-quinocarbocyanine
D-4-QC	1,1'-Diethyl-4,4'-quinocarbocyanine(cryptocyanine)
D-2-QDC	1,1'-Diethyl-2,2'-quinodicarbocyanine
D-4-QDC	1,1'-Diethyl-4,4'-quinodicarbocyanine
11-Br-Dm-2-QDC	1,1'-Dimethyl-11-bromo-2,2'-quinodicarbocyanine
11-Br-D-2-QDC	1,1'-Diethyl-11-bromo-2,2'-quinodicarbocyanine
11-Br-D-4-QDC	1,1'-Diethyl-11-bromo-4,4'-quinodicarbocyanine
D-2-QTC	1,1'-Diethyl-2,2'-quinotricarbocyanine
D-4-QTC	1,1'-Diethyl-4,4'-quinotricarbocyanine

TUNING CURVES

Figures 3.1.15 to 3.1.21 are tuning curves showing the laser wavelength range and relative output of various organic dyes. They indicate some of the pump and dye laser systems commercially available. A representative sampling of the types of laser emission possible is given but is by no means complete. Other types of pump sources not covered include copper vapor, excimer, and ruby lasers.

The tuning curves have been supplied by and reproduced with the permission of the Candela Corporation, 96 South Avenue, Natick, MA 01760; Coherent Inc., 3210 Porter Drive, Palo Alto, CA 94304; Molectron Corporation, 177 North Wolfe Road, Sunnyvale, CA 94086; the Phase-R Company, Box G-2, Old Bay Road, New Durham, NH 03855; Quanta-Ray, 2134 Old Middlefield Way, Mountain View, CA 94043; Quantel International, 928 Benecia Avenue, Sunnyvale, CA 94086; and Spectra-Physics, 1250 West Middlefield Road, Mountain View, CA 94042.

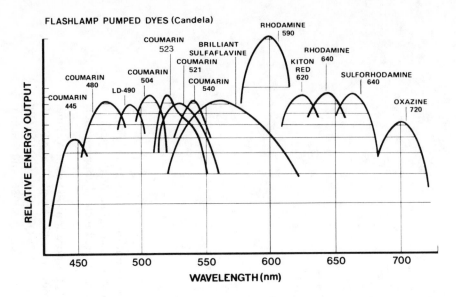

FIGURE 3.1.15.   Relative energy outputs of various organic dyes used in dye lasers excited by flashlamps[104] (diagram courtesy of Candela Corporation, Natick, Mass.).

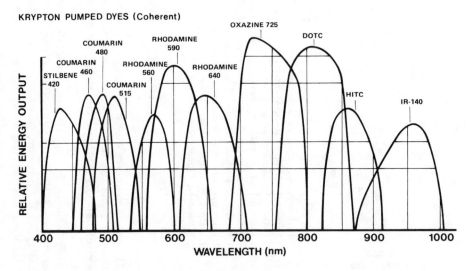

FIGURE 3.1.16.   Relative energy outputs of various organic dyes used in dye lasers excited by krypton lasers[103] (diagram courtesy of Coherent, Inc., Palo Alto, Calif.).

NITROGEN PUMPED DYES (Molectron)

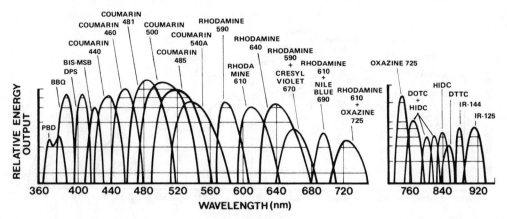

FIGURE 3.1.17. Relative energy outputs of various organic dyes used in dye lasers excited by nitrogen lasers[50] (diagram courtesy of Molectron Corporation, Mountain View, Calif.).

COAXIAL FLASHLAMP PUMPED DYES (Phase-R)

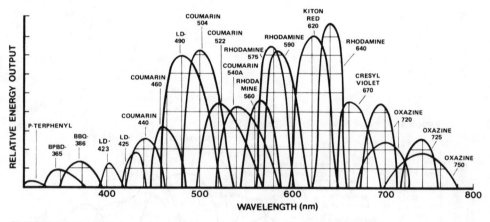

FIGURE 3.1.18. Relative energy outputs of various organic dyes used in dye lasers excited by coaxial flashlamps[48] (diagram courtesy of Phase-R Company, New Durham, N.H.).

Nd:YAG PUMPED LASER DYES (Quanta-Ray)

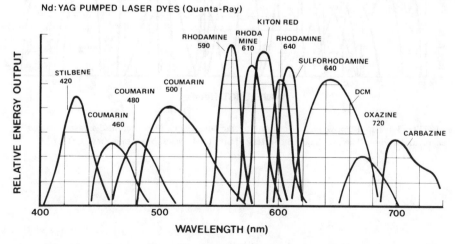

FIGURE 3.1.19. Relative energy outputs of various organic dyes used in dye lasers excited by Nd:YAG lasers[94] (diagram courtesy of Quanta-Ray, Mountain View, Calif.).

Nd:YAG PUMPED LASER DYES (Quantel International)

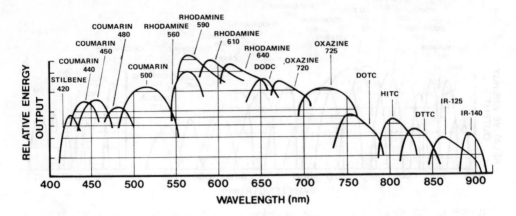

FIGURE 3.1.20.   Relative energy outputs of various organic dyes used in dye lasers excited by Nd:YAG lasers[90] (diagram courtesy of Quantel International, Sunnyvale, Calif.).

ARGON & KRYPTON PUMPED DYES (Spectra Physics)

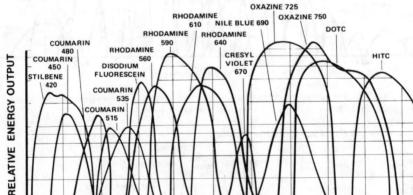

FIGURE 3.1.21.   Relative energy outputs of various organic dyes used in dye lasers excited by argon and krypton[62] lasers (diagram courtesy of Spectra-Physics, Mountain View, Calif.).

REFERENCES

1. Sorokin, P. P. and Lankard, J. R., *IBM J. Res. Dev.*, 10, 162, 1966.
2. Drexhage, K. H., Structure and properties of laser dyes, in *Dye Lasers*, Vol. 1, Schafer, F. P., Ed., Springer-Verlag, Berlin, 1973, 144 and references therein.
3. Hammond, P. R., Fletcher, A. N., Bliss, D. E., Henry, R. A., Atkins, R. L., and Moore, D. W., Search for efficient, near uv lasing dyes. III. Monocyclic and miscellaneous dyes, *Appl. Phys.*, 9, 67, 1976.
4. Drexhage, K. H., "Design of Laser Dyes," VII. International Quantum Electronics Conference, Montreal, Canada, 1972, paper B.5.; see also References 2 and 5.
5. Drexhage, K. H., What's ahead in laser dyes, *Laser Focus*, 9, 35, 1973.
6. Schimitschek, E. J., Trias, J. A., Taylor, M., and Celto, J. E., New improved laser dyes for the blue-green spectral region, *IEEE J. Quantum Electron.*, QE-9, 781, 1973.
7. Schimitschek, E. J., Trias, J. A., Hammond, P. R., and Atkins, R. L., Laser performance and stability of fluorinated coumarin dyes, *Opt. Commun.*, 11, 352, 1974.
8. Drexhage, K. H., Erickson, G. R., Hawks, G. H., and Reynolds, G. A., Water soluble coumarin-dyes for flashlamp-pumped dye lasers, *Opt. Commun.*, 15, 399, 1975.
9. Henry, R. A., U.S. Patent 961578, 1979.
10. Reynolds, G. A. and Drexhage, K. H., New coumarin dyes with rigidized structures for flashlamp-pumped dye lasers, *Opt. Commun.*, 13(3), 222, 1975.
11. Hammond, P. R., Fletcher, A. N., Henry, R. A., and Atkins, R. L., Search for efficient, near uv lasing dyes, *Appl. Phys.*, 8, 311, 1975.
12. Hammond, P. R., Fletcher, A. N., Henry, R. A., and Atkins, R. L., Search for efficient, near uv lasing dyes. II. Aza substitution in bicyclic dyes, *Appl. Phys.*, 8, 315, 1975.
13. Schimitschek, E. J., Trias, J. A., Hammond, P. R., Henry, R. A., and Atkins, R. L., New laser dyes with blue-green emissions, *Opt. Commun.*, 16, 313, 1976.
14. Fletcher, A. N., Fine, D. A., and Bliss, D. E., Laser dye stability. I. Flash frequency and fluid filtration efforts, *Appl. Phys.*, 12, 99, 1977.
15. Fletcher, A. N., Laser dye stability. II. Input energy per flash, dye concentration, and mirror reflectivity efforts, *Appl. Phys.*, 12, 327, 1977.
16. Fletcher, A. N., Laser dye stability. III. Bicyclic dyes in ethanol, *Appl. Phys.*, 14, 295, 1977.
17. Fletcher, A. N., Laser dye stability. IV. Photodegradation relationships for bicyclic dyes in alcohol solutions, *Appl. Phys.*, 16, 93, 1978.
18. Fletcher, A. N. and Bliss, D. E., Laser dye stability. V. Efforts of chemical substituents of bicyclic dyes upon photodegradation parameters, *Appl. Phys.*, 16, 289, 1978.
19. Neister, S. E., *Opt. Spectra*, 11, 34, 1977.
20. Morton, R. G., Mack, M. E., and Itzkan, I., Efficient cavity dumped dye laser, *Appl. Opt.*, 17, 3268, 1978.
21. Winters, B. H., Mandelberg, H. I., and Mahr, W. B., Photochemical products in coumarin dyes, *Appl. Phys. Lett.*, 25, 723, 1974.
22. Rulliere, C. and Joussat-Dubien, J., Dye laser action at 330 nm using benzoxazole: a new class of lasing dyes, *Opt. Commun.*, 24(1), 38, 1978.
23. Rulliere, C., Morand, J. P., and de Witte, O., KrF laser pumps new dyes in the 3500 Å spectral range, *Opt. Commun.*, 20(3), 339, 1977.
24. Ziegler, L. D. and Hudson, B. S., Tuning ranges of 266 nm pumped dyes in the near uv, *Opt. Commun.*, 32(1), 119, 1980.
25. Ducasse, L., Rayez, J. C., and Rulliere, C., Substitution effects enhancing the lasing ability of organic compounds, *Chem. Phys. Lett.*, 57(4), 547, 1978.
26. Huffer, W., Schieder, R., Telle, H., Raue, R., and Brinkwerth, W., CW dye laser emission down to the near uv, *Opt. Commun.*, 28(3), 353, 1979.
27. Beterov, I. M., Ishchenko, V. N., Kogan, B. Ya., Krasovitskii, B. M., and Chernenko, A. A., Stimulated emission from 2-phenyl-5(4-difluoromethylsulfonylphenyl)oxazole pumped with nitrogen laser radiation, *Sov. J. Quantum Electron.*, 7(2), 246, 1977.
28. Majewski, W. and Krasinski, J., Laser properties of fluorescent brightening agents, *Opt. Commun.*, 18(3), 255, 1976.
29. Telle, H., Brinkmann, U., and Raue, R., Laser properties of bis-styryle compounds, *Opt. Commun.*, 24(3), 248, 1978.
30. Takakusa, M. and Itoh, U., The 3-CN-4-MU dye laser in ethanol containing various amounts of water, acid or alkali, *Opt. Commun.*, 26(3), 401, 1978.
31. Telle, H., Brinkmann, U., and Raue, R., Laser properties of triazinyl stilbene compounds, *Opt. Commun.*, 24, 33, 1978.
32. Kuhl, J., Telle, H., Scheider, R., and Brinkmann, U., New efficient and stable laser dyes for cw operation in the blue and violet spectral range, *Opt. Commun.*, 24, 251, 1978.

33. **Gacoin, P., Bokobza, A., Bos, F., LeBris, M. T., and Hayat, G.**, New class of high-efficiency laser dyes: the quinoxalinones, Conference on Laser and Electrooptical Systems, San Diego, Febuary 7 to 9, 1978, 56, WEE6.

34. **Basting, D., Schafer, F. P., and Steyer, B.**, New laser dyes, *Appl. Phys.,* 3, 81, 1974.

35. **See Reference 8.**

36. **Lee, L. A. and Robb, R. A.**, "Water soluble blue-green lasing dyes for flashlamp-pumped dye lasers", IEEE/OSA Conference on Laser Engineering and Applications, Washington, D.C., May 30 to June 1, 1979; *IEEE J. Quant. Electron.* to be published.

37. **Schafer, F. P., Bor, Zs., Luttke, W., and Liphardt, B.**, Bifluorophoric laser dyes with intramolecular energy transfer, *Chem. Phys. Lett.,* 56, 455, 1978.

38. **Zapka, W. and Brackmann, U.**, Shorter dye laser wavelengths from substituted p-terphenyl, *Appl. Phys.,* 20, 283, 1979.

39. **Bryon, D. A.**, McDonnell Douglas Astronautics Company, private communication, 1979.

40. **Scabo, A.**, National Research Council of Canada, private communication, 1980, Jessop, P. E. and Scabo, A., Single frequency cw dye laser operation in the 690-700 nmgap; *J. Quantum Electron.,* to be published.

41. **Wayashita, M., Kasamatsu, M., Kashiwagi, H., and Machida, K.**, The selective excitation of lithium isotopes by intra cavity nonlinear absorption in a cw dye laser, *Opt. Commun.,* 26(3), 343, 1978.

42. **Lill, E., Schneider, S., and Dorr, F.**, Passive mode-locking of a flashlamp-pumped fluorol 7GA dye laser in the green spectral region, *Opt. Commun.,* 20(2), 223, 1977.

43. **Fahfu, Ho**, University of Pennsylvania, private communication, 1979.

44. **Jarett, S.**, Spectra Physics, private communication, 1980.

45. **Profitt, W.**, Coherent, private communication, 1980.

46. **Godard, B. and de Witte, O.**, Efficient laser emission in para-terphenyl tunable between 323 and 364 nm, *Opt. Commun.,* 19(3), 325, 1976.

47. **Furumoto, H. W. and Ceccon, H. L.**, Ultraviolet organic liquid lasers, *IEEE J. Quantum Electron.,* QE-6, 262, 1970.

48. **Phase-R Company,** Box G-2, New Durham, N.H. 03855.

49. **Dunning, F. B. and Stebbings, R. F.**, The efficient generation of tunable near UV radiation using an N_2 pumped dye laser, *Opt. Commun.,* 11(2), 112, 1974.

50. **Molectron Corporation,** 177 N. Wolfe Road, Sunnyvale, Calif. 94086.

51. **Myer, J. A., Itzkan, I., and Kierstead, E.**, Dye lasers in the ultraviolet, *Nature (London),* 225, 544, 1970.

52. **Furumoto, H. W. and Ceccon, H. L.**, Flashlamp pumped organic scintillator lasers, *J. Appl. Phys.,* 40, 4204, 1969.

53. **Hammond, P. R., Fletcher, A. N., Henry, R. A., Atkins, R. L., and Moore, D. W.**, Near-ultraviolet lasing dyes, Part 1: Search for new dyes and summation of results; Fletcher, A. N., Near-ultraviolet lasing dyes, Part 2: Effects of coaxial flashlamp excitation, NWC TP 5768, 1975; Fletcher, A. N., Laser dye stability, Part 3: Bicyclic dyes in ethanol, *Appl. Phys.,* 14, 295, 1977; Fletcher, A. N. and Bliss, D. E., Laser dye stability, Part 5: Effect of chemical substituents of bicyclic dyes upon photodegradation parameters, *Appl. Phys.,* 16, 289, 1978.

54. **Williamson, A.**, private communication, 1977.

55. **Kittrell, C.**, private communication, 1977.

56. **Marling, J. B., Hawley, J. H., Liston, E. M., and Grant, W. B.**, Lasing characteristics of seventeen visible-wavelength dyes using a coaxial-flashlamp-pumped laser, *Appl. Optics,* 13(10), 2317, 1974. a. With Rhodamine 6G.

57. **Chromatix,** 560 Oak Meade Parkway, Sunnyvale, Calif. 94086.

58. **Tuccio, S. A., Drexhage, K. H., and Reynolds, G. A.**, CW laser emission from coumarin dyes in the blue and green, *Opt. Commun.,* 7(3), 248, 1973.

59. **Yarborough, J. M.**, CW laser emission spanning the visible spectrum, *Appl. Phys. Lett.,* 24(12), 629, 1974.

60. **Srinivasan, R., von Gutfield, R. J., Angadiyavar, C. S., and Tynan, E. E.**, Photochemical studies on organic lasers, Air Force materials laboratory, Wright-Patterson Air Force Base, Dayton, Ohio, AFML-TR-74-110, 1974. a. With Rhodamine 6G.

61. **Srinivasan, R.**, New materials for flash-pumped organic lasers, *IEEE J. Quantum Electron.,* QE-5, 552, 1969.

62. **Spectra-physics,** 1250 W. Middlefield Road, Mountain View, Calif. 94042.

63. **Ledbetter, J. W.**, private communication, 1977.

64. **Drexhage, K. H. and Reynolds, G. A.**, New highly efficient laser dyes, VII Int. Quantum Electronics Conf., Paper F.1, San Francisco, Calif. 1974; see also References 2 and 5.

65. **Roullard, F. P.**, private communication, 1976.

66. **Blazej, D.**, private communication, 1977.

67. **Lambropoulos, M.,** Fluorol 7GA: An efficient yellow-green dye for flashlamp-pumped lasers, *Opt. Commun.,* 15(1), 35, 1975.

68. **Hargrove, R. S. and Kan, T.,** Efficient, high average power dye amplifiers pumped by copper vapor lasers, *IEEE J. Quantum Electron.,* QE-13, 28D, 1977.

69. **Drake, J. M., Steppel, R. N., and Young, D.,** Kiton red s and rhodamine b. The spectroscopy and laser performance of red laser dyes, *Chem. Phys. Lett.,* 35(2), 181, 1975.

70. **Woodruff, S. and Ahlgren, D.,** private communication, 1977.

71. **Marling, J. B., Wood, L. L., and Gregg, D. W.,** Long pulse dye laser across the visible spectrum, *IEEE J. Quantum Electron.,* QE-7, 498, 1971.

72. **McDonald, J.,** private communication, 1974.

73. **Kato, K.,** A high-power dye laser at 6700-7700 Å, *Opt. Commun.,* 19(1), 18, 1976.

74. **Oettinger, P. E. and Dewey, C. F.,** Lasing efficiency and photochemical stability of infrared laser dyes in the 710-1080 nm region, *IEEE J. Quantum Electron.,* QE-12(2), 95, 1976.

75. **Romanek, K. M., Hildebrand, O., and Gobel, E.,** High power CW dye laser emission in the near IR from 685 nm to 965 nm, Opt. Commun., 21(1), 16, 1977; *Spectra-Physics Laser Review,* 4(1), April, 1977.

76. **Miyazoe, Y. and Maeda, M.,** Polymethine dye lasers, *Opto Electronics,* 2(4), 227, 1970.

77. **Donzel, A. and Weisbuch, C.,** CW dye laser emission in the range 7540-8880Å, *Opt. Commun.,* 17(2), 153, 1976.

78. **Maeda, M. and Miyazoe, Y.,** Flashlamp-excited organic liquid laser in the range from 342 to 889 nm, *Jpn. J. Appl. Phys.,* 11(5), 692, 1972.

79. **Loth, C. and Gacoin, P.,** Improvement of infrared flashlamp-pumped dye laser solution with a double effect additive, *Opt. Commun.,* 15(2), 179, 1975.

80. **Tomin, V. I., Alcock, A. J., Sarjeant, W. J., and Leopold, K. E.,** Tunable, narrow bandwidth, 2 MW dye laser pumped by a KrF* discharge laser, *Opt. Commun.,* 28(3), 336, 1979.

81. **Tomin, V. I., Alcock, A. J., Sarjeant, W. J., and Leopold, K. E.,** Some characteristics of efficient dye laser emission obtained by pumping at 248 nm with a high-power KrF* discharge laser, *Opt. Commun.,* 26(3), 396, 1978.

82. **McKee, T. J., Stoicheff, B. P., and Wallace, S. C.,** Tunable, coherent radiation in the lyman-region [1210-1290Å] using magnesium vapor, *Opt. Lett.,* 3(6), 207, 1978.

83. **McKee, T. J. and James, D. J.,** Characterization of dye laser pumping using a high-power KrF excimer laser at 248 nm, *Can. J. Phys.,* Sept. 1979.

84. **Fehrenback, G. W., Gruntz, K. J., and Ulbrich, R. G.,** Subpicosecond light pulses from a synchronously mode-locked dye laser with composite gain and absorber medium, *Appl. Phys. Lett.,* 33(2), 159, 1978.

85. **Halstead, J. A. and Reeves, R. R.,** Mixed solvent systems for optimizing output from a pulsed dye laser, *Opt. Commun.,* 27(2), 273, 1978.

86. **Collins, C. B. Taylor, K. N., and Lee, F. W.,** Dyes pumped by the nitrogen ion laser, *Opt. Commun.,* 26(1), 101, 1978.

87. **Holton, G.,** private communication, 1978.

88. **Blazy, J.,** private communication, 1978.

89. **Eckstein, N. J., Ferguson, A. I., Hansch, T. W., Minard, C. A., and Chan, C. K.,** Production of deep blue tunable picosecond light pulses by synchronous pumping of a dye laser, *Opt. Commun.,* 27(3), 466, 1978.

90. **Quantel International,** 928 Benecia Avenue, Sunnyvale, Calif., 94086.

91. **Green, W. R.,** private communication, 1977.

92. **Hartig, W.,** A high power dye-laser pumped by the second harmonic of a Nd-YAG laser, *Opt. Commun.,* 27(3), 447, 1978.

93. **a. Shirley, J.,** private communication, 1977; b. **Hall, R. J., Shirley, J. A., and Eckbreth, A. C.,** Coherent anti-stokes raman spectroscopy: Spectra of water vapor in flames, *Opt. Lett.,* 4, 87, 1979.

94. **Quanta-Ray,** 1250 Charleston Rd., Mountain View, Calif. 94043.

95. **J. K. Lasers Ltd.,** Somers Road, Rugby, Warwickshire, U.K.

96. **Kato, K.,** 3547-Å Pumped high power dye laser in the blue and violet, *IEEE J. Quantum Electron.,* QE-11, 373, 1975.

97. **Passner, A. and Venkatesan, T.,** Inexpensive, pulsed, tunable ir dye laser pumped by a driven dye laser, *Rev. Sci. Instrum.,* 49(10), 1413, 1978.

98. **Allain, J. Y.,** High energy pulsed dye lasers for atmospheric sounding, *Appl. Optics,* 18(3), 287, 1979.

99. **Mialocq, J. C. and Goujon, P.,** Tunable blue picosecond pulses from a flashlamp-pumped dye laser, *Appl. Phys. Lett.,* 33(9), 819, 1978.

100. **Mahon, R., McIlrath, T. J., and Koopman, D. W.,** High-power TEM_{oo} tunable laser system, *Appl. Optics,* 18(6), 891, 1979.

101. **Kato, K.,** Near infrared dye laser pumped by a carbazine 122 dye laser, *IEEE J. Quantum Electron.,* QE-12, 442, 1976.
102. **Drell, P.,** private communication, 1978.
103. **Coherent Inc.,** 3210 Porter Dr., Palo Alto, Calif., 94304.
104. **Candela Corporation,** 96 South Ave., Natick, Mass., 01760.
105. **Kuhl, J., Lambrich, R., and von der Linde, D.,** Generation of near-infrared picosecond pulses by mode locked synchronous pumping of a jet-stream dye laser, *Appl. Phys. Lett.,* 31(10), 657, 1977.
106. **Bennett, J.,** private communication, 1978 (Noted increase in efficiency by addition of R590).
107. **Hammond, P. R.,** Laser dye DCM, spectral properties, synthesis and comparison with other dyes in the red, *Opt. Commun.,* Preprint.
108. **Marling, J. B., Gregg, D. W., Thomas, S. J.,** *IEEE J. Quantum Electron.,* QE-5, 570, 1970.
109. **Petty, B. W., Morris, K.,** *Opt. Quantum Electron.,* 8(4), 371, 1976.
110. **Marowsky, G., Cordray, R., Tittel, F. K., Wilson, W. L., and Collins, C. B.,** Intense laser emission from electron-beam - pumped ternary mixtures of Ar, N_2, and POPOP vapor, *Appl. Phys. Lett.,* 33(1), 59, 1978.
111. **Smith, P. W., Liao, P. F., Schank, C. V., Gustafson, T. K., Lin, C., and Maloney, P. J.,** Optically excited organic dye vapor laser, *Appl. Phys. Lett.,* 25(3), 144, 1974.
112. **Schafer, F. P., Schmidt, W., and Marth, K.,** New dye lasers covering the visible spectrum, *Appl. Phys. Lett.,* 24A, 280, 1967.
113. **Kato, K.,** Efficient ultraviolet generation of 2073-2174 Å in KB$_5$ O$_8$/4H$_2$O, *IEEE J. Quantum Electron.,* QE-13(7), 544, 1977.
114. **Azuma, K., Nakagawa, O., Segawa, Y., Aoyagi, Y., and Namba, S.,** A tunable picosecond UV dye laser pumped by the third harmonic of a neodymium-doped YAG laser, *Jpn. J. Appl. Phys.,* 18(1), 209, 1979.
115. **Decker, C. D. and Tittel, F. K.,** Broadly tunable, narrow linewidth dye laser emission in the near infrared, *Opt. Commun.,* 7(2), 155, 1973.
116. **Moore, C. A. and Decker, C. D.,** Power-scaling effects in dye lasers under high power laser excitation, *J. Appl. Phys.,* 49(1), 47, 1978.
117. **Kato, K.,** Nd:YAG laser pumped infrared dye laser, *IEEE J. Quantum Electron.,* QE-14, 7, 1978.
118. **Kato, K.,** Broadly tunable dye laser emission to 12850 Å, *Appl. Phys. Lett.,* 33(6), 509, 1978.
119. **Ferrario, A.,** A 13 MW peak power dye laser tunable in the 1.1 μm range, *Opt. Commun.,* 30(1), 83, 1979.
120. **Ferrario, A.,** A picosecond dye laser tunable in the 1.1 μ region, *Opt. Commun.,* 30(1), 85, 1979.
121. **Decker, C. D.,** Excited state absorption and laser emission from infrared dyes optically pumped at 532 nm, *Appl. Phys. Lett.,* 27(11), 607, 1975.
122. **Fouassier, J. P., Longnot, D. J., and Faure, J.,** Photoisomerization processes in the IR-140 laser dye, *Opt. Commun.,* 23(3), 393, 1977.
123. **Hirth, A., Faure, J., and Longnot, D.,** Quenching effects in flashlamp-excited polymethine dye lasers, *Opt. Commun.,* 8, (4), 318, 1973.
124. **Hildebrand, O.,** Nitrogen laser excitation of polymethine dyes for emission wavelength up to 9500 Å, *Opt. Commun.,* 10(4), 310, 1974.
125. **Leduc, M. and Weisbach, C.,** CW dye laser emission beyond 1000 nm, *Opt. Commun.,* 26, (11), 78, 1978.
126. **Romanek, K. and Hildebrand, O.,** Dyes for CW infrared output, *Laser Focus,* 50, 1977; Donzel, A. and Weisbach, C., Focus on science, *Coherent,* 1, 1, 1977.
127. **Donzel, A. and Weisbach, C.,** CW dye laser emission in the range 7540-8880 Å, *Opt. Commun.,* 17(2), 153, 1976.
128. **Webb, J. P., Webster, F. G., and Plourde, B. E.,** Sixteen new infrared laser dyes excited by a simple, linear flashlamp, *IEEE J. Quantum Electron.,* QE-11, 114, 1975.
129. **Leduc, M.,** Synchronous pumping of dye lasers up to 1095 nm, *Opt. Commun.,* 31(1), 66, 1979.
130. **Ammann, E. O., Decker, C. D., and Falk, J.,** High-peak-power 532 nm pumped dye laser, *IEEE J. Quantum Electron.,* QE-10, 463, 1974.
131. **Oudar, J. L., Kupock, Ph. J., and Chemba, D. S.,** Medium infrared tunable down conversion of a YAG-pumped infrared dye laser in gallium selenide, *Opt. Commun.,* 29(1), 119, 1979.
132. **Miyazoe, Y. and Mitsuo, M.,** Stimulated emission from 19 polymethine dyes-laser action over the continuous range 710-1060 mu, *Appl. Phys. Lett.,* 12, 206, 1968.

3.2 INORGANIC LIQUID LASERS

Harold Samelson

The subject of inorganic liquid lasers is divided in two parts based on the nature of the environment around the rare-earth ion, the compound in which the gain transition is found. In one category, the chelates, the rare-earth ion is part of a complex with, usually, bidentate ligands such as β di-ketonate and carboxylate ions or organic phosphate. Generally, organic solvents are employed. In the second category, the rare-earth ion is in an inorganic complex usually with oxyhalides and halides of the heavier elements such as phosphorus, sulfur, selenium, zirconium, tin, etc. The primary characteristic of these solvent systems is that they are aprotic — they have no hydrogen atoms. These, then, are called "aprotic liquid lasers" and the general title of "inorganic liquid lasers" includes both types and serves to distinguish them from organic or dye lasers.

The basis for the intense luminescence of the chelate lasers is the intense absorption in the ligand component and the efficient transfer of the absorbed energy to the rare-earth ion. Some quenching does occur at this stage because of the high energy vibrations characteristic of the hydrogen-containing organic components and not too large energy gap between the emitting state of the rare earth ion and its ground manifold. Some improvement results if the hydrogens are replaced by deuterium or heavier elements. In the case of the important laser ion, neodymium, the fluorescence is almost completely quenched in the usual chelate systems and it requires almost heroic efforts to bring the fluorescence efficiency to only a few percent. In this case, the aprotic solvent system has to be used. With aprotic solvents, however, the excitation is pumped directly into the levels of the rare earth ion.

The two inorganic laser systems are rather different in their functioning and their properties. Because of this, the data and discussion on them are given separately.

3.2.1. RARE EARTH CHELATE LASERS

INTRODUCTION

The first suggestion that organic materials might be useful in laser applications was made by Rautian and Sobelman.[1] Although their analysis was principally for organics as the active species, they make reference to complexes of rare earth ions. A few years later this concept was experimentally realized using europium benzoylacetonate.[2]

In subsequent development, the ions used in this type of laser were Eu^{+3}, Tb^{+3}, and Nd^{+3}. For europium and terbium the conventional, bidentate β-diketonate types of ligands have been employed while acetate and phosphate ligands were applied in the case of Nd^{+3}. The neodymium chelate laser is excited by direct pumping into the levels of the rare earth ion. This will be discussed in more detail in the section on aprotic liquid lasers. The excitation of the other chelate lasers is accomplished by energy transfer as illustrated in Figure 3.2.1.1. The most effective absorption is in the singlet absorption band of the β-diketone ligand. Because of the heavy element effect, this energy is efficiently transferred to the triplet state of the ligand by an intersystem crossing transition. From this state, it is again efficiently transferred to the rare-earth ion levels if the triplet state of the ligand is higher in energy than the emitting state of the ion.[3-7] Whether or not such excitation will result in laser action depends on the chemistry of the rare earth complex (ligand, solvent, compensating cation) and the cell or cavity arrangement.

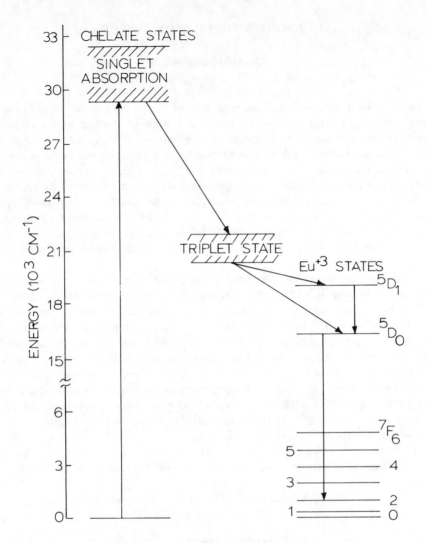

FIGURE 3.2.1.1.   Energy level system for a europium chelate. The ^5D states of the ion are excited by the transfer of the energy absorbed in the ground state — singlet transition (30,000 cm^{-1}) via the triplet state (21,000 cm^{-1}). The precise location of these states will vary with ligand. The laser transition is between the ^5D_o state and the $^7F_2$ state as indicated.

PREPARATION AND SOLUTION CHEMISTRY OF RARE EARTH CHELATES

There are many ways to prepare the chelate materials. A typical preparation is as follows:[8] The β-diketone is dissolved in an appropriate amount of 95% ethanol and an equivalent amount of base (say, piperidine) is added to form the salt of the enolate ion. An ethanol solution of EuCl$_3$ containing ¼ the number of moles of β-diketone is slowly added to the first solution with stirring. The precipitate, which may form immediately, after some time or may require the removal of solvent, is separated by filtration, washed several times with ethanol and dried by flowing dry nitrogen through the powder. Other techniques may be used depending on the ligand and the base.[9-15]

The compound resulting from this type of synthesis usually is a tetrakis chelate with the available sp^{3}d^4 orbitals of the rare earth used in the bonding. In some cases, the coordination is nine-fold rather than eight-fold and the remaining d orbital of europium is used. It is possible that the coordination is always nine-fold with the electrons of a solvent molecule occupying the ninth orbital. In europium complexes, the most intense emissions are found in the ^5D_o - $^7F_2$ and ^5D_o - $^7F_4$ transitions and one component of the former is the laser transition. In the case of Tb^{+3}, because of the inversion of the energy levels, lasing occurs in one component of the $^5D_4$ - $^7F_5$ transition. In Nd^{+3}, lasing occurs in the $^4F_{3/2}$ - $^4I_{11/2}$ transition.[10,11,16-19] Such detailed spectroscopic analysis forms the basis for understanding the structure and solution chemistry of the chelate.

The maximum number of bidentate ligands that can be accommodated by the rare-earth ion is four. The uninegative property of the ligand requires that a positive cation be associated with the coordinate anion. (The major exceptions to this are the two Nd^{+3} complexes tris perfluoro propionate, 1, 10 phenanthroline and tris deutero tributyl phosphate.) The tetrakis chelate, in solution, undergoes a dissociative reaction.

$$ReK_4^- = REK_3 + K^- \tag{1}$$

where K^- is the ligand and RE the rare earth ion. The extent to which this occurs depends on the solvent and the cation. In the case of the solvent, the dependence seems to reside in the strength of the solvent coordination compared to the ligand-rare earth ion bond strength. The dependence on the cation arises from the equilibrium:

$$C^+ + K^- = CK \tag{2}$$

where C^+ is the cation. If, in a particular solvent, the cation is combined with the ligand to form a neutral species, this equilibrium will have to be taken into account in establishing the ultimate concentration of tetrakis chelate in solution. The importance of the tetrakis form is that, to this point, only the tetrakis species of europium chelate has been shown to lase. The interaction of these effects is discussed in References 8, 16, 17, 20 through 25, and in several review articles 23, 26, 27.

The material components used in chelate lasers are presented in Table 3.2.1.1, Ligands; Table 3.2.1.2, Cations; and Table 3.2.1.3, Solvents.

CELLS FOR LIQUID LASERS

Liquid lasers, in general, require a device for containing the gain medium in some shape so that it can be a part of a resonant cavity. The approach that has been used with these intrinsically low gain substances is to create a region that the liquid fills and in which it resembles a crystal. Typical examples are illustrated in Figures 3.2.1.2a-d. Those shown in a and b are used with stationary liquids. The wider diameter (4 mm) cells were the first ones used but because of the intense absorption of the ligands, the capillary-type cells (b) give better results. The cells shown in (c) and (d) are for circulating liquids. Variations of these approaches are discussed in References 22, 24, 28 through 31. In addition to the cell, reflecting mirrors are required to complete the resonator. In the early cuvettes of the type shown in Figures 3.2.1.2a, b, these were usually dielectric mirrors on the interior surfaces of the pistons. In later type cuvettes such as in Figures 3.2.1.2b, c, they were deposited on the external surfaces of the cell windows or, in some cases, external to the cuvette.

The final component of the laser is the excitation system consisting of flash lamps and enclosure. These have ranged from single lamps in elliptical or spherical enclosures to multilamp, close-coupled arrangements. A number of good systems can be found in References 17, 22, 29, and 31.

Most liquid lasers are subject to significant thermal effects because of the large tem-

Table 3.2.1.1
LIGANDS

Name	Formula	Symbol
Benzoylacetonate	$C_6H_5COCHCOCH_3^-$	B
Dibenzoylmethide	$C_6H_5COCHCOC_6H_5^-$	D
Benzoyltrifluoroacetonate	$C_6H_5COCHCOCF_3^-$	BTF
Deuterated benzoyltrifluoroacetonate	$C_6D_5COCHCOCF_3^-$	BTF_{d5}
Thenoyltrifluoroacetonate	$C_4H_3SCOCHCOCF_3^-$	TTF
Trifluoroacetylacetonate	$CF_3COCHCOCH_3^-$	TFA
α-Naphthoyltrifluoroacetonate	$C_{10}H_7COCHCOCF_3^-$	NTF
o-Halobenzoyltrifluoroacetonate[a]	$C_6H_4XCOCHCOCF_3^-$	o-XBTF
m-Halobenzoyltrifluoroacetonate[a]	$C_6H_4XCOCHCOCF_3^-$	m-XBTF
p-Halo benzoyltrifluoroacetonate[a]	$C_6H_4XCOCHCOCF_3^-$	p-XBTF
m,p-Dichlorobenzoyltrifluoroacetonate	$C_6H_3Cl_2COCHCOCF_3$	$3,4Cl_2BTF$
p-R benzoyltrifluoroacetonate[b]	$C_6H_4RCOCHCOCF_3^-$	p-RBTF
Pentafluoropropionate and 1,10 phenanthroline	$CF_3CF_2COO^-$ and $C_{12}H_8N_2$	PFP1,10PHEN
Deuterotributyl phosphate	$(CD_3CD_2CD_2CD_2O)_3PO^-$	TBP_{d27}

[a] In the case of *o-* and *m-* substituents the halogens F, Cl, and Br have been used. In *p-* substitutions I has also been used.

[b] CH_3O-(methoxy) and CF_3 (trifluoromethyl) have been used as R.

Table 3.2.1.2
CATIONS

Name	Formula	Symbol
Piperidinium	$C_5H_{12}N^+$	P^+
Deuterated piperidinium	$C_5D_{10}NH_2^+$	P^+_{d10}
Sodium	Na^+	Na^+
Morpholinium	$C_4H_8ONH_2$	M^+
Imidazolium	$C_3H_5N_2^+$	I^+
Pyrrolidonium	$C_4H_8NO^+$	Pyo^+
Pyrrolidinium	$C_4H_{10}N^+$	Pyi^+
Pyridinium	$C_5H_5NH^+$	Pyr^+
2,4,6 Trimethylpyridinium	$(CH_3)_3C_5H_2NH^+$	Me_3Pyr^+
Isoquinolinium	$C_9H_8N^+$	Iq^+

Table 3.2.1.2 (continued)
CATIONS

Name	Formula	Symbol
Ammonium	$NH_4{}^+$	$NH_4{}^+$
Dimethylammonium	$(CH_3)_2NH_2{}^+$	Me_2A^+
Tetramethylammonium	$(CH_3)_4N^+$	Me_4A^+
Diethylammonium	$(C_2H_5)_2NH_2{}^+$	Et_2A^+
Triethylammonium	$(C_2H_5)_3NH^+$	Et_3A^+
Tetraethylammonium	$(C_2H_5)_4N^+$	Et_4A^+
n-Butylammonium	$(C_4H_9)NH_3{}^+$	BA^+
Tetra n-butylammonium	$(C_4H_9)_4N^+$	B_4A^+
Tetra n-propylammonium	$(C_3H_7)_4N^+$	P_4A^+
2 Hydroxyethylammonium	$HO(CH_2)_2NH_3{}^+$	EOA^+
Benzylammonium	$C_7H_{10}N^+$	BeA^+
Dibenzylammonium	$(C_7H_7)_2NH_2{}^+$	Be_2A^+
Tetramethylguanidinium	$(N(C_2H_5)_2)_2CNH_2{}^+$	Me_4G^+

(Benzylammonium structure: benzene ring—$CH_2NH_3{}^+$)

Table 3.2.1.3
SOLVENTS

Name	Formula	Symbol
Ethanol	C_2H_5OH	E
Methanol	CH_3OH	M
Acetonitrile	CH_3CN	A
Dimethyl sulfoxide	$(CH_3)_2SO$	DMSO
Deuterated dimethyl sulfoxide	$(CD_3)_2SO$	DMSOd6
Dimethyl formamide	$(CH_3)_2NCHO$	DMF
Ethoxyproprionitrile	$CH_3CH(OC_2H_5)CN$	EN
Ethoxyethanol	$C_2H_5O(CH_2)_2OH$	EO
Carbon tetrachloride	CCl_4	CCL
Hexafluorobenzene	C_6F_6	HFB
p-Dioxane	$C_4H_8O_2$	p-D
Polymethylmethacrylate		PMMA
Ethanol-methanol 3:1		EM
Ethanol-methanol-dimethyl formamide		
7.5:2.5:1		DMFA1
3:1:1		DMFA2
9:3:2		DMFA3
Acetonitrile-proprionitrile-butyronitrile		N
1:1:1		
Ethoxyproprionitrile-acetonitrile-ethoxy-ethanol		ENAEO
2:1:1		

(p-Dioxane structure shown with $\overset{H_2}{C}$—$\overset{H_2}{C}$ at top, O on each side, $\underset{H_2}{C}$—$\underset{H_2}{C}$ at bottom)

perature coefficient of the index of refraction (water at 4° C is an exceptional case) and the nonuniformity of the pump absorption. These problems are discussed by Winston and Gudmundsen[32] and Rideal.[33] Dynamic loss determinations in a chelate laser were made by De Witte and Meyer.[34]

LASER RESULTS

The results using chelated rare earth ions for lasers are given in Table 3.2.1.4. Information relating to the transition involved is not given in the table but is stated in Section

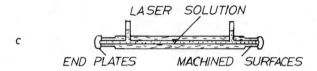

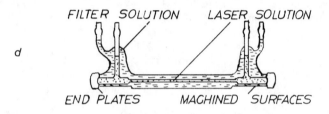

FIGURE 3.2.1.2.   Examples of cells used for chelate liquid lasers. (a) Piston cell with a wide base, usually about 4 mm. (b) Capillary piston type cell. The bore in these is about 1 mm or less. (c) Capillary cell with fixed windows and side arms. This type of cell can be used in static or circulatory systems. (d) A variation of a cell for a circulatory laser system with a provision for cooling the laser liquid and filtering the excitation radiation.

Table 3.2.1.4
CHELATE LASERS

RE Ion	Ligand	Cation	Solvent	Conc. (M)	Temp. (K)	Wavelength (nm)	Ref.
Eu^{+3}	B	P$^+$	EM	0.0018—0.05	110—150	612.9	4, 15-18, 35-39
			E	0.01—0.015	77—123	613	28, 38
			A	0.33	300	612.3	39[a]
		NH$_4^+$	EM	0.015	123	613	9
		Na$^+$	EM	0.01—0.02	123—133	611.1	20, 40
			DMF	0.02	133	611.1	40
		P$^+$	DMFA2	0.01	—	—	24
			N	0.01	—	—	24
	D	P$^+$	DMFA3	0.015	128	612	21
			DMFA2	0.01	—	—	8, 24
			N	0.01	—	—	24
	BTF	P$^+$	DMFA2	—	168	611.9	8
			A	0.01	298	611.9	41
		I$^+$	A	0.0075	238	611.8	42
		Pyo$^+$	A	0.0015—0.01	300	611.8	43
		P$^+$	ENAEO	0.0135	168	—	5
		P$^+_{d10}$	ENEAO	0.0135	168	—	5
	BTF$_{d5}$	P$^+$	ENEAO	0.0135	168	—	5
		P$^+_{d10}$	ENEAO	0.0135	168	—	5

Table 3.2.1.4 (continued)
CHELATE LASERS

RE Ion	Ligand	Cation	Solvent	Conc. (M)	Temp. (K)	Wavelength (nm)	Ref.
	TTF	P$^+$	PMMA	0.0001	77		48
			A	0.33	300	612.3	39^a
		Me$_2$A$^+$	A	0.005	238	612.5	42
	TFA	Me$_2$A$^+$	A	0.01	238	611.9	42
		NH$_4$$^+$	EM	0.015	123	612.2	9
	NTF	P$^+$	A	0.01	253	611.8	22
	o-ClBTF	Me$_2$A$^+$	A	0.0075	300	611.8	7, 30, 44
	m-ClBTF	Me$_2$A$^+$	A	0.0075	300	611.7	44
	p-ClBTF	Me$_2$A$^+$	A	0.0075	300	611.7	44
	o-FBTF	Me$_2$A$^+$	A	0.0075	300	611.7	44
	m-FBTF	Me$_2$A$^+$	A	0.0075	300	611.8	44
	p-FBTF	Me$_2$A$^+$	A	0.0075	300	611.7	44
		Me$_3$Pyr$^+$	ENAEO	—	168	—	7
	o-BrBTF	Me$_2$A$^+$	A	0.0075	300	611.7	44
	m-BrBTF	Me$_2$A$^+$	A	0.0075	300	611.7	44
	p-BrBTF	Me$_2$A$^+$	A	0.0075	300	611.7	44
	p-IBTF	Me$_3$Pyr$^+$	ENEAO	—	168	—	7
	p-CF$_3$BTF	Me$_3$Pyr$^+$	ENEAO	—	168	—	7
	p-CH$_3$OBTF	P$^+$	ENEAO	—	168	—	7
	m,p-Cl$_2$BTF	P$^+$	ENEAO	—	168	—	7
Tb^{+3}	TTF	P$^+$	PMMA	—	77	545	45
	TFA	—^{b}	A	0.0025	300	547	26
		—^{b}	D	0.0025	300	547	26
Nd^{+3}	PFP1,10PHEN	—	DMSO$_{d6}$	0.2	300	1057	46
	TBP$_{d27}$	—^{c}	HFB	0.033	300	1054	47
			CCL	0.033	300	1054	47

^{a} The lasers listed in this table were all excited by flashlamp discharge. Those given in Ref. 34 were excited by a coumarin dye laser.
^{b} The active component of this laser is the dihydrate of a tris chelate and thus requires no cation.
^{c} In these lasers the active material seems to be a simple tris compound.

IIB. In all cases but two, excitation was accomplished through the use of a flashtube arrangement.

REFERENCES

1. **Rautian, S. G. and Sobelman, I. I.,** Remarks on negative absorption, *Opt. and Spectr.,* 10, 65—66, 1961.
2. **Lempicki, A. and Samelson, H.,** Optical maser action in europium benzoylacetonate, *Phys. Lett.,* 4, 133—135, 1963.
3. **Crosby, G. A., Whan, R. E., and Alire, R. M.,** Intramolecular energy transfer in rare earth chelates. Role of the triplet state, *J. Chem. Phys.,* 34, 743—748, 1961.
4. **Bykov, V. P.,** Intramolecular energy transfer and quantum generators, *J. Exptl. Theoret. Phys. (U.S.S.R.),* 43, 1634—1635, 1962.
5. **Ross, D. L., Blanc, J., and Pressley, R. J.,** Deuterium isotope effect on the performance of europium chelate lasers, *App. Phys. Lett.,* 8, 101—102, 1966.
6. **Shionoya, S., Matsuda, Y., Morita, M., and Makishima, S.,** Energy transfer in the luminescence process of rare earth chelates, in *Proc. Int. Conf. on Luminescence,* Publishing House of the Hungarian Academy of Sciences, Budapest, 1966, 1709—1713.
7. **Ross, D. L. and Blanc, J.,** Europium chelates as laser materials, in *Advances in Chemistry Series,* Number 71, American Chemical Society, Washington, D. C., 1967, chap. 12.

8. **Brecher, C., Samelson, H., and Lempicki, A.,** Laser phenomena in europium chelates. III: spectroscopic effects of chemical composition and molecular structure, *J. Chem. Phys.*, 42, 1081—1096, 1965.

9. **Nehrich, R. B., Schimitschek, E. J., and Tras, J. A.,** Laser action in europium chelates prepared with NH$_3$, *Phys. Lett.*, 12, 198—199, 1964.

10. **Nugent, L. J., Bhaumik, M. L., George, S., and Lee, S. M.,** Ligand field spectra of some new laser chelates, *J. Chem. Phys.*, 41, 1305—1312, 1964.

11. **Lempicki, A., Samelson, H., and Brecher, C.,** Europium chelates as laser materials, in *Proc. IV Rare Earth Conf.*, Phoenix, Ariz., 1964, 351—361.

12. **Charles, R. G. and Perotto, A. M.,** Rare earth dibenzoylmethides preparation, dehydration and thermal stability, *J. Inorg. Nucl. Chem.*, 26, 373—389, 1964.

13. **Charles, R. G.,** Europium mixed ligand complexes derived from dibenzoylmethane and carboxylate anions, *J. Inorg. Nucl. Chem.*, 26, 2195—2199, 1964.

14. **Charles, R. G.,** Rare earth salicylaldehyde chelates, *J. Inorg. Nucl. Chem.*, 26, 2298—2300, 1964.

15. **Charles, R. G. and Ohlmann, R. C.,** Europium thenoyl trifluoro acetonate, *J. Inorg. Nucl. Chem.*, 27, 255—259, 1965.

16. **Metlay, M.,** Fluorescence lifetime of the europium dibenzoylmethides, *J. Phys. Chem.*, 39, 491—492, 1963.

17. **Bhaumik, M. L., Fletcher, P. C., Nugent, L. J., Lee, S. M., Higa, S., Telk, C. L., and Weinberg, M.,** Laser emission from a europium benzoylacetonate alcohol solution, *J. Phys. Chem.*, 68, 1490—1493, 1964.

18. **Ohlmann, R. C. and Charles, R. G.,** Fluorescence properties of europium dibenzoylmethide and its complexes with Lewis bases, *J. Chem. Phys.*, 41, 3131—3133, 1964.

19. **Charles, R. G. and Ohlmann, R. C.,** Europium dibenzoylmethide adducts, *J. Inorg. Nucl. Chem.*, 27, 119—127, 1965.

20. **Samelson, H., Brophy, V. A., Brecher, C., and Lempicki, A.,** Shift of laser emission of europium benzoylacetonate by inorganic ions, *J. Chem. Phys.*, 41, 3998—4000, 1964.

21. **Schimitschek, E. J. and Nehrich, R. B.,** Laser action in europium dibenzoylmethide, *J. Appl. Phys.*, 35, 2786—2787, 1964.

22. **Riedel, E. P. and Charles, R. G.,** Spectroscopic and laser properties of europium naphthoyltrifluoroacetonate in solution, *J. Chem. Phys.*, 42, 1908—1914, 1966.

23. **Derkacheva, L. D., Peregudov, G. V., and Sokolovskaya, A. I.,** Rare earth chelate lasers, *Sov. Phys. Uspekhi*, 10, 91—99, 1967.

24. **Samelson, H., Brecher, C., and Lempicki, A.,** Europium chelate lasers, *J. Chim. Phys.*, 64, 165—172, 1967.

25. **Riedel, E. P. and Charles, R. G.,** Effect of organic cations on the laser threshold of europium tetrakis benzoyl trifluoroacetonate, *J. Appl. Phys.*, 36, 3954—3955, 1965.

26. **Bjorklund, S., Kellermeyer, G., Hurt, C. R., McAvoy, N., and Filipescu, N.,** Laser action from terbium trifluoroacetylacetonate in p-dioxane and acetonitrile at room temperature, *Appl. Phys. Lett.*, 10, 160—162, 1967.

27. **Lempicki, A. and Samelson, H.,** Organic laser systems, in *Lasers*, Levine, A. K., Ed., Marcel Dekker, New York, 1966, 181—252.

28. **Schimitschek, E. J.,** Stimulated emission in rare earth chelate (europium benzoylacetonate) in a capillary tube, *Appl. Phys. Lett.*, 3, 117—118, 1963.

29. **Schimitschek, E. J. and Schumacher, E. R.,** Capillary quartz cell for liquid laser research, *Rev. Sci. Inst.*, 35, 521, 1964.

30. **Schimitschek, E. J., Nehrich, R. B., and Trias, J. A.,** Recirculating liquid laser, *Appl. Phys. Lett.*, 9, 103—104, 1966.

31. **Bjorklund, S., Kellermeyer, G. L., McAvoy, N., and Filipescu, N.,** Liquid laser cavities, *J. Sci. Instrum.*, 44, 947—948, 1967.

32. **Winston, H. and Gudmundsen, R. A.,** Refractive gradient effects in proposed liquid lasers, *Appl. Opt.*, 3, 143—146, 1964.

33. **Riedel, E. P.,** Light scattering in a solution of europium benzoylacetonate during optical pumping, *Appl. Phys. Lett.*, 5, 162—165, 1964.

34. **De Witte, O. and Meyer, Y.,** Etude del'amplification optique dans les lasers a chelate, *J. Chim. Phys.*, 64, 186—190, 1967.

35. **Aristov, A. V., Maslyukov, Yu. S., and Reznikova, I. I.,** Luminescence of a europium chelate solution under intense pulsed excitation, *Opt. Spectr.*, 21, 286—287, 1966.

36. **Lempicki, A., Samelson, H., and Brecher, C.,** Laser phenomena in europium chelates. IV. Characteristics of the europium benzoylacetonate laser, *J. Chem. Phys.*, 41, 1214—1224, 1964.

37. **Lempicki, A., Samelson, H., and Brecher, C.,** Laser action in rare earth chelates, in *Applied Optics Supplement 2 of Chemical Lasers*, 1965, 205—213.

38. Aristov, A. V. and Maslyukov, Yu. S., Stimulated emission in europium benzoylacetonate solutions, *J. Appl. Spectr.*, 8, 431—433, 1968.
39. Malashkevich, G. E. and Kuznetsova, V. V., Laser excited lasing in solutions of some europium chelates, *J. Appl. Spectr.*, 22, 170—172, 1975.
40. Meyer, Y., Astier, R., and Simon, J., Emission stimulee a 6111Å dans le benzoylacetonate d'europium active au sodium, *Compt. Rend.*, 259, 4604—4607, 1964.
41. Samelson, H., Lempicki, A., Brecher, C., and Brophy, V., Room temperature operation of a europium chelate liquid laser, *Appl. Phys. Lett.*, 5, 173—174, 1964.
42. Schimitschek, E. J., Nehrich, R. B., and Trias, J. A., Laser action in fluorinated europium chelates in acetonitrile, *J. Chem. Phys.*, 42, 788—790, 1965.
43. Schimitschek, E. J., Trias, J. A., and Nehrich, R. B., Stimulated emission in an europium chelate solution at room temperature, *J. Appl. Phys.*, 36, 867—868, 1965.
44. Schimitschek, E. J., Nehrich, R. B., and Trias, J. A., Fluorescence properties and stimulated emission in substituted europium chelates, *J. Chim. Phys.*, 64, 173—182, 1967.
45. Huffman, E. H., Stimulated optical emission of a Tb^{3+} chelate in a vinylic resin matrix, *Phys. Lett.*, 7, 237—238, 1963.
46. Heller, A., Fluorescence and room temperature laser action of trivalent neodymium in an organic liquid solution, *J. Am. Chem. Soc.*, 89, 167—169, 1967.
47. Goryaeva, E. M., Shablya, A. V., and Serov, A. P., Luminescence and stimulated emission for solutions of complexes of neodymium nitrate with perdeuterotributylphosphate, *J. Appl. Spectr.*, 28, 55—59, 1976.
48. Wolff, N. E. and Pressley, R. J., Optical laser action in an Eu^{+3}-containing organic matrix, *Appl. Phys. Lett.*, 2, 152—154, 1963.

3.2.2 APROTIC LIQUID LASERS

INTRODUCTION

The quenching of the fluorescence of rare earth ions is related to the vibrational frequencies found in the environment of the ion. Generally, the higher these frequencies are, the more readily is the ion fluorescence quenched. With this as a guide, Heller[1] was able to make a chelate laser with neodymium as the active ion by replacing nearest-neighbor hydrogen atoms with deuterium and fluorine, thus lowering the vibrational frequencies. As an alternative solution to this problem, Heller[2] was able to dissolve Nd^3 in the aprotic solvent selenium oxychloride, $SeOCl_2$, thereby achieving a high fluorescence yield and producing the first aprotic liquid laser.

Subsequent to this, a number of aprotic solvent systems were developed and other rare earth ions were incorporated into these solvents. The most successful, and apparently only, laser ion was Nd^{+3}. In the following sections, information relating to the chemistry of these solutions, the laser apparatus used, the laser systems developed and various phenomena encountered is presented.

THE CHEMISTRY OF APROTIC LASER SOLUTIONS WITH Nd^{+3}

The Materials

The solvent materials used in aprotic liquid lasers are given in Table 3.2.2.1 and the solvent systems derived from them are listed in Table 3.2.2.2 in the order of their discovery and their development. Those based on $SeOCl_2$ are hazardous and corrosive while those utilizing $POCl_2$ are less so. Solvents 11 and 12 have been developed more recently. The addition of $SOCl_2$ to make the ternary solvent system results in a marked reduction in the viscosity of the solution. The most recent solvent system, the bromide mixture (12) is reported to be much less toxic.

Preparative Procedures and Chemistry

The solvent systems listed in Table 3.2.2.2 break down into five classes. In the first four cases, the solvent is an oxyhalide and a halide capable of combining chemically with additional halide ions. The latter are called Lewis acids since they combine with the halide ion, a Lewis base. This is a generalization of the customary acid-base concept of aquo or protic chemistry to the aprotic case. The oxyhalide, called the donor molecule, can be viewed as dissociating in a manner analogous to water to provide the acid ion $SeOCl^+$ and the base Cl^-. The Lewis acid or acceptor $SnCl_4$ combines with Cl^- to form the anion $SnCl_6^{2-}$. In these systems, the neodymium compound can be viewed as a base. The four oxyhalide systems are

a. $SeOCl_2$, $SnCl_4$ or $SbCl_5$
b. $POCl_3$, $SnCl_4$
c. $POCl_3$, $ZrCl_4$
d. $POCl_3$, $SnCl_4$, $SOCl_2$

The fifth system is the bromide system.

e. PBr_3, $AlBr_3$, $SbBr_3$

The preparative procedures for a, b, and c are taken from Brecher and French,[10] that for (d) from Alekseev et al.[15] and for (e) from Bondarev et al.[16]

Table 3.2.2.1

PROPERTIES OF MATERIALS USED IN APROTIC LIQUID LASERS

Name of compound	Formula	Molecular weight	Density g/cm³	Refractive index	Melt. pt. °C	Boil. pt. °C	Dielectric const.
Selenium oxychloride	$SeOCl_2$	165.9	2.42	1.651	8.5	176.4	46.2
Phosphorous oxychloride	$POCl_3$	153.4	1.68	1.46	2	105.3	13.9
Phosphorous sulfochloride	$PSCl_3$	169.4					
Vanadium oxychloride	$VOCl_3$	173.8	1.83		−77	126.7	
Phosphorous tribromide	PBr_3	270.8	2.85	1.697	−40	172.9	32
Tin tetrachloride	$SnCl_4$	260.5	2.23		−33	114.1	
Antimony pentachloride	$SbCl_5$	299.1	2.34	1.601	2.8	140	0.8
Zirconium tetrachloride	$ZrCl_4$	233.1	2.80		300 (subl)		
Aluminum trichloride	$AlCl_3$	133.3	2.44		190 (2.5 at)	182.7 (752 mm)	
Aluminum tribromide	$AlBr_3$	266.7	3.01		97.5	263.3 (747 mm)	
Titanium tetrachloride	$TiCl_4$	189.7	1.73	1.61	−30	136.4	
Boron tribromide	BBr_3	250.6	2.65	1.553	−46.1	90.1 (740 mm)	
Antimony tribromide	$SbBr_3$	361.5	4.15		96.6	280	
Thionyl chloride	$SOCl_2$	119	1.61	1.527	−105	78.8	

Table 3.2.2.2
LIQUID LASER SOLVENT SYSTEMS

No.	Components	Fluorescence lifetime μsec	Ref.
1	$SeOCl_2 - SnCl_4$	280	2-8
2	$SeOCl_2 - SbCl_5$	275-300	2,3,5,6
3	$POCl_3 - SnCl_4$	245-370	4,6,9-11
4	$POCl_3 - SnCl_4 - D_2O$	200	12
5	$POCl_3 - ZrCl_4$	300-400	6,10,13
6	$POCl_3 - AlCl_3$	330	6
7	$POCl_3 - TiCl_4$	—	13,14
8	$POCl_3 - BBr_3$	—	13
9	$VOCl_3 - SnCl_4$	20	6
10	$PSCl_3 - AlCl_3$	100	6
11	$POCl_3 - SOCl_2 - SnCl_4$	>200	15
12	$PBr_3 - SbBr_3 - AlBr_3$	230	16
13	YAG	240	25
14	Glass, ED-2	310	25

a. $SeOCl_2 - SnCl_4 - Nd^{+3}$

Dissolve pure, anhydrous Nd_2O_3 in an anhydrous 5:1 mixture of $SeOCl_2 - SnCl_4$. This solution is distilled at a pressure of 40 mm until the boiling point of pure $SeOCl_2$ (90°C) is reached and about one third of the volume has been removed. The solution is then reconstituted to the desired concentration and acidity ($SnCl_4$) by adding appropriate amounts of anhydrous $SeOCl_2$ and $SnCl_4$. In this procedure, $NdCl_3$ can be used in place of Nd_2O_3 and $SbCl_5$ in place of $SnCl_4$. The preparative method outlined above avoids many of the cumbersome details inherent in other[2-8] methods. The chemical equations are

$$NdCl_3 + 3SeOCl^+ = Nd^{+3} + 3SeOCl_2 \qquad (1)$$
$$Nd_2O_3 + 6SeOCl^+ = 2Nd^{+3} + 3SeOCl_2 + 3SeO_2 \qquad (2)$$
$$SnCl_4 + 2SeOCl_2 = SnCl_6^{2-} + 2SeOCl^+ \qquad (3)$$
$$SbCl_5 + SeOCl_2 = SbCl_6^- + SeOCl^+ \qquad (4)$$

b. $POCl_3 - SnCl_4 - Nd^{+3}$

This is prepared in the same manner as the solution in (a) except that water in the mole ratio of 1:10 is added to the solvent mixture to facilitate the solution of the neodymium salt. The protic contamination thus introduced is removed by distilling (116°C) at atmospheric pressure until the volume has been reduced by about two thirds. The solution is then reconstituted by the addition of the appropriate pure anhydrous liquids. The overall chemical reaction for this procedure is, nominally:

$$Nd_2O_3 + 3SnCl_4 + 6POCl_3 = 2Nd^{+3} + 3SnCl_6^{2-} + 3P_2O_3Cl_4 \qquad (5)$$

However, the chemistry is more complex and involves compounds of the form:

$$Nd(PO_2Cl_2 \cdot SnCl_4)_{3-n}(SnCl_4POCl_3)_n$$

$$(6)$$

$$(n = 0,1,2)$$

The water added enters into a reaction of the type:

$$H_2O + POCl_3 = HPO_2Cl_2 + HCl \qquad (7)$$

and the $PO_2Cl_2^-$ ion plays a key role in solubilizing the neodymium salts.

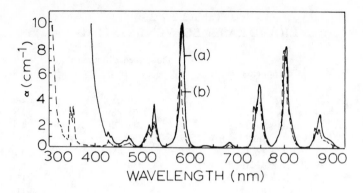

FIGURE 3.2.2.1.   Absorption spectra of 0.3 M Nd^{+3} laser solutions: (a) SeOCl$_2$ – SnCl$_4$ (b) POCl$_3$ – SnCl$_4$

c. POCl$_3$ – ZrCl$_4$ – Nd^{+3}

The neodymium salt most conveniently used for the preparation is Nd(CF$_3$COO)$_3$ prepared by recrystallization from a solution of Nd$_2$O$_3$ in aqueous trifluoroacetic acid. The water is removed from the salt by heating under vacuum. This anhydrous salt is dissolved in a mixture of POCl$_3$ – ZrCl$_4$ and the solution is distilled to remove traces of protic contamination and reconstituted. The nominal equation in this case is

$$2Nd(CF_3COO)_3 + ZrCl_4 + 12POCl_3 = Nd_2(ZrCl_6)_3 + 6P_2O_3Cl_4 + 6CF_3COCl \qquad (8)$$

In this preparation, the trifluoroacetate serves to generate the needed PO$_2$Cl$_2^-$ ion.

d. POCl$_3$ – SnCl$_4$ – SOCl$_2$ – Nd^{+3}

This solution[15] is prepared by starting with the standard POCl$_3$ – SnCl$_4$ – Nd^{+3} laser solution and distilling off an azeotropic mixture of POCl$_3$SnCl$_4$. The original volume is then reconstituted by adding anhydrous SOCl$_2$. This can be done to the point at which solvent is 70% by volume of SOCl$_2$.

e. PBr$_3$ – AlBr$_3$ – SbBr$_3$ – Nd^{+3}

Details on this preparation are limited. Bondarev et al.[16] simply state that AlBr$_3$ and SbBr$_3$ were added to PBr$_3$ forming chemical compounds because this improved the solubility of neodymium ions. The neodymium ions were dissolved in the form of anhydrous salts.

Spectroscopy and Other Physical Properties

The absorption and emission spectra have the same general appearance and shape as do those of the neodymium ion in glasses. Figure 3.2.2.1 compares the absorption of 0.3 M (1.8 × 10^{20} cm^{-3}) Nd^{+3} solutions in SeOCl$_2$:SnCl$_4$ and POCl$_3$:SnCl$_4$ while Figure 3.2.2.2 compares the former with POCl$_3$:SnCl$_4$ and with Schott laser glass LG 55. The emission spectra at various concentrations in SeOCl$_2$ are shown in Figure 3.2.2.3. Figure 3.2.2.4 compares the emission spectra of an 0.3 M solution of Nd^{+3} in SeOCl$_2$:SnCl$_4$ and POCl$_3$:SnCl$_4$ and Figure 3.2.2.5 shows the emission corresponding to the samples whose absorption is shown in Figure 3.2.2.2. The peak absorbance of the various pump bands in the three solvents a, b, and c is summarized in Table 3.2.2.3. Finally, the absorption cross section of the laser transition and other physical properties of the laser solutions are given in Table 3.2.2.4.

The cross section of the liquid systems lies between that of a typical glass and that of YAG as is seen from Table 3.2.2.4. On this basis, it would be expected that the

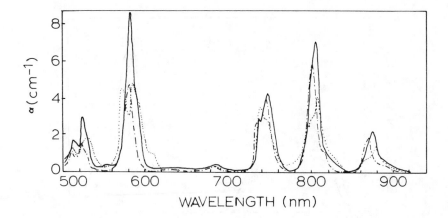

FIGURE 3.2.2.2.   Absorption spectra of Nd^{+3} in SeOCl$_2$ − SnCl$_4$, POCl$_3$ − ZrCl$_4$ and LG 55 glass: − SeOCl$_2$; −·−· POCl$_3$; ······ Glass.

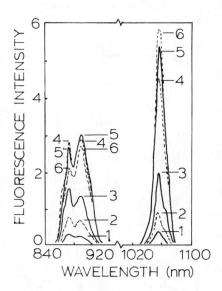

FIGURE 3.2.2.3.   Fluorescence spectra of Nd^{+3} in SeOCl$_2$ − SnCl$_4$ laser solutions at various concentrations.

No.	Concentration
1	6×10^{18} cm^{-3} (0.01 N)
2	18×10^{19} cm^{-3} (0.05 N)
3	6×10^{19} cm^{-3} (0.1 N)
4	18×10^{20} cm^{-3} (0.3 N)
5	3×10^{20} cm^{-3} (0.5 N)
6	6×10^{20} cm^{-3} (1.0 N)

liquid systems will have good efficiencies and energy storage characteristics and will probably find their principal application in pulsed operation.

The fluorescence lifetimes for the liquid systems is comparable to that of the solid state systems. For the liquids, the highest lifetimes found, for a given solvent, are almost equal to those calculated from the branching ratios and the cross-section data.

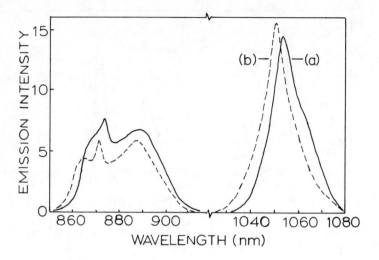

FIGURE 3.2.2.4.   Emission spectra of 0.03 M Nd^{+3} laser solutions: (a) SeOCl$_2$ − SnCl$_4$ (b) POCl$_3$ − SnCl$_4$

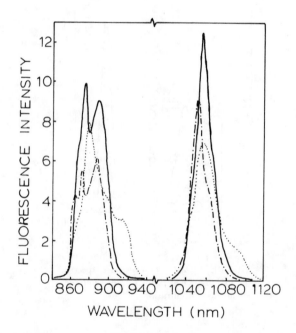

FIGURE 3.2.2.5.   Emission spectra of Nd^{+3} in different solvent systems: − SeOCl$_2$; −·− POCl$_3$; ····· Glass.

This implies that the quantum efficiencies for the neodymium fluorescence in the liquid hosts is close to unity.

Overall, these spectroscopic characteristics indicate that the performance of the liquid laser will be between that in YAG and in the conventional glasses. Indeed, this is what is found. The major differences will reside in the ability of the respective materials to handle the thermal load in the various applications. In general, this, too, is what has been observed.

Structure

The broad emission and absorption spectra raise the issue of the nature of the site

Table 3.2.2.3
PEAK ABSORBANCE OF Nd^{+3} PUMP BANDS
IN SOME APROTIC SOLVENTS

	Peak absorbance (ℓ mol^{-1} cm^{-1})	
Peak wavelength (nm)	Solvent No. 1	Solvent No. 3,5
350	—	11.7
520	11.7	6.7
580	33.3	2.0
750	16.7	16.7
800	26.7	26.7
870	10.0	5.0

Table 3.2.2.4
SOME PROPERTIES OF LASER SOLUTIONS

Solvent no.	Absorption cross section (cm^2 × 10^{20})	Refractive index	Viscosity (cp)[24,25]	Loss Passive (%/cm)	Loss Dynamic (%/cm)
1	7.7[17] 7.8[18]	1.667	10—30		0.12[25,41]
3	9.6[18] 5.6, 8.3, 8.5[9] 8.1[20] 10.0[21] 7.8[22] 4.8[23]	1.488	5—6		2.[25,41]
5	9.6[4,24]	1.487	4—5	0.16[7]	0.5[25,41]
YAG	88[25]	1.823	—		
Glass, ED-2	4.5[25]	1.556	—		

and the nature of the line broadening. Brecher and French[10] using chemical and spectroscopic arguments were able to show that the structure of the emitting site in the solvents $SeOCl_2 - SnCl_4$, $POCl_3 - SnCl_4$, and $POCl_3 - ZrCl_4$ were different and were able to deduce structures consistent with the eight-fold coordination number of the rare earth ion and the chemical behavior of the solutions. More recently Bonch-Bruyevich et al.[26] were able to show that there were two sites in the $POCl_3 - SnCl_4$ solvent. Andreeva et al.[27] deduced four types of centers in this solvent and Levkin and Ral'chenko[28] showed that one of these was the same as that found in neodymium pentaphosphate. Gilyarov et al.[29] propose some possible structures for these sites.

Other Rare Earth Ions

The preparation and spectroscopic properties of trivalent Pr, Sm, Eu, Tb, Dy, Ho, Er, Tm, and Yb in $SeOCl_2 - SnCl_4$ and $- SbCl_5$ were studied by Heller.[31] Kato and Shimoda[32] studied Tb^{+3} in $POCl_3 - SnCl_4$ and concluded that the symmetry of the terbium ion was T_d. Watanabe[33] looked into Pr^{+3} in $SeOCl_2 - SnCl_4$ and $POCl_3 - TiCl_3$ and concluded that the quantum efficiency of the $^3P_0$ level was too low for laser action. Jezowska-Trzebiatoeska et al.[34] calculated a quantum efficiency of 0.05 for Er^{+3} in $POCl_3 - ZrCl_4$ and concluded that this ion would not show laser action. Friedman

Table 3.2.2.5

DATA FOR STATIC LIQUID LASER SYSTEMS

Solvent	Conc. (m/l)	Wavelength (nm)	Lifetime (μsec)	Cell Dia.	Length (cm)	R_{out} %	Threshold (J)	Input (J)	Output (J)	Ref.
1	0.5	1056	110	0.4	5-15	95	<5	1000	1	36
	0.1	1056	180	0.6	9	90		1250	0.2	8
	0.3	1330	180			99.5	270			38
	0.3	1056	240				2			18
	0.3	1055	230	0.64	15	45	200	1500	2.5	39
				0.95	15	45	450	1500	4.5	39
				1.25	15	45	650	1750	7.5	39
		1058	83	0.8	13	95	65	500	0.1	37
	0.23		240			70	160	500	1.6	40
	0.3	1056	220							6
	0.3	1056	280	0.64	15	85	10	200	0.7	41
2	0.3	1056	255				2			18
	0.1-0.5		270-300							6
3	0.1	1060		1.2	12.5	90	3300			9
	1%[a]	1055	120-180	0.5	8	84		600	0.2	14
	0.1		240	0.6	10	70		375	0.75	40
	0.34	1052	360	0.6	10	45	125	500	1.6	42
	0.3	1056	310	0.6	15	85	85	200	1	41
				0.96	24	4		14700	300	43
				2.6	24	4		45900	1275 (?)	43
5	0.2		400							6
	0.1	1053					<30	600	2.5	13
	0.3	1056	330	1.25	25	52	100	4000	100	25
	0.3	1056	300	0.8	15	4	450	1500	4	25
	0.3	1056	330	1.6	25	55	180	1000	28	41
	0.3	1056	330	2.5	70	48	1500	4000	40	45
6	0.2		330							6
7	0.1	1053					<30	600	2.5	13
	1.0%[a]	1054	140	0.5	8.0	92	32	340	0.3	14
8	0.2									13
10	0.2		100				?			6
11	0.18-0.7		210—230							15
12	0.17	1066	230	1.8	10	86	20	100	0.4	16

[a]   Solutions made with 1 wt. % Nd_2O_3.

and Bell[35] extended the use of aprotic solvents $POCl_3 - ZrCl_4$ to the actinide Am^{+3} and found only a weak fluorescence having a lifetime too short for them to measure.

LASER SYSTEMS

Long-Pulse, Static

There are many examples relating to this mode of operation and these are listed in Table 3.2.2.5. The principal variables in this table are the solvent system, cell size, and output. A given output performance has to be interpreted in terms of the various system entries in the table.

Long-Pulse, Circulating

This mode of operation takes advantage of the liquid nature of the gain medium. The characteristics of the various systems built for this mode are listed in Table 3.2.2.6.

Table 3.2.2.6
DATA FOR CIRCULATORY LIQUID LASER SYSTEMS

Solvent	Conc. (m/l)	Circul'n velocity (l/sec)	Reynold's number	Cell Dia. (cm)	Length (cm)	Mode	Rep rate (pps)	Pulse thresh. (J)	Input	Output	Ref.
3	0.3	0						364	1.8 kJ	8.6 J	
		0.037	880	1.2	40	Single		500	1.8 kJ	9.3 J	48
		0.111	2630			Shot		500	1.8 kJ	7.5 J	
5	0.3	0		1.2	15	Single		300	2 kJ	17 J	41
		Laminar				Shot		400	2 kJ	6 J	
	0.3	0.7	5000	2.2	25	Single		100	4 kJ	100	25
							5		20 kW	425 W	
	0.3	1—2		2.2	30		30		30 kW	336 W	46

Table 3.2.2.7
DATA FOR Q-SWITCHED LIQUID LASERS

Solvent	Conc. (m/l)	Pulse width (nsec)	Mode	Rep rate (pps)	Cell Dia. (cm)	Length (cm)	R_{out} %	Peak power (MW)	Input	Output	Ref.
1		80	Dye		0.8	13		0.5			37
	0.3		Dye		0.64	15	72	170	810 J		51
	0.1	30	Dye + Prism					1			59
2		8	Dye		1.0	30	1	180	1500 J		50
		11	E. Opt					50	2000 J		
3	2%[a]	12	Dye		0.5	8	40	10			
5	0.3	50	Prism		0.8	15		80			44
	0.3	25	Pockels		0.9	35	80		1000 J	0.5 J	52
	0.3	50	Prism	5	2.2	30	55		1000 W	3 W	46
	0.3	20	Pockels	10	1.4	16.5	30	100	900 W	4.5 W	53
	0.3	13	Pockels		0.75	15.5	28	50	500 J		57
	0.3	20			0.8	15	25				61

[a] Solution made with 2 wt. % Nd_2O_3.

Very high single pulse energies (100 J) have been obtained as well as high average powers.

Q-Switched

The results from this mode of operation are presented in Table 3.2.2.7. Two of these systems[46,53] have been circulatory and repetitively pulsed.

Mode-Locked

Mode locking is a frequent concomitant to passive Q-switching with a saturable absorber such as the Kodak 9740 and 9860 dyes. It was first observed by Samelson and Lempicki[51] in $SeOCl_2 - SnCl_4$. These were broad pulses, about 2 nsec in half width. Two photon fluorescence indicated a complex pulse structure. More careful experiments by Alfano and Shapiro[55] in a 25 cm cell filled with Nd^{+3} $POCl_3$-$ZrCl_4$ produced clean pulse trains of microsecond length with individual pulses 3 psec long and having a peak power of 1 GW. A glass laser under comparable conditions resulted in pulse trains several hundred nsec long but of higher peak power. Selden[67] observed that, in the $POCl_3$-$ZrCl_4$ solvent, the pulse train is produced in a single transverse mode and the mode depends on the thermal lens character produced in the liquid. Fill[54,56]

Table 3.2.2.8
DATA FOR LIQUID LASER AMPLIFIERS

Solvent	Oscillator	Mode of operation	Cell Dia. (cm)	Length (cm)	Pump energy (J)	Gain	Ref.
1	Nd glass LG 55	Circulated				25	41
		Random spike input	1.4	16.5	2000	Small sig.	
	Nd glass silicate	Static				1.2	59
		Q-switched input	0.8	33	4900	Sat'd	
	Nd-SeOCl₂-SnCl₄					2.1	59
	Nd glass silicate	Static	1.0	30	4900	2.0	60
		Q-switched 10 MW 30 nsec					
5	Nd-POCl₃-ZrCl₄	Static	0.8	30		50	44
		Q-switched 5 MW					
	Nd glass	Static	0.9	33	1500	12.6	62
	Nd POCl₃	Q-switched			1500	63	
	Nd-POCl₃	Static	0.8	30	4000	20	61
		Q-switched 5—10 MW					
	Nd glass	Static	0.75	15.5	+ 1000	9	63,64
		mode locked 5 psec					
	Nd POCl₃	Static	0.9	22.8	1760	18	65
		Q-switched 11 nsec 0.5j ∿ 50 MW				(500 MW)	
	NdPOCl₃	Static 10 nsec 0.5 j	2.5	70	12000	100 (3 GW)	45,66

observed mode locking in $POCl_3 - AlCl_3$ using a saturable dye and an acousto-optic shutter.

Amplifier

The liquid gain medium has also been used as an amplifier in a single pass, nonregenerative manner. The oscillators used to drive the amplifier have been glass, YAG, and the liquid itself, all of them doped with Nd^{+3}. These oscillators emit at different wavelengths within the bandwidth of the amplifier and the gain is modified accordingly. Pertinent data on amplifiers is given in Table 3.2.2.8.

OTHER PHENOMENA

Nonlinear Scattering

In condensed phase lasers, the medium in which the oscillation develops plays an important role. In the liquid, Rayleigh, Brillouin, and Raman scattering can give rise to losses. Rayleigh and Brillouin scattering usually takes place at frequencies within the linewidth of the laser transition but Raman scattering falls outside this linewidth and stimulated Raman scattering (SRS) can be a limiting factor in the generation of laser power. The relevant information is given in Table 3.2.2.9.

In a single pass amplifier, 70 cm long, Green et al.[69] found a 10% conversion to the Raman frequency while in the oscillator mode, with the same cell, only an 8% conversion was experienced.[45] Andreou[66] points out that in a laser operating in a single transverse mode the conversion to Raman light could become complete.

Self Q-Switching and Self Mode-Locking[72]

Under conditions of low feedback, output reflectivities less than about 50% the output of aprotic, Nd^{+3} liquid lasers begins to show some giant spikes in the midst of

Table 3.2.2.9

NONLINEAR PARAMETERS OF LIQUID LASER SOLUTIONS

Parameter	$POCl_3:ZrCl_4$	$SeOCl_2:SnCl_4$	Ref.
Raman shift	488 cm^{-1}	386 cm^{-1}	69,71
	450 cm^{-1}		68
Brillouin triplet spacing	0.22 cm^{-1}	0.30 cm^{-1}	70
Stimulated Brillouin shift	0.12 cm^{-1}	0.11 cm^{-1}	71
Gain coefficient at 1060 nm (calculated)	1.6×10^{-3} cm/MW	Same	71
Forward scattering cross section	1.6×10^{-30} cm^2	4.9×10^{-30} cm^2	71

the random spike type of output characteristic of the free running laser. These giant spikes are of megawatt power and of duration usually less than 50 nsec. As the output reflectivity continues to decrease to that of a single output window (about 4%) and then to the point where both windows provide the only feedback, the self-Q-switched spikes dominate and frequently have the character of a high order transverse mode. The most reasonable physical basis for this phenomenon seems to be the formation of a thermal phase grating.[73] This arises from the heat due to the transition between the terminal laser level and the ground state when there is a standing wave in the liquid gain medium.

Distortion

The temperature dependence of the refractive index of liquids is generally 10^2 to 10^3 times greater than that for solids. For $SeOCl_2$ and $POCl_3$ based laser solutions, including those with $SOCl_2$, dn/dT is between $-(5 \text{ to } 7) \times 10^{-4}$ °C^{-1}.[25,55] For the PBr$_3$-based solvent it is -1×10^{-3}.[72] When the solution in a cylindrical cell has a radial temperature gradient, it behaves as a lens. This problem was first considered by Winston and Gudmundsen[75] and Riedel.[76] In effect, a higher temperature near the laser cell wall leads to the formation of a positive lens.[47,77-79] The situation is complicated since the cell wall itself becomes heated by the excitation radiation. The external cooling by the water jacket serves to maintain the wall temperature so that the isothermal profile of the circulating laser liquid can be maintained.[25] The turbulent flow results in thermal variations on a microscale so that even a macroscopically uniform medium will have only a limited optical quality. This and the related thermal problems are discussed in depth in Reference 25.

APPARATUS

Much of the apparatus used in conjunction with aprotic lasers is similar to that used with solid state lasers, either crystalline or glass. This relates to power supplies, cooling systems, and flash enclosures. For the latter, close coupling with aluminum wrapping is satisfactory for simple experimental systems. For more complex ones, modified radiant heating furnaces proved satisfactory. The active medium was located at the common focus and the flash lamps at the other focus of multiple elliptical systems.

Unique to the liquid laser, however, are the circulatory system and the cells. The circulatory arrangement for a three laser head system is shown in Figure 3.2.2.6. The pump requirements are exceedingly severe because of the corrosive nature of the laser liquid and its sensitivity to water contamination. For this reason, magnetically driven pumps constructed of nickel,[46] Carpenter 20 stainless steel,[25] or teflon and glass[6] have been most successful. Heat exchangers have varied from simple chemical condensers to nickel shell and tube[25] arrangements and filters are readily available in the form of fritted quartz cylinders. The free surface is a standpipe with a liquid-dry gas interface

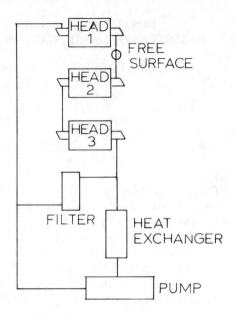

FIGURE 3.2.2.6. Schematic diagram of a circulatory liquid laser system showing the relative placement of the various components.

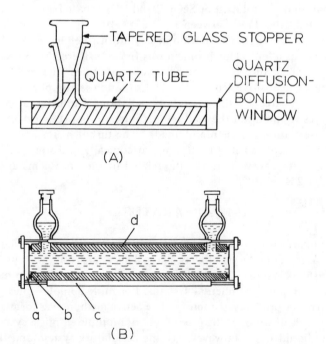

FIGURE 3.2.2.7. Types of liquid laser cells for use in static systems: (A) quartz, diffusion-bonded laser cell (B) demountable laser cell: a) cell windows, b) teflon-coated, silicone O-ring, c) invar tie rods, d) cell body.

that serves as a means for eliminating trapped gas and establishing a reference pressure. A detailed description can be found in Reference 25.

Typical cells are illustrated in Figure 3.2.2.7 to Figure 3.2.2.9. These are described

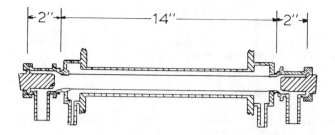

FIGURE 3.2.2.8.   Assembly of a demountable cell for a circulatory liquid laser showing external jacket for water cooling.

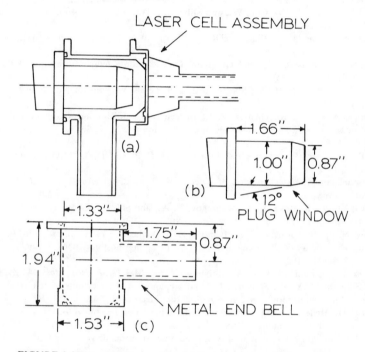

FIGURE 3.2.2.9.   Detail showing nickel plenum chamber at the end of the laser cell for circulatory system: (a) detail of plenum chamber assembly at the end of the laser cell, (b) detail of plug window for the laser cell, (c) nickel end bell (cell inlet tube shown at right).

in detail in Reference 25. The first of these in Figure 3.2.2.7A is the simplest and the end windows can be heat-sealed on or cemented on with a number of adhesives.[45] A more convenient arrangement is the demountable cell in Figure 3.2.2.7B. Such cells are easy to clean and dry and, in general, easier to use since there is some adjustment for the parallelism of the end windows. The expansion volumes serve as a means for filling the cells and dissipating the thermal shock arising from the excitation. The final two figures are for larger cells used in the laser system of Figure 3.2.2.6. These cells are designed to meet the hydrodynamic requirements and are discussed in detail in Reference 25. Other cell designs can be found in References 20, 47, and 74.

REVIEW ARTICLES

Reviews of the field can be found in References 80 and 81; the most complete survey is in Reference 25.

REFERENCES

1. **Heller, A.**, Fluorescence and room temperature laser action of trivalent neodymium in an organic liquid solution, *J. Am. Chem. Soc.*, 89, 167—169, 1967.
2. **Heller, A.**, A high gain, room-temperature liquid laser: trivalent neodymium in selenium oxychloride, *Appl. Phys. Lett.*, 9, 106—108, 1966.
3. **Heller, A.**, Liquid lasers. Preparative techniques for selenium oxychloride based laser solutions, *J. Am. Chem. Soc.*, 90, 3711—3712, 1968.
4. **Brecher, C. and French, K. W.**, Comparison of aprotic solvents for Nd^{+3} liquid laser systems: SeOCl$_2$ and POCl$_3$, *J. Phys. Chem.*, 73, 1785—1789, 1969.
5. **Weichselgartner, H.**, Zur preparation laseraktiver flüssigkeiten mit höhen fluoreszenz lebensdauern, *Z. Naturforsch.*, 24a, 1665—1666, 1969.
6. **Weichselgartner, H. and Perchermeier, J.**, Anorganischer flüssigkeits laser, *Z. Naturforsch.*, 25a, 1244—1247, 1970.
7. **Brecher, C., French, K., Watson, W., and Miller, D.**, Transmission losses in aprotic liquid lasers, *J. Appl. Phys.*, 41, 4578—4581, 1970.
8. **Kato, D. and Shimoda, K.**, Liquid SeOCl$_2$:Nd^{+3} laser of high quality, *Jpn. J. Appl. Phys.*, 7, 548, 1968.
9. **Blumenthal, N., Ellis, C. B., and Grafstein, D.**, New room temperature liquid laser: Nd(III) in POCl$_3$-SnCl$_4$, *J. Chem. Phys.*, 48, 5726, 1968.
10. **Brecher, C. and French, K. W.**, Spectroscopy and chemistry of aprotic Nd^{+3} laser liquids, *J. Phys. Chem.*, 77, 1370—1377, 1973.
11. **Polyanova, A. G. and Ral'chenko, V. I.**, Influence of oxidizing agents on the characteristics of spectral luminescence of laser materials of composition POCl$_3$ + SnCl$_4$: Nd, *J. Appl. Spect.*, 20, 41—43, 1974.
12. **Brun, P. and Caro, P.**, Laser effect in a solution of POCl$_3$, D$_2$O, Nd$_2$O$_3$, *Compt. Rend.*, 274, 1072—1074, 1972.
13. **Schimitschek, E. J.**, Laser emission of a neodymium salt dissolved in POCl$_3$, *J. Appl. Phys.*, 39, 6120—6121, 1968.
14. **Voronko, Yu. K., Krotova, L. V., Sychugov, V. A., and Shipulo, G. P.**, Lasers with liquid active materials based on POCl$_3$:Nd^{+3}, *J. Appl. Spect.*, 10, 168—170, 1969.
15. **Alekseev, N. E., Zhabotinski, M. E., Ivanova, E. B., Malashko, Ya. I., and Rudnitskii, Y. P.**, Effect of thionyl chloride on the laser characteristics of the liquid phosphor POCl$_3$ − SnCl$_4$ − Nd^{+3}, *Inorg. Mater.*, 9, 215—217, 1973.
16. **Bondarev, A. S., Buchenkov, V. A., Volyukin, V. M., Mak, A. A., Pogodaev, A. K., Przhevaskii, A. K., Sidorenko, Yu. K., Soms, L. N., and Stepanov, A. I.**, New low toxicity inorganic Nd^{+3} − activated liquid medium for lasers, *Sov. J. Quantum Electron.*, 6, 202—204, 1976.
17. **Samelson, H., Heller, A., and Brecher, C.**, Determination of the absorption cross section of the laser transition of the Nd^{+3} ion in the Nd^{+3}:SeOCl2 system, *J. Opt. Soc. Am.*, 58, 1054—1056, 1968.
18. **Heller, A.**, Liquid lasers — design of neodymium based inorganic systems, *J. Molec. Spectroscopy*, 28, 101—117, 1968.
19. **Collier, F., Girard, G., Michon, M., and Pocholle, J. P.**, Amplification cross section of the 1,052 μ transition of Nd^{+3} in POCl$_3$ − SnCl$_4$(H$_2$O) system by three different methods, *IEEE J. Quantum Electron.*, QE-7, 519—522, 1971.
20. **Grigoryants, V. V., Zhabotinski, M. E., and Markushev, V. M.**, Lasing and spectral line characteristics in phosphate glasses and inorganic liquids with neodymium, *IEEE J. Quantum Electron.*, QE-8, 196—198, 1972.
21. **Boling, N. L. and Dube, G.**, Comments on "Amplification cross section of the 1.052 μ transition of Nd^{+3} in POCl$_3$ − SnCl$_4$(H$_2$O) system at three different methods", *IEEE J. Quantum Electron.*, QE-8, 388—389, 1972.
22. **Birnbaum, M. and Gelbwachs, J. A.**, Stimulated emission cross section of Nd^{+3} at 1.06 μ in POCl$_3$, YAG, CaWO4,ED-2 glass and LG55 glass, *J. Appl. Phys.*, 43, 2335—2338, 1972.
23. **Yanush, O. V., Karapetyan, G. O., Mosichev, V. I., Sinyuta, S. A., and Chinyakov, S. V.**, Quantum yield and amplification cross section for stimulated emission for solutions of neodymium in phosphorous oxychloride, *J. Appl. Spectr.*, 24, 444—450, 1976.
24. **Kato, D.**, Measurement of the stimulated emission cross section of the Nd^{+3}:POCl$_3$:ZrCl$_4$ liquid laser, *IEEE J. Quantum Electron.*, QE-8, 529—530, 1972.
25. **Samelson, H. and Kocher, R.**, Final Technical Report, High Energy Liquid Laser, Contract N00014-68-C-0110, 1974.
26. **Bonch-Bruyevich, A. M., Kaporskii, L. N., and Kalabushkin, O. I.**, Inorganic liquid lasers, *Opt. Technol.*, 40, 770—781, 1973.

27. Andreeva, T. K., Zhabotinsky, M. E., Levkin, L. V., and Ralchenko, V. I., Spectral luminescence study of solutions of active Nd^{+3} centers in $POCl_3$ + $SnCl_4$, *Opt. Spectr.*, 37, 529—532, 1974.

28. Levkin, L. V. and Ral'chenko, V. I., Structure of active complexes in $POCl_3$-$SnCl_4$-Nd^{+3}, *Sov. J. Quantum Electron.*, 5, 176—179, 1975.

29. Gilyarov, O. N., Zhabotinsky, M. E., Kulikovskii, B. N., Lebedev, V. G., Levkin, L. V., and Ral'chenko, V. I., Changes in the active center in the liquid laser material ($POCl_3$ + $SnCl_4$):Nd^{+3} in relation to temperature, *Inorg. Mater.*, 13, 1784—1787, 1977.

30. Gilyarov, O. N., Zhabotinsky, M. E., Kulikovskii, B. N., Lebedev, V. G., Levkin, L. V., and Ral'chenko, V. I., Changes in the active center in the liquid laser material ($POCl_3$ + $SnCl_4$):Nd^{+3} and their relations with the crystallization process, *Inorg. Mater.*, 13, 1788—1790, 1977.

31. Heller, A., Fluorescence, absorption an energy transfer of rare earth ion solutions in selenium oxychloride, *J. Molec. Spectroscopy*, 28, 208—232, 1968.

32. Kato, D. and Shimoda, K., Fluorescence spectrum of Tb^{+3} in $POCl_3$ - $SnCl_4$ liquid, *Jpn. J. Appl. Phys.*, 9, 581—582, 1970.

33. Watanabe, A., The spectroscopy of Pr^{+3} ions in aprotic solvents, *Can. J. Phys.*, 52, 868—875, 1974.

34. Jezowska-Trzebiatoeska, B., Ryba-Romanowski, W., and Mazurak, Z., Radiative transition probabilities within 4-f configurations of Nd^{+3} and Er^{+3} in $POCl_3$:$ZrCl_4$, *Chem. Phys. Lett.*, 43, 417—419, 1976.

35. Friedman, H. A. and Bell, J. T., A search for laser phenomena in the actinides: studies of the investigations of Am^{+3} in liquid $POCl_3$, *J. Inorg. Nucl. Chem.*, 34, 3928—3930, 1972.

36. Lempicki, A. and Heller, A., Characteristics of the Nd^{+3}: $SeOCl_2$ liquid laser, *App. Phys. Lett.*, 9, 108—110, 1966.

37. Yamaguchi, G., Endo, F., Murakawa, S., Okamura, S., and Yamanaka, C., Room temperature, Q-switched liquid laser ($SeOCl_2$-Nd^{+3}), *Jpn. J. Appl. Phys.*, 7, 179, 1968.

38. Heller, A. and Brophy, V., Liquid lasers: stimulated emission of Nd^{+3} in selenium oxychloride solutions in the $^4F_{3/2}$ to $^4I_{13/2}$ transition, *J. Appl. Phys.*, 39, 6120—6121, 1968.

39. Samelson, H., Lempicki, A., and Brophy, V., Output properties of the Nd^{+3}:$SeOCl_2$ liquid laser, *IEEE J. Quantum Electron.*, QE-4, 849—855, 1968.

40. Le Sergent, C., Michon, M., Rousseau, S., Collier, F., Dubost, H., and Raoult, G., Characteristics of the laser emission obtained with the solution $POCl_3$, $SnCl_4$, Nd_2O_3, *Compt. Rend.*, 268, 1501—1503, 1969.

41. Samelson, H., Kocher, R., Waszak, T., and Kellner, S., Oscillator and amplifier characteristics of lasers based on Nd^{+3} dissolved in aprotic solvents, *J. Appl. Phys.*, 41, 2459—2469, 1970.

42. Collier, F., Michon, M., and Le Sergent, C., Paramètres laser du système liquide Nd^{+3}-$POCl_3$-$SnCl_4(H_2O)$ compares à ceux du YAG et du verre dopé au néodyme, *Compt. Rend.*, 272, 945—947, 1971.

43. Zaretskii, A. I., Vladimirova, S. I., Kirillov, G. A., Kormes, S. B., Negiva, V. R., and Sukharov, S. A., Some characteristics of a $POCl_3$ + $SnCl_4$ + Nd^{+3} inorganic liquid laser, *Sov. J. Quantum Electron.*, 4, 646—648, 1974.

44. Ueda, K., Hongyo, M., Sasaki, T., and Yamanaka, C., High power Nd^{+3} $POCl_3$ liquid laser system, *IEEE J. Quantum Electron.*, QE-7, 291, 1971.

45. Green, M., Andreou, D., Little, V. I., and Selden, A. C., A multigigawatt liquid laser amplifier, *J. Phys. D*, 9, 701—707, 1976.

46. Fahlen, T. S., High average power Q-switched liquid laser, *IEEE J. Quantum Electron.*, QE-9, 493—496, 1973.

47. Watson, W., Reich, S., Lempicki, A., and Lech, J., A circulating liquid laser system, *IEEE J. Quantum Electron.*, QE-4, 842—849, 1968.

48. Malyshev, B. N., Karnaukh, N. P., Paramonova, N. A., and Kulikovskii, B. N., Spatial and energy characteristics of a $POCl_3SnCl_4Nd$ circulating liquid laser, *Sov. J. Quantum Electron.*, 1, 103—104, 1971.

49. Aristov, A. V., Batyaev, I. M., Lubimov, E. I., Maslyukov, Yu. S., and Cherkasov, A. S., Stimulated emission in solutions of neodymium in heavy-atom inorganic solvents, *Opt. Spect.*, 26, 365, 1969.

50. Lang, R. S., Die erzeugung von reisen impulsen durch einen aktiv und passiv geschalteten anorganischen neodym-flüssigkeits laser, *Z. Naturforsch.*, 25a, 1354—1355, 1970.

51. Samelson, H. and Lempicki, A., Q switching and mode locking of Nd^{+3}:$SeOCl_2$ liquid laser, *J. Appl. Phys.*, 39, 6115—6116, 1968.

52. Brinkschulte, H., Fill, E., and Lang, R., Spectral output properties of an inorganic liquid laser, *J. Appl. Phys.*, 43, 1807—1811, 1972.

53. Brinkschulte, H., Perchermeier, J., and Schimitschek, E. J., A repetitively pulsed, Q-switched, inorganic liquid laser, *J. Phys. D*, 7, 1361—1368, 1974.

54. Fill, E. E., Subnanosecond pulses from an Nd liquid laser, *J. Appl. Phys.*, 41, 4749—4750, 1970.

55. **Alfano, R. R. and Shapiro, S. L.**, Picosecond pulse emission from a mode locked Nd⁺³ POCl₃ liquid laser, *Opt. Commun.*, 2, 90—92, 1970.
56. **Fill, E. E.**, Mode locking experiments with an Nd-POCl₃ liquid laser, *Jpn. J. Appl. Phys.*, 9, 1542—1543, 1970.
57. **Andreou, D., Little, V., Selden, A. C., and Katzenstein, J.**, Output characteristics of a Q-switched laser system, Nd⁺³:POCl₃:ZrCl₄, *J. Phys. D*, 5, 59—63, 1972.
58. **Andreou, D. and Little, V. I.**, The spiking behavior of a laser having combined liquid and glass active media, *J. Phys. D*, 6, 390—394, 1973.
59. **Yamanaka, C., Yamanaka, T., Yamaguchi, G., Sasaki, T., and Nakai, S.**, Tandem amplifier systems of glass and SeOCl₂ liquid lasers doped with neodymium, *Nachrichten Tech. Fachberichte*, 35, 791—795, 1968.
60. **Sasaki, T., Yamanaka, T., Yamaguchi, G., and Yamanaka, C.**, A construction of the high power laser amplifier using glass and selenium oxychloride doped with Nd⁺³, *Jpn. J. Appl. Phys.*, 8, 1037—1045, 1969.
61. **Hongyo, M., Sasaki, T., Ngao, Y., Ueda, K., and Yamanaka, C.**, High power ⁻Nd⁺³POCl₃ liquid laser system, *IEEE J. Quantum Electron.*, QE-8, 192—196, 1972.
62. **Fill, E. E.**, Ein Nd-POCl₃ laser verstärker, *Z. Angew. Phys.*, 32, 356—358, 1972.
63. **Andreou, D., Selden, A. C., and Little, V. I.**, Amplification of mode locked trains with a liquid laser amplifier, Nd⁺³:POCl₃:ZrCl₄, *J. Phys. D*, 5, 1405—1417, 1972.
64. **Andreou, D. and Little, V. I.**, The effect of frequency shifts on the power gain of a laser amplifier, *Opt. Commun.*, 6, 180—184, 1972.
65. **Andreou, D.**, A high power liquid laser amplifier, *J. Phys. D*, 7, 1073—1077, 1974.
66. **Andreou, D.**, On the growth of stimulated Raman scattering in amplifying media, *Phys. Lett.*, 57A, 250—252, 1976.
67. **Selden, A. C.**, Transverse modes in a liquid laser, *Opt. Commun.*, 5, 62, 1972.
68. **Lang, R. S., Baumbacher, H., and Fill, E. E.**, Stimulated Raman scattering in an inorganic liquid laser, *Phys. Lett.*, 32A, 433—434, 1970.
69. **Green, M., Andreou, D., Little, V. I., and Selden, A. C.**, Stimulated Raman scattering from a multigigawatt liquid laser amplifier, *J. Appl. Phys.*, 46, 4854—4856, 1975.
70. **Pappalardo, R. and Lempicki, A.**, Brillouin and Rayleigh scattering in aprotic laser solutions containing neodymium, *J. Appl. Phys.*, 43, 1699—1708, 1972.
71. **Alfano, R. R., Lempicki, A., and Shapiro, S. L.**, Non-linear effects in inorganic liquid lasers, *IEEE J. Quantum Electron.*, QE-7, 416—424, 1971.
72. **Samelson, H., Lempicki, A., and Brophy, V.**, Self Q-switching of the Nd⁺³ SeOCl₂ liquid laser, *J. Appl. Phys.*, 39, 4029—4030, 1968.
73. **Selden, A. C.**, On self Q-switching the Nd:liquid laser, *Opt. Commun.*, 6, 415—417, 1972.
74. **Kaporskii, L. N. and Kalabushkin, O. I.**, Cuvettes for inorganic liquid lasers, *Sov. J. Opt. Technol.*, 42, 180—182, 1975.
75. **Winston, H. and Gudmundsen, R. A.**, Refractive index gradient effects in proposed liquid lasers, *Appl. Opt.*, 3, 143—6, 1964.
76. **Riedel, E. P.**, Light scattering in a solution of europium benzoylacetonate during optical pumping, *Appl. Phys. Lett.*, 5, 162—165, 1964.
77. **Malyshev, B. N. and Salyuk, V. A.**, Effect of the lens formed in the active element on the radiation from a liquid laser, *Sov. Phys. Tech. Phys.*, 16, 1331—1336, 1972.
78. **Rubinov, A. N. and Anufrik, S. S.**, Dynamic compensation of the thermal distortion in a liquid laser cavity, *J. Appl. Spectr.*, 17, 858—860, 1972.
79. **Zaretskii, A. I., Kirillov, G. A., Kormer, S. B., and Sukharov, S. A.**, Dependence of the divergence of the radiation emitted by laser oscillators and amplifiers on the optical homogeneity of the medium, *Sov. J.Quantum Electron.*, 4, 649—652, 1974.
80. **Batyaev, I. M.**, Liquid lasers based on lanthanide complexes, *Russ. Chem. Rev.*, 40, 622—631, 1971.

Section 4
Other Lasers

4.1 FREE ELECTRON LASERS

4.1.1 INFRARED AND VISIBLE LASERS

Donald Prosnitz

INTRODUCTION

In the Free Electron Laser (FEL), gain is generated by the interaction of photons with a relativistic electron beam. A freely propagating electron does not interact with an electromagnetic field. To obtain gain the electrons and photons must interact within a perturbing environment that permits the simultaneous conservation of energy and momentum; spontaneous emission from the electron is then possible. The synchrotron radiation that occurs when the trajectory of a high-energy electron is bent by a magnetic field is an example of one such process. The process that generates gain may be viewed as stimulated scattering, as stimulated "free-free" transitions between continuum states of the perturbed electron-photon system, or as the inverse of the interaction that accelerates electrons in an accelerator. If the velocity distribution of the electrons in the beam is carefully selected, the radiation emitted by each electron adds coherently to the radiation from other electrons in the beam. The wavelength of maximum gain is primarily a function of the energy of the beam. With a minimum of constraints, the operation of an FEL should be possible at any wavelength from millimeter wavelengths into the visible and near ultraviolet.

FEL CLASSIFICATION

A free-free electron transition can occur only in an environment that permits the simultaneous conservation of energy and momentum in an electron-photon interaction. Free Electron Lasers can be classified on the basis of the different structures that have been suggested to provide the proper environment. These are summarized in Table 4.1.1.1. (These FELS are reviewed in References 1 to 6.) Relevant references to both experimental and theoretical studies of these devices (along with typical FEL configurations) are included in Table 4.1.1.1.

- The Bremsstrahlung FELs[1-6,39,44] use an external field (often a periodic static magnetic field) to force the electrons to undergo rapid oscillations and radiate energy into a narrow frequency band. The laser wavelength and the electron velocity are found by requiring that the imposed external field and the laser field have the same frequency in the electron's rest frame (Equation 7 below). As a result, if the electron velocity is close to the speed of light, long wavelength (cm) imposed fields can be used to build FELs operating in the visible region of the spectrum.
- The Smith-Purcell laser[4,5,120] relies on the fact that, if an electron beam passes close to a diffraction grating, image charges are induced on the grating that should radiate at a wavelength given by

$$\ell\lambda_s = d\left[\frac{1}{\beta} - \cos\Theta\right] \tag{1}$$

where λ_s is the laser wavelength, β the ratio of the electron velocity to the speed of light, θ the angle between light emission and the plane of the grating, d the

Table 4.1.1.
FEL CONFIGURATIONS

Bremsstrahlung		References	
Compton		Experimental	Theoretical
Magnetic		7-15	16-80
Electromagnetic		147-150	81-84
Electrostatic		85-86	87
Raman		88-97	98-102
Cerenkov		103-105	106-112
Smith-Purcell		113-116	117-121
Limited Interaction			122-124
Optical Klystron		125	126-129 151-153

grating spacing, and l an integer (harmonic number). (Definitions of all variables are given in Table 4.1.1.2) If the electrons are relativistic ($\beta \lesssim 1$) and θ is close to 0°, visible radiation is emitted by the electrons even if the grating spacing is orders of magnitude larger than the laser wavelength. When placed in a suitable cavity, the device will radiate coherently.

- The Cerenkov laser[1-5] is based on the fact that the phase velocity of an electromagnetic wave in a dielectric medium may be matched to the velocity of a relativistic electron, thereby permitting significant energy exchange between photons and electrons.

- The limited interaction "FEL"[1,124] takes advantage of the uncertainty principle, hence, energy and momentum need not be strictly conserved if the electron-photon interaction distance is short. This interaction would be most useful in an optical klystron.

- An optical klystron[151,153] consists of three sections: a velocity modulator, a drift region to allow the electrons to bunch, and a wiggler to force the bunched electrons to radiate. The modulator and wiggler may be chosen from the structures discussed above. Momentum dependent path lengths (using strong magnetic fields) have been suggested as a way to shorten the drift region in a relativistic optical klystron.

Included in Table 4.1.1.1 are references to experiments which have demonstrated many of the principles essential to FEL operation. The most important of these are summarized in Table 4.1.1.3 (with the exception of Bremsstrahlung-Raman Experiments, see Section 4.1.2 this volume) and discussed at the end of this article.

We have arbitrarily excluded the gyrotron from Table 4.1.1.1 because it is not read-

Table 4.1.1.2
SYMBOLS

B_w	Wiggler magnetic field amplitude
b_w	rms normalized wiggler field, $eB_w/\sqrt{2}\,mc$) for linear field
c	Speed of light
d	Grating spacing
e	Electron charge ($e > 0$)
E_s	Laser electric field magnitude
e_s	Normalized electric field, $eE_s/\sqrt{2}\,mc^2$)
I	Electron-beam current
I_0	$ec/r_0 = 1.7 \times 10^4$ Amperes
k_s	Laser wave number
k_w	Wiggler wave number
l	Harmonic number (integer)
L	Amplifier length
m	Electron rest mass
n	Electron density, lab frame
N	Number of wiggler periods
P_{laser}	Laser power
P_{syn}	Spontaneous synchrotron power radiated in a storage ring $= ESC\,mc^2\,\dfrac{2}{3}\,\dfrac{Cr_0}{\rho_2}\,\gamma^4$ (ϱ = ring radius)
P_{eb}	Electron beam power
r	Electron-beam radius
r_0	Classical electron radius
Z_o	Impedance of free space
β	$[1 - 1/\gamma^2]^{1/2}$
γ	Electron energy in units of electron rest mass
γ_o	Initial electron energy
γ_r	Resonant electron energy
$\Delta\gamma$	Full width of electron energy distribution
ε	(Electron beam emittance)/π
ε_o	Permittivity of free space
Θ	Angle between light emission and plane of a Smith Purcell grating
Σ	Larger of electron beam or laser beam cross section
ϕ_s	Phase of laser field
ω_s	Laser frequency
ω_p	Plasma frequency — $(ne^2/m\varepsilon_o)^{1/2}$
Ω	Synchrotron frequency
Ψ	Phase of the electron in the wiggler field relative to the laser field

ily scalable to short wavelengths. References 130 to 136 give a summary of past and projected gyrotron performance, along with a complete list of references.

As is apparent from Table 4.1.1.1, the Bremsstrahlung FELs have received the most attention. In particular, the magnetic Bremsstrahlung FEL is the most likely to achieve high power and (possibly) high efficiency. The magnetic Bremsstrahlung FEL may be separated into two operating regions, depending on whether or not collective effects (Raman regime) or single-particle effects (Compton regime) play the dominant role in the interaction.

Hasegawa[30] and Kroll[35] have shown that collective effects must be considered whenever

$$\left(\frac{\omega_p}{\omega_s}\right) \gamma^{1/2} \gg \frac{\Delta\gamma}{\gamma} \tag{2}$$

where ω_p is the electron plasma frequency,* ω_s the laser frequency, γ the electron energy in units of its rest mass energy, and $\Delta\gamma$ the energy spread of the electron beam.

---

* Editors note: In Section 4.1.2, ω_p is the frame-invariant electron plasma frequency given by $\omega_p = [nC^2/\gamma m\varepsilon_0]^{1/2}$. In this notation Equation 2 becomes $(\omega_p/\omega)\gamma \gg \Delta\gamma/\gamma$.

Table 4.1.1.3
COMPTON FEL EXPERIMENTS

Wavelength (μm)	Beam energy (MeV)	Beam current (A)	Wiggler field (kG, peak)	Wiggler period (cm)	Wiggler length (m)		Ref.
3.3—3.4	43	1.3—2.6 (peak)	2.0	3.3	5.3	Oscillator, 130 kW peak, 5 W avg. during macro pulse	9, 147
10.6	24	0.07 (peak)	2.4	3.3	5.3	Amplifier, 7% gain	11
0.488	150—240	0.01 (peak)	4	4	0.96	FEL on ACO storage ring, measured gain of 4.3×10^{-4}/pass	144, 149
0.6	350	—	3.0	10	0.3 (2 wigglers)	Optical klystrons on VEPP3 storage ring; modulator — buncher-radiator	125
0.44	100	—	7.9	4	1	Checked polarization, first attempted use of coherent wiggler	13
0.17—0.5	100—175	0.03 (avg.)	0.36	4	0.8	Synchrotron radiation from PAKHRA synchrotron	7, 8
0.5	70—800	0.1 (avg.)	0.25—1.0	14	0.7	Studied angular dependence and polarization of radiation from wiggler on SIRIUS synchrotron	10, 12, 14
10^{-3}	3×10^3	0.2 (avg.)	0.5—2.8	6.1	1.95	Measured X-ray spectrum from wiggler installed on SPEAR storage ring	150
10.6	19	0.1 (avg.)	2.6	2.3	2.3	Tapered Wiggler Amplifier, electron deceleration observed	148

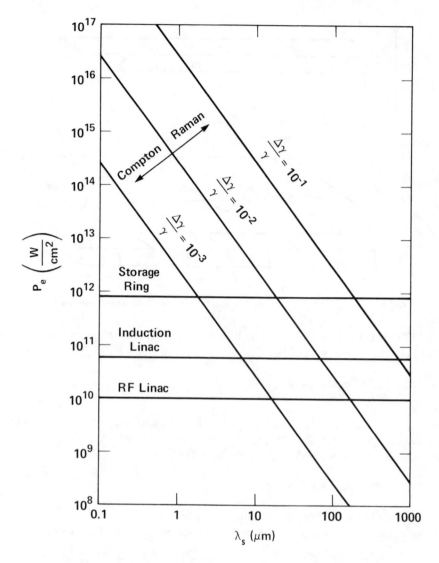

FIGURE 4.1.1.1.   Compton and Raman operating regimes of the magnetic Brems-strahlung FEL.

Equation 2 is plotted in Figure 4.1.1.1, with electron-beam power density replacing $\gamma \omega_p{}^2$. The operating region of typical electron beam machines is also indicated in Figure 4.1.1.1. The storage ring[137] and rf linac[138] both operate at high voltages, implying that they are most suitable for visible FELs (Equation 7, below), while current induction linacs[139] are low-voltage devices and are most suitable for far-infrared FELs. Future, high-voltage induction linacs may be useful for visible FELs. It is clear that FELs operating in the near-infrared and visible-wavelength regions will operate in the Compton Region, while far-infrared and millimeter wave lasers will be Raman devices. (The latter are covered in Section 4.1.2 of this volume.) The remainder of this article concentrates on the Compton FEL.

SMALL SIGNAL GAIN OF THE COMPTON FEL

The operational principle of the Compton FEL is shown in Figure 4.1.1.2. As an electron beam travels along the Z axis, it is given a Y velocity component by the mag-

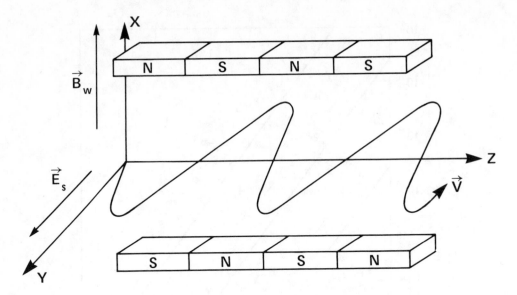

FIGURE 4.1.1.2.   Electron photon interaction in the magnetic Bremsstrahlung FEL.

netic field (helical fields are also used). Energy transfer between the electron and the laser field may then be described by

$$\frac{d\gamma}{dZ} = \frac{-e_s b_w}{k_w \gamma} \sin \Psi \tag{3}$$

where e_s and b_w are the normalized laser electric field and wiggler magnetic field, respectively (see Table 4.1.1.2), k_w is the wiggler-wave vector, and ψ is the relative phase of the electron's transverse velocity and the laser field,

$$\Psi = (k_w + k_s)Z - \omega_s t + \phi_s \tag{4}$$

k_s and ϕ_s are the laser wave vector and phase, respectively.

Colson[1,25] has shown that ψ is governed by the simple pendulum equation

$$\frac{d^2 \Psi}{dZ^2} = -\Omega^2 \sin \Psi \tag{5a}$$

in which

$$\Omega^2 \equiv \frac{2e_s b_w}{\gamma^2} \tag{5b}$$

The periodic nature of ψ, when combined with Equation 3, would seem to indicate that electrons first gain and then lose energy to the field, with no net energy extracted. However, gain is possible under appropriate conditions. In the small signal regime defined by

$$\frac{\gamma_0 - \gamma_r}{\gamma_r} \gtrsim \frac{\Omega}{k_w} \tag{6}$$

where

$$\gamma_r^2 \equiv \frac{k_s}{2k_w} \left[1 + \frac{b_w^2}{k_w^2} \right] \tag{7}$$

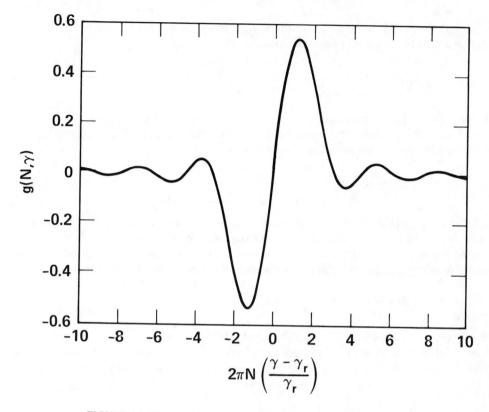

FIGURE 4.1.1.3.   Gain function for the fixed parameter Compton FEL.

Equations 3 to 5 may be solved and averaged over all initial electron phases to give the small signal FEL gain.[25,49]

$$G = 2\pi^2 \left(\frac{b_w}{k_w}\right)^2 \frac{\lambda_w^2}{\Sigma} \frac{I}{I_0} \left(\frac{N}{\beta\gamma}\right)^3 g\left(2\pi N \frac{\gamma_0 - \gamma_r}{\gamma_r}\right) \qquad (8a)$$

$$g(x) = \frac{-d}{dx} \frac{\sin^2 x}{x^2} \qquad (8b)$$

where λ_w is the wiggler wavelength, I the electron beam current, I_0 the Alfven current $(1.7 \times 10^4$ A), N the number of wiggler periods, and Σ the larger of the e-beam or laser-beam cross section.

Since $\gamma_r \approx \gamma_0$, Equation 7 determines the operating wavelength of the FEL $(2\pi/k_s)$ in terms of the electron energy, wiggler period, and field strength. It can be seen from the plot of g in Figure 4.1.1.3 that there is an optimum combination of length $(N\lambda_w)$ and initial energy that is given by

$$N = 0.207 \left(\frac{\gamma_r}{\gamma_0 - \gamma_r}\right) \qquad (9)$$

when g is maximized and equal to 0.54.

If the wiggler is made longer, some of the energy originally lost by the electrons is regained, while in a shorter wiggler, maximum energy is not extracted. It should also be noted that if the laser is propagating in free space (as opposed to an optical waveguide), the cross-sectional area and the wiggler length ($N\lambda_w$) are related by diffraction, so that $\Sigma \approx \lambda_w^2 N$. The gain becomes saturated when

$$\frac{\Omega}{k_W} > \frac{\gamma - \gamma_r}{\gamma_r} \tag{10}$$

If we assume that at saturation no more energy can be extracted from the FEL, we find from the solutions of Equations 3 to 9 that the fraction of energy that can be extracted from the electron beam is approximately[49]

$$<\frac{\gamma_f - \gamma_i}{\gamma_i}> \text{avg} \approx \frac{1}{2N} \tag{11}$$

Maximum gain (Equation 8, large N) and maximum energy extraction (Equation 11, small N) conflict, and practical FEL design requires a compromise.

An examination of Figure 4.1.1.3 reveals that the maximum gain predicted by Equation 8 requires that the electron beam has an energy spread $\Delta\gamma_o$ (FWHM) given by

$$2\pi N \frac{\Delta\gamma}{\gamma} \stackrel{\sim}{<} 1.5 \text{ or } \frac{\Delta\gamma}{\gamma} < \frac{1}{4N} \tag{12}$$

Furthermore, a periodic Y-directed magnetic field has a transverse variation (because $\nabla \cdot B = 0$) that must be kept in order to preserve electron-photon synchronism (Equation 7). This variation is minimized by limiting the electron-beam radius[6,49-51] as follows:

$$r^2 < \frac{1 + \left(b_w^2/k_w^2\right)}{b_w^2} \left(\frac{1}{2N}\right) \tag{13}$$

(Wiggler magnet designs are discussed in References 140 to 143.)

The electron beam will also have a transverse momentum component due to finite emittance. This too must be limited[49-51] such that

$$\gamma \epsilon << b_w r^2 \tag{14}$$

Because $1/2N$ may be thought of as the homogeneous broadening ($\Delta\omega_s/\omega_s = 2\Delta\gamma/\gamma$) due to the finite length of the laser, conditions (12) to (14) are often described by saying that, for maximum gain, the inhomogeneous broadening must be kept less than the homogeneous broadening.[51]

Operation of the Compton FEL as an oscillator requires a knowledge of the gain saturation. If the FEL is inserted in a linear electron accelerator (rf or induction linac) "fresh electrons" are continually inserted into the oscillator, and saturation is simply a result of the growing E field (Equation 10). The gain in this region has been calculated by several authors,[55-64] and peak gain in the saturated regime is found to occur slightly away from the optimum conditions predicted in Equation 9. If the FEL is placed in an electron storage ring, an additional saturation mechanism that appears is due to the progressive deterioration of the electron beam each time it passes through the wiggler. This tendency is counterbalanced by synchrotron-radiation damping of the beam. An equilibrium condition is reached[51] when

$$P_{laser} \approx \frac{1}{2N} P_{syn} \tag{15}$$

where P_{laser} and P_{syn} are the FEL power and incoherent synchrotron power, respectively.

Table 4.1.1.4
FEL SMALL SIGNAL GAIN EQUATIONS

Wavelength

$$\lambda_s = \frac{\lambda_w}{2\gamma_r^2}\left[1 + \frac{b_w^2}{k_w^2}\right]$$

Saturation

$$\frac{\Omega}{k_w} = \left[\frac{2e_s b_w}{k_w^2 \gamma_r^2}\right]^{1/2} > \frac{\gamma - \gamma_r}{\gamma_r}$$

Gain

$$G = 2\pi^2 \left[\frac{b_w}{k_w}\right]^2 \frac{\lambda_w^2}{\Sigma} \frac{I}{I_0} \left[\frac{N}{\beta_0 \gamma_r}\right]^3 g\left(2\pi N \frac{\gamma_0 - \gamma_r}{\gamma_r}\right) \qquad g(x) = \frac{-d}{dx} \frac{\sin^2 x}{x^2}$$

Maximum gain

$$G = 10.66 \left[\frac{b_w}{k_w}\right]^2 \frac{\lambda_w^2}{\Sigma} \frac{I}{I_0} \left[\frac{N}{\beta_0 \gamma_r}\right]^3$$

$$\text{when } N = 0.207\left(\frac{\gamma_0 - \gamma_r}{\gamma_r}\right)^{-1}$$

Maximum energy extraction (single pass)

$$\frac{\gamma_f - \gamma_i}{\gamma_i} = \frac{1}{2N}$$

Laser power radiated in storage ring configuration

$$P_\varrho = \frac{1}{2N} P_{syn}$$

Maximum electron-beam radius

$$r^2 < \frac{1 + b_w^2/k_w^2}{b_w^2} \frac{1}{2N}$$

Maximum emittance

$$\gamma\epsilon \ll b_w r^2$$

Laser output (but not efficiency) can be increased by inserting a special structure into the storage ring to increase the synchrotron power (P_{syn}) and damping. A typical storage ring configured in this manner might produce more than 5 kW average power from 0.3 to 30 μm with a 2% electron to photon conversion efficiency.[19] The rf structure of the electron beam introduces multiple modes into the FEL oscillator with minimum linewidth determined by the pulse structure of the electron source. This problem has been studied by several authors.[55-68]

The basic equations governing FEL operation in the small signal regime are summarized in Table 4.1.1.4.

FEL Enhancement Techniques

Equations 11 and 15 imply that FEL efficiency is at best a few percent. Several authors have suggested ways of increasing the FEL gain and/or efficiency.[69-80,83,84] These include imposing a strong axial magnetic field;[74,78] adding an additional low-frequency pump source;[73,84] building a two-stage, low-voltage, visible FEL with electrostatic energy recovery;[83] building a gain expanded magnet;[77] building an isochronous storage ring;[72] and building a tapered wiggler.[70,75,76,78-80]

The gain-expanded magnet is designed so that each electron has an equilibrium orbit that precisely matches its energy to the magnetic field strength required for synchronism. Consequently, the adverse effects of electron beam energy spread are reduced. Much higher efficiencies in a storage ring may then be possible.

The isochronous storage-ring FEL and the tapered-wiggler FEL both operate in the saturated regime. In this situation, electrons are trapped in potential (pondermotive) wells, and they form electron bunches. The isochronous storage ring is designed to ensure that the bunches are preserved as the electrons circle the ring. The phase of the bunches relative to the phase of the laser is optimized at the input of the wiggler for maximum energy extraction. The ring is designed so that this phase is preserved each time the electrons circle the ring. Such a device might operate with high efficiency at the multiple kilowatt level.

The tapered-wiggler FEL also traps electrons in bunches, but it then relies on changing the parameters of the wiggler to decelerate the bunches and extract energy from the electron beam. The gain in this device is given by

$$\frac{dE_s}{dZ} \approx \frac{b_w}{\sqrt{2}\,k_w} \frac{Z_0}{\Sigma} I \frac{\sin \Psi_r}{\gamma} \tag{16}$$

where E_s is the peak laser electric field, Z_0 the impedance of free space, $\sin \psi_r$ characterizes the wiggler taper according to

$$\sin \Psi_r = \frac{-k_w}{b_w} \frac{\gamma_r}{e_s} \frac{d\gamma_r}{dZ} \tag{17}$$

and $d\gamma_r/dZ$ is the rate of deceleration (with γ_r determined from Equation 7). One must have $(1/\gamma_r)(d\gamma_r/dZ)(1/k_w) \ll 1$ to preserve the electron traps (often called buckets). It is projected that 30 to 40% of the electron energy can be converted to photon energy in a single-pass amplifier before the traps collapse.[76] This device might have wide application where high-power, short-wavelength lasers are required.

One problem with operating the FEL in the saturated regime is that large field strengths are required to create potential wells that are large enough to trap an electron beam with energy spread. The required field strength is obtained from Equation 10 with a slight modification due to the decelerating buckets.[76]

$$\frac{\Omega}{k_w} \left\{ \cos\Psi_r - (\Pi/2 - \Psi_r)\sin\Psi_r \right\}^{1/2} \geq \frac{|\gamma - \gamma_r|}{\gamma_r} \tag{18}$$

or, if $\Psi_r = 0.4$ and $b_w/k_w \approx 1$,

$$\frac{P_\varrho}{\Sigma_\varrho} > \frac{3 \times 10^{10}}{\lambda_s^2 \,(\mu m)} \left(\frac{\Delta\gamma}{2\gamma}\right)^4 \, W/cm^2 \tag{19}$$

where λ_s is expressed in microns and $\Delta\gamma/2\gamma$ as a percentage. Thus, 3×10^{10} W/cm² at 1 μm are required to trap electrons with a 2% energy spread. About 40% of the electrons are trapped with the field given by Equation 19.[76] This device is clearly useful only where high power is required.

EXPERIMENTAL TESTS OF FEL CONCEPTS

There have been several experiments designed to test FEL concepts (most are listed in Table 4.1.1.1.). Of particular note is the Ubitron,[15] a microwave (11 cm) device based on a magnetic wiggler that achieved 13% efficiency at 900-kW output. This device is the nonrelativistic analogue of the Bremsstrahlung FEL. (Phillips[15] even suggested using tapered wigglers.) Many microwave devices have also been built that use the Smith-Purcell effect. Some of these, such as the Oratron[115] and Ledatron,[114] have achieved outputs of 0 .1 W to several watts in high Q cavities. Several far-infrared magnetic Bremsstrahlung experiments have been performed. (These are discussed in the following chapter.) Only a few experiments have used relativistic electron beams to produce coherent infrared and visible radiation, and these are summarized in Table 4.1.1.3. The first three experiments are the only relativistic Compton FELs that have exhibited gain. The same helical wiggler[141] was used for the 10.6 μm gain experiment and the 3.4 μm oscillator experiment, graphically illustrating the tunability of the FEL. The Stanford group has made time-resolved measurements of the optical and electron momentum spectra and the total optical power produced by their 3.4 μm FEL oscillator. The electron source was a high quality, low emittance super-conducting linac.

The low gain produced in the ACO storage ring FEL precluded operation as an oscillator. It is anticipated that this advice will be converted to an optical klystron yielding higher gains[151] and possibly permitting laser oscillation. Similarly, the next experiment is an optical klystron which has been installed on the VEPP3 storage ring in Novosibirsk with no results reported yet.

The following four experiments were not configured as lasers and did not illustrate gain, but they all demonstrated the enhanced spontaneous emission produced by electrons traversing a longer wiggler. Interference effects resulting from the multiple period wigglers were observed. These experiments investigated the spectrum, polarization, and angular distribution of the radiation produced.

The final experiment listed is primarily concerned with demonstrating that tapered wigglers will permit one to extract a large amount of optical energy from an electron beam. This experiment has thus far demonstrated net loss of electron energy (parts of the beam were decelerated more than 6%), but no attempt has yet been made to measure laser gain.

ACKNOWLEDGMENT

Work performed under the auspices of the U.S. Department of Energy by the Lawrence Livermore National Laboratory under contract number W-7405-ENG-48.

REFERENCES

1. **Jacobs, S. F., Pilloff, H. S., Sargent M., III, Scully, M. O., Spitzer, R.,** *Free-Electron Generators of Coherent Radiation,* Addison-Wesley, Reading, Mass., 1980, 112.
2. **Jacobs, S. F., Sargent, M., III, and Scully, M. O.,** *Novel Sources of Coherent Radiation,* Addison-Wesley, Reading, Mass., 1978, 113.
3. **Gover, A. and Yariv, A.,** Collective and single-electron interactions of electron beams with electromagnetic waves, and free-electron lasers, *Appl. Phys.,* 16, 121, 1978.
4. **Gover, A. and Livni, Z.,** Operation regimes of Cerenkov — Smith-Purcell free electron lasers and T. W. Amplifiers, *Opt. Commun.,* 26, 375, 1978.
5. **Bratman, V. L, Ginzburg, N. S., and Petelin, M. I.,** Common properties of free electron lasers, *Opt. Commun.,* 30, 409, 1979.
6. **Lawson, J. D.,** Coherent interaction between waves and particle streams, *Part. Accelerators,* 10, 73, 1980.
7. **Alferov, D. F., Bashmakov, Y. A., Belovintsev, K. A., Bessonov, E. G., and Cherenkov, P. A.,** The ondulator as a source of electromagnetic radiation, *Part. Accelerators,* 9, 223, 1979.
8. **Alferov, D. F., Bashmakov, Y. A., Belovintsev, K. A., Bessonov, E. G., and Cherenkov, P. A.,** Observation of undulating radiation with the "Pakhra" synchrotron, *JETP Lett.,* 26, 385, 1977.
9. **Deacon, D. A. G., Elias, L. R., Madey, J. M. J., Ramian, G. J., Schwettman, H. A., and Smith, T. I.,** First operation of a free-electron laser, *Phys. Rev. Lett.,* 38, 892, 1977.
10. **Didenko, A. N., Kozhevnikov, A. V., Medvedev, A. F., Nikitin, M. M., and Epp, V. Ya.,** Radiation from relativistic electrons in a magnetic wiggler, *Sov. Phys. JETP,* 49, 973, 1979.
11. **Elias, L. R., Fairbank, W. M., Madey, J. M. J., Schwettman, H. A., and Smith, T. I.,** Observation of stimulated emission of radiation by relativistic electrons in a spatially periodic transverse magnetic field, *Phys. Rev. Lett.,* 36, 717, 1976.
12. **Medvedev, A. F., Nikitin, M. M., and Epp, V. Ya.,** Undulator emission from relativistic electrons, *Sov. Tech. Phys. Lett.,* 5, 327, 1979.
13. **Motz, H., Thon, W., and Whitehurst, R. N.,** Experiments on radiation by fast electron beams, *J. Appl. Phys.,* 24, 826, 1953.
14. **Nikitin, M. M., Medvedev, A. F., and Moiseev, M. B.,** Interference of synchrotron radiation, *Sov. Tech. Phys. Lett.,* 5, 347, 1979.
15. **Phillips, R. M.,** The Ubitron, a high-power traveling-wave tube based on a periodic beam interaction in unloaded waveguide, *IRE Trans. Electron Dev.,* 17, 231, 1960.
16. **Alferov, D. F., Bashmakov, Y. A., Bessonov, E. G., and Govorkov, B. B.,** Emission of polarized quasimonochromatic rays by ultrarelativistic electrons in a transverse periodic magnetic field, *Soc. J. Nucl. Phys.,* 27, 514, 1978.
17. **Al-Abawi, H. A., Hopf, F. A., and Meystre, P.,** Electron dynamics in a free-electron laser, *Phys. Rev. A,* 16, 666, 1977.
18. **Baier, V. N. and Milstein, A. I.,** To the theory of a free-electron laser, *Phys. Lett.,* 65A, 319, 1978.
19. **Bambini, R., Dattoli, G., Letardi, T., Marino, A., Renieri, A., and Vignola, G.,** Leda-F storage ring dedicated to the free electron laser operation, preliminary design, *IEEE Trans. Nucl. Sci.,* 26, 3836, 1979.
20. **Bambini, A. and Renieri, A.,** The free electron laser: a single-particle classical model, *Lettere Al Nuovo Cimento,* 21, 399, 1978.
21. **Bambini, A. and Stenholm, S.,** Quantum description of free electrons in the laser, *Opt. Commun.,* 30, 391, 1979.
22. **Bambini, A. and Stenholm, S.,** The momentum distribution in the free electron laser, *Opt. Commun.,* 25, 244, 1978.
23. **Becker, W.,** On the frequency of a free-electron laser, *Phys. Lett.,* 65A, 317, 1978.
24. **Cocke, W. J.,** Stimulated emission and absorption in classical systems, *Phys. Rev. A,* 17, 1713, 1978.
25. **Colson, W. B.,** One-body electron dynamics in a free electron laser, *Phys. Lett.,* 64A, 190, 1977.
26. **Colson, W. B.,** Theory of a free electron laser, *Phys. Lett.,* 59A, 187, 1976.
27. **Deacon, D. A. G. and Madey, J. M. J.,** Boltzmann simulation of a storage ring laser, *Appl. Phys.,* 19, 295, 1979.
28. **Elias, L. R., Madey, J. M. J., and Smith, T. I.,** Monte Carlo analysis of a free electron laser in a storage ring (to be published).
29. **Godwin, R. P.,** Synchrotron radiation as a light source, in *Springer Tracts in Modern Physics,* Springer-Verlag, Berlin, 1969, 51.
30. **Hasegawa, A.,** Free electron laser, *Bell Syst. Tech. J.,* 57, 3069, 1978.
31. **Hasegawa, A., Mina, K., Sprangle, P., Szu, H. H., and Granatstein, V. C.,** Limitation in growth time of stimulated compton scattering in X-ray regime, *App. Phys. Lett.,* 29, 542, 9/76.

32. Hopf, F. A., Meystre, P., and Scully, M. O., and Louisell, W. H., Classical theory of a free-electron laser, *Opt. Commun.*, 18, 413, 1976.

33. Kincaid, B. M., A short-period helical wiggler as an improved source of synchrotron radiation, *J. Appl. Phys.*, 48, 2684, 1977.

34. Kolomenskii, A. A. and Lebedev, A. N., Stimulated undulator radiation from relativistic electrons and the physical processes in an "electron laser," *Sov. J. Quantum Electron.*, 70, 879, 1978.

35. Kroll, N. M. and McMullin, W. A., Stimulated emission from relativistic electrons passing through a spatially periodic transverse magnetic field, *Phys. Rev. A*, 17, 300, 1978.

36. Louisell, W. H., Lam, J. F., and Copeland, D. A., Effect of space charge on free-electron-laser gain, *Phys. Rev. A*, 18, 655, 1978.

37. Madey, J. M. J., Stimulated Emission of Radiation in Periodically Deflected Electron Beam, U.S. Patent 3, 822, 410, 1974.

38. Madey, J. M. J., Deacon, D. A. G., and Smith, T. I., Free-electron lasers, boundary deformation, and the Robinson-Liouville Theorem, *J. Appl. Phys.*, 50, 7875, 1979.

39. Madey, J. M. J., Stimulated emission of bremsstrahlung in a periodic magnetic field, *J. Appl. Phys.*, 42, 1906, 1971.

40. Mayer, G., Collisionless Landau damping and stimulated Compton scattering, *Opt. Commun.*, 20, 200, 1977.

41. Medvedev, A. F., Nikitin, M. M., and Epp, V. Ya., Use of undulator radiation to measure the angular spread of electron velocities, *Sov. Tech. Phys. Lett.*, 5, 144, 1979.

42. Motz, H. and Nakamura, M., Proc. of the Symposium on Millimeter Waves, Microwave Res. Inst. Symposium, Series IX, 155, 1960.

43. Motz, H. and Nakamura, M., Radiation of an electron in an infinitely long waveguide, *Ann. Phys.*, 7, 84, 1959.

44. Motz, H., Applications of the radiation from fast electron beams, *J. Appl. Phys.*, 22, 527, 1951.

45. Motz, H., Is the free-electron laser a laser?, *Phys. Lett.*, 71A, 41, 1979.

46. Namiot, V. A., Coherent X radiation from relativistic electrons, *Sov. Tech. Phys. Lett.*, 5, 509, 1979.

47. Palmer, R. B., Interaction of relativistic particles and free electromagnetic waves in the presence of a static helical magnet, *J. Appl. Phys.*, 43, 3014, 1972.

48. Pantell, R. H., Soncini, G., and Puthoff, H. E., Stimulated photon-electron scattering, *IEEE J. Quantum Electron.*, QE-4, 905, 1968.

49. Pellegrini, C., The free electron laser and its possible developments, *IEEE Trans. Nucl. Sci.*, NS-26, 3791, 1979.

50. Renieri, A., Free electron laser amplifier operation, *IEEE Trans. Nucl. Sci.*, NS-26, 3827, 1979.

51. Renieri, A., Storage ring operation of the free-electron laser: the amplifier, *Nuovo Cimento*, 53 B, 160, 1979.

52. Sukhatme, V. P. and Wolff, P. A., Stimulated Compton scattering as a radiation source — theoretical limitations, *J. Appl. Phys.*, 44, 2331, 1973.

53. Vainshtein, L. A., Type-O relativistic electron devices. I. Linear theory, *Sov. Phys. Tech. Phys.*, 24, 625, 197 5.

54. Winick, H. and Bienenstock, A., Synchrotron radiation research, *Ann. Rev. Nucl. Part. Sci.*, 28, 33, 1978.

55. Dubrovskii, V. A., Lerner, N. B., and Tsikin, B. G., Theory of a Compton laser, *Sov. J. Quantum Electron.*, 5, 1248, 1975.

56. Fedorov, M. V. and McIver, J. K., Saturation in the classical theory of the free-electron laser, *Opt. Acta*, 26, 1121, 1979.

57. Fedorov, M. V. and McIver, J. K., The classical theory of saturation in the free-electron laser, *Kvantoraya Elektron.*, 7, 1980.

58. Fedorov, M. V. and McIver, J. K., The Quantum theory of stimulated processes in the free electron laser in the region of strong field, *Zh. Eksp. Teor. Fiz.*, 76, 1996, 1979.

59. Hopf, F. A., Meystre, P., Scully, M. O., and Louisell, W. H., Strong-signal theory of a free-electron laser, *Phys. Rev. Lett.*, 37, 1342, 1976.

60. Louisell, W. H., Lam, J. F., Copeland, D. A., and Colson, W. B., "Exact" classical electron dynamic approach for a free-electron laser amplifier, *Phys. Rev. A*, 19, 288, 1979.

61. McIver, J. K. and Fedorov, M. V., Theory of the free-electron laser, *Sov. Tech. Phys. Lett.*, 5, 248, 1979.

62. McIver, J. K. and Fedorov, M. V., Quantum theory of stimulated processes in a free-electron laser in a strong field, *Sov. Phys. JETP*, 49, 1012, 1979.

63. Planner, C. W., Numerical solutions to the non-linear phase equation for the free electron laser, *Phys. Lett.*, 67A, 263, 1978.

64. Sprangle, P. and Smith, R.A., Theory of free-electron lasers, *Phys. Rev. A*, 21, 293, 1980.

65. Al-Abawi, H., Hopf, F. A., Moore, G. T., and Scully, M. O., Coherent transients in the free-electron laser: laser lethargy and coherence brightening, *Opt. Commun.*, 30, 235, 1979.

66. Dattoli, G. and Renieri, A., Classical multimode theory of the free electron laser, *Lettere Al Nuovo Cimento.*, 24, 121, 1979.

67. Moore, G. T. and Scully, M. O., Coherent dynamics of a free-electron laser with arbitrary magnet geometry. I. General formalism, *Phys. Rev. A*, 21, 2000, 1980.

68. Bonifacio, R., Meystre, P., Moore, G. T., and Scully, M. O., Coherent dynamics of a free-electron laser with arbitrary magnet geometry. II. Conservation laws, small-signal theory, and gain-spread relations, *Phys. Rev. A*, 21, 2009, 1980.

69. Boscolo, I., Brautt, G., Clauser, T., and Stagno, V., Free-electron lasers and masers on curved paths, *Appl. Phys.*, 19, 47, 1979.

70. Brau, C. A., Small signal gain of free electron lasers with nonuniform wigglers, *IEEE J. Quantum Electron.*, QE-16, 300, 1980.

71. Cocke, W. J., Increasing the output of the free-electron laser by using dielectric effects, *Opt. Commun.*, 28, 123, 1979.

72. Deacon, D. A. G. and Madey, J. M. J., Isochronous storage-ring laser: a possible solution to the electron heating problem in recirculating free-electron lasers, *Phys. Rev. Lett.*, 44, 449, 1980.

73. Dubrovskii, V. A. and Tsikin, B. G., Stimulation of scattering in a Compton laser by decelerated waves (semiclassical theory), *Sov. J. Quantum Electron.*, 7, 832, 1977.

74. Friedland, L and Hirshfield, J. L., Free-electron laser with a strong axial magnetic field, *Phys. Rev. Lett.*, 44, 1456, 1980.

75. Kroll, N. M., Morton, P., and Rosenbluth, M. N., *Free Electron Lasers With Variable Parameter Wigglers*, Tech. Rep. JSR-79-01, SRI International, Arlington, Va., 1980.

76. Prosnitz, D., Szoke, A., and Neil, V. K., High-gain, free-electron laser amplifiers: design considerations and simulation, *Phys. Rev. A*, 24, 1436, 1981.

77. Smith, T. I., Madey, J. M. J., Elias, L. R., and Deacon, D. A. G., Reducing the sensitivity of a free-electron laser to electron energy, *J. Appl. Phys.*, 50, 4580, 1979.

78. Sprangle, P. and Granatstein, V. L., Enhanced gain of a free-electron laser, *Phys. Rev. A*, 17, 1792, 1978.

79. Sprangle, P., Tang, Cha-Mei, and Manheimer, W. M., Nonlinear formulation and efficiency enhancement of free-electron lasers, *Phys. Rev. Lett.*, 43, 1932, 1979.

80. Sprangle, P., Tang, Cha-Mei, and Manheimer, W. M., Nonlinear theory of free-electron lasers and efficiency enhancement, *Phys. Rev. A*, 21, 302, 1980.

81. Bratman, V. L., Ginzburg, N. S., and Petelin, M. I., Energy feasibility of a relativistic compton laser, *JETP Lett.*, 28, 190, 1978.

82. Chan, Y. W., Proposed free-electron laser stimulated by traveling microwave radiation, *Phys. Rev. Lett.*, 42, 92, 1979.

83. Elias, L. R., High-power, cw, efficient, tunable (uv through ir) free-electron laser using low-energy electron beams, *Phys. Rev. Lett.*, 42, 977, 1979.

84. Fedorov, M. V. and McIver, J. K., Multiphoton stimulated compton scattering, *Opt. Commun.*, 32, 179, 1980.

85. Alguard, M. J., Swent, R. L., Pantell, R. H., Berman, B. L., Bloom, S. D., and Datz, S., Observation of radiation from channeled positrons, *Phys. Rev. Lett.*, 42, 1148, 1979.

86. Swent, R. L., Pantell, R. H., Alguard, M. J., Berman, B. L., Bloom, S. D., and Datz, S., Observation of channeling radiation from relativistic electrons, *Phys. Rev. Lett.*, 43, 1723, 1979.

87. Bekefi, G. and Shefer, R. E., Stimulated Raman scattering by an intensive relativistic electron beam subjected to a rippled electric field, *J. Appl. Phys.*, 50, 5158, 1979.

88. Boehmer, H., Buzzi, J. M., Doucet, H. J., Etlicher, B., Lamain, H., and Rouille, C., Resonance effect on relativistic electron beam propagation for collective free electron laser, *Bull. Am. Phys. Soc.*, 24, 1066, 1979.

89. Bohmer, H., Munch, J., and Caponi, M. Z., A free electron laser experiment, *IEEE Trans. Nuc. Sci.*, 26, 79, 3830, 1979.

90. Efthimion, P. C. and Schlesinger, S. P., Stimulated Raman scattering by an intense relativistic electron beam in a long rippled magnetic field, *Phys. Rev. A*, 16, 633 1977.

91. Friedman, M. and Herndon, M., Emission of coherent microwave radiation from a relativistic electron beam propagating in a spatially modulated field, *Phys. Fluids*, 16, 1982, 1973.

92. Friedman, M. and Herndon, M., Generation of intense infrared radiation from an electron beam propagating through a rippled magnetic field, *Appl. Phys. Lett.*, 22, 658, 1973.

93. Gilgenbach, R. M., Marshall, T. C., and Schlesinger, S. P., Spectral properties of stimulated Raman radiation from an intense relativistic electron beam, *Phys. Fluids*, 22, 971, 1979.

94. Granatstein, V. L., Schlesinger, S. P., Herndon, M., Parker, R. K., and Pasour, J. A., Production of megawatt submillimeter pulses by stimulated magneto-Raman scattering, *Appl. Phys. Lett.*, 30, 384, 1977.

95. Marshall, T. C., Talmadge, S., and Efthimion, P., High-power millimeter radiation from an intense relativistic electron-beam device, *Appl. Phys. Lett.*, 31, 320, 1977.

96. McDermott, D. B., Marshall, T. C., Schlesinger, S. P., Parker, R. K., and Granatstein, V. L., High-power free-electron laser based on stimulated Raman backscattering, *Phys. Rev. Lett.*, 41, 1368, 1978.

97. Zhukov, P. G., Ivanov, V. S., Rabinovich, M. S., Raizer, M. D., and Rukhadze, A. A., Stimulated Compton scattering by a relativistic electron beam, *Sov. Phys. JETP*, 49, 1045, 1979.

98. Sprangle, P., Granatstein, V. L., and Baker, L., Stimulated collective scattering from a magnetized relativistic electron beam, *Phys. Rev. A*, 12, 1697, 1975.

99. Lin, A. T. and Dawson, J. M., High-efficiency free-electron laser, *Phys. Rev. Lett.*, 42, 1670, 1979.

100. Kwan, T. and Godfrey, B. B., Simulations of free electron laser, *IEEE Trans. Nucl. Sci.*, NS-26, 3833, 1979.

101. Kwan, T., Dawson, J. M., and Lin, A. T., Free electron laser, *Phys. Fluids*, 20, 581, 1977.

102. Granatstein, V. L. and Sprangle, P., Mechanisms for coherent scattering of electromagnetic waves from relativistic electron beams, *IEEE Trans. Microwave Theory Tech.*, MTT-25, 545, 1977.

103. Piestrup, M. A., Rothbart, G. B., Fleming, R. N., and Pantell, R. H., Momentum modulation of a free electron beam with a laser, *J. Appl. Phys.*, 46, 132, 1975.

104. Piestrup, M. A., Powell, R. A., Rothbart, G. B., Chen, C. K., and Pantell, R. H., Cerenkov radiation as a light source for the 2000-620-A spectral range, *Appl. Phys. Lett.*, 28, 92, 1976.

105. Walsh, J. E., Marshall, T. C., and Schlesinger, S. P., Generation of coherent Cerenkov radiation with an intense relativistic electron beam, *Phys. Rev.*, 20, 709, 1977.

106. Andreev, Yu. A., Davydovskii, V. Ya., and Danilenko, V. N., Resonant energy transfer from electrons to a slow electromagnetic wave, *Sov. Phys. Tech. Phys.*, 24, 885, 1979.

107. Dekker, H., A theory of cooperative effects in stimulated Cerenkov radiation, *Physica*, 90C, 283, 1977.

108. Soln, J., Differential equations with respect to a coupling constant: an approach to Cerenkov and stimulated radiations, *Phys. Rev. D*, 18, 2140, 1978.

109. Walsh, J. E., Marshall, T. C., Mross, M. R., and Schlesinger, S. P., Relativistic electron-beam-generated coherent submillimeter wavelength Cerenkov radiation, *IEEE Trans. Microwave Theory Tech.*, MTT-25, 1977.

110. Kalashnikova, Y. S., Parametric resonance in Cerenkov radiation in a medium excited by an external field, *Sov. Phys. Tech. Phys.*, 22, 127, 1977.

111. Schneider, S. and Spitzer, R., Interaction of electrons and pump fields at superluminal electron velocities, in *Free Electron Generators of Coherent Radiation*, Addison-Wesley, Reading, Mass., 1980, 323.

112. Kroll, N. M., Relativistic synchrotron radiation in a medium and its implications for SESR, in *Free Electron Generators of Coherent Radiation*, Addison-Wesley, Reading, Mass., 1980, 355.

113. Korneenkov, V. K., Petrushin, A. A., Skrynnik, B. K., and Shestopalov, V. P., Diffractive-radiation generator with a spherocylindrical open resonator, *Sov. Radio Phys.*, 20, 290, 1977.

114. Mizuno, K., Ono, S., and Shibata, Y., Two different mode interactions in an electron tube with a Fabry-Perot resonator — The Ledatron, *IEEE Trans. Electron Devices*, Ed-20, 749, 1973.

115. Rusin, F. S. and Bogomolov, G. D., Generation of electromagnetic oscillations in an open resonator, *JETP Lett.*, 4, 160, 1966.

116. Smith, S. J. and Purcell, E. M., Visible light from localized surface charges moving across a grating, *Phys. Rev.*, 92, 1069, 1953.

117. Bekefi, G., Electrically pumped relativistic free-electron wave generators, *J. Appl. Phys.*, 51, 3081, 1980.

118. Stroke, G. W., Diffraction gratings, in *Encyclopedia of Physics*, XXIX, Springer-Verlag, New York, 426, 1967.

119. Leavitt, R. P., Wortman, D. E., and Morrison, C. A., The Orotron — a free-electron laser using the Smith-Purcell effect, *Appl. Phys. Lett.*, 35, 363, 1979.

120. Wachtel, J. M., Free-electron lasers using the Smith-Purcell effect, *J. Appl. Phys.*, 50, N. 1, 1979.

121. Yariv, A., and Shih, Chun-Ching, Amplification of radiation by relativistic electrons in spatially periodic optical waveguides, *Opt. Commun.*, 24, 233, 1978.

122. Fradkin, D. M., Radiation reaction as a mechanism for increasing energy-momentum of a particle interacting with a laser field, *Phys. Rev. Lett.*, 42, 1209, 1979.

123. Edighoffer, J. A. and Pantell, R. H., Energy exchange between free electrons and light in vacuum, *J. Appl. Phys.*, 50, 6120, 1979.

124. Pantell, R. H. and Piestrup, M. A., Free-electron momentum modulation by means of limited interaction length with light, *Appl. Phys. Lett.*, 32, 781, 1978.

125. Atamonov, A S., Vinokurov, N. A., Veblyi, P. D., Gluskin, E. S., Kornuykhin, G. A., Nochubei, V. A., Kulipanov, G. N., Litvinenko, V. N., Menzentsev, N. A., and Skrinskii, A. N., The First Experiments with an Optical Klystron Installed on the VEPP-3 Storage Ring, (unpublished) Institute of Nuclear Physics, Novosibirsk, USSR.

126. **Chen, C. K., Sheppard, J. C., Piestrup, M. A., and Pantell, R. H.,** Analysis of bunching of an electron beam at optical wavelengths, *J. Appl. Phys.,* 49, 41, 1978.

127. **Shih, C. and Yariv, A.,** Electron rebunching and radiation gain in two element free electron lasers, in *Free Electron Generators of Coherent Radiation,* Addison-Wesley, Reading, Mass., 1980, 473.

128. **Vinokurov, N. A. and Skrinskii, A. N.,** Oscillator Klystron in the Optical Band Using Ultrarelativistic Electrons, Preprint IWP 77-59, Institute of Nuclear Physics, Siberian Branch, USSR Academy of Sciences, 1977.

129. **Vinokurov, N. A. and Skrinskii, A. N.,** Limiting Power of an Optical Klystron Installed on an Electron Storage Ring, Preprint IWP 77-67, Institute of Nuclear Physics, Siberian Branch, USSR Academy of Sciences, 1977.

130. **Budker, G. I., Karliner, M. M., Makarov, I. G., Morosov, S. N., Nezhevenko, O. A., Ostreiko, G. N., and Shekhtman, I. A.,** The Gyrocon — an efficient relativistic high-power VHF generator, *Part. Accel.,* 10, 41, 1979.

131. **Chu, K. R., Read, M. E., and Ganguly, A. K.,** Methods of efficiency enhancement and scaling for the gyrotron oscillator, *IEEE Trans. Microwave Theory Tech.,* MTT-28, 1980.

132. **Flyagin, V. A., Gaponov, A. V., Petelin, M. I., and Yulpatov, V. K.,** The Gyrotron, *IEEE Trans. Microwave Theory Tech.,* MTT-25, 514, 1977.

133. **Hirshfield, J. L. and Granatstein, V. L,,** The electron cyclotron maser — an historical survey, *IEEE Trans. Microwave Theory Tech.,* MTT-25, 522, 1977.

134. **Sprangle, P. and Smith, R. A.,** The nonlinear theory of efficiency enhancement in the electron cyclotron maser (Gyrotron), *J. Appl. Phys.,* 51, 3001, 1980.

135. **Seftor, J. L., Granatstein, V. L., Chu, K. R., Sprangle, P., and Read, M.,** "The electron cyclotron maser as a high power amplifier of millimeter waves," *IEEE J. Quantum Electron.,* QE-15, 848, 1979.

136. **Read, M. E., Gilgenbach, R. M., Lucey, Jr., R. F., Chu, K. R., Drobot, A. T., and Granatstein, V. L.,** Spatial and temporal coherence of a 35 GHz gyromonotron using the TE_{01} circular mode, *IEEE Trans. Microwave Theory Tech.,* MTT-28, 875, 1980.

137. **Krinski, S., Blumberg, L. Bittner, J., Galayda, R., Hess, P. R., Schuchman, J. C., and Steenberten, A. Van,** Design status of the 2.5 Gev. national synchrotron light source X-ray ring, *IEEE Trans. Nucl. Sci.,* NS26, 3806 1979.

138. 40 MeV, 250 A/cm² peak current.

139. **Paul, A. C., Neil, V. K., Craly, G. D., and Fessenden, T. J.,** Characteristics of the ETA gun, UCRL #84065, 1980, to be published.

140. **Blewett, J. P. and Chasman, R.,** Obits and Fields in the Helical Wiggler, *J. Appl. Phys.,* 48, 2692, 1977.

141. **Elias, L. R. and Madey, J. M.,** Superconducting Helically wound magnet for the free-electron laser, *Rev. Sci. Inst.,* 50, 1339, 1979.

142. **Poole, M. W. and Walker, R. P.,** Some limitations on the design of plane periodic electromagnets for undulators and free electron lasers, Daresburg Laboratory Report DL/SCI/0215A, 1980, to be published.

143. **Winick, H. and Knight, T.,** Wiggler Magnets — a collection of materials presented at the Wiggler Workshop held at SLAC (March 21—23, 1977) and other material relating to Wiggler Magnets, SSRP Report No. 77105, 1977.

144. **Deacon, D. A. G., Madey, J. M. J., Robinson, K. E., Bazin, C., Brillardon, M., Elleaume, P., Farge, Y., Ortega, J. M., Petroff, Y., and Yelghe, M. F.,** Gain measurement on the ACO storage ring, *IEEE Trans. Nucl. Sci.,* NS-28, 3142, 1981.

145. **Coisson, R.,** Angular-spectral distribution and polarization of synchrotron radiation from a "short" magnet, *Phys. Rev. A,* 20, 524, 1979.

146. **Winick, H. and Donlach, S.,** *Synchrotron Radiation Research,* Plenum Press, New York, 1980.

147. **Eckstein, J. N., Madey, J. M. J., Robinson, K., and Smith, T. I.,** Additional Experimental Results from the Stanford 3 micron FEL, presented at the ONR Workshop on Free Electron Lasers, Sun Valley, Idaho, June 22—25, 1981.

148. **Slater, J.,** Status of the FEL Experiment at Mathematical Sciences Northwest, presented at the ONR Workshop on Free Electron Lasers, Sun Valley, Idaho, June 22—25, 1981.

149. **Bazin, C., Billardon, M., Deacon, D., Farge, Y., Ortega, J. M., Perot, J., Petroff, Y., and Velghe, M.,** First results of a superconducting undulator on the ACO storage ring, *J. Phys. Lett.,* 41, L-547, 1980.

150. **Halbach, K., Chin, J., Hoyer, E., Winick, H., Cronin, R., Yang, J., and Zambre, Y.,** A permanent magnet undulator for SPEAR, *IEEE Trans. Nucl. Sci.,* NS-28, 3136, 1981.

151. **Shih, C-C. and Yariv, A.,** two-element free-electron lasers, *Optics Lett.,* 5, 76, 1980.

152. **Stagno, V., Brautti, G., Clauser, T., and Boscolo, I.,** Coherent radiation from electrons in transverse periodic fields, *Nuovo Cimento,* 56, 219, 1980.

153. **Boscolo, I. and Stagno, V.,** The converter and the transverse optical klystron, *Nuovo Cimento,* 58, 267, 1980.

4.1 FREE ELECTRON LASERS

4.1.2 MILLIMETER AND SUBMILLIMETER LASERS

V. L. Granatstein, R. K. Parker, and P. A. Sprangle

TWO CLASSES OF FREE ELECTRON LASERS

An overlay of accelerator technology and interaction physics strongly suggests that two distinct classes of free electron lasers will evolve:

R. F. Accelerators and Single Electron Interactions

Lasers such as those described above in Section 4.1.1 are based on high energy, low current electron beams produced in rf accelerators (Linacs, microtrons, and storage rings). They can be expected to operate in the region of parameter space where single particle electron dynamics will predominate. Such lasers will have modest gain and will be of practical interest as oscillators rather than amplifiers. Efficiency for conversion of electron kinetic energy to photon energy will also be modest and high overall device efficiency will depend on techniques for recovering the unused electron energy. The high energy characteristics of rf accelerators favors their use for free electron lasers at shorter wavelengths from the near infrared to the ultraviolet. The microstructure of the electron beam produced in rf accelerators ($\tau \lesssim 10$ psec) will limit the coherence of the radiation.

Pulse-Line Accelerators and Collective Interactions

The relativistic electron beams produced by pulse-line accelerators (induction Linacs,[1] IREB accelerators,[2] and radial line accelerators[3]) can have sufficient intensity to shift the laser interaction to the regime where collective electron oscillations participate in the wave amplification process. Free electron lasers with collective interactions can have sufficient gain so that practical amplifiers can be developed in addition to oscillators. Also, there is potential for high efficiency in converting electron kinetic energy to photon energy. The electron energy characteristic of pulse-line accelerators (roughly 1 to 10 MeV) is more modest and lasers employing these accelerators are expected to operate in the wavelength range extending from millimeters to the near infrared. Electron beam pulse duration (10 nsec to 10 μsec) is much longer than with rf accelerators, so that linewidth of the radiation may be compatible with good coherence even with operations in the millimeter wave regime.

BASIC PROCESSES IN COLLECTIVE INTERACTION LASERS: STIMULATED RAMAN SCATTERING

Collective free electron lasers amplify coherent radiation in a medium consisting of a dense, cold stream of relativistic electrons which are wiggled transversely either by a periodic wiggler magnet or by a strong electromagnetic wave. In the rest frame of the electron beam, the wiggler field appears to be a pump electromagnetic wave at frequency ω_o' and wave number k_o' (rest frame quantities denoted by primes). The incident pump wave scatters into a counterpropagating scattered electromagnetic wave (ω_s', k_s') and an electron density wave at the frequency, ω_e'. Both the backscattered e.m. wave and the plasma wave grow exponentially.

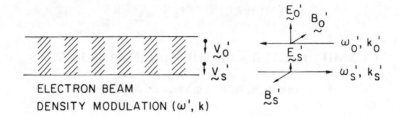

FIGURE 4.1.2.1.   Stimulated scattering in the electron beam rest frame.

The rest frame geometry is shown in Figure 4.1.2.1. The incident pump e.m. wave has a transverse electric field $E_o' \, \hat{e}_y$ which excites a transverse oscillation of the electrons with velocity $\underline{v}_o' = -e_y \, eE_o' \, (1 - v_o^2/C^2)^{1/2}/m\omega_o'$ where e is the magnitude of electron charge and m is the nonrelativistic electron mass. In the presence of an incipient scattered wave with magnetic field $B_s' \, \hat{e}_y$, an axial force $e \, v_o' \, B_s' \, e_z$ is exerted on the electrons. The coupling between the incident e.m. wave and the scattered e.m. wave thus produces a radiation pressure force (pondermotive force) which leads to a low frequency density modulation of the electrons. The complete expression for the pondermotive force is $\underline{F} = -e \, (\underline{v}_o' \times \underline{B}_s' + \underline{v}_s' \times \underline{B}_o')$. The frequency and wave number of the electron density modulation satisfies the following conservation laws

$$\omega_e' = \omega_o' - \omega_s' \text{ and } k_e' = k_o' + k_s' \qquad (1)$$

where k_e', k_o', and k_s' are positive real quantities denoting wave number magnitudes.

The growth of the density modulation gives increasing coherence to the scattering process, resulting in a growing scattered wave which in turn increases the density modulation still further. Thus, there is a feedback mechanism in this process which may result in an instability and exponential growth of both the scattered wave and the density modulation.

If the electron distribution is sufficiently dense and cold, it has a natural frequency of collective oscillation at the plasma frequency, ω_p. In that case the stimulated scattering process will be greatly strengthened by synergism with the growth of plasma waves when $\omega_e' = \omega_p$. The frequency of the scattered wave is then displaced from the incident frequency by a characteristic frequency of the medium (viz; $\omega_s' = \omega_o' - \omega_p$) and the process is called Stimulated Raman Scattering.

A number of theoretical analyses of stimulated Raman scattering of e.m. radiation from relativistic electron beams have appeared in the literature.[4-11] The papers by Sprangle and Drobot[10] and Sprangle and Smith[11] treat the nonlinear regime. Particle simulation codes have also been used to study the nonlinear problem.[12] Recently, both particle simulation studies[13] and analytical work[14] have shown substantial efficiency enhancement by appropriate contouring of wiggler period and amplitude. Several review articles covering collective free electron lasers have also appeared in the literature.[15-19] We note especially the review article by Sprangle et al.[17] from which we have adapted many of the expressions for growth rate and efficiency which follows. In general, the expressions presented assume that the electron beam is highly relativistic.

THE DOPPLER-SHIFTED LASER FREQUENCY

Equation 1 gives the relationship between the wave frequencies in the rest frame of the electron beam. In the remainder of this paper, for the convenience of the reader all expressions will be given in the laboratory frame where the electrons are streaming with velocity v_z.

In the beam frame, the stimulated scattering is a Stokes process with the frequency of the scattered output wave being smaller than the incident pump wave frequency. However, in transforming from the beam frame to the laboratory frame, the pump frequency is downshifted while the backscattered wave frequency is upshifted. This double doppler shift results in a scattered output wave in the laboratory frame at much shorter wavelength than the wavelength of the wiggler pump.

For a scattered e.m. wave propagating at the speed of light in the same direction as the electron beam, we have the following expression for output frequency

$$\omega_s = \gamma_z (1 + \beta_z) [\gamma_z \beta_z \kappa_o c - \omega_p] \tag{2}$$

where

$$\kappa_o = 2\pi/\ell \tag{3a}$$

when the pump wave is a magnetostatic wiggler of period ℓ, and

$$\kappa_o = \omega_o (1 + v_z/v_{ph})/v_z \tag{3b}$$

when the pump is an electromagnetic wave propagating counter to the streaming electrons with phase velocity v_{ph}, and frequency ω_o. In Equations 2 and 3, $\beta_z = v_z/c$, $\gamma_z = (1 - \beta_z^2)^{-1/2}$, and the frame invariant plasma frequency

$$\omega_p = (e^2 n/\epsilon_o m \gamma)^{1/2} \tag{4}$$

where n is the electron density, $\gamma = (1 - v_z^2/c^2 - v_\perp^2/c^2)$, $v\perp$ is the transverse electron velocity, and we are employing relationalized m.k.s. units.

For the case of a highly relativistic beam, the laser output frequency of Equation 2 may be as large as

$$\mathrm{MAX}(\omega_s) = 2\gamma_z^2 (2\pi/\ell) c \tag{5a}$$

for a magnetostatic wiggler pump. For an electromagnetic pump wave with $v_{ph} = c$,

$$\mathrm{MAX}(\omega_s) = 4\gamma_z^2 \omega_o \tag{5b}$$

The relationship between the frequencies and wave numbers of the pump wave, the scattered output wave, and the idler wave at ω_p are shown in the Stokes diagrams, Figure 4.1.2.2. Figure 4.1.2.2a shows the beam rest frame relationships, while Figure 4.1.2.2b shows the laboratory frame relationships. Figure 4.1.2.2 depicts the case where the pump wave is a zero-frequency magnetostatic ripple in the laboratory frame, and the scattered wave propagates at the speed of light.

GROWTH RATE OF THE OUTPUT SCATTERED WAVE

As discussed above in above, the interaction of a cold dense electron beam with a wiggler field can result in an instability in which the scattered output wave grows exponentially. In that case, an electromagnetic wave at ω_s will be amplified as it propagates along the z axis colinear with the electron beam; the wave amplitude will increase as $e^{\Gamma z}$, where Γ is the amplitude growth rate.

The expressions presented below for the growth rate were derived[7,17] assuming that the pump wave is RHCP in the case of an electromagnetic wiggler, and helical in the

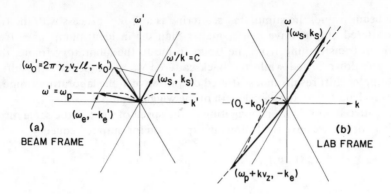

FIGURE 4.1.2.2.   Stokes diagrams for free electron laser based on stimulated Raman scattering for case where pump wave is a magnetostatic wiggler. — — Dispersion curves for space charge waves.

case of a magnetostatic wiggler. It was also assumed that for an interaction of length L, $\Gamma L \gg 1$ so that end effects are not important in the wave amplification process.

The wiggler field excites a transverse electron velocity with magnitude given by

$$v_\perp = \frac{e\,B_w}{m\,\gamma\,\kappa_0} \tag{6}$$

where for a magnetostatic wiggler with transverse magnetic field
$\underset{\sim}{b}_r = b_r\,(\hat{e}_x \cos 2\pi z/\ell + \hat{e}_y \sin 2\pi z/\ell)$

$$B_w = b_r \tag{7a}$$

while for an electromagnetic wiggler with electric field
$\underset{\sim}{E}_o = E_o\,[\hat{e}_x \cos (k_o z + \omega_o t) + \hat{e}_y \sin (k_o z + \omega_o t)],$

$$\underset{\sim}{E}_o = E_0\,[\hat{e}_x \cos (k_0 z + \omega_0 t) + \hat{e}_y \sin (k_0 z + \omega_0 t)],$$
$$B_w = E_0\,(1 + v_z/v_{ph})/v_z \tag{7b}$$

It should be noted that the relationship between the streaming electron energy, γ_z, and the total electron energy γ depends on the magnitude of transverse velocity, $v_\perp$ as

$$\gamma_z = \gamma/(1 + \gamma^2\,v_\perp^2/c^2)^{1/2} \tag{8}$$

Thus, when a magnetostatic wiggler is used to excite a strong $v_\perp$ the streaming energy and the doppler shift decrease; a magnetostatic wiggler does not increase the total energy.

When the electron beam is propagated along a strong uniform magnetic guide field of strength $\overline{B}$, the transverse velocity excited by the wiggler may be enhanced[5] by resonance effects near the cyclotron frequency $\overline{\Omega} = c\,\overline{B}/m\gamma$. In that case the right hand side of Equation 6 should be multiplied by the enhancement factor $v_z\,\kappa_o/(v_z\kappa_o - \overline{\Omega})$; of course, in a physical system, the quantity $v_z\kappa_o - \overline{\Omega}$ cannot be made arbitrarily small[5] and its magnitude has been calculated[5] for the case where the pump wave is a normal mode of the system as $v_z\kappa_o - \overline{\Omega} = \frac{1}{2}(\overline{\Omega}^2 + 4\,\omega_p^2)^{1/2} - \frac{1}{2}\,\overline{\Omega}$.

The amplitude growth rate in the stimulated Raman scattering regime is given by

$$\Gamma_R = 0.5\,(\gamma_z\,\omega_p/c\kappa_0)^{1/2}\,\beta_\perp\,\Gamma^{1/2}\,\kappa_0 \tag{9a}$$

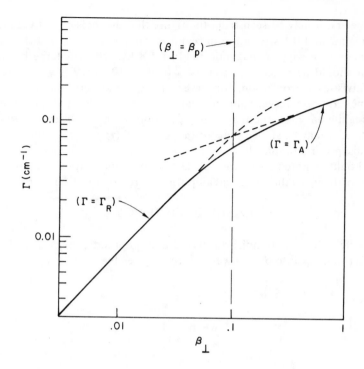

FIGURE 4.1.2.3.   Wave amplitude growth rate as a function of transverse electron velocity.

where $\beta_\perp = v_\perp/c$ and F is the filling factor i.e., the ratio of the electron beam area to
the area of the electromagnetic beam being amplified at ω_s with the limitation F $\leqslant$ 1.
Equation 9a is valid when space charge forces dominate the pondermotive forces requiring

$$\beta_\perp \ll \beta_p \qquad\qquad (9b)$$

where

$$\beta_p = 4\,\omega_p^{1/2}\,(\gamma_z^3\,c\,\kappa_0\,F)^{-1/2} \qquad\qquad (9c)$$

When the wiggler field is very strong, the pondermotive forces may modify the
plasma wave dispersion. In that case the wave amplification process is altered. The
pondermotive wave dominates the collective space charge wave. The expression for
amplitude growth rate becomes

$$\Gamma_A = \frac{\sqrt{3}}{2}\,\Gamma_R\,(\beta_\perp/\beta_p)^{-1/3} \qquad\qquad (10a)$$

Validity of Equation 10a requires

$$\beta_\perp \gg \beta_p \qquad\qquad (10b)$$

which has been called the "strong pump regime".

Growth rate is plotted as a function of $\beta_\perp$ in Figure 4.1.2.3; the case of a magnetostatic wiggler has been assumed and free electron laser parameters have been taken as
$\gamma = 5$, $\ell = 3$ cm, F = 1, and electron density n $= 10^{11}$ cm^{-3} in Figure 4.1.2.3. For

these parameters, it may be seen from the figure that the division between the Raman scattering regime and the strong pump regime occurs at $\beta_\perp = \beta_p = 0.1$; this value of $\beta_\perp$ corresponds to a wiggler magnetic field of 1.8 kG which is large but attainable. However, it should be noted that a density of n = 10^{11} cm^{-3} is very modest for an intense relativistic electron beam; for larger values of n, operation in the strong scattering regime requires a larger value of wiggler magnetic field, and the normal Raman scattering regime has been the regime of operation in intense beam experiments conducted to date. It may be shown that for $(\omega_p/c\,\varkappa_o) > F\,(\gamma^2 - 1)/(4\gamma)^2$ the strong scattering regime disappears completely.

Last, it should be noted in Figure 4.1.2.3 that amplitude exponentiating lengths are on the order of 10 cm so that practical amplifiers of reasonable length are possible.

LASER EFFICIENCY

For a wiggler with axially uniform period and amplitude, efficiency of converting electron kinetic energy into output wave energy is given by

$$\eta_R = \omega_p/\gamma_z\,c\,\varkappa_o \tag{11}$$

when $\beta_\perp \ll \beta_p$

and by

$$\eta_A = \eta_R\,(\beta_\perp/\beta_p)^{2/3} \tag{12}$$

when $\beta_\perp \gg \beta_p$

The expression for efficiency in the Raman regime may be expressed in terms of the laser output frequency ω_s and shown to have an upper bound given by

$$\eta_R \lesssim 2\,\gamma\,\omega_p/\omega_s$$

where we have used $\omega_s = 2\,\gamma_z^{2-}\,c\,\varkappa_o$ and $\gamma_z \lesssim \gamma$. This upper bound efficiency is plotted as a function of ω_s in Figure 4.1.2.4 for values of accelerator voltage and current density appropriate to intense relativistic electron beam accelerators. (In present intense beam accelerators current densities of 10^4 A/cm^2 are routinely achieved while J = 10^5 A/cm^2 is a reasonable estimate of what could be attained with careful design.)

Note that good efficiency, on the order of 10%, is predicted in Figure 4.1.2.4 for wavelengths longer than about 10 μm. It should also be stressed that recent theoretical studies have indicated the possibility of greatly increasing efficiency above the values predicted by Equations 11 and 12 by such techniques as decreasing the wiggler period and increasing the wiggler magnetic field in the region where the extraction of energy from the electrons is beginning to saturate.[14]

Transfer of energy from the electrons to the output e.m. wave saturates when the electrons become trapped in the minima of the total longitudinal wave potential (i.e., the pondermotive plus space-charge waves), and the electrons are no longer free to maintain a phase relationship conducive to energy loss. This process in a uniform wiggler leads to the expressions for saturated efficiency of energy transfer given by Equations 11 and 12. However, when the wiggler period is reduced the phase velocity of the total longitudinal wave is also reduced and the trapped electrons are forced to slow down and give up more energy to the output radiation.

UPPER BOUND ESTIMATES ON SINGLE PASS EFFICIENCY IN A UNIFORM WIGGLER F.E.L.

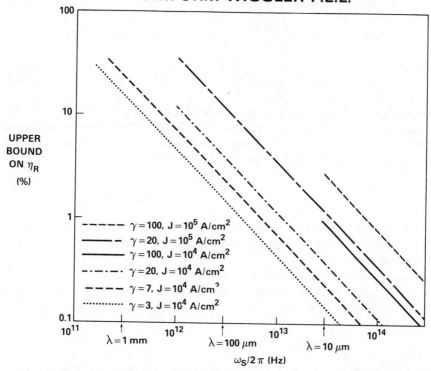

FIGURE 4.1.2.4. Upper bound estimates of single-pass, unenhanced, Raman FEL efficiency as a function of output frequency. Several examples of electron current density and accelerator voltage are given.

LIMITATIONS IMPOSED BY ELECTRON VELOCITY SPREAD, AND THE REQUIREMENT FOR HIGH QUALITY ACCELERATORS

In order that the preceding expressions for gain and efficiency be valid the electron beam must have a sufficiently small axial velocity spread, Δv, at a given time and spatial position. If Δv is excessive, bunching of the electrons becomes smeared out and stimulated scattering tends to take place from individual electrons (stimulated Compton scattering) rather than from a density wave in the electron beam. The condition for validity of the expressions for gain and efficiency in the Raman and strong scattering regimes is

$$\Delta v/c \ll \eta/\gamma_z^2 \tag{13}$$

where η is given by Equations 11 and 12. If Equation 13 is not satisfied and stimulated Compton scattering holds, both the gain and the efficiency of the stimulated scattering process fall rapidly as Δv increases, and in general are much below the values of gain and efficiency given by the expression in sections D and E above.

The measure of electron beam quality which will determine if the inequality in Equation 13 is satisfied is emittance, ε, defined for a solid streaming electron beam of radius R as

$$\epsilon = \pi R \langle \Theta^2 \rangle^{1/2} \tag{14}$$

where $\langle \theta^2 \rangle^{1/2}$ is the r.m.s. spread in the angle between the electron trajectories and the axis. Clearly a spread in orientation of the electron trajectories implies a spread in the velocity components. For small values of $\langle \theta \rangle^{1/2}$, the spread in axial velocity is related to emittance as

$$\Delta v / c = \epsilon^2 \, \beta / 2\pi^2 \, R^2 \qquad (15)$$

Then Equation 13 and Equation 15 imply that for R = 3 mm, η = 10% and $\gamma_z \approx \gamma$ = 4, the emittance must satisfy $\epsilon \beta_z \gamma_z \ll 130$ mrad·cm ; this is a realizable but not a trivial requirement for an intense electron beam.

Aside from variations of Δv at a given time and spatial position, there can also be systematic variations of Δv in time and space which also will degrade free electron laser performance. Variation with time arises from fractional fluctuations in accelerator voltage, $\Delta V/V$. The inequality in Equation 13 implies that

$$\frac{\Delta V}{V} \ll \eta \qquad (16)$$

in order that stimulated Raman scattering occur. Marshall et al.[18] have shown that if $\Delta V/V > \eta$, the free electron laser radiation will be absorbed during a portion of the accelerator current pulse.

A systematic spacewise variation in electron energy occurs because of space charge effects in an intense electron beam. For a solid beam on axis the potential variation across the beam radius is

$$\Delta V = I / 4\pi \, \epsilon_o \, v_z \qquad (17)$$

where I is the beam current, and ϵ_o is the permittivity of free space. For a thin annular beam of thickness t and radius a,

$$\Delta V = (I / 4\pi \, \epsilon_o \, v_z) \, (t/a) \qquad (18)$$

In order for stimulated Raman scattering to occur ΔV due to electrostatic effects must satisfy the inequality in Equation 16; this will place a limitation on beam current. For example, if η = 10% and V = 3 MV then from Equation 16 and Equation 17 one has the requirement for a solid beam that I $\ll$ 10 kA; of course, it is clear from the form of Equation 18 that larger total currents are permitted for annular electron beams. Last, there will also be a potential variation across the beam radius due to the transverse gradient of the wiggler magnetic field[19,30] which is necessary to satisfy $\nabla \cdot \underset{\sim}{b}_r$ = $\nabla \times \underset{\sim}{b}_r$ = 0. The shear due to the transverse dependence of the wiggler field opposes the shear due to spare charge effects.

EXPERIMENTS

Data on laboratory studies of stimulated scattering from intense relativistic electron beams which resulted in millimeter and submillimeter radiation are summarized in Table 4.1.2.1; the experiments listed were carried out by the Naval Research Laboratory,[20,23] Columbia University,[21-23] or the Lebedev Institute.[24] Experimental studies are also in progress at M.I.T. where a periodic electrostatic wiggler is being employed,[25] the Ecole Polytechnique,[26] TRW Inc.,[27] and the Hebrew University of Jerusalem.[28]

The intense-electron-beam studies listed in Table 4.1.2.1 are characterized by electron energy in the range 0.7 to 2 MeV and corresponding output wavelengths that

Table 4.1.2.1

Experimental configuration	Wiggler fields	Electron beam	Output radiation	Ref.
Amplified spontaneous emission	Electromagnetic wave $\ell = 2$ cm $E_o = 4 \times 10^4$ V/cm $L = 0.3$ m	2 MeV, 30 kA a = 1.8 cm, t = 2 mm $n = 3 \times 10^{12}$ cm^{-3} $\Delta V/V$ (electrostatic) = 4%	$\lambda_s = 400 \, \mu$m 1 MW $G_L = \int_o^L$ $e\Delta^z \, dz > 2$	20
Amplified spontaneous emission	Magnetostatic wiggler $\ell = 0.6$ cm $b_r = 0.5$ kG $L = 0.36$ m	0.86 MeV, 5 kA a = 1 cm, t = 1.5 mm $n = 10^{12}$ cm^{-3} $\Delta V/V$ (elec- trostatic) = 2%	$\lambda_s = 1.5$ mm 8 MW	21, 22
Oscillator with optical cavity; mirror trans. = 2%	Magnetostatic wiggler $\ell = 0.8$ cm $b_r = 0.4$ kG L = 0.4 m	1.2 MeV, 25 kA a = 2.2 cm, t = 1 mm $n = 4 \times 10^{12}$ cm^{-3} $\Delta V/V$ (electrostatic) = 3%	$\lambda_s = 400 \, \mu$m $\Delta\lambda_s/\lambda_s = 2\%$ 1 MW	23
Amplified spontaneous emission	Electromagnetic wave $\ell = 3.2$ cm $E \approx 10^5$ V/cm $L = 0.6$ m	0.7 MeV, 4.5 kA a = 1 cm, t = 1.5 mm $n = 1.4 \times 10^{12}$ $\Delta V/V$ (elec- trostatic) = 4%	$\lambda_s = 3.2$ mm 20 W	24

range from 3 mm down to 400 μm. The current is large (4.5 to 30 kA) making possible the participation of collective electron beam modes in the stimulated scattering process and resulting in large gain; amplified spontaneous emission was demonstrated in relatively short interaction lengths (30 to 60 cm) and in one experiment the single pass amplitude gain $G_L > 2$. As yet, there is no continuous power capability in the intense beam studies, the experiments being characterized by a single electron pulse of 10 to 50 nsec in duration.

Two types of pump waves were investigated: a magnetostatic wiggler (Examples 2 and 3 in Table 4.1.2.1) and a powerful electromagnetic wave (Examples 1 and 4 in Table 4.1.2.1). In the experiments which used an electromagnetic pump wave, it was generated by a portion of the same electron beam in which the stimulated backscattering occurred. The pump wave with wavelength 2 to 3 cm was generated in a region of the beam far downstream from the cathode near the output end of the experiment. "Conventional" processes were used to generate the pump wave (i.e., the electron cyclotron maser instability in example 1, and the usual BWO instability in a tube with periodically rippled walls in Example 4). The pump wave was then made to propagate upstream; when it encountered the cold streaming electrons near the cathode, stimulated scattering occurred, resulting in backscattered millimeter and submillimeter radiation. In addition to the wiggler field, the intense beam experiments typically had an externally imposed uniform axial magnetic field that was large enough so that magnetic resonances in the output power were observed.

Theory predicts that intense-beam free-electron lasers with output in the millimeter and submillimeter will have saturated efficiencies on the order of several to several tens of percent (when operated as traveling wave amplifiers). The efficiency measured for the free electron laser oscillators listed in Table 4.1.2.1 has been lower by one to two orders of magnitude. A more detailed discussion of one of the experiments may shed light on this apparent discrepancy. Figure 4.1.2.5 is a schematic of the free electron laser oscillator corresponding to example 3 in Table 4.1.2.1. The free electron laser oscillator was created by passing a 1.2 MV electron beam of 25 kA through a spatially periodic, linearly polarized, magnetostatic field which had a ripple amplitude

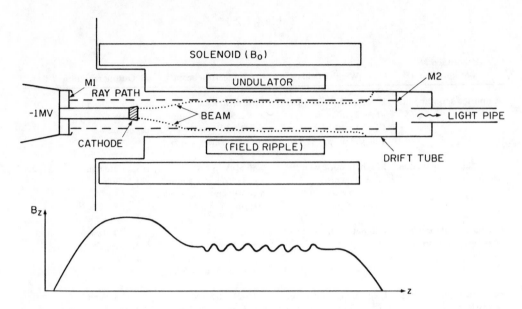

FIGURE 4.1.2.5.   Experimental arrangement in study of a Raman regime free electron laser with quasi-optical cavity. M_1 and M_2 designate mirrors. (From McDermott, D. B., Marshall, T. C., Schlesinger, S. P., Parker, R. K., and Granatstein, V. L., *Phys. Rev. Lett.*, 41, 1368, 1978. With permission.)

B_o = 400 G in the radial field and a period of ℓ = 8 mm. As shown in Figure 4.1.2.5, an annular electron beam was expanded radially by passing it through an adiabatic reduction in the confining magnetic field. Beam expansion not only reduced the transverse electron energy spread but also placed the beam within the resonant volume of the Fabry-Perot cavity formed by the two mirrors. However, the short beam duration (30 nsec) and the long cavity length (1.5 m) limited the feedback radiation to a maximum of three interactive passes. The double-doppler shifted radiation was diffraction coupled from an aperture in the output mirror. The peak output power was measured to be 1 MW at 400 μm, corresponding to an efficiency of 0.02%.

This study was instructive and did result not only in an impressive output power, but also in a narrowing of the linewidth of the output radiation to $\Delta\lambda_s/\lambda_s \sim$ 2% compared with $\Delta\lambda_s/\lambda_s >$ 10% in the case of the superradiant oscillators. However, variations in accelerator voltage during the "constant" portion of the accelerator waveform ($\Delta V/V \approx 5 - 10\%$) may well have caused periodic reversal in the energy transfer from beam to wave. Moreover, computer studies of the electron trajectories within the interaction volume have shown that the annular electron beam is composed of an outer layer with large velocity spread and an inner layer with small velocity spread[29] Calculations using the linear theory of Raman scattering and the results of the electron trajectory analysis indicate that the growth in the scattered wave amplitude was limited to approximately one exponential increase during a single pass through the 40 cm ripple length. Since feedback was limited to only three passes by the short beam duration, it is highly unlikely that the process became saturated.

An improved experiment in which the stimulated Raman scattering process could reach saturation requires an electron beam of improved quality as well as a much stronger wiggler magnet. As this volume goes to press, a new superradiant experiment has been reported on [31,32]that makes use of a 1 kA, 1.35 MeV, 6 mm diameter solid electron beam of greatly reduced velocity spread ($\Delta v/C <$ 0.1%). The 3 cm period wiggler was variable up to b_r = 4 kG, and its effect was enhanced by gyroresonance with a uniform axial magnetic field. The measured Raman amplitude growth rate was

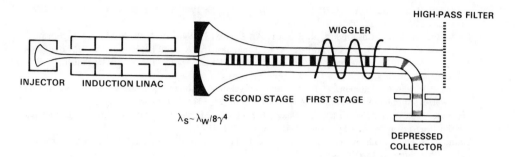

FIGURE 4.1.2.6.   Conceptual operation of a two-stage Raman free electron laser with output in the near infrared. An electron accelerator producing a beam with normalized emittance $\varepsilon\beta\gamma \leqslant 10$ mrad. cm would be required for operation in the Raman regime at $\lambda_s \sim 10$ μm; this represents a substantial improvement over the present state of the art in induction linacs and other high current accelerators. Energy recovery by employing a collector which is depressed in potential is depicted as a means of improving overall laser efficiency.

$\Gamma_R = 0.12$ cm^{-1} or 8 e-folds of the wave amplitude in a 63 cm interaction length. The measured efficiency was $\eta_R = 2.5\%$ corresponding to an output power of 35 MW at a wavelength of 4 mm.

Finally, one should consider for future experimentation the possibility of two-stage free electron lasers. Free-electron laser operation at short wavelengths has the disadvantage of requiring very high electron kinetic energy. Accelerator voltage requirements could be reduced for intense beam systems by using a two-stage approach. To illustrate this technique, a near-infrared radiation source is considered. In a two-stage laser, two consecutive and distinct scattering interactions take place within a single electron beam. The output radiation from the first stage, in which the pump is a circularly polarized static magnetic field, is reflected back on the beam and used as an electromagnetic pump wave in the second stage. The final wavelength of the output radiation, from the second stage, is $\lambda \approx \ell/8\gamma^4$ instead of $\ell/2\gamma^2$ as would be the case in a single-stage device. Hence, in a two-stage laser, far shorter output wavelengths can be realized for the same electron kinetic energy. For example, a 3 MeV electron beam with a total current of 7 kA is passed in the first stage through a wiggler with a period of 2 cm. This interaction would produce radiative power pulse of 3.6 GW at a wavelength of 340 μm. Reflection of this intermediate frequency back on the beam would result in a second scattering, upshifting the output radiation to a wavelength of 10 μm. With efficiency enhancement the output power could be 200 MW for an efficiency of $\sim 1\%$. Sequential acceleration of the electron beam during the second scattering interaction would have a similar effect on efficiency enhancement as decreasing the period in a wiggler magnet. This two-stage free electron laser concept is depicted in Figure 4.1.2.6.

REFERENCES

1. Christofilos, N. C., Hester, R. E., Lamb, W. A. S., Reagen, D. D., Sherwood, W. A., and Wright, R. E., *Rev. Sci. Instrum.*, 35, 886, 1964; Avery, R., Behrsing, G., Chupp, W. W., Faltens, A., Hartwig, E. C., Hernandez, H. P., MacDonald, C., Meneghetti, J. R., Nemetz, R. G., Popenuch, W., Salig, W., and Vanecek, D., *IEEE Trans. Nucl. Sci.*, NS-18, 479, 1971; and Hester, R. E., Bubp, D. G., Clark, J. C., Chesterman, A. W., Cook, E. G., Dexter, W. L., Fessenden, T. J., Reginato, L. L., Yokota, T. T., and Faltens, A. A., *IEEE Trans. Nucl. Sci.*, NS-26, 4180, 1979.

2. **Bernstein, B. and Smith, I.**, *IEEE Trans. Nucl. Sci.*, NS-20, 1973; **Parker, R. K. and Ury, M.**, *IEEE Trans. Nucl. Sci.*, NS-22, 983, 1975.

3. **Pavlovskii, A. I. and Bosamykin, V. S.**, *Sov. At. Energy*, 37, 942, 1974; Pavlovskii, A. I., Bosamykin, V. S., Kuleshov, G. D., Gerasimov, A. I., Tananakin, V. A., and Klementev, A. P., *Sov. Phys. Dokl.*, 20, 441, 1975; **Eccleshall, D. and Temperley, J. K.**, *J. Appl. Phys.*, 49, 1978; **Smith, I.**, *Rev. Sci. Instrum.*, 50, 714, 1979.

4. **Sprangle, P. and Granatstein, V. L.**, Stimulated cyclotron resonance scattering and the production of powerful submillimeter radiation, *Appl. Phys. Lett.*, 25, 377—379, 1974.

5. **Sprangle, P., Granatstein, V. L., and Baker, L.**, Stimulated collective scattering from a magnetized relativistic electron beam, *Phys. Rev.*, A-12, 1697—1701, 1975.

6. **Uhm, H. S. and Davidson, R. C.**, Self-consistent Vlasov description of the free electron laser instability, *Phys. Fluids*, 23, 2076—2084, 1980.

7. **Kroll, N. M. and McMullin, W. A.**, Stimulated emission from relativistic electrons passing through a spatially periodic transverse magnetic feld, *Phys. Rev.*, A-17, 300—308, 1978.

8. **Hasegawa, A.**, Free electron laser, *Bell Syst. Tech. J.*, 57, 3069—3089, 1978.

9. **Bernstein, I. B. and Hirshfield, J. L.**, Amplification on a relativistic electron beam in a spatially periodic transverse magnetic field, *Phys. Rev.*, A-20, 1661—1670, 1979; Bernstein, I. B. and Friedland, L., Theory of the free electron laser in combined helical pump and axial guide fields, *Phys. Rev.*, A23, 816—823, 1981.

10. **Sprangle, P. and Drobot, A. T.**, Stimulated backscattering from relativistic unmagnetized electron beams, *J. Appl. Phys.*, 50, 2652—2661, 1979.

11. **Sprangle, P. and Smith, R. A.**, The theory of free electron lasers, *Phys. Rev.*, A21, 293—301, 1980.

12. **Kwan, T., Dawson, J. M., and Lin, A. T.**, Free electron laser, *Phys. Fluids*, 20, 581—588, 1977.

13. **Lin, A. T. and Dawson, J. M.**, High efficiency free electron laser, *Phys. Rev. Lett.*, 42, 1670—1673, 1979.

14. **Sprangle, P., Tang, C. M., and Manheimer, W. M.**, Non-linear theory of free electron lasers and efficiency enhancement, *Phys. Rev.*, A-21, 302—318, 1980; Non-linear formulation and efficiency enhancement of free electron lasers, *Phys. Rev. Lett.*, 43, 1932—1936, 1979.

15. **Granatstein, V. L. and Sprangle, P.**, Mechanism for coherent scattering of electromagnetic waves from relativistic electron beams, *IEEE Trans. Microwave Theory Tech.*, MTT-25, 545—550, 1977.

16. **Gover, A. and Yariv, A.**, Collective and single-electron interactions of electron beams with electromagnetic waves, and free-electron lasers, *Appl. Phys.*, 16, 121—133, 1978.

17. **Sprangle, P., Smith, R. A., and Granatstein, V. L.**, Free electron lasers and stimulated scattering from relativistic electron beams, *Infrared and Millimeter Waves*, Vol. 1, Button, K. J., Ed., Academic Press, New York, 1979, 279—327.

18. **Marshall T. C., Schlesinger, S. P., and McDermot, D. B.**, The free electron laser: a high power submillimeter radiation source, in *Advances in Electronics and Electron Physics*, Vol. 53, Marton, L., Ed., Academic Press, New York, (to be published).

19. **Bratman, V. L., Ginzburg, M. S., and Petelin, M. I.**, Ubitrons and scattrons, in *Relativistic High Current Electronics*, Gaponov, A. V., Ed., Academy of Sciences of the U.S.S.R., Institute of Applied Physics, Gorki'i, 1979, 217—148 (in Russian).

20. **Granatstein, V. L., Schlesinger, S. P., Herndon, M., Parker, R. K., and Pasour, J. A.**, Production of megawatt submilimeter pulses by stimulated Magneto-Raman scattering, *Appl. Phys. Lett.*, 30, 384—386, 1977.

21. **Marshall, T. C., Talmadge, S., and Efthimion, P.**, High-power millimeter radiation from an intense relativistic electron-beam device, *Appl. Phys. Lett.*, 31, 320—322, 1977.

22. **Gilgenbach, R. M., Marshall, T. C., and Schlesinger, S. P.**, Spectral properties of stimulated Raman radiation from an intense relativistic electron beam, *Phys. Fluids*, 22, 971—977, 1979.

23. **McDermott, D. B., Marshall, T. C., Schlesinger, S. P., Parker, R. K., and Granatstein, V. L.**, High-power free-electron laser based on stimulated Raman backscattering, *Phys. Rev. Lett.*, 41, 1368, 1978.

24. **Zhukov, P. G., Ivanov, V. S., Rabinovich, M. S., Raizer, M. D., and Ruchadze, A. A.**, Stimulated Compton scattering from relativistic electron beam, Proceedings of the Third International Topical Conference on Higher Power Electron and Ion Beam Research and Technology, Vol. 1., Novosibirsk, 1979, 705—714.

25. **Shefer, R. E., Jacobs, K. D., and Bekefi, G.**, Quasistatic pump for free electron lasers, *Bull. Am. Phys. Soc.*, 24, 1067, 1979.

26. **Boehmer, H., Buzzi, J. M., Doucet, H. J., Etlicher, B., Lamain, H., and Rouille, C.**, Resonance effect on relativistic electron beam propagation for collective free electron laser, *Bull. Am. Phys. Soc.*, 24, 1066, 1979.

27. **Boehmer, H., Munch, J., and Caponi, M. Z.**, Free electron laser experiment with a spatially varying pump amplitude, *Bull. Am. Phys. Soc.*, 24, 1066, 1979.

28. Private discussions with J. Hirshfield.
29. Jackson, R. H., Parker, R. K., and Granatstein, V. L., Beam quality studies for intense beam free electron lasers, Digest of Fourth International Conference on Infrared and Millimeter Waves and Their Applications, Miami, 1979, IEEE Cat. No. 79 CH 1384-7 MTT, IEEE, Piscataway, N.J., 1979, 96—97.
30. Sprangle, P. and Tang, C. M., Three-Dimensional, Non-Linear Theory of the Free Electron Laser, Naval Research Laboratory Memorandum Report 4280, 1980.
31. Parker, R. K., Jackson, R. H., Gold, S. H., Freund, H. P., Granatstein, V. L., Efthimion, P. C., Herndon, M., and Kinkead, A. K., Axial magnetic field effects in a millimeter wave Raman free electron laser, Phys. Rev. LeH., to be published.
32. Gold, S. H., Jackson, R. H., Parker, R. K., Granatstein, V. L., Efthimion, P. C., Freund, H. P., Herndon, M., and Kinkead, A. K., A millimeter free electron laser experiment based on a cold electron beam, Bull. Am. Phys. Soc., 26, 846, 1981.

4.2 X-RAY LASERS

Raymond C. Elton

INTRODUCTION

Research toward advancing lasing to the X-ray spectral regions is in an early and progressive state, such that the usual categorization and tabulation typified by a handbook article is not possible. Instead, this represents an overview of the field, somewhat abbreviated since a comprehensive review[1] with 268 references published in 1976 is generally available to the reader seeking more detail. In addition, various useful literature searches and compilations are available.[2] The references here will be limited to articles of more general guidance, as well as to those necessary to support specific points of emphasis and milestones.

The challenge of inventing and developing X-ray lasers may be approached by

1.   Adapting familiar X-ray sources to lasing action
2.   Extending proven ion laser processes progressively toward shorter wavelengths, perhaps through isoelectronic extrapolation
3.   Discovering new pumping and emission processes more appropriate to the task

With potential applications[3] in the vacuum-ultraviolet spectral region seemingly limited as compared to those for the penetrating X-ray region, early thoughts were directed toward making the big leap to the X-ray and perhaps γ-ray regions. Formidable pumping problems were projected. Meanwhile advancements into the ultraviolet regions, accompanied by rising uses and interests as specific devices have emerged, seem to indicate that the more reasonable approach is the continued systematic advance toward shorter wavelengths. Indeed, over the past 12 years the so-called short-wavelength "barrier" has been pushed from 200 nm into the vacuum region — first near 100 nm, and presently it appears that 60 nm has been reached.[4] These advances have been achieved both with cavities and in the amplified spontaneous emission (ASE) single pass mode, where the latter requires considerably higher gain.

The emphasis in the following is on the extreme-ultraviolet (EUV or XUV) portion of the vacuum-ultraviolet (VUV) spectral region which includes soft X-rays (as defined by Samson[5]), i.e., wavelengths between 0.2 and 100 nm as indicated also in Figure 4.2.1. Some early work at longer wavelengths is included for historical perspective.

BASIC TASKS

The basic problems in extrapolating laser concepts to extremely short wavelengths were early understood on the basis of a rapid decline in surface reflectivity[6] for efficient cavity operation, leading to consideration of single-pass ASE gain $I/I_o = \exp(\alpha L)$ over a length L, where the small-signal gain coefficient α at the center of a line of width $\Delta\nu$ for a transition of probability $A_{\mu\varrho}$ is given by[1,7]

$$\alpha \approx \xi \ \frac{\lambda^2}{8\pi\Delta\nu} \ A_{u\varrho}N_u \left[1 - \frac{N_\varrho/g_\varrho}{N_u/g_u} \right] \qquad (1)$$

The upper and lower state densities N_u, N_ℓ and statistical weights g_u, g_ℓ enter as shown and ξ is a numerical factor dependent on line shape. Although $A_u\ell$ may be expected to scale as $\sim f\lambda^{-2}$, and $\Delta\nu$ as λ^{-1} for Doppler broadening, a linear wavelength (λ) scaling

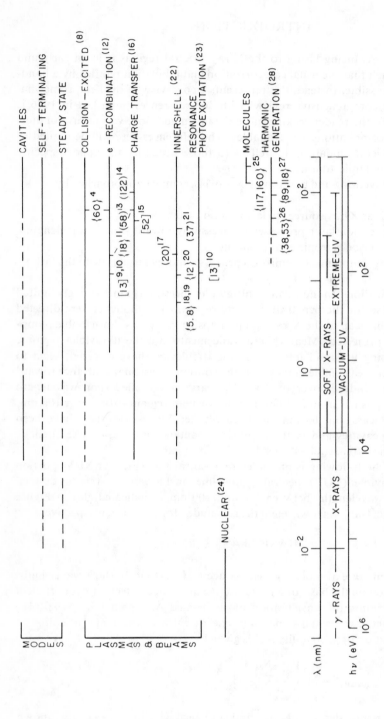

FIGURE 4.2.1. Diagram categorizing various operating modes and pumping processes for achieving gain, as well as for harmonic generation. The wavelength regions are somewhat arbitrary but follow Samson's definition[5] for the vacuum-UV region. The numbers on the bars refer to milestone wavelengths in nm, where () indicates modeling, [] indicates population inversions and {} indicates emission either as gain or harmonics. The dashed lines indicate projections into spectral regions of less certainty. References are indicated by superscripts as in the text. The emphasis here and in the text is on wavelengths shorter than 100 nm; molecular hydrogen is included for historical perspective.

for α cannot be explicit until a specific scheme is invoked. Nevertheless, the problem of achieving high gain at short λ breaks down into needs for transitions of large oscillator strength, f, small line width, and large upper state density, along with a significant degree of inversion [bracketed in Equation 1] as evidenced from fluorescent measurements by a ratio $N_u g_l / N_l g_u$ significantly exceeding unity.

Some magnitude relations useful for scaling such important parameters as the pump power line density P in W/cm^2, ($\div$ PL, where p is the volume power density), the mean line density H in cm^{-2} ($\div$ NL, where N is the volume density), and the pump rate R in sec^{-1} vs. wavelength λ in nm can be obtained for a gain product of $\alpha L = 5$ from a series of approximations. For pumping by free-electron collisions (recombination, excitation and dielectronic capture) as well as by photoexcitation and possibly ion-ion collisions, two requirements become approximately:

$$\eta \ \lambda^2 > 10^{20} \ cm^{-1} \ ; \ R\lambda > 10^{10} \ sec^{-1} \ cm \qquad (2)$$

where the second follows from the density set by the first. For atom-ion resonance charge transfer collisions of higher cross section, the same ''threshold'' conditions become:

$$\eta \ \lambda > 10^{19} \ cm^{-2} \ ; \ R\lambda^2 > 10^{11} \ sec^{-1} \ cm \qquad (3)$$

i.e., the density could be lower with an associated higher pump rate. For all cases the minimum pump power line density in the plasma itself scales as

$$\mathcal{P}\lambda^4 > 10^{15} \ W - cm \qquad (4)$$

These relations imply that pulsed pumping may already be technically feasible to wavelengths as short as 1 nm, if the steady-state inversions are present.

Schemes in which the inversion self-terminates in times $\sim A^{-1}_{u\ell}$, i.e., when the final state density is not depleted, are technically limited for the shortest wavelengths, as indicated in Figure 4.2.1. Indeed, for this reason an air of pessimism prevailed for some years in anticipation of the need for femtosecond (10^{-15} sec) pumping pulse risetimes for such nonsteady-state inversions in the X-ray region, with coherence lengths as short as 10^{-5} cm for X-rays. This would require at least extremely delicate traveling-wave pumping and synchronization for a finite-size cavity, even if one could be constructed to withstand the X-radiation with high efficiency.

PROGRESS

A number of promising approaches that have evolved are indicated in Figure 4.2.1 for the spectral regions of interest. Included are milestone wavelengths and references, separately designated where modeling, population inversions, or output has been reported. A quick glance at the references will indicate that while some ideas emerged in the latter 1960s including ASE amplification on molecules such as H_2, most modeling for shorter wavelengths was published in the mid-1970s, with population inversions and harmonic or gain output reported in the late 1970s and early 1980s. Some important general categories are described further in the following paragraphs, following a somewhat historical vein for continuity.

Innershell Processes

Perhaps it is to the credit of a report[29] of X-ray lasing on the Cu Kα transition in $CuSO_4$ in the early 1970s that stimulation of new thinking began along the lines of cm-long gain lengths with corresponding quasi-cw inversion lifetimes of at least 30 ps

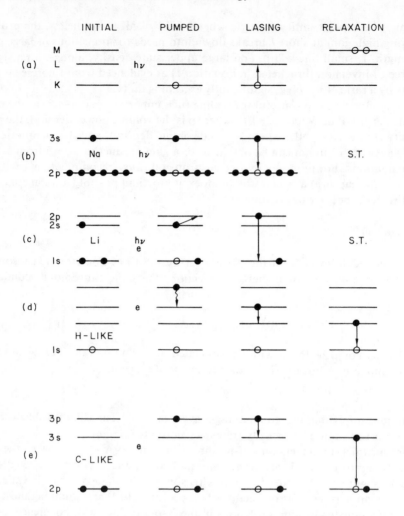

FIGURE 4.2.2.   Simple diagrams illustrating various lasing schemes discussed in the text. Open circles refer to normally occupied holes; closed circles refer to bound electrons. Pumping is either by photons (hν) or by electrons (e). The relaxation process indicated in the last column assures steady-state inversion if it proceeds at a sufficient rate to prevent self-termination (S.T.) in an absorbing state.

duration for single-pass gain. While these reports were not substantiated as being associated with X-rays and a physical process could not be identified,[29] some excellent ideas emerged during the ensuing period concerning, for example, possible quasi-cw schemes involving "traditional" innershell X-ray transitions, for which a pumped K-vacancy state would lase to an L-vacancy state which would more rapidly deplete by, e.g., an Auger transition to perhaps a shifted NMM-vacancy state, so that reabsorption would be minimized.[22] This is illustrated in Figure 4.2.2a. This scheme is not unlike the ultraviolet excimer lasers where the final laser state selfdestructs at a rapid rate, generating the required final stable states. This method becomes rapidly complex as do most innershell schemes when all possible decay modes are considered. A thorough numerical analysis based on improved atomic parameters would be useful but complicated. In addition, it requires a powerful well-tuned pump source to selectively and efficiently pump the K-shell without perturbing the outer shell populations. As such, serious experiments have not been pursued, even though it continues to indicate promising gains for the 4-20 keV X-ray region with narrow lines in a cold medium.

The Auger decay complications and branching competition is avoided in a selfterminating (0.4 nsec) innershell scheme[21] involving the selective removal of L-shell 2p electrons from sodium atoms (or sodium-like ions) with possible lasing on a 3s-2p transition (37 nm for neutral sodium). This interesting scheme, illustrated in Figure 4.2.2b, also awaits a powerful and wavelength-selective short-pulse pumping source.

Plasma Ions

Auger transitions can also be avoided for short wavelength K-radiation by the removal of outer electrons by using, for example, hydrogenic and helium-like one- and two-electron ions such as created at high density in plasmas. Such systems are also quite amenable to analysis and compatible with other programs. For example, the requirements of high peak power density, small dimensions, short lifetimes, high densities, and high temperatures for ionization lead naturally to an association with fusion pellet plasmas.

The use of long-lived metastable states, often associated with helium-like and beryllium-like ions, comes to mind for lasing as a means of extending the pumped-state lifetimes. Unfortunately, for the same conditions, the gain achievable is also decreased by the lowered decay probability, which is an offsetting effect. Since both pump power and gain scale as $N_u A_u$, no less power is required and the density must rise, resulting in additional collisional effects. One variation[17] (Figure 4.2.2c) which has received some experimental verification in a plasma flow experiment is the transfer of metastable stored-electrons to nearby "allowed" decaying states at a rate more rapid than that of the subsequent decay, using absorption of intense long-wavelength emission. The physical analyses for $(Li^+)^*$ ions created by selective VUV photoionization or by electron capture (see below) with subsequent lasing on a K-transition at $\lambda \approx 20$ nm appears promising.

More fundamental than the pumping requirement is the achievement of significant inverted population densities between laser levels. This can be studied quantitatively in fluorescence. Indeed, much progress has been made in demonstrating inversions in highly stripped ions by what is thought to be electron capture (Figure 4.2.2d), either of free electrons when very rapid cooling is effected or by charge transfer of bound electrons, and associated mixing or cascading among excited states. Inversions have now been reported[9-11,15,30,31] for levels extending from $n = 2$ through 6 under various conditions and for various elements ranging from helium to aluminum, inferring the possibility of lasing at wavelengths as short as 13 nm if the proper conditions can be continued into the dense gain region. There appears to be still much to be learned about the optimum plasma conditions required, but the significance of demonstrating a quasi-cw inversion for such short wavelengths should not be minimized. How this will ultimately relate to K-shell inversions is not presently known, but it is not beyond reason to speculate that a quasi-cw laser based on upper level transitions could provide the selective pump source required for the sophisticated innershell or the metastable-transfer schemes mentioned above.

Related to this possibility is again the idea of extrapolating proven ion-laser transitions isoelectronically toward shorter wavelengths, where the 3p-3s outershell transitions in the beryllium-like to neon-like sequences are prime candidates.[8] Here the upper p-level does not decay readily to the ground state as does the final s-level (see Figure 4.2.2e). Indeed, np-ns ion laser transitions now cover the visible-UV spectral region. Numerical modeling based on electron-collisional pumping has been carried out which indicates the promise of ASE for the vacuum-UV region. Even more significant are recent experimental results where significant gain in a cavity is reported near 60 nm for calcium ions generated by a high-power laser;[4] the wavelength region suggests a

3p-3s transition. The pumping mechanism remains to be ascertained, since the quasi-cw inversion could be independent of whether the electrons arrive from lower levels by excitation or by capture and downward cascade followed by reionization.

OTHER CONSIDERATIONS

Gain Verification

An important problem faced by experiments in this field is that of proving convincingly the presence of gain, particularly in the ASE mode near threshold.[19] Optical depth effects on intense lines can be misleading, for example. Also variations of lasant length methods to demonstrate exponential scaling can give misleading results from plasma alteration. Here cavities, even if inefficient, prove helpful as demonstrated in the Lebedev experiments[4] where multiple passes results in increased utilization of pumped quasi-cw inversions and in enhanced collimation. Thus cavities will probably prove very useful in demonstrating, developing, and researching higher-gain ASE devices.

Coherence

Coherence is minimal in ASE devices. A system approach conceived[32] for eventual coherent laser operation in the short wavelength region involves the use of ASE devices as amplifiers of weak but coherent beams, generated as harmonics[28] of high power long wavelength lasers. For example, the 20th and 28th harmonics of a Nd-glass laser at 53 and 38 nm wavelength have been demonstrated[26] for this purpose. The coupling problems will of course await the exact configuration of both the generator and the amplifier and will not be trivial.

SUMMARY

In summary, early pessimism toward X-ray lasers has given way to cautious optimism with the demonstration of quasi-cw inversions and some gain in the extreme ultraviolet spectral region. A synergism between the analyses and modeling and definitive experiments appears to be emerging. With present understanding and technology it is possible to predict lasing to about 1 nm wavelength, well within the soft X-ray region. However, activity at the present must be considered at a rather low ebb, reflecting the somewhat academic attitude that surrounds this area of research without an apparent urgent need. Perhaps progress will appropriately follow the advances in inertial fusion toward high temperatures and densities for pumping, or perhaps entirely new concepts involving for example crystal lattices[33] will emerge. Lasing at even shorter wavelengths in the γ-ray region is envisioned also as reviewed currently by Baldwin,[24] and appears at present to be in the thought stage of X-ray lasers of the early 1970s. Relevant gain experiments there will be difficult and expensive, but some preliminary exploratory experiments already seem reasonable.

REFERENCES

1. Waynant, R. W. and Elton, R. C., Review of short wavelength laser research, *Proc. IEEE*, 64, 1059, 1976; Elton, R. C., Recent advances in X-ray laser research, in *Advances in X-ray Analysis*, Vol. 21, Plenum Press, New York, 1978.

2. **Carrigan, B.,** X-ray Lasers, a Bibliography with Abstracts 1969—78, Rep. NTIS/PS-79/0011, National Technical Information Service, Department of Commerce, Springfield, Va., 1979; X-ray Lasers, Rep. IJ22-013, July 1979, Smithsonian Science Information Exchange, Inc., Washington, D.C., 1979; **Mauk, S. C.,** X-ray Lasers (citations from the International Aerospace Abstracts Data Base), Sept. 1979, Report NTIS/PS-79/0863/5WH, National Technical Information Service, Department of Commerce, Springfield, Va.; **Baldwin, G. C.,** Bibliography of Graser Research, Report No. LA-7783-MS, Los Alamos Scientific Laboratory, Los Alamos, N.M., April 1979; **Kelly, R. L. and Palumbo, L. J.,** *Atomic and Ionic Emission Lines below 2000A,* Naval Research Laboratory Rep. No. 7599, U.S. Govt. Printing Office, Washington, D.C., 1973, revision in press.

3. **Jorna, S., Bailey, N. A., Galligan, J., Gardner, C. G., Hirth, J., Mueller, R., Shirley, D. A., Smith, W., Sullivan, P. A., Thomas, J. K.,** and **Trammell, G. T.,** X-ray Laser Applications Study, Report No. PD-LJ-76-132, Physical Dynamics, Inc., La Jolla, Calif., 1976.

4. **Ilyukhin, A. A., Peregudov, G. V., Ragozin, E. N., Sobel'man, I. I.,** and **Chrikob, V. A.,** Concerning the problem of lasers for the far ultraviolet $\lambda \sim 500$—700 A, *Sov. Phys. JETP Lett.,* 25, 535, 1977; **Vinogradov, A. V., Sobel'man, I. I.,** and **Yukov, E. A.,** On the problem of short wavelength lasers, *J. Physique, Colloque C4 Suppl.,* 39, C4, 1978.

5. **Samson, J. A.,** *Techniques of Vacuum Ultraviolet Spectroscopy,* John Wiley & Sons, New York, 1967, 2.

6. **Hass, G.** and **Hunter, W. R.,** The use of evaporated films for space applications, in *Physics of Thin Films,* Vol. 10, Academic Press, New York, 1978.

7. **Elton, R. C.** and **Dixon, R. H.,** X-ray laser research, guidelines and progress, *Ann. N.Y. Acad. Sci.,* 267, 3, 1976.

8. **Elton, R. C.,** Extension of 3p-3s ion lasers into the vacuum ultraviolet region, *Appl. Opt.,* 14, 97, 1975; **Palumbo, L. J.** and **Elton, R. C.,** Short wavelength laser calculations for electron pumping in carbon-like and helium-like ions, *J. Opt. Soc. Am.,* 67, 480, 1977.

9. **Bhagavatula, V. A.** and **Yaakobi, B.,** Direct observation of population inversion between Al^{+11} levels in a laser-produced plasma, *Opt. Commun.,* 24, 331, 1978.

10. **Bhagavatula, V. A.,** Experimental evidence for soft X-ray population inversion by resonant photoexcitation in multicompent laser plasmas, *Appl. Phys. Lett.,* 33, 726, 1978.

11. **Irons, F. E.** and **Peacock, N. J.,** Experimental evidence for population inversion in C^{5+} in an expanding laser-produced plasma, *J. Phys. B.,* 7, 1109, 1974; **Dewhurst, R. J., Jacoby, D., Pert, G. J.,** and **Ramsden, S. A.,** Observation of a population inversion in a possible extreme ultraviolet lasing system, *Phys. Rev. Lett.,* 97, 1265, 1976; **Jacoby, D., Pert, G. J., Ramsden, S. A., Shorrock, L.,** and **Tallents, G. J.,** Observation of gain in a possible extreme ultraviolet lasing system, *Opt. Commun.,* 37, 193, 1981.

12. **Gudzenko, L. I.** and **Shelepin, L. A.,** Radiation enhancement in a recombining plasma, *Sov. Phys.-Doklady,* 10, 147, 1965; **Gordiets, B. F., Gudzenko, L. I.,** and **Shelepin, L. A.,** Relaxation processes and amplification of radiation in a dense plasma, *Sov. Phys. JETP,* 28, 489, 1969.

13. **Louisell, W. H., Scully, M. O.,** and **McKnight, W. B.,** A soft X-ray laser utilizing charge exchange, *Opt. Commun.,* 9, 246, 1973; Analysis of a soft-X-ray laser with charge-exchange excitation, *Phys. Rev. A,* 11, 989, 1975.

14. **Copeland, D. A., Mahr, H.,** and **Tang, C. L.,** Threshold and rate equation considerations for a H^+-Cs charge-exchange laser, *IEEE J. Quantum Electron.,* QE-12, 665, 1976.

15. **Dixon, R. H.** and **Elton, R. C.,** Resonance charge transfer and population inversion following C^{5+} and C^{6+} interactions with carbon atoms in a laser-generated plasma, *Phys. Rev. Lett.,* 38, 1072, 1977; **Dixon, R. H., Seely, J. F.,** and **Elton, R. C.,** Intensity inversion in the Balmer spectrum of C^{5+}, *Phys. Rev. Lett.,* 40, 122, 1978.

16. **Vinogradov, A. V.** and **Sobel'man, I. I.,** The problem of laser radiation sources in the far ultraviolet and X-ray regions, *Sov. Phys. JETP,* 36, 1115, 1973.

17. **Mahr, H.** and **Roeder, U.,** Use of metastable ions for a soft X-ray laser, *Opt. Commun.,* 10, 227, 1974; **Vekhov, A. A., Makhov, V. N., Nikolaev, F. A.,** and **Rozanov, V. B.,** Possibility of using metastable helium-like ions in generation of ultrashort X-ray stimulated radiation, *Sov. J. Quantum Electron.,* 5, 718, 1975.

18. **Zerikhin, A. N., Koshelev, K. N., Kryukov, P. G., Letokhov, V. S.,** and **Chekalin, S. V.,** Observation of intensity anomalies at 58-78 Å in Cl VII transitions in two-stage plasma heating by ultrashort laser pulses, *Sov. Phys. JETP Lett.,* 25, 300, 1977; also *Sov. J. Quantum Electron.,* 11, 48, 1981.

19. **Elton, R. C.** and **Dixon, R. H.,** Gain-verification problem in extreme ultraviolet lasing, *Opt. Lett.,* 2, 100, 1978.

20. **Jaegle, P., Jamelot, G., Carillon, A., Sureau, A.,** and **Dhez, P.,** Superradiant line in the soft X-ray range, *Phys. Rev. Lett.,* 33, 1070, 1974.

21. **Duguay, M. A.** and **Rentzepis, P. M.,** Some approaches to vacuum-uv and X-ray lasers, *Appl. Phys. Lett.,* 10, 350, 1967.

22. **Elton, R. C.**, Quasi-stationary population inversion on Kα transitions, *Appl. Opt.*, 14, 2243, 1975.
23. **Vinogradov, A. V., Sobel'man, I. I., and Yukov, E. A.**, Possibility of constructing a far-ultraviolet laser utilizing transitions in multiply charged ions in an inhomogeneous plasma, *Sov. J. Quantum Electron.*, 5, 59, 1975.
24. **Baldwin, G. C.**, On the feasibility of grasers, in *Laser Interaction and Related Plasma Phenomena*, Vol. 4A, Schwarz, H. J. and Hora, H., Eds., Plenum Press, N.Y. 1977; **Baldwin, G. C., Solem, J. C., and Gol'danskii, V. I.**, A review of grasers, *Rev. Mod. Phys.*, in press.
25. **Waynant, R. W., Shipman, J. D., Elton, R. C., and Ali, A. W.**, Vacuum ultraviolet laser emission from molecular hydrogen, *Appl. Phys. Lett.*, 17, 383, 1970; Laser emission in the vacuum ultraviolet from molecular hydrogen, *Proc. IEEE*, 59, 679, 1971; **Waynant, R. W.**, Observations of gain by stimulated emission in the Werner band of molecular hydrogen, *Phys. Rev. Lett.*, 28, 533, 1972.
26. **Reintjes, J., Eckhardt, R. C., She, C. Y., Karangelen, N. E., Elton, R. C., and Andrews, R. A.**, Generation of coherent radiation at 53.2 nm by fifth-harmonic conversion, *Phys. Rev. Lett.*, 37, 1540, 1976; Seventh harmonic conversion of mode-locked laser pulses to 38.0 nm, *Appl. Phys. Lett.*, 30, 480, 1977.
27. **Kung, A. H., Young, J. F., Bjorklund, G. C., and Harris, S. E.**, Generation of vacuum ultraviolet radiation in phase-matched Cd vapor, *Phys. Rev. Lett.*, 29, 985, 1972; Generation of 1182 Å radiation in phase-matched mixtures of inert gases, *Appl. Phys. Lett.*, 22, 301, 1973; errata 28, 294, 1976; **Harris, S. E., Young, J. F., Kung, A. H., Bloom, D. M., and Bjorklund, G. C.**, Generation of ultraviolet and vacuum ultraviolet radiation, in *Laser Spectroscopy*, Brewer, R. G. and Mooradian, A., Eds., Plenum Press, New York, 1973, 59.
28. **Reintjes, J. F., She, C. Y., and Eckardt, R. C.**, Generation of coherent radiation in the xuv by fifth- and seventh-order frequency conversion in rare gases, *IEEE J. Quantum Electron.*, QE-14, 581, 1978.
29. **Krepros, J. G., Eyring, E. M., and Cagle, F. W., Jr.**, Experimental evidence of an X-ray laser, *Proc. Natl. Acad. Sci.*, *U.S.*, 69, 1744, 1972; **Bradford, J. N., Elton, R. C., Lee, T. N., Andrews, R. A., Palumbo, L. J., and Eckardt, R. C.**, Further comments on collimated X-ray emission from laser-heated CuSO$_4$-doped gelatin, *Appl. Opt.*, 12, 1095, 1973.
30. **Elton, R. C., Lee, T. N., Dixon, R. H., Hedden, J. D., and Seely, J. F.**, Short wavelength population inversions associated with charge transfer in laser-produced plasmas, in *Laser Interaction and Related Plasma Phenomena*, Vol. 5, Schwarz, H. J. and Hora, H., Eds., Plenum Press, New York, 1980.
31. **Suckewer, S., Hawryluk, R. J., Okabayaski, M., and Schmidt, J. A.**, Observation of inverted population levels in the FM-1 spherator, *Appl. Phys. Lett.*, 29, 537, 1976; **Sato, K., Shiho, M., Hosokawa, M., Sugawara, H., Oda, T., and Sasaki, T.**, Population inversions inferred from intensity measurements in stationary recombining He plasma, *Phys. Rev. Lett.*, 39, 1074, 1977.
32. **Andrews, R. A., Reintjes, J., Eckardt, R. C., Dixon, R. H., Waynant, R. W., Lee, T. N., Palumbo, L. J., Lehmberg, R., DeRosa, J., and Jones, W.**, ARPA/NRL X-ray Laser Program Technical Report, NRL Memorandum Rep. No. 3130, Naval Research Laboratory, Washington, D.C., September 1975.
33. **Das Gupta, K.**, Non-linear increases in Bragg peak and narrowing of X-ray lines, *Phys. Lett.*, 46A, 179, 1973; Nondivergent radiation of discrete frequencies in continuous X-ray spectrum, *Phys. Rev. Lett.*, 33, 1415, 1974.

Section 5
Masers

5.1 MASERS

Adrian E. Popa

INTRODUCTION

The foundation for the concept of a quantum mechanical amplifier of electromagnetic radiation was laid by Einstein in 1917 when he concluded that a quantized atomic system in thermodynamic equilibrium exhibits not two but three radiative processes: absorption, spontaneous emission, and induced or stimulated emission.[1] However, under normal physical conditions, the first two processes dominate, and the degree of stimulated emission is negligible. Seventeen years later, the field of microwave spectroscopy was born with the pioneering experiments of Cleeton and Williams.[2] They succeeded in reducing the dimensions of a magnetostatic spark gap until it oscillated at 24 GHz and then used the dispersive scatter from a 3-ft-wide diffraction grating to scan through the inversion spectrum of ammonia gas held at atmospheric pressure in a large rubber cell.

Although some microwave spectroscopy was conducted during World War II, there were no publications until 1946, when the components and techniques developed for radar came into general use.[3] By 1953, the microwave spectrum of many molecules and atoms had been measured, and many new sophisticated experimental techniques had come into general use. The state of the art in 1952, a year before the first operational maser, has been documented by Gordy et al.,[4] who cite 591 references to work in the field between 1946 and 1952. Some of the key experiments of this period include Bloch's study of nuclear magnetic resonance in water,[5] Lamb and Rutherford's use of an atomic beam and a hot wire detector to measure the hyperfine structure of the hydrogen atom,[6] and Purcell and Pound's evidence of stimulated emission when a sudden reversal of magnetic field created a population inversion in lithium fluoride.[7] The first detection of a molecular beam by microwave methods was accomplished by Johnson and Strandberg when they propagated a 24-GHz signal through a beam of ammonia molecules to obtain a factor of six reduction in the Doppler broadening of the inversion spectrum.[8]

During the early 1950s, the concept of using stimulated emission for coherent microwave amplification originated independently at Columbia University,[9] the University of Maryland,[10] and the Lebedev Institute in Moscow.[11] In 1954, Gordon, Zeiger, and Townes[9] published details of the first successful molecular microwave oscillator, which, in a subsequent publication, they named a maser, an acronym for microwave amplification by stimulated emission of radiation.[12] To sustain oscillation, the Columbia maser focused a state-selected ammonia beam into a microwave cavity tuned to the 23.8-GHz hyperfine inversion frequency. When the beam flux was high enough to enable the stimulated emission signal to overcome the losses in the cavity, an oscillation of such high spectral purity was produced that it immediately suggested the maser might be used as a molecular clock. Also, some previously unresolved magnetic hyperfine structure was detected in the ammonia inversion spectrum, suggesting that the device could also be used as a high-resolution spectrometer. Unfortunately, the narrow bandwidth and limited tunability of the ammonia-beam maser detracted from its use as an amplifier.

The continued search for a low-noise microwave amplifier led Bloembergen at Harvard to propose a maser scheme based on the three-level excitation of paramagnetic ions in a crystal cooled to near absolute zero.[13] In 1957 during an attempt to establish the feasibility of Bloembergen's idea, Scovil, Feher, and Seidel operated the first successful three-level solid-state maser at the Bell Telephone Laboratories (BTL) using lanthanum ethyl sulfate doped with the Gd^{3+} paramagnetic ion.[14] Continuous amplification and oscillation were obtained with the BTL device at 9 GHz using a 17.5-GHz pump frequency.

Between 1956 and the development of the first operating laser in 1960, many maser materials, frequencies, and circuits were reported. During this period, several two-level solid-state masers were also demonstrated;[15-18] these devices could only amplify pulses, however, and were never put to practical use. The first ruby maser was reported by Makhov et al. when they achieved 20-dB gain at 9.2 GHz using a 24.2-GHz pump.[19]

For the past two decades, the ruby maser has remained the mainstay in low-noise amplifiers for space communications and astronomy. Numerous pumping techniques and circuits have been developed for signal frequencies ranging from 300 MHz to 75 GHz. A detailed survey of this technology can be found in Siegman's books.[20,21]

In 1960, Ramsey's group at Harvard reported operation of an atomic hydrogen maser operating at 1420 MHz.[22-24] This device, which uses a storage bulb to eliminate first-order Doppler shifts, offers unprecedented stability. The hydrogen maser was the first maser to be flown in space and continues to be refined for use as a ground and spaceborne frequency standard.[25-27]

TYPES OF MASERS

In thermodynamic equilibrium, the number of particles in a lower discrete energy level N_n is related to the number of particles in a higher discrete energy level N_m by the Boltzmann distribution condition:

$$\frac{N_m}{N_n} = e^{-(E_m-E_n)/kT} = e^{-h\nu/kT} \tag{1}$$

where E_m and E_n are the respective energies of the two levels; k is Boltzmann's constant; T is the absolute temperature; h is Planck's constant; and ν is the frequency of the characteristic radiation emitted or absorbed during the particle's transition between energy levels.

The characteristic wavelength for radiation from particles in thermodynamic equilibrium at room temperature (T = 300 K) is about 50 μm. For room-temperature particles characterized by transitions at optical wavelengths ($h\nu/kT \gg 1$), the number of particles in a higher energy level will be quite small as compared with the number at ground level. In contrast, for room-temperature particles with transitions at microwave lengths ($h\nu/kT \ll 1$), thermal energy dominates, and the ratio N_m/N_n is close to unity with slightly more atoms in the ground energy level. The fact that these transitions have energies of less than 50 eV demonstrates that masers, in contrast to lasers, use transitions that are less energetic than those required for orbital changes.

Under standard conditions, particles stimulated by microwave radiation to make transitions from a lower to a higher energy level absorb energy from the wave, while particles making the opposite transition (which is equally probable) give energy to the wave. The slight predominance of particles in the lower energy level required by the Boltzmann distribution results in a net absorption of microwave energy. In a maser device, an artificial imbalance must be created to invert to normal Boltzmann condition so that more particles give energy to the wave than absorb energy from it. This condition is generally called a population inversion.

Gas Beam Masers

The first maser operated by Townes and his students was the gas beam device shown

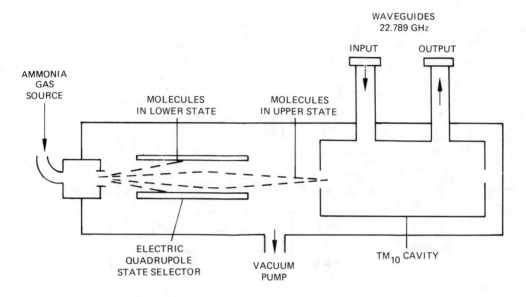

FIGURE 5.1.1.   Schematic of an ammonia beam maser.

schematically in Figure 5.1.1.[9,12] An ammonia beam formed by a pressure drop through a tubular collimator passes through an electric quadrupole field. The electric field focuses molecules in the upper inversion state into a cavity and defocuses molecules in the lower inversion state. Since a population inversion ($N_m > N_n$) is created in the cavity, downward transitions to the lower inversion state predominate and hence energy is given up to the cavity. The cavity is tuned to the 23.870-GHz hyperfine inversion frequency of ammonia ($N^{14}H_3$), and, when the beam flux is sufficient to provide enough stimulated radiation to overcome cavity losses, the maser oscillates. If the device is operated at a beam flux below the threshold of oscillation, signals injected into the cavity at the hyperfine inversion frequency will be amplified. To operate the maser as a high-resolution spectrometer, rf power from a tunable signal source can be injected into the cavity. Then, when the external oscillator passes through the molecular resonance frequency, molecular resonances will be observed as sharp increases in output power level. Eventually, all of the ammonia molecules will be removed by a vacuum pump.

In the atomic hydrogen maser, shown schematically in Figure 5.1.2, an atomic hydrogen beam is produced by a glow discharge in a dissociator. The beam passes through an inhomogeneous magnetic field, which focuses upper energy atoms into a Teflon-coated quartz storage bulb located in a TE_{011}-mode cavity. The cavity is tuned to 1.420 GHz, the hyperfine transition frequency of hydrogen. The hydrogen atom spends most of its time in free space in the storage bulb, and the effects of wall collisions are small because the electric polarizability is low. The effect of the first-order Doppler shift is nearly eliminated because the average velocity of the atom in the storage bulb is close to zero. The long interaction time of the confined hydrogen atom produces line Qs as high as 10^9 and stabilities near one part in 10^{14} for averaging times from 1 to 10^5 sec. Several experimental results for beam masers are presented in Table 5.1.1.

Optically Pumped Gas Masers

A typical optically pumped gas maser is shown schematically in Figure 5.1.3. The device consists of a microwave cavity filled with Rb^{87} vapor and a buffer gas, usually

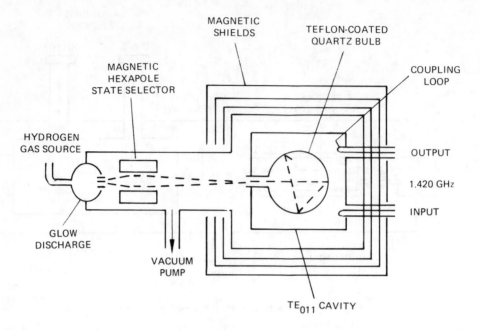

FIGURE 5.1.2.   Schematic of an atomic hydrogen maser.

Table 5.1.1
GAS BEAM MASERS

Molecule/atom	Frequency (GHz)	Output power (W)	Temperature (K)	Ref.
CH_2O	14.4886	3×10^{-11}	300	30
	72.838			
	72.409			
CH DO	16.0381	—	Dry ice	30
$C_{12}H_2O_{16}$	72.8	3×10^{-12}	300	31
D	0.327384	—	300	32
DCN	72.4 (3 lines)	—	300	33
	144.8 (2 lines)	—	300	
H	1.420405	1×10^{-13}	300	22—24
HCN	82.2 (5 lines)	—	300	33
	177.4 (3 lines)	—		34
HDO	10.2782	—	Salt ice	30
HDS	11.2838	—	Dry ice	30
$N^{14}H_3$	23.870	5×10^{-14}	300	9, 12
$N^{15}H_3$	22.789	2×10^{-12}	300	35

nitrogen. When the vapor in the cavity is illuminated with rubidium resonance radiation filtered by a cell containing Rb^{85}, amplification or oscillation can occur at 6.835 GHz, the ground-state hyperfine frequency. Output powers are on the order of 10^{-10} W with a frequency stability of a few parts in 10^{12} for 1-sec averaging times. Because changes in pump lamp intensity or gas pressure can cause a frequency shift, the device is not a primary frequency standard. However, it can be used as a compact, portable secondary atomic frequency standard. Table 5.1.2 presents data for three different optically pumped maser configurations.

Solid State Masers

The solid-state maser uses electron-paramagnetic resonance (EPR) energy levels in

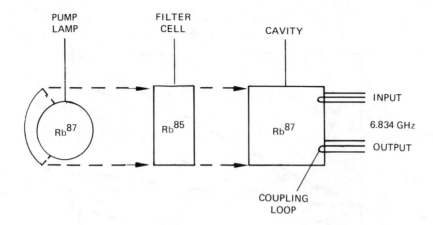

FIGURE 5.1.3.   Schematic of an optically pumped rubidium maser.

Table 5.1.2
OPTICALLY PUMPED GAS MASERS

Molecule/atom	Frequency	Output power (W)	Temperature (K)	Ref.
He³	103 kHz	5×10^{-14}	300	36
Rb⁸⁵	3.0359 GHz	1×10^{-10}	300	37
Rb⁸⁷	6.8347 GHz	1×10^{-10}	300	38

atoms that individually possess permanent magnetic dipolar moments. For example, the lowest orbital energy levels of the Cr^{3+} ion are actually four relatively closely spaced Zeeman levels that depend on the direction and strength of the magnetic field relative to the symmetry axis of the crystal. Figure 5.1.4 is a plot of the energy spacing between Zeeman levels for the Cr^{3+} ions in a ruby crystal as a function of the magnetic field for three different angles (ϕ) between the magnetic field and the C axis of the crystal.

In the popular three-level scheme shown schematically in Figure 5.1.5, the normal Boltzmann distribution for thermal equilibrium is modified by applying energy at the pump frequency. If the applied pump energy is strong enough, it will dominate the relaxation processes, causing the population of Level 3 to increase beyond thermal equilibrium and the population of Level 1 to decrease below thermal equilibrium. When the population of Level 3 is about equal to the population of Level 1, the transition is saturated, and a continuous population inversion is maintained between Level 2 and Level 1. If a small signal is then applied to the crystal at the signal frequency, the signal will be amplified by maser action using energy delivered to the wave.

A simple three-level reflection cavity maser is shown schematically in Figure 5.1.6. The signal frequency and pump frequency are applied to a paramagnetic crystal located in a dual-mode cavity that is resonant at both frequencies. Energy at the signal frequency is amplified in the crystal and reflected to the output port. To prevent oscillation, a ferrite circulator is used to maintain isolation between the signal at the input port and the amplified signal at the output port. In practice, the entire maser cavity is cooled to near liquid helium temperatures to enhance the ratio of the maser signal to the noise of spontaneous emission. Tables 5.1.3 and 5.1.4 present several different solid-state maser materials and pumping techniques.

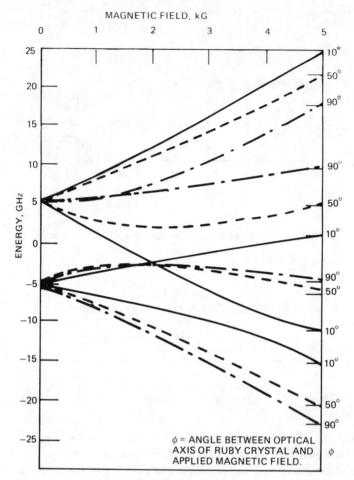

FIGURE 5.1.4. Energy spacing between Zeeman levels for Cr^{3+} ions in a ruby crystal as a function of magnetic field for three different angles (ϕ) between the magnetic field and the C axis of the crystal.

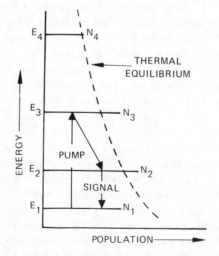

FIGURE 5.1.5. Relative populations of the first four energy levels as a function of energy spacing for thermal equilibrium and for a population inversion.

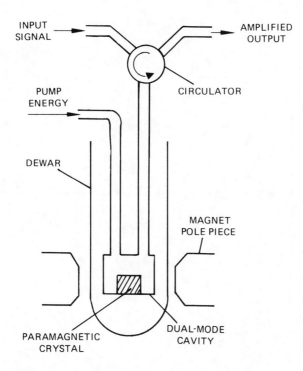

FIGURE 5.1.6.   Schematic of a three-level reflection cavity maser.

PRACTICAL MASER DESIGNS

A Hydrogen Maser Oscillator Design

A design for a spaceborne hydrogen maser atomic clock for possible use in the NAV-STAR/Global Positioning System (GPS) is being developed at Hughes Research Laboratories under a contract from the Naval Research Laboratory.[26] The goals of the program are to achieve one part in 10^{14} stability over a 10-day period in a 15-in.-diameter, 30-in.-long package that consumes less than 100 W of power from the spacecraft bus. The operational temperature range of the maser in the spacecraft will be $20°C \pm 10°C$, and an operating life greater than five years is required.

The operation of the spaceborne hydrogen maser can be followed by referring to the device layout shown in Figure 5.1.7 and the energy level diagram in Figure 5.1.8. A 10-year supply of hydrogen gas held at high pressure in an external 6-in.-diameter spherical tank (not shown) is delivered through a pressure-reduction system to a spherical quartz dissociator mounted in a metal ceramic envelope. Atomic hydrogen created by an rf discharge in the dissociator passes through a beam-forming glass collimator and into the inhomogeneous magnetic field of a hexapole permanent magnet. This state-selecting magnetic field focuses atoms in the (F = 1, m = 0) and (F = 1, m = 1) states into the aperture of a Teflon-coated quartz storage bulb. The bulb is mounted in the center of a high-Q (5×10^4) cylindrical TE_{011}-mode microwave cavity made from silver-plated quartz. The cavity is tuned to 1,420,405,751 Hz, the (F = 1, m = 0) → (F = 0, m = 0) hyperfine transition frequency of hydrogen. There is little interaction between the atoms and the Teflon surface, and the atoms are able to spend up to 1 sec in the storage bulb, producing oscillations with a spectral linewidth of 1 Hz. The effect of first-order Doppler shift is greatly reduced by the fact that the velocity of the atom in the bulb, when suitably averaged, is close to zero. A small coupling loop delivers about 10^{-13} W of maser energy to a low-noise receiving system.

Table 5.1.3
TWO-LEVEL SOLID STATE MASERS

Material (dopant)	Signal frequency (GHz)	Pump frequency (GHz)	Gain (dB)	Bandwidth (MHz)	Gain bandwidth product (MHz)	Magnetic field	Temperature (K)	Comments	Ref.
Al_2O_3 (Cr^{3+}, 0.1%)	9	9	—	—	—	500 G 1.6 msec pulse	1.4	Ruby 10 Hz pulse rate	15
$Bi_{92}Sb_8$	0.8—4	dc	Oscillator 20 μW	250	—	1—8 kG	1.8—20.4	Maserlike emission	16
MgO (Neutron irradiation)	9	9	20 Oscillator 12 mW	—	—	Swept	4.2	10-μsec pulses	17
SiO_2 (Neutron irradiation)	9	9	8—21	—	5×10^6	Swept	4.2	10-μsec pulse	17
Si^{28} (P)	9	9	Oscillator 2.5 μW	—	—	—	1.2	—	18

Table 5.1.4
MULTIPLE-LEVEL SOLID STATE MASERS

Material (Dopant)	Signal frequency (GHz)	Pump frequency (GHz)	Gain (dB)	Bandwidth (MHz)	Gain bandwidth product (MHz)	Magnetic field	Temperature (K)	Comments	Ref.
Al_2O_3 (Cr^{3+})	9.22	22.4	20	—	—	4 kG	4.2	First ruby maser	19
(Cr^{3+})	5.75–6.1	18.9–19.5	23	25	—	4 kG	1.5	First traveling-wave maser, ruby	39
(Cr_2O_3)	9.3	23.5	—	—	14	4 kG	77–195	Highest-temperature solid-state maser	40
(Cr^{3+})	75	12.7	—	—	—	30 kOe	4.2	Pulsed-field ruby	41
(Fe^{3+}, Cr)	9.375	25	Oscillator	—	—	1.2 kOe	1.8	Parametric amplification	42
(Cr_2O_3)	22.4	6.93.4 nm (Optical)	—	—	—	6.5 kOe	4.2	Optically pumped ruby flashlamp	43
(Fe^{3+})	12.03	31.34	3–10	—	—	0	4.2	Powdered material zerofield	44
$BeAl_2Si_{16}O_{18}$ (Cr^{3+})	10	58.5	16	20	126	1.9 kG	4.2	First emerald	45
CaF_2 (Tm^{+2})	9.2–11.75	Mercury lamp	—	—	—	8 kOe	1.4–4	CW optical pumping	46
$K_3C_0(CN)_6$ ($K_3Cr(CN)_6$) 0.5%	1.375	8	0–10	—	—	—	2	For radio astronomy	47
$K_3C_0(CN)_6$ ($K_3Cr(CN)_6$) 0.5%	0.3	5.35	10	0.1	1	600 Oe	1.6	Lowest-frequency ruby	48

Table 5.1.4 (continued)
MULTIPLE-LEVEL SOLID STATE MASERS

Material (Dopant)	Signal frequency (GHz)	Pump frequency (GHz)	Gain (dB)	Bandwidth (MHz)	Gain bandwidth product (MHz)	Magnetic field	Temperature (K)	Comments	Ref.
$K_3C_0(CN)_6$ $(K_3Cr(CN)_6)_{0.5\%}$	2.8	9.4	37	0.025	—	3 kOe	1.25	Detailed hardware description	49
La $(C_2H_5SO_4)_3$ 0.9 H_2O (Gd^{3+}, Cr^{3+})	9.06	17.52	—	—	—	2.5 kOe	1.2	First 3-level solid-state maser	14
(Gd^{3+}, Dr^{3+})	6–6.3	11.7–12.3	12	25	—	1.8 kG	1.6	First traveling wave	50
TiO_2	8.44	35.85	2.4–3.8	20	—	—	1.4	First rutile traveling wave	51
(Cr^{3+})	22–24	70	3	—	—	—	1.4	First rutile traveling wave	51
(Cr^{3+})	4.25	0.15 & 46.5	Oscillator	—	—	—	—	Parametric gain in inverted medium	52
(Fe^{3+})	10.64	51.6	—	84	—	0.5–2.7 kG	—	Powdered material	53
(Fe^{3+})	12.05	31.29	30	6.6	200	0	—	First zero field	44
(Fe^{3+})	96.3	62.5	Oscillator	—	—	7.3 kG	2.1	5 levels with inverted pump millimeter wavelengths	53
(Fe^{3+})	70	118	20	10	—	4.6 kG	4.2	5 levels with inverted pump millimeter wavelengths	53

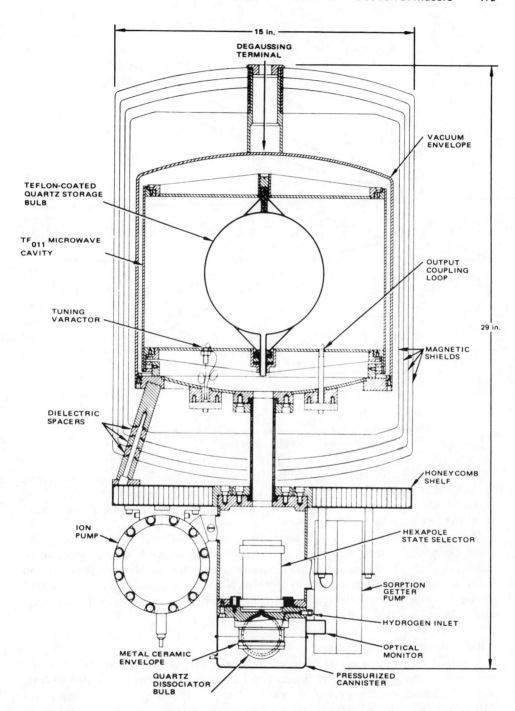

FIGURE 5.1.7.   Layout of a hydrogen maser oscillator designed by Hughes Research Laboratories for use in the Global Positioning System.

To conserve power and extend life, the major portion of the spent hydrogen is selectively collected by a passive sorption getter pump. A small ion pump is used as a pressure gauge and to pump background contaminant gases. The maser vacuum envelope is maintained at pressures below 10^{-6} Torr. A small solenoid maintains a 100-μG quantizing field in the cavity, and four concentric magnetic shields are used to atten-

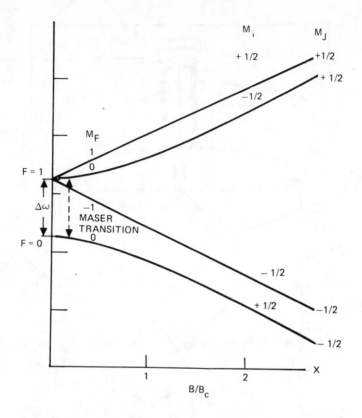

FIGURE 5.1.8.   Energy-level diagram of atomic hydrogen as a function
of magnetic field normalized to the critical field.

uate external magnetic fields. A thermal control system, consisting of servoed heaters and a varactor tuner located in the cavity, holds the cavity frequency to within 1 Hz of the atomic transition frequency to reduce coupled-Q pulling of the atomic line to less than one part in 10^{14}.

The maser receiver and thermal servos are under the control of a microprocessor-based diagnostic system. A frequency synthesizer programmable in microhertz increments is included in the design to correct the clock output signals for relativistic time shifts on the order of a few parts in 10^{10} which are incurred in the 12,000-mile GPS orbits. The goal of the GPS system is to achieve world-wide three-dimensional navigational fixes to within 10 m.

A Ruby Maser Amplifier Design

A reflected-wave ruby maser designed to operate between 18.3 GHz and 26.6 GHz was recently developed at the Jet Propulsion Laboratory (JPL) of the California Institute of Technology under a contract from the National Radio Astronomy Observatory (NRAO).[28] The reflected-wave maser employs a unique design in which four stages are cascaded via ten circulator junctions operating at 4.6 K. The four rubies, each 15-cm long, are biased by a single superconducting magnet, and the ten circulators are biased in pairs by five sets of permanent magnets. Each stage has approximately 10 dB of electronic gain and 9.5 dB of net gain, excluding circulator losses. The resulting amplifier (including circulator losses) has a tuning range of 18.3 to 26.6 GHz, at least 30 dB of net gain, up to 240 MHz of 3-dB bandwidth, and a noise temperature of 13 ± 2 K.

Figure 5.1.9 depicts the various techniques used to achieve the wide bandwidth and

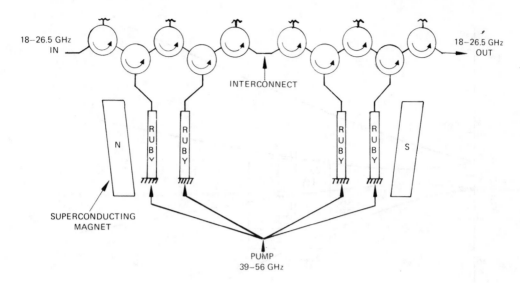

FIGURE 5.1.9.   Schematic of the JPL/NARO K-band ruby maser amplifier.[28] (From Moore, C. R. and Clauss, R. C., *IEEE Trans. Microwave Theory Tech.,* MTT-27, 249, 1979. With permission.)

tuning range. The maser microwave structure consists of a ruby-filled waveguide. Amplification occurs as the signal travels down and back through the ruby. A microwave circulator is used to separate the incoming and outgoing signals. The tuning range is limited by only the bandwidth of the circulator. The pump power is injected at the shorted end by the ruby-filled guide, where the pump guide appears as a waveguide beyond cutoff to signal frequencies. The magnetic field biasing the ruby is tapered linearly along the length of the ruby, giving rise to maser material linewidth spreading. Gain over this broadened linewidth is accomplished by using frequency modulation to spectrally distribute the pump energy. The frequency is modulated at a rate with a period much shorter than the pump transition spin relaxation time.

A pink ruby (0.05% Cr^{3+} in Al_2O_3) was chosen as the active material. The applicable energy level diagram, and the various requirements for wide-band operation, are shown in Figure 5.1.10. As implied by the double-pump orientation, the four ground-state energy levels are symmetrical. The 1 to 3 and 2 to 4 transitions can be pumped simultaneously because they occur at the same frequency. This results in an inversion of the 2 to 3 transition, which corresponds to the signal frequency.

Preliminary results with the reflected-wave ruby maser mounted on the NARO 43-m antenna in a Cassegrain configuration show the following:

- System noise temperature (including atmospheric contributors) of 73 K at 22.23 GHz, 50 K below 20 GHz, and 60 K above 24 GHz during clear weather. The accumulated contributions to system noise temperature from the feed horn, waveguide components, maser, and follow-up receiver are no greater than 30 K.
- Total power variation equivalent to 0.15 K peak to peak for periods of tens of minutes during clear weather.
- Gain stability (as measured with a calibration noise signal) of better than 4% ($\pm$ 0.1 dB) for 5 hr with the telescope tracking sources. Figure 5.1.11 presents the gain and absorption of the four stages near 22 GHz.

Table 5.1.5 presents a performance summary of other masers developed for the National Aeronautics and Space Administration's Deep-Space Network (DSN).[29] These systems are used for communication and control of unmanned spacecraft traveling at

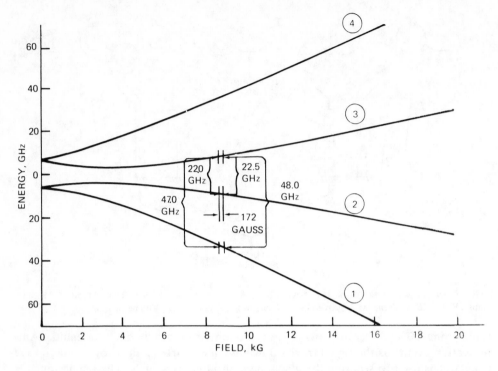

FIGURE 5.1.10.   Energy-level diagram for the JPL/NARO K-band ruby maser amplifier.[28] (From Moore, C. R. and Clauss, R. C., *IEEE Trans. Microwave Theory Tech.*, MTT-27, 249, 1979. With permission.)

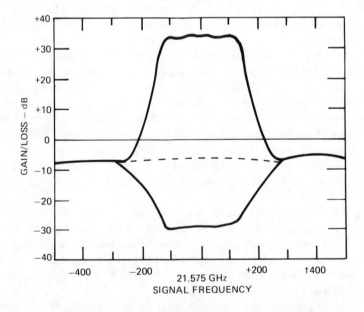

FIGURE 5.1.11.   Gain and absorption as a function of frequency for the JPL/NARO K-band ruby maser amplifier.[28] (From Moore, C. R. and Clauss, R. C., *IEEE Trans. Microwave Theory Tech.*, MTT-27, 249, 1979. With permission.)

interplanetary ranges, planetary radar astronomy, very long base-line interferometry (VLBI), and pulsar and other astronomical observations.

Table 5.1.5
ANTENNA MOUNTED PERFORMANCE SUMMARY OF DSN MASER AMPLIFIERS

Frequency (MHz)	Maser Type, Refrigerator, Magnet Type[a]	Net gain (dB)	Half-power bandwidth (MHz)	Noise temperature (K)	Installation Month, Year (First System)	Removal Month, Year (Last System)
960	Cavity, OCR, PM	20	0.75	22, 30[e]	September 1960	June 1965
960	Cavity, CCR, PM	20	0.75	17	March 1962	May 1962
2295	TWM, CCR, PM	35	17	9	March 1962	May 1962
2270–2300	TWM, CCR, PM[b,c]	35	17	9	March 1964	September 1971
2260–2310	TWM, CCR, PM	45	45	5(8)	October 1970	Currently in use
2388	Cavity, OCR, PM	20	2.5	25	February 1961	May 1961
2388	Dual cavity, OCR, PM	34	2.5	18	August 1962	June 1963
2388	TWM, CCR, PM	40	12	8	September 1963	February 1966
2270–2400	TWM, CCR, PM	50(35)	14	5.5(7)	March 1966	Currently in use
2240–2420	TWM, CCR, PM	45(30)	16	4.2(6)	September 1967	Currently in use
2240–2420	TWM, CCR, SCM	45(27)	16	4.2(6)	January 1972	Currently in use
8450	Multiple cavity OCR, PM[d]	33	15	18	October 1964	July 1966
8370–8520	TWM, CCR, PM	45(30)	17	18(23)	November 1966	November 1968
7750–8750	TWM, CCR, SCM	45	20(17)	6.5(10.5)	January 1973	Currently in use
7600–8900	TWM, CCR, PM	42(30)	17	7(13)	February 1970	November 1972
14,300–16,300	TWM, CCR, SCM	48(30)	17	8.5(13)	September 1971	Currently in use

[a] OCR — open-cycle liquid–helium dewar refrigeration.   PM — permanent magnet.   SCM — superconducting magnet.
    TWM — traveling wave maser.   CCR — closed-cycle refrigerator

[b] TWMs purchased from Airborne Instruments.

[c] CCRs purchased from Arthur D. Little, Inc.

[d] System purchased from Hughes Research Laboratories.

[e] First system was 30°K; later systems were 22°K.

From Reid, M. S., Clauss, R. C., Bathker, D. A., and Stelzreid, C. T., *Proc. IEEE*, 61, 1330, 1973. With permission.

REFERENCES

1. **Einstein, A.,** Zur Quantentheorie der Strahlung., *Phys. Z.,* 18, 121, 1917.
2. **Cleeton, C. W. and Williams, N. H.,** Electromagnetic waves of 1.1 cm wavelength in the absorption spectrum of ammonia, *Phys. Rev.,* 45, 234, 1934.
3. **Van Vleck, J. H. and Weisskopf, V. F.,** On the shape of collision-broadened lines, *Revs. Mod. Phys.,* 17, 227, 1945.
4. **Gordy, W., Smith, W. L., and Trambarrulo, R. F.,** *Microwave Spectroscopy,* John Wiley & Sons, New York, 1953.
5. **Bloch, F., Hansen, W. W., and Packard, M.,** The nuclear induction experiment, *Phys. Rev.,* 70, 474, 1946.
6. **Lamb, W. E. and Rutherford, R. L.,** Fine structure of the hydrogen atom, Part 1, *Phys. Rev.,* 79, 549, 1950.
7. **Purcell, E. M. and Pound, R. V.,** A nuclear spin system at negative temperature, *Phys. Rev.,* 81, 279, 1951.
8. **Johnson, H. R. and Strandberg, M. W. P.,** Beam system for reduction of Doppler broadening of a microwave absorption line, *Phys. Rev.,* 85, 503, 1952.
9. **Gordon, J. P., Zeiger, H. J., and Townes, C. H.,** Molecular microwave oscillator and new hyperfine structure in the microwave spectrum of NH_3, *Phys. Rev.,* 95, 282, 1954.
10. **Weber, J.,** Amplification of microwave radiation by substances not in thermal equilibrium, Trans. IRE., *PGED,* 3, 1, 1953.
11. **Basov, N. G. and Prokhovov, A. M.,** 3-level gas oscillator, *JETP,* 27, 431, 1954.
12. **Gordon, J. P., Zeiger, H. G., and Townes, C. H.,** The maser; a new type of microwave amplifier, frequency standard and spectrometer, *Phys. Rev.,* 99, 1264, 1955.
13. **Bloembergen, N.,** Proposal for a new type solid state maser, *Phys. Rev.,* 104, 2, 324.
14. **Scovil, H. E. D., Feher, G., and Seidel, H.,** Operation of a solid state maser, *Phys. Rev.,* 105, 762, 1957.
15. **Hoskins, B. H.,** Two level maser materials, *J. Appl. Phys.,* 30, 397, 1959.
16. **Nanney, C. A. and George, E. V.,** Coherent microwave radiation from BiSb alloys, *Phys. Rev. Lett.,* 22, 1062, 1969.
17. **Chester, P. F., Wagner, P. E., and Castle, J. G.,** Two level solid-state maser, *Phys. Rev.,* 110, 281, 1958.
18. **Feher, G., Gordon, J. P., Buehler, E., Gere, E. A., and Thurmond, C. D.,** Spontaneous emission of radiation from an electron spin system, *Phys. Rev.,* 109, 221, 1958.
19. **Makhov, G., Kikuchi, C., Lambe, J., and Terhune, R. W.,** Maser action in ruby, *Phys. Rev.,* 109, 1399, 1958.
20. **Siegman, A. E.,** *Microwave Solid State Masers,* McGraw-Hill, New York, 1964.
21. **Siegman, A. E.,** *An Introduction to Lasers and Masers,* McGraw-Hill, New York, 1971.
22. **Kleppner, D., Goldenberg, H. M., and Ramsey, N. F.,** Atomic hydrogen maser, *Phys. Rev. Lett.,* 5, 361, 1960.
23. **Kleppner, D., Goldenberg, H. M., and Ramsey, N. F.,** Theory of the hydrogen maser, *Phys. Rev.,* 126, 603, 1962.
24. **Kleppner, D., Berg, H. C., Crampton, S. B., and Ramsey, N. F.,** Hydrogen maser principles and techniques, *Phys. Rev.,* 138, 4A, 1965.
25. **Vessot, R. F. C., Levine, M. W., Mattison, E. M., and Hoffman, T. E.,** Spaceborne hydrogen maser design, Proc. 8th Annu. precise time and time interval (PTTI) applications and planning meeting, *NASA Tech. Memo.,* NASA-X-814-77-149, 277, 1976.
26. **Popa, A. E., Wang, H. T. M., Bridges, W. B., Etter, J. E., Schnelker, D., Goodwin, F. E., and Dials, M.,** A space-borne hydrogen maser design, Proceedings of the ninth annual precise time and time interval (PTTI) applications and planning meeting, *NASA Tech. Memo.,* NASA TM 78104, 403, 1978.
27. **Hellwig, H. W.,** Atomic frequency standards: a survey, *Proc. IEEE,* 63, 212, 1975.
28. **Moore, C. R. and Clauss, R. C.,** A reflected-wave ruby maser with K-band tuning range and large instantaneous bandwidth, *IEEE Trans. Microwave Theory Tech.,* MTT-27, 249, 1979.
29. **Reid, M. S., Clauss, R. C., Bathker, D. A., and Stelzreid, C. T.,** Low-noise microwave receiving systems in a worldwide network of large antennas, *Proc. IEEE,* 61, 1330, 1973.
30. **Thaddeus, P., Krishner, L. C., and Loubser, J. H. N.,** Hyperfine structure in the microwave spectrum of HDO, HDS, CH_2O, and CHDO: Beam-maser spectroscopy on asymmetric-top molecules, *J. Chem. Phys.,* 40, 257, 1964.
31. **Krupnov, A. F. and Kvortsov, V. A.,** Four millimeter beam maser depending on the 1_{01}—0_{00} transition in the CH_2O molecule, *JETP,* 18, 74, 1964.

32. **Wineland, D. J. and Ramsey, N. F.,** Atomic deuterium maser, *Phys. Rev. A.* 5, 821, 1972.

33. **DeLucia, F. and Gordy, W.,** Molecular-beam maser for the shorter millimeter wave region: spectral constants of HCN and DCN, *Phys. Rev.,* 187, 58, 1969.

34. **Marcuse, D.,** Maser oscillation observed from HCN maser at 88.6 kMc., *Proc. IRE.,* 49, 1706, 1961.

35. **De Prins, J.,** $N^{15}H_3$ double beam maser at a frequency standard, *IRE Trans.,* 111, 200, 1962.

36. **Robinson, H. G. and Myint, T.,** ^{3}He nuclear Zeeman maser, *Appl. Phys. Lett.,* 5, 116, 1964.

37. **Davidovits, P. and Novick, R.,** The optically pumped rubidium maser, *Proc. IEEE,* 54, 155, 1966.

38. **Davidovits, P. and Stern, W. A.,** A field-independent optically pumped Rb^{87} maser oscillator, *Appl. Phys. Lett.,* 6, 20, 1965.

39. **Schulz-Du Bois, E. O., Scovil, H. E. D., and De Grasse, R. W.,** Use of active material in three-level solid-state masers, *Bell Syst. Tech. J.,* 38, 355, 1959.

40. **Maiman, T. H.,** Maser behavior temperature and concentration effects, *J. Appl. Phys.,* 31, 222, 1960.

41. **Momo, L. R., Myers, R. A., and Foner, S.,** Pulsed field millimeter wave maser, *J. Appl. Phys.,* 31, 443, 1960.

42. **Mornienko, L. S. and Prokhorov, A. M.,** A paramagnetic amplifier and generator using Fe^{3+} ions in corundum, *JETP,* 36, 919, 1959.

43. **Devor, D. P., D'Haenens, I. J., and Asawa, C. K.,** Microwave generation in ruby due to population inversion produced by optical absorption, *Phys. Rev. Lett.,* 8, 432, 1962.

44. **Nagy, A. W. and Friedman, G. F.,** A no field powder maser, *Proc. IRE,* 51, 7, 1963.

45. **Goodwin, F. E.,** Maser action in emerald, *J. Appl. Phys.,* 32, 1624, 1961.

46. **Sabisky, E. S. and Anderson, C. H.,** A solid-state cw optically pumped microwave maser, *Appl. Phys. Lett.,* 8, 798, 1968.

47. **Artman, J. O., Bloembergen, N., and Shipiro, S.,** Operation of a three-level solid-state maser at 21 cm, *Phys. Rev.,* 109, 1392, 1958.

48. **Kingston, R. H.,** A UHF solid-state maser, *Proc. IRE,* 46, 916, 1958.

49. **McWhorter, A. L. and Meyer, J. W.,** Solid state maser amplifier, *Phys. Rev.,* 109, 312, 1958.

50. **De Grasse, R. W., Schulz-Du Bois, E. O., and Scovil, H. E. D.,** The three-level solid-state traveling-wave maser, *Bell Syst. Tech. J.,* 38, 305, 1959.

51. **Sabisky, E. S. and Gerritsen, H. J.,** A traveling wave maser using chromium-doped rutile, Proc. IRE, 49, 1329, 1961, *J. Appl. Phys.,* 33, 1450, 1962.

52. **Dathe, G., Steiner, K. H., Roth, D., and Schollmeier, G.,** Parametric amplification and oscillation using an inverted maser material, *IEEE J. Quantum Electron.,* QE-5, 12, 623.

53. **Hughes, W. E.,** A new type of powder maser, *Proc. IEEE,* 58, 480, 1970.

5.2 MASER ACTION IN NATURE

James M. Moran

INTRODUCTION

Gaseous masers occur naturally in astrophysical environments at microwave frequencies where appropriate energy sources, such as stars, are available to create population inversions, path lengths are large enough for sufficient amplification, and densities are low enough so that collisions do not thermalize the population levels. Astronomers study masers in these natural settings for the information they can provide about the excitation and dynamics of the gas. These cosmic masers operate in circumstellar clouds having dimensions of $\sim 10^{16}$ cm at densities of $\sim 10^9$ cm^{-3}. In contrast to laboratory devices, cosmic masers have no walls, end mirrors, or significant feedback. Hence, mode theory is unimportant in describing their operation. Cosmic masers exhibit only a small amount of temporal and spatial coherence, properties normally associated with man-made lasers.

Maser action was proposed to explain the unusual emission spectra of OH first observed in 1965 at a wavelength of 18 cm toward regions of ionized hydrogen (H II regions) in our galaxy.[1-6] Some typical spectra of a cosmic maser are shown in Figure 5.2.1. The radiation fields from these sources appear to be Gaussian random processes,[7] but the spectra contain many narrow features arising from clouds of gas having diameters of about $\sim 10^{14}$ cm at slightly different velocities within a region having an overall diameter of $\sim 10^{16}$ cm (see Figure 5.2.2). The brightness temperature of a blackbody giving the same radiated power can be as high as 10^{15} K although the kinetic temperature deduced from the linewidth is typically ~ 100 K. The spectral features are often highly polarized and sometimes change significantly over weeks or months. There is little similarity in the spectra among different transitions observed in the same source which suggests an unusual excitation mechanism. Shklovskii[8] first proposed that many of these masers may form in the envelopes of protostars or very young stars. Barrett[9] and Turner[10] describe the early development of research on cosmic masers. Reviews of the subject have been written by Cook,[11] Kegel,[12] Litvak,[13] Moran,[14] and Strel'nitskii.[15] This review emphasizes the basic physics of cosmic masers, reports the research results available until mid-1980, and provides an extensive bibliography.

This review describes the spectacular maser emission observed towards molecular clouds in our galaxy. This maser emission is spectacular because there is both population inversion and sufficient gain (e^{20} - e^{30}) to achieve very high brightness. There are other examples of population inversion in astrophysical objects where the gain is small. The hydrogen gas in H II regions has a significant population in the high Rydberg states. There may be population inversions between adjacent levels so that enhanced stimulated emission in recombination lines at radio wavelengths is observed.[164] Weak natural laser emission may occur in the 10 μ CO bands in the atmosphere of Mars.[165]

MODELS

A simple model for the geometry of a cosmic maser is a tube — a long filament of gas throughout which the velocity of the molecules is constant to within the linewidth. The equation of radiative transfer through the tube at the center frequency of the maser transition, is

$$\frac{dI}{dz} = \frac{h\nu}{4\pi\Delta\nu} B\Delta NI \qquad (1)$$

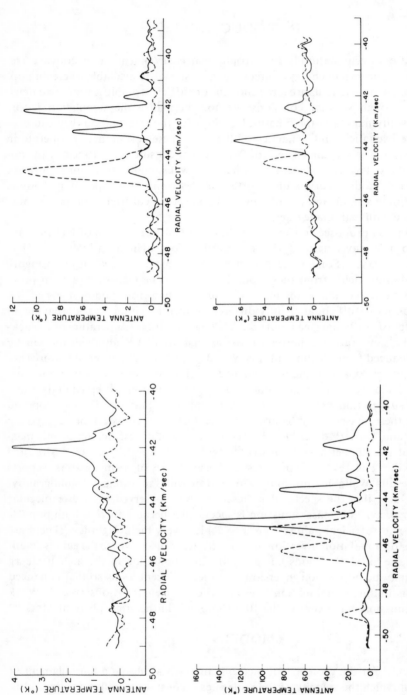

FIGURE 5.2.1.   The spectra (left to right, top to bottom) of the $1\rightarrow2$, $1\rightarrow1$, $2\rightarrow2$, and $2\rightarrow1$ transitions in the ground state ($^2\pi_{3/2}'$ $J = 3/2$) of OH at 18 cm wavelength for the maser source W3(OH).[53] The solid lines are the spectra in right circular polarization and the broken lines are the spectra in left circular polarization. T_A is proportional to received flux density where $1\ K \cong 2 \times 10^{-26}\ Wm^{-2}Hz^{-1}$. The velocity axis is calculated from the Doppler equation and is based on the rest frequency listed in Table 5.2.4 and the conventional local standard of rest. These spectra, typical of cosmic OH masers, contain many narrow, highly polarized components with little similarity among spectra in different transitions. The angular structure of the features is shown in Figure 5.2.2.

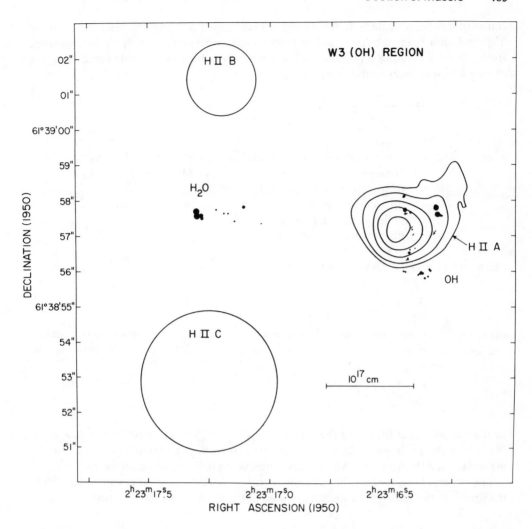

FIGURE 5.2.2.   The region near the masers in W3(OH). H IIA, H IIB, and H IIC are compact regions of ionized hydrogen which are thought to be excited by newly formed massive stars still invisible because of their dust cocoons.[38] Contours of intensity of the radio emission at 6 cm wavelength are shown for H IIA. Infrared emission from the dust cocoon associated with H IIA has been detected.[159] The areas of the maser spots are proportional to emitted flux density. The OH spots[90] are typically 2×10^{14} cm (0!005) in diameter and the H_2O spots[86] are typically 10^{13} cm (0 !! 0003) in diameter. OH and H_2O masers usually appear in the same region but they are rarely coincident. The maps of the OH and H_2O maser spots were made with very long baseline interferometers with antennas across the United States. (From Reid, M. J., Haschick, A. D., Burke, B. F., Moran, J. M., Johnston, K. J., and Swenson, G. W., *Astrophys. J.*, 239, 89-111, 1980. With permission.)

where $I =$ specific intensity ($Wm^{-2}Hz^{-1}$ steradian^{-1})
        $z =$ distance along the axis of tube toward the observer
        $h =$ Planck's constant
        $\nu =$ maser transition frequency
       $\Delta\nu =$ linewidth
       $B =$ Einstein coefficient for stimulated emission
      $\Delta N =$ population density difference between the upper and lower levels of masing transition $= N_2 - N_1$

The statistical weights are assumed to be unity and the contribution of spontaneous emission has been neglected. An energy source, called a pump, causes a net transfer of population from the lower maser level to the upper maser level, via intermediate levels, thereby creating a population inversion, ΔN, which can be calculated from the equation of statistical equilibrium, given by

$$\Delta N = \frac{\Delta N_o}{1 + \frac{(C+W)}{P}} \tag{2}$$

where ΔN_o =   population inversion established by the pump in the absence of collisions and stimulated emission = $N_1 \Delta P / P$
$\quad\quad\quad$ P = $\quad$ mean pump rate into the maser levels
$\quad\quad\quad$ ΔP = $\quad$ difference in pump rates to the upper and lower levels
$\quad\quad\quad$ C = $\quad$ collision rate across the maser levels
$\quad\quad\quad$ W = $\quad$ microwave emission rate

The microwave emission rate per molecule is

$$W = BI \frac{\Omega_m}{4\pi} \tag{3}$$

where Ω_m is the angle into which the maser emission is beamed. For a tube of radius ϱ and length L, strong emission occurs only along rays which traverse the entire length of the tube, so that

$$\Omega_m \cong \pi \left(\frac{\rho}{L}\right)^2 \tag{4}$$

In the astronomical literature the maser is said to be "unsaturated" when $P > (C + W)$. When the radiation field is strong enough to reduce the population inversion the maser is called "saturated." Most cosmic masers are believed to be saturated.

The observable quantities are the flux density, S, and the solid angle subtended by the source at the Earth, Ω_s, which are related to the intensity by the equation

$$S = I\Omega_s = \pi I \left(\frac{\rho}{R}\right)^2 \tag{5}$$

where R is the distance between the earth and the maser. The in tensity can be converted to the temperature of an equivalent blackbody, or brightness temperature, T_B, using the Rayleigh-Jeans approximation to the Planck equation,

$$T_B = \frac{\lambda^2}{2k} I \tag{6}$$

where k is Planck's constant and λ is the wavelength.

If the maser is unsaturated, the brightness temperature, from Equations 1, 2, and 6, will be

$$T_B = T_o e^{\alpha L} \tag{7}$$

where

$$\alpha = \frac{Bh}{4\pi} \left(\frac{\nu}{\Delta\nu}\right) \Delta N_o \tag{8}$$

and T_o is the temperature of a background source being amplified, or, in the case of no external input, the magnitude of the excitation temperature. Saturation is reached when W = (P + C). For P>>C, the saturation intensity, I_s, is therefore

$$I_s = \frac{4\pi P}{B\Omega_m} \tag{9}$$

at which point the brightness temperature, from Equations 3, 4, and 6, is

$$T_B = 4 \left(\frac{h\nu}{k}\right) \left(\frac{L}{\rho}\right)^2 \left(\frac{P}{A}\right) \tag{10}$$

A is the Einstein A coefficient of the maser transition, equal to $2h\nu^3 B/c^2$ where c is the speed of light. $A = 2 \times 10^{-9}$ sec^{-1} for the $6_{16}-5_{23}$ transition of H_2O. The length of a cosmic maser is not observable. However, masers occur in clusters (see Figure 5.2.2) and a reasonable estimate of L is the distance between the masers. Typically, L/ϱ is less than ~ 25. P is limited by the Einstein A coefficients linking the maser levels to other rotational levels which are ~ 1 sec^{-1}. Since $h\nu/k = 1$ K^{-1}, the highest that T_B can be for an unsaturated maser is about 10^{12} K. Masers that have T_B greater than this value, which includes most known water vapor masers, must be saturated.

When the maser is saturated the solution to equation (1), using Equations 2, 3, and 8, is

$$I = I_s + \alpha I_s (L - L_s) \tag{11}$$

where L_s is the path length at which saturation occurs. For a well-saturated maser, where $L_s<<L$ and $I_s<<I$, the intensity is

$$I \sim \alpha I_s L \tag{12}$$

or

$$I \sim \frac{h\nu\Delta N_o PL^3}{\Delta\nu\pi\rho^2} \tag{13}$$

The total radiated power is

$$P_R \sim \pi\rho^2 \, I\Omega_m \Delta\nu = h\nu\Delta N_o PV \tag{14}$$

where V is the volume of the maser. Using Equations 5 and 13, the population inversion density is given by the equation

$$\Delta N_o = \frac{SR^2\Delta\nu}{h\nu L^3 P} \tag{15}$$

Parameters for a typical H_2O maser are $S = 10^{-19}$ ergs sec^{-1} cm^{-2} Hz^{-1}; $\Delta\nu = 5 \times 10^4$ Hz, $\nu = 2 \times 10^{10}$ Hz, $L = 5 \times 10^{14}$ cm; $\varrho = 10^{13}$ cm, $P = 1$ sec^{-1}, and $R = 10^{22}$ cm. With these values the population inversion density, ΔN_o, is 30 cm^{-3}. The total gas density, which is predominantly composed of molecular hydrogen can be estimated by the equation

$$N_{H_2} = \Delta N_o \left(\frac{N_1}{\Delta N_o}\right)\left(\frac{N_{H_2O}}{N_1}\right)\left(\frac{N_{H_2}}{N_{H_2O}}\right) \tag{16}$$

Table 5.2.1
TYPICAL PARAMETERS FOR
SPHERICAL MASERS IN YOUNG
STELLAR OBJECTS[a]

Quantity	H_2O maser	OH maser
Decay rate (sec^{-1})	1	10^{-1}
Pump rate (sec^{-1})	10^{-1}	10^{-2}
Pump efficiency	10^{-2}	10^{-2}
Hydrogen density (cm^{-3})	10^9	10^8
OH or H_2O density (cm^{-3})	3×10^4	10^2
ΔN_o (cm^{-3})	30	0.1
Kinetic temperature (K)	10^3	10^2
Excitation temperature (K)	-45	-4
α^{-1} (cm)	5×10^{12}	2×10^{13}
T_B(K)	6×10^{14}	6×10^{11}
Apparent diameter (cm)	4×10^{13}	1×10^{14}
True diameter (cm)	2×10^{15}	2×10^{15}

[a] These parameters are taken from Goldreich, P. and
Keeley, D. A.[17] A wide range of parameter values is possible.

where $N_1/\Delta N_o \sim 50$; N_{H_2O}/N_1, the ratio of total H_2O density to density in the lower maser level, ~ 50; and $N_{H_2}/N_{H_2O} \sim 3 \times 10^4$. Hence, $N_{H_2} \sim 3 \times 10^9$ cm^{-3} or 10^{-13} g cm^{-3}. The radiated power P_R, of this maser is $\sim 6 \times 10^{26}$ ergs sec^{-1}; the flux density at the output face of the maser is 2 ergs sec^{-1} cm^{-2}; the rms electric field strength is about 1 vm^{-1} (3×10^{-5} esu). The flux density from this maser model is relatively low because the maser is highly beamed.

In addition to tubular masers, spherical masers have been analyzed by Litvak[16] and by Goldreich and Keeley.[17] The longest amplification paths occur along rays that pass through the center of the sphere. Saturation occurs at the ends of these paths. Hence spherical masers saturate first at the outer surface and retain an unsaturated core. An observer will predominantly see rays that pass through the unsaturated core and hence the size of the actual masing cloud may be much larger than the apparent size. This is analogous to the tube geometry where the amplification path length may be much larger than the observed cross section. A consistent set of parameters for a spherical maser model is given in Table 5.2.1. The actual geometries and global structures of maser sources are discussed by Elitzur and de Jong[18] and Elmegreen and Morris.[19]

The line profile of a maser narrows during unsaturated growth by the factor $\sqrt{\alpha L}$, which is typically ~ 5. For saturated growth the line broadens back to its thermal width unless IR trapping is important.[20,21]

MASER CHARACTERISTICS

In the following sections the properties of masers are discussed. References according to topic are provided in Table 5.2.2.

Molecular Species

Radio emission has been identified from about 50 species of atoms and molecules (not including isotopes).[22] The general characteristics of these clouds are described by Turner[23] and by Zuckerman and Palmer.[24] The gross differences between normal molecular clouds and masers are listed in Table 5.2.3, and differences in their spectra are shown in Figure 5.2.3. Only 3 molecules, OH, H_2O, and SiO, have been found to have

Table 5.2.2
SELECTED REFERENCES ON COSMIC MASERS

Maser type	Spectra	Polarization	Time variations	Structure and positions	Pumping theory
OH (late-type stars)	104-107	51-52	63,161	80-84	94-96
OH (YSO)[a]	1-4,108-117	2,4,53-61	67	37,90,146-148	92,93,97
H_2O (late-type stars)	118,119	46,47	64,65	78-79	98
H_2O (YSO)	27,120-130	46,47	68-73	35,36,85-89,102,149-152	91,92,99,100
SiO	131-138	48	66,142,143	77,153,154	101,102
CH_3OH	139-141		144,145	141,155,156	

[a] YSO = Young Stellar Object.

Table 5.2.3
CHARACTERISTICS OF MASER[a] AND NONMASER MOLECULAR CLOUDS

Characteristic	Maser	Nonmaser[b]
Linewidth (km/sec)	0.1 - 5	2 - 20
Number of distinct spectral features	1 - 500	1 - 3
Polarization (%)	0 - 100	None known
Size (cm)	$10^{12} - 10^{15c}$	$10^{18} - 10^{21}$
T_B (K)	$10^9 - 10^{15}$	10 -150
Time scale of variability (sec)	$10^5 - 10^8$	$> 10^8$
Density (cm^{-3})	$10^7 - 10^{11}$	$10^1 - 10^6$

[a] OH, H_2O, and SiO masers.
[b] See Turner[23] for more details.
[c] Masers usually appear in clusters having diameters of 10^{14} to 10^{17} cm.

strong maser action with $T_B > 10^9$ K. There is one source that shows moderate maser action in CH_3OH where $T_B \sim 10^4$ K. Several other molecules appear to show weak maser action.[25,26] The molecular transitions that are known to have strong maser action are listed in Table 5.2.4. Energy level diagrams for the four significant maser species are shown in Figures 5.2.4 through 5.2.7.

Location

A few powerful masers have ben detected in nearby galaxies.[27] All other known cosmic masers are located in our galaxy at distances from the sun of $\sim 10^2$ to 10^4 parsec (1 parsec $= 3 \times 10^{18}$ cm). Most of them are associated with circumstellar envelopes around either cool evolved stars, called late-type stars, or young stellar objects. The characteristics of these two types of maser are given in Tables 5.2.5 and 5.2.6. The late-type stars that have maser emission are giants or supergiants having surface temperatures of about 2000 K and luminosities of 10^4 times that of the sun. Radiation pressure drives a large flow of material from them with mass loss rates of about 10^{-8} to 10^{-5} solar-masses/year.[28] This flow creates an extended circumstellar envelope of dust and gas suitable for maser emission. The physical properties of these envelopes are described by Goldreich and Scoville,[29] the maser properties by Snyder,[30] and maser models by Kwok[31] and by Elitzur et al.[32] About 300 masers in this class have been discovered.[33]

Masers of the second class, those associated with young stellar objects, are more

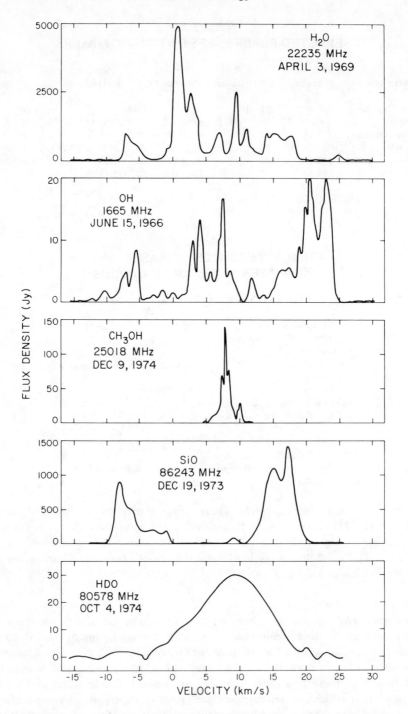

FIGURE 5.2.3.   The Spectra of 4 masers in the Orion Nebula — H_2O,[71] OH,[110] SiO,[134] and CH_3OH[144] — and one nonmaser, HDO.[160] The maser spectra are essentially free of measurement noise. The ripples in the HDO spectrum are due to noise and probably only one spectral component is present. The top four spectra show a principal characteristic of cosmic maser emission — a multiplicity of narrow features. Data taken from Moran.[14]

enigmatic because they are not usually associated with any visible objects. Most of these masers are near very compact H II regions of diameter $\sim 10^{17}$cm which are excited

Table 5.2.4

PARAMETERS OF OBSERVED MASER TRANSITIONS

Molecule	Transition	Frequency[a] (MHz)	λ^b (cm)	E/hc[c] (cm⁻¹)	E/k (K)	A[d] (sec⁻¹)	Ref.
OH	$^2\pi_{3/2}$ J = 3/2, F = 1 → 2	1612.2310	18.6	0	0	1.30×10^{-11}	104
	$^2\pi_{3/2}$ J = 3/2, F = 1 → 1	1665.4018	18.0	0	0	7.18×10^{-11}	112
	$^2\pi_{3/2}$ J = 3/2, F = 2 → 2	1667.3590	18.0	0	0	7.78×10^{-11}	4
	$^2\pi_{3/2}$ J = 3/2, F = 2 → 1	1720.5300	17.4	0	0	9.50×10^{-12}	108
	$^2\pi_{1/2}$ J = 1/2, F = 0 → 1	4660.42	6.4	126	181	1.08×10^{-9}	113
	$^2\pi_{1/2}$ J = 1/2, F = 1 → 0	4765.562	6.3	1 26	181	3.89×10^{-10}	114
	$^2\pi_{3/2}$ J = 5/2, F = 2 → 2	6030.747	5.0	84	120	1.55×10^{-9}	115
	$^2\pi_{3/2}$ J = 5/2, F = 3 → 3	6035.092	5.0	84	120	1.57×10^{-9}	116
	$^2\pi_{3/2}$ J = 7/2, F = 4 → 4	13441.4173	2.2	184	265	9.34×10^{-9}	117
H_2O	$^1_\Sigma$ 6_{16} → 5_{23}	22235.080[e]	1.35	447	644	1.91×10^{-9}	121
S,O	$^1_\Sigma$ v = 3, J = 1 → 0	42519.3	0.70	3654	5257	2.88×10^{-6}	131
	v = 2, J = 1 → 0	42820.54	0.70	2448	3522	2.93×10^{-6}	132
	v = 1, J = 1 → 0	43122.03	0.70	1230	1770	3.00×10^{-6}	134
	v = 1, J = 2 → 1	86243.35	0.35	1232	1773	2.87×10^{-5}	137
	v = 1, J = 3 → 2	129363.26	0.23	1234	1776	1.04×10^{-4}	138
CH_3OH	$^1_\Sigma$ 3_2 → 3_1 E[f]	24928.7	1.20	24	35	7.19×10^{-8}	139
	4_2 → 4_1 E	24933.468	1.20	31	45	7.77×10^{-8}	140
	2_2 → 2_1 E	24934.382	1.20	19	27	5.76×10^{-8}	139
	5_2 → 5_1 E	24959.080	1.20	39	56	8.08×10^{-8}	140
	6_2 → 6_1 E	25018.123	1.20	49	71	8.30×10^{-8}	140
	7_2 → 7_1 E	25124.873	1.19	61	87	8.51×10^{-8}	140
	8_2 → 8_1 E	25294.411	1.18	74	106	8.76×10^{-8}	140
	10_2 → 10_1 E	25878.2	1.15	102	147	9.52×10^{-8}	141

[a] Transition frequencies from Lovas, Snyder, and Johnson.[22]
[b] Transition wavelength.
[c] Energy of lower transition level above ground state.
[d] Transition probability, OH values from Destombes et al.[162]
[e] Weighted mean frequency of six hyperfine transitions. See Kukolich.[157]
[f] Transition nomenclature ($J_k \rightarrow J'_{k'}$) described by Lees.[158]

by newly formed hot stars (T $\sim$3 × 10⁴K) of high mass ($\sim$30 × solar mass). They are frequently found near the interface between the ionization front surrounding an old H II region and adjacent molecular clouds.[34] The stars are formed when a shock front, preceding the ionization front, triggers the collapse of local density inhomogeneities in the molecular cloud. During the collapse period, material falls inward onto an accretion core until nuclear burning begins and the star begins to radiate. Radiation pressure from the new star eventually blows off the remaining placental material and ionizes the gas. The time scale for these events is less than 10⁶ years. The ionized gas can be seen by free-free radiation in the microwave continuum. The heated dust creates an infrared source but visible radiation from the star itself is initially obscured by the dust. Figure 5.2.2 shows a typical situation. OH and H_2O masers usually occur near each other but not in the same volume of gas,[35,36] suggesting that these two types of masers may occur at different stages of stellar evolution. The evolutionary sequence is discussed by Evans et al.[37] and Habing and Israel.[38] Downes and Genzel[39] discuss masers in this class in detail.

Spectra

The spectra of most masers are complex, containing up to several hundred components which are presumably due to radiation from cloudlets having slightly different velocities and hence different frequencies. The velocity structure of masers in the late-

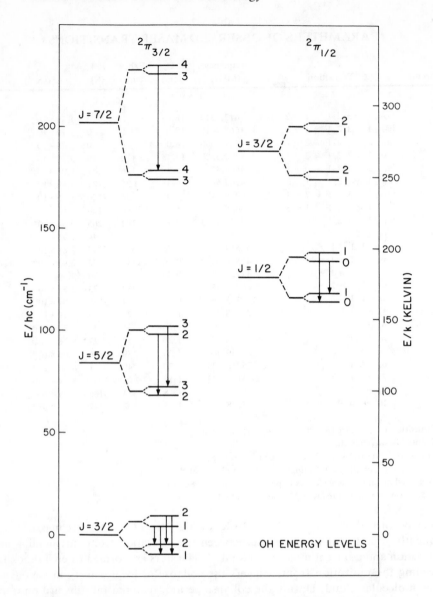

FIGURE 5.2.4.   Part of the rotational spectrum of OH. The rotational ladder has two branches due to spin splitting. Each rotational level is divided into four sublevels (shown highly exaggerated) because of the interaction of the unpaired electron and the rotation of the molecule (lambda doubling) and because of the hyperfine interaction between the hydrogen nuclear magnetic moment and the molecular magnetic moment. The number to the right of each energy level is the total angular momentum quantum number F. The known maser transitions are indicated by the downward arrows.

type stars generally extends over ∿50 km sec⁻¹ and this spread can be reasonably explained by the stellar winds. The velocities of H_2O masers toward young stellar objects sometimes extend over a range of 500 km sec⁻¹.[40] Some authors have argued that these velocity or frequency displacements could be due to solitons,[42] Raman scattering,[43] Stark effect,[44] or Langmuir scattering.[45] None of these effects has been verified. Transverse motions of some masers have been measured with interferometry techniques[163] which strongly suggests that most of the frequency shifts are due to kinematic motions.[41] Small velocity shifts due to Zeeman effect and hyperfine splitting almost certainly exist.

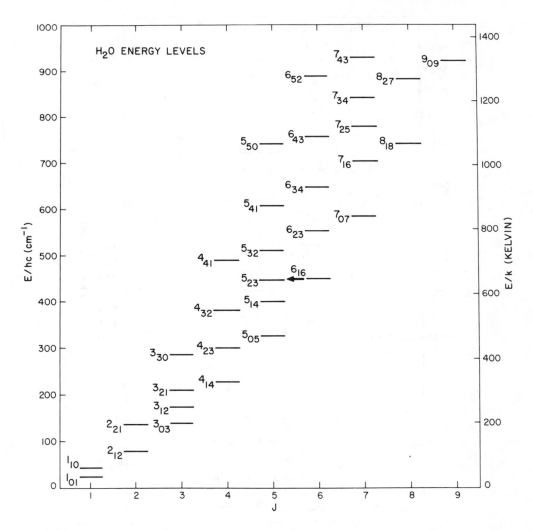

FIGURE 5.2.5.   Part of the rotational energy level spectrum for H_2O, an asymmetrical rotator.[100] Only the ortho branch (K_{-1}, K_{+1} = odd, even $\leftrightarrow$ even, odd) is shown. The microwave maser transition is due to a chance proximity of the 6_{16} and 5_{23} levels.

Polarization

Cosmic maser emission is significantly polarized. H_2O[46,47] and SiO[48] masers show linear polarization in the range of 0 to 50%. None shows circular polarization. SiO and H_2O are nonparamagnetic and field strengths of $\sim$100 Gauss, about four orders of magnitude greater than expected in the maser environment, would be required to produce a Zeeman splitting of the lines equal to the Doppler widths. Hence, theory[49,50] predicts linear polarization, but not circular polarization, if certain conditions are met.

OH masers in late type stars generally show little polarization[51,52] but OH masers in young stellar objects show a large degree of polarization.[4,53-57] The observed spectra often contain many 100% circularly polarized features but only a few linearly polarized ones. The Zeeman pattern for the OH transitions[58] at 1665 and 6035 MHz contains two σ components, of opposite circular polarization, for the magnetic field along the line of sight, and three components, two σ and one π component, all linearly polarized, for the magnetic field perpendicular to the line of sight. The separation between the σ components is 0.6 km sec^{-1} mG^{-1} (3.3 KHz mG^{-1}) for the 1665 MHz transition and 0.06 km sec^{-1} mG^{-1} for the 6035 MHz transition. The observed polarization is almost

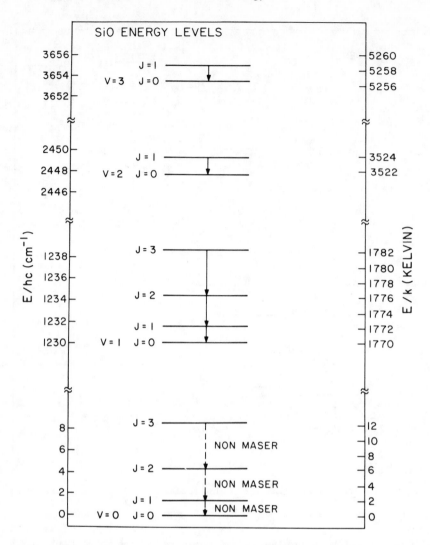

FIGURE 5.2.6.   Some of the rotational and vibrational energy levels of SiO, a simple diatomic molecule. The known masing and nonmasing transitions are shown.

certainly caused by the Zeeman effect.[61] The absence of linearly polarized components is explained as due to internal Faraday rotation or to cross relaxation among the magnetic sublevel populations.[50] Although the Zeeman pairs are not completely obvious in the spectra (see Figure 5.2.1), they can be identified readily by their spatial coincidence in the interferometer maps. The magnetic fields deduced from these measurements are ∿10 mG.[59-61] This is about the expected value[62] if the field is frozen into the medium as it collapses from the interstellar medium where the density is 1 molecule cm^{-3} and the field is 10^{-6} G, to the maser cloud where the density is about 10^9 molecules cm^{-3}.

Time Variations

Maser emission is variable on time scales of days to years. For masers in the envelopes of late-type stars the emission generally varies approximately in-phase with the optical luminosity of the star.[63-66] This strongly suggests that the stellar radiation pumps the maser. In masers toward young stellar objects the variations appear to be random[67-72] with individual features appearing and vanishing over a period of several

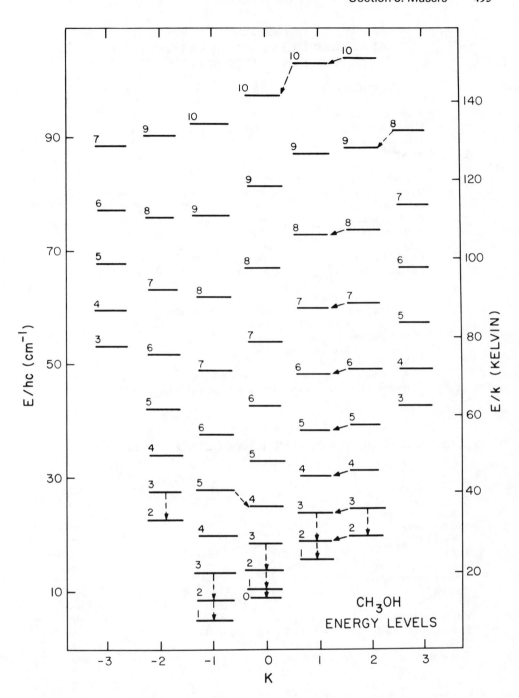

FIGURE 5.2.7.   Part of the rotational energy spectrum of CH_3OH showing only the E stack.[158] Observed masing transitions are shown as solid arrows, nonmasing as dashed arrows.

months such that in many sources the spectra are completely different after about one year. Most of these changes are probably due to turbulent motions that alter the amplification path length of constant velocity. There is one well-sampled case of a variation that showed a sudden rise and slow decay discovered by Burke et al.[73] They modeled this as a sudden influx of pump energy that slowly diffused through the masing cloud. Mathematical theories of cosmic maser variations are available.[74,75]

Table 5.2.5
CHARACTERISTICS OF MASERS ASSOCIATED WITH LATE-TYPE STARS

Quantity	H_2O	OH	SiO
Transitions observed	1	3	5
Number known[a]	100	130[b]	60
Linewidth (km/sec)	1-2	1-2	1-2
$T_K(K)$[c]	400-1500	400-1500	250-3500
Number of spectral features[d]	1-10	1-10	1-10
Velocity range (km/sec)	5-50	5-80	2-15
Polarization (%)	None	Small	Linear (0-50)
Lifetime of feature (sec)	$>10^7$	$>10^7$	$>10^7$
Spot size (cm)	10^{14}	10^{15}	10^{14}
$T_B(K)$	10^{11}-10^{12}	10^9-10^{11}	10^{11}
Cluster size (cm)	10^{15}	2×10^{15}	10^{14}-10^{15}
Power (ergs/sec)[e]	10^{24}-10^{28}	10^{24}-10^{28}	10^{26}-10^{27}

[a] Complete list in Engels.[33]

[b] 145 more sources are known whose spectra resemble those seen toward late-type stars but for which there is no optical or IR object seen, possibly because of obscuration.

[c] Assuming no line narrowing or mass motions.

[d] More for super giants.

[e] Assuming isotropic radiation. 10^{24} ergssec^{-1} is the sensitivity limit for present radio telescopes for masers at a distance of 100 pc.

Table 5.2.6
CHARACTERISTICS OF MASERS ASSOCIATED WITH YOUNG STELLAR OBJECTS

Quantity	H_2O	OH	SiO	CH_3OH
Transitions observed	1	9	4	8
Number known	170	~100	1[a]	1[b]
Linewidth (km/sec)	0.5-2	0.1-1	1-2	0.5-2
$T_K(K)$[c]	100-1500	5-500	3500	150
Number of spectral features	1-200	1-50	~5	~10
Velocity range (km/sec)	1-400	1-30	25	4
Polarization (%)	Linear (0-20)	Linear (0-100) Circular (0-100)	None	None known
Lifetime of feature[d] (sec)	10^6-10^8	10^7-10^8	10^7	$>10^7$
Spot size (cm)	10^{13}-10^{14}	10^{14}-10^{15}	10^{14}	$\sim10^{16}$
$T_B(K)$	10^{13}-10^{15}	10^{12}-10^{13}	10^9	10^3-10^4
Cluster size (cm)	10^{16}-10^{17}	10^{16}-10^{17}	$<10^{15}$	3×10^{17}
Power[e] (ergs/sec)	10^{25}-10^{33}	10^{25}-10^{30}	10^{29}	10^{28}

[a] The only source known is in Orion A. The classification of the SiO maser is controversial.

[b] The only source known is in Orion A.

[c] Assuming no line narrowing or mass motions.

[d] There are some cases of shorter time scales.

[e] Assuming isotropic radiation. 10^{25} ergs/sec is the sensitivity limit for present radio telescopes for masers at a distance of 300 parsecs.

Structure

Masers appear as a cluster of bright spots of size 0 ʺ 0002 to 1″, spread over a region having a diameter of 0 ʺ 2 to 2″ corresponding to 10^{14} to 10^{17} cm. Because of the small size most of the structural information comes from radio interferometry, in particular, very long baseline interferometry.[76] Observation of H_2O masers at a resolution of 0 0002 is routinely achieved with a network of antennas located in Simeiz, USSR; Effelsberg, Germany; Onsala, Sweden; Westford, Mass.; Green Bank, W. Va.; and Big Pine, Calif. Lower resolution observations of OH masers are done with arrays in the U.S. and Europe. Precise astrometric work can be done with interferometers having baselines of a few kilometers.[36]

In the envelopes of late-type stars the different molecular species mase at different distances from the exciting star. SiO masers, which require 1500 K of excitation in the $v = 1$ level, form near the stellar surface[77] at a radius of $\sim 10^{14}$ cm. The water masers[78,79] form at ~ 10 stellar radii and the OH masers[80-84] form at ~ 100 stellar radii.

Water vapor masers around young stellar objects are often clumped in groups suggesting the presence of several exciting stars.[85,86] The mass flow from these stars interacts with the surrounding molecular cloud and creates a complex distribution of masers.[87-89] An example of OH maser emission in a collapsing envelope may be W3(OH),[90] shown in Figure 5.2.2.

Dynamics

Maps of the relative distribution of maser spots have been made to an accuracy of 50 microarcseconds (μas) for several H_2O masers on many occasions over a period of several years.[163] For a maser at 1 kiloparsec (kpc) from the sun, a motion of 50 μas in one year corresponds to a velocity of 0.3 km sec^{-1}. From a comparison of the maser maps, proper motions of the spots, and hence three dimensional velocity vectors, were measured for about 30 spots in each source. For the maser in one source, Orion A, the motions clearly defined a radial outflow of material from a central exciting star. In modeling this flow the distance to the source was determined, essentially by a comparison of radial velocities and proper motions.

In masers where the velocity vectors cannot be modeled in any simple way, but can be assumed to be random variables, the distance to the source can be estimated by the method of "statistical parallax". That is, if the dispersions in the radial and transverse velocities are equal, then the distance is given by

$$D = \frac{\sigma_R}{\sigma_\theta} \tag{17}$$

where σ_R is the dispersion in radial velocity and σ_θ is the dispersion in angular velocity calculated from the proper motions. The accuracy of this method is about

$$\frac{\Delta D}{D} = \frac{1}{2\sqrt{N}} \sqrt{\frac{1}{1 - \left(\frac{\sigma_M}{\sigma_A}\right)^2}} \tag{18}$$

where

$$N = \quad \text{number of maser spots}$$
$$\sigma_M = \quad \text{rms measurement error in transverse velocity}$$
$$\sigma_A = \quad \text{velocity dispersion in transverse velocity}$$

Since N is typically 30 to 100 and $\sigma_M \ll \sigma_A$, statistically parallax measurements on masers provides a primary method of determining distances within the galaxy to an accuracy of $\sim 10\%$. There are several masers in external galaxies. When measurement accuracies improve to ~ 10 μas it will be possible to measure the distances to these galaxies to an accuracy of $\sim 10\%$. Currently, indirect techniques are only accurate to

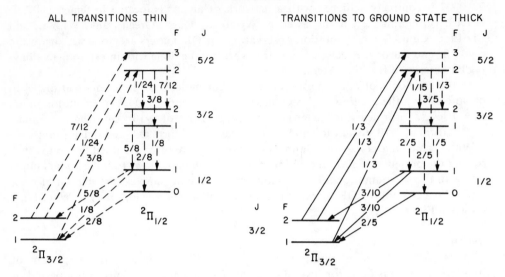

FIGURE 5.2.8.   The population flow following the absorption of a 35 μ photon by an OH molecule in the $^2\pi_{3/2}$ J = 3/2 level. The total absorption rate out of the ground state is normalized to unity. The transition rates shown are proportional to the Einstein A coefficients modified for the effects of radiation trapping. For the case on the left, where all the transitions are optically thin, there is no net transfer of population between the hyperfine levels of the Λ doublets. For the case on the right, transitions with the solid arrows are optically thick, and there is a net transfer of population to the upper level of the lambda doublet. Because of the symmetry between upper and lower halves of the Λ doublets, only one half of each doublet is shown. This pump cycle is described in detail by Elitzur et al.[95]

40%. Hence, the measurements of proper motions in masers may contribute to the refinement of the distance scale of the universe.

Pumping

Goldreich and Kwan[91] and Litvak[13,92] have reviewed the general problem of pumping cosmis masers. Kegel[93] has studied the range of physical conditions required for pumping OH. Many other pump models have been proposed.[94-102] Several conditions are necessary for population inversion. Maser clouds, which are predominantly composed of H_2 gas, must have densities of less than 10^{11} hydrogen molecules cm^{-3} lest the population be thermalized by collisions. Nonthermodynamic equilibrium conditions must be established. This can happen in many ways near a strong radiation source in clouds consisting of gas and dust. The kinetic temperature of the gas, the thermal temperature of the dust, and the radiation temperature of the radiation field may be different, thereby providing heat reservoirs for the pump cycle. Many lines are usually involved in the pump cycle and the overlapping of doppler broadended lines, some from other molecules, may be important in establishing population inversion. There are two types of pumping mechanisms: radiative pumps, where the heat source is from radiative absorptions, and collisional pumps, where the heat source is from collisions. Most radiative pumps involve near or far IR radiation for excitation to higher rotational levels. The pump source can be either inside or outside the maser cloud. A typical pump cycle for OH masers in late-type stars is shown in Figure 5.2.8. It requires a source of 35 μ radiation. The masing region must be optically thick at the interladder transition wavelengths of 35 and 80 μ but optically thin at the intraladder transition wavelengths of 110 and 175 μ. Detailed pump calculations involve a simultaneous solution of the radiative transfer equation and the equations of statistical equilibrium for the populations of many energy levels.

Most proposed pumping mechanisms require the emission of at least one infrared photon per maser photon. The number of infrared photons available for the pump cycle is

$$N_{IR} = 4\pi R^2 \ \frac{S_p \Delta\nu_p \Omega_p}{h\nu_p} \tag{19}$$

where S_p =   infrared flux density
$\Delta\nu_p$ =   linewidth of pumping transition
Ω_p =   solid angle subtended by maser cloud seen from the source
ν_p =   pump frequency
R =   distance between observer and source

The number of maser photons is

$$N_m = 4\pi R^2 \ \frac{S_m \Delta\nu_m \Omega_m}{h\nu_m} \tag{20}$$

where S_m =   maser flux density
ν_m =   maser transition
$\Delta\nu_m$ =   maser linewidth
Ω_m =   solid angle into which the maser radiates

If $f_m = \Delta\nu_m/\nu_m$ and $f_p = \Delta\nu_p/\nu_p$, then the condition $N_{IR} > N_m$ gives

$$S_p > S_m \left(\frac{\Omega_m}{\Omega_p}\right)\left(\frac{f_m}{f_p}\right) \tag{21}$$

The ratios (Ω_m/Ω_p) and $f_m/f_p)$ are unknown, but for a maser close to the pump source with saturated amplification they may both be close to unity. Hence $S_p > S_m$. Figure 5.2.9 shows the correlation between pump flux and masing flux towards a group of late-type stars. There is probably adequate power to pump the masers. Many of the most powerful water vapor masers are not closely associated with IR sources and explaining how they are pumped is difficult.[103]

Conclusions

Maser emission occurs primarily in the circumstellar envelopes of evolved stars or newly formed stars. Several hundred masers have been found in four molecular species. Mapping these masers may be the best technique for determining the kinematics of the material around the exciting stars and the magnetic field distribution. Because of the large number of spots in each maser, for which proper motions can be measured, the distance of the maser can be determined by the method of statistical parallax.

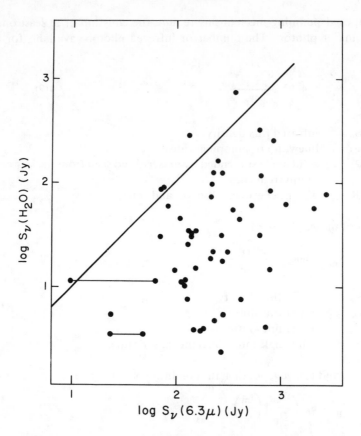

FIGURE 5.2.9.   The peak H_2O maser flux density versus the infrared flux density at 6.3 μm, the wavelength of a vibrational transition in H_2O which may be important in the maser pump cycle.[98] The diagonal line shows where the number of 6.3 μm infrared and H_2O maser photons are equal. The horizontal lines show the variability of the maser flux density of some of the stars. Generally there is adequate radiation from the stars to pump the masers. These data are taken from Kleinmann et al.[118]

REFERENCES

1. **Weaver, H., Williams, D. R. W., Dieter, N. H., and Lum, W. T.,** Observations of a strong unidentified microwave line and of emission from the OH molecule, *Nature (London),* 208, 29—31, 1965.
2. **Weinreb, S., Meeks, M. L., Carter, J. C., Barrett, A. H., and Rogers, A. E. E.,** Observations of polarized OH emission, *Nature (London),* 208, 440—441, 1965.
3. **McGee, R. X., Robinson, B. J., Gardner, F. F., and Bolton, J. G.,** Anomalous intensity ratios of the interstellar lines of OH in absorption and emission, *Nature (London),* 208, 1193—1195, 1965.
4. **Palmer, P. and Zuckerman, B.,** Observations of galactic OH, *Astrophys. J.,* 148, 727—744, 1967.
5. **Litvak, M. M., McWhorter, A. L., Meeks, M. L., and Zeiger, H. J.,** Maser model for the interstellar OH microwave emission, *Phys. Rev. Lett.,* 17, 821—826, 1966.
6. **Perkins, F., Gold, T., and Salpeter, E. E.,** Maser action in interstellar OH, *Astrophys. J.,* 145, 361—366, 1966.
7. **Evans, N. J., Hills, R. E., Rydbeck, O. E. H., and Kollberg, E.,** Statistics of the radiation from astronomical masers, *Phys. Rev. A,* 6, 1643—1647, 1972.
8. **Shklovskii, I. S.,** *Stars: Their Birth, Life, and Death,* W. H. Freeman, San Francisco, 1978.
9. **Barrett, A. H.,** Radio signals from hydroxyl radicals, *Sci. Am.,* 219, 36—44, 1969.
10. **Turner, B. E.,** Anomalous emission from interstellar hydroxyl and water, *R. Astron. Soc. Can.,* 64, 221—237, 282—304, 1970.

11. **Cook, A. H.**, *Celestial Masers,* Cambridge University Press, New Rochelle, N.Y., 1977.

12. **Kegel, W. H.**, Cosmic masers, in *Problems in Stellar Atmospheres and Envelops,* Bascheck, B., Kegel, W. H., and Traving, G., Eds., Springer-Verlag, Berlin, 1975, 257—299.

13. **Litvak, M. M.**, Coherent molecular radiation, in *Annu. Rev. of Astronomy and Astrophysics,* Vol. 12, Burbidge, G. R., Ed., Annual Reviews, Palo Alto, Calif., 1974, 97—112.

14. **Moran, J. M.**, Radio observations of galactic masers, in *Frontiers of Astrophysics,* Avrett, E., Ed., Harvard University Press, Cambridge, Mass., 1976, 385—437.

15. **Strel'nitskii, V. S.**, Cosmic masers, *Sov. Phys.-USP,* 17, 507—527, 1975.

16. **Litvak, M. M.**, Radiative transport in interstellar masers, *Astrophys. J.,* 182, 711—730, 1973.

17. **Goldreich, P. and Keeley, D. A.**, Astrophysical masers. I. Source size and saturation, *Astrophys. J.,* 174, 517—525, 1972.

18. **Elitzur, M. and de Jong, T.**, A model for the maser sources associated with H II regions, *Astron. Astrophys.,* 67, 323—332, 1978.

19. **Elmegreen, B. G. and Morris, M.**, Disk structure in protostellar H_2O maser sources, *Astrophys. J.,* 229, 593—603, 1979.

20. **Goldreich, P. and Kwan, J.**, Astrophysical masers. IV. Line widths, *Astrophys. J.,* 190, 27—34, 1974.

21. **Litvak, M. M.**, Linewidths of a Gaussian broadband signal in a saturated two-level system, *Phys. Rev. A,* 2, 2107—2115, 1970.

22. **Lovas, F. J., Snyder, L. E., and Johnson, D. R.**, Recommended rest frequencies for observed interstellar molecular transitions, *Astrophys. J. Suppl. Ser.,* 41, 451—480, 1979.

23. **Turner, B. E.**, General physical characteristics of the interstellar molecular gas, in *The Large Scale Characteristics of the Galaxy,* (IAU Symp. 84), Burton, W. B., Ed., D. Reidel, Dordrecht, the Netherlands, 1979, 257—270.

24. **Zuckerman, B. and Palmer, P.**, Radio radiation from interstellar molecules, in *Annu. Rev. of Astronomy and Astrophysics,* Vol. 12, Burbidge, G., Ed., Annual Reviews, Palo Alto, Calif., 1974, 279—313.

25. **Forster, J. R., Goss, W. M., Wilson, T. L., Downes, D., and Dickel, H. R.**, A formaldehyde maser in NGC7538, *Astron. Astrophys.,* 84, L1—3, 1980.

26. **Broten, N. W., MacLeod, T., Oka, T., Avery, L. W., Brooks, J. W., McGee, R. X., and Newton, L. M.**, Evidence for weak maser action in interstellar cyanodiacetylene, *Astrophys. J.,* 209, L143—147, 1976.

27. **Churchwell, E., Witzel, A., Huchtmeier, W., Pauliny-Toth, I., Roland, J., and Sieber, W.**, Detection of H_2O maser emission in the galaxy M33, *Astron. Astrophys.,* 54, 969—971, 1977.

28. **Reimers, D.**, On the absolute scale of mass-loss in red giants, *Astron. Astrophys.,* 61, 217—224, 1977.

29. **Goldreich, P. and Scoville, N.**, OH-IR Stars. I. Physical properties of circumstellar envelopes, *Astrophys. J.,* 205, 144—154, 1976.

30. **Snyder, L. E.**, Observational characteristics of masers associated with stars, in *Interstellar Molecules,* IAU Symp. 87, Andrew, B., Ed., D. Reidel, Dordrecht, the Netherlands, 1980, 525—533.

31. **Kwok, S.**, A study of the velocity pattern of maser emission from infrared stars, *J. R. Astron. Soc. Can.,* 70, 49—66, 1976.

32. **Elitzur, M., Goldreich, P., and Scoville, N.**, OH-IR Stars. II. A model for the 1612 MHz masers, *Astrophys. J.,* 205, 384—396, 1976.

33. **Engels, D.**, Catalogue of late-type stars with OH, H_2O, or SiO maser emission, *Astron. Astrophys. Suppl.,* 36, 337—345, 1979.

34. **Elmegreen, B. G. and Lada, C. J.**, Sequential formation of subgroups in OB associations, *Astrophys. J.,* 214, 725—741, 1977.

35. **Mader, G. L., Johnston, K. J., and Moran, J. M.**, The spatial distribution of the OH and H_2O masers associated with W3(OH), W49N and W51, *Astrophys. J.,* 224, 115—124, 1978.

36. **Forster, J. R., Welch, W. J., Wright, M. C. H., and Baudry, A.**, Accurate Interferometer Positions of H_2O Masers, *Astrophys. J.,* 221, 137—144, 1978.

37. **Evans, N. J., Beckwith, S., Brown, R. L., and Gilmore, W.**, Type I OH Masers: A study of positions, polarization, nearby water masers, and radio continuum and infrared properties, *Astrophys. J.,* 227, 450—465, 1979.

38. **Habing, H. J. and Israel, F. P.**, Compact HII regions and OB star formation, in *Annu. Rev. Astronomy and Astrophysics,* Vol. 17, Burbidge, G., Ed., Annual Reviews, Palo Alto, Calif., 1979, pp. 345—385.

39. **Downes, D. and Genzel, R.**, Observations of masers in regions of star formation, in *Interstellar Molecules,* IAU Symp. 87, Andrew, B., Ed., D. Reidel, Dordrecht, the Netherlands, 1980, 565—577.

40. **Heckman, T. M. and Sullivan, W. T.**, The puzzle of the high velocity water vapor features in W49, *Astrophys. Lett.,* 17, 105—112, 1976.

41. Strel'nitskii, V. S. and Syunyaev, R. A., Nature of the high velocities of H_2O sources in W49, *Sov. Astron.-AJ*, 16, 579—584, 1973.

42. Montes, C., Variability of intensity of interstellar maser lines due to induced Compton scattering, *Astrophys. J.*, 216, 329—345, 1977.

43. Fernandez, J. C. and Reinisch, G., Generation of very-high-velocity satellite features through stimulated Raman scattering of the 22.2 GHz H_2O maser lines in compact HII plasma regions. One dimensional model, *Astron. Astrophys.*, 67, 163—174, 1978.

44. Slysh, V. I., Resonance stark effect in OH and H_2O interstellar masers, *Astrophys. Lett.*, 14, 213—216, 1973.

45. Burdjuzha, V. V., Charugin, V. M., and Tomozov, V. M., The role of plasma effects in generating high velocity and symmetric spectral features in galactic masers, *Astron. Astrophys.*, 79, 306—311, 1979.

46. Bologna, J. M., Johnston, K. J., Knowles, S. H., Mango, S. A., and Sloanaker, R. M., Observations of the polarization characteristics of galactic H_2O sources, *Astrophys. J.*, 199, 86—91, 1975.

47. Knowles, S. H. and Batchelor, R. A., Linear polarization of 22-GHz water vapor line emission in southern sources, *Mon. Not. R. Astron. Soc.*, 184, 107—117, 1978.

48. Troland, T. H., Heiles, C., Johnson, D. R., and Clark, F. O., Polarization Properties of the 86.2 GHz v = 1, J = 2 → 1 SiO Maser, *Astrophys. J.*, 232, 143—157, 1979.

49. Goldreich, P., Keeley, D. A., and Kwan, J. Y., Astrophysical masers. II. Polarization properties, *Astrophys. J.*, 179, 111—134, 1973.

50. Goldreich, P., Keeley, D. A., and Kwan, J. Y., Astrophysical masers. III. Trapped infrared lines and cross relaxation, *Astrophys. J.*, 182, 55—66, 1973.

51. Wilson, W. J., Barrett, A. H., and Moran, J. M., OH emission associated with infrared stars, *Astrophys. J.*, 160, 545—571, 1970.

52. Reid, M. J., Moran, J. M., Leach, R. W., Ball, J. A., Johnston, K. J., Spencer, J. H., and Swenson, G. W., Stellar OH masers and magnetic fields: VLBI observations of U Orionis and IRC +10420, *Astrophys. J.*, 227, L89—92, 1979.

53. Barrett, A. H. and Rogers, A. E. E., Observations of circularly polarized OH emission and narrow spectral features, *Nature (London)*, 210, 188—190, 1966.

54. Ball, J. A. and Meeks, M. L., Observations of galactic OH emission, *Astrophys. J.*, 153, 577—594, 1968.

55. Manchester, R. N., Robinson, B. J., and Goss, W. M., 18 cm Observations of galactic OH from longitudes 128° to 300°, *Aust. J. Phys.*, 23, 751—775, 1970.

56. Coles, W. A., Rumsey, V. H., and Welch, W. J., Polarization and time variation of OH sources, *Astrophys. J.*, 154, L61—65, 1968.

57. Chaisson, E. J. and Beichman, C. A., Further evidence for magnetism in the orion region, *Astrophys. J.*, 199, L39—42, 1975.

58. Davies, R. D., Magnetic fields in OH masers clouds, in *Galactic Radio Astronomy*, Kerr, F. J. and Simonson, S. C., Eds., D. Reidel, Dordrecht, the Netherlands, 1974, 275—292.

59. Lo, K. Y., Walker, R. C., Burke, B. F., Moran, J. M., Johnston, K. J., and Ewing, M. S., Evidence for Zeeman splitting in 1720 MHz OH line emission, *Astrophys. J.*, 202, 650—654, 1975.

60. Hansen, S. S., Moran, J. M., Reid, M. J., Johnston, K. J., Spencer, J. H., and Walker, R. C., The hydroxyl maser in the Orion nebula, *Astrophys. J.*, 218, L65—69, 1977.

61. Moran, J. M., Reid, M. J., Lada, C. J., Yen, J. L., Johnston, K. J., and Spencer, J. H., Evidence for the Zeeman effect in the OH maser emission from W3(OH), *Astrophys. J.*, 224, L67—71, 1978.

62. Mouschovias, T. Ch., Formation of stars and planetary systems in magnetic interstellar clouds, in *Protostars and Planets*, Gehrels, T., Ed., University of Arizona Press, Tucson, 1978, 209—242.

63. Harvey, P. M., Bechis, K. P., Wilson, W. J., and Ball, J. A., Time variations in the OH microwave and infrared emission from late-type stars, *Astrophys. J. Suppl.*, 27, 331—357, 1974.

64. Schwartz, P. R., Harvey, P. M., and Barrett, A. H., Time variation of the H_2O maser and infrared continuum in late-type stars, *Astrophys. J.*, 187, 491—496, 1974.

65. Cox, G. C. and Parker, E. A., Time variations of stellar water masers, *Mon. Not. R. Astron. Soc.*, 186, 197—215, 1979.

66. Hjalmarson, A. and Olofsson, H., Time variability of the R Leonis, O Ceti, and Orion A SiO (v = 1, J = 2 − 1) masers, *Astrophys. J.*, 234, L199—204, 1979.

67. Sullivan, W. T. and Kerstholt, J. H., Time variations in 18-cm OH emission profiles over the period 1965-1972, *Astron. Astrophys.*, 51, 427—450, 1976.

68. White, G. J., Correlated variability in the W49 water vapour maser source, *Mon. Not. R. Astron. Soc.*, 186, 377—381, 1979.

69. Gammon, R. H., Correlated variability in W49 (H_2O), *Astron. Astrophys.*, 50, 71—77, 1976.

70. Little, L. T., White, G. J., and Riley, P. W., Time variations of interstellar water masers: strong sources in HII regions, *Mon. Not. R. Astron. Soc.*, 180, 639—656, 1977.

71. Sullivan, W. T., Variations in frequency and intensity of 1.35-cm H_2O emission profiles in galactic HII regions, *Astrophys. J.*, 166, 321—332, 1971.

72. White, G. J. and Macdonald, G. H., Time variations of interstellar water masers in HII regions, *Mon. Not. R. Astron. Soc.*, 188, 745—764, 1979.

73. Burke, B. F., Giuffrida, T. S., and Haschick, A. D., Maser time variations, *Astrophys. J.*, 226, L21—24, 1978.

74. Bettwieser, E., Remarks on time variations and radiative stability of the celestial masers, *Astron. Astrophys.*, 72, 97—103, 1979.

75. Salem, M. and Middleton, M. S., Time variability of astrophysical masers, *Mon. Not. R. Astron. Soc.*, 183, 491—500, 1978.

76. Moran, J. M., Spectral line analysis of very long baseline interferometric data, *Proc. IEEE*, 1236—1242, 1973.

77. Moran, J. M., Ball, J. A., Predmore, C. R., Lane, A. P., Huguenin, G. R., Reid, M. J., and Hansen, S. S., VLBI observations of SiO masers at a wavelength of 7 millimeters in late-type stars, *Astrophys. J.*, 231, L67—71, 1979.

78. Rosen, B. R., Moran, J. M., Reid, M. J., Walker, R. C., Burke, B. F., Johnston, K. J., and Spencer, J. H., Observations of OH and H_2O microwave maser emission from VY Canis Majoris, *Astrophys. J.*, 222, 132—139, 1978.

79. Spencer, J. H., Johnston, K. J., Moran, J. M., Reid, M. J., and Walker, R. C., The structure of H_2O masers associated with late-type stars, *Astrophys. J.*, 230, 449—455, 1979.

80. Moran, J. M., Ball, J. A., Yen, J. L., Schwartz, P. R., Johnston, K. J., and Knowles, S. H., Very long baseline interferometric observations of OH masers associated with infrared stars, *Astrophys. J.*, 211, 160—169, 1977.

81. Reid, M. J. and Muhleman, D. O., Very long baseline interferometric observations of the hydroxyl masers in VY Canis Majoris, *Astrophys. J.*, 220, 229—238, 1978.

82. Benson, J. M. and Mutel, R. L., Multibaseline VLBI observations of the 1612 MHz OH masers toward NML Cygni and VY Canis Majoris, *Astrophys. J.*, 233, 119—126, 1979.

83. Reid, M. J., Muhleman, D. O., Moran, J. M., Johnston, K. J., and Schwartz, P. R., The structure of the stellar hydroxyl masers, *Astrophys. J.*, 214, 60—77, 1977.

84. Masheder, M. R. W., Booth, R. S., and Davies, R. D., The structure of four 1612 MHz OH emission sources, *Mon. Not. R. Astron. Soc.*, 166, 561—583, 1974.

85. Walker, R. C., Johnston, K. J., Burke, B. F., and Spencer, J. H., VLBI observations of high-velocity H_2O emission in W49N, *Astrophys. J.*, 211, L135—138, 1977.

86. Walker, R. C., Burke, B. F., Haschick, A. D., Crane, P. C., Moran, J. M., Johnston, K. J., Lo, K. Y., Yen, J. L., Broten, N. W., Legg, T. H., Greisen, E. W., and Hansen, S. S., VLBI aperture synthesis observations of H_2O masers associated with molecular clouds, *Astrophys. J.*, 226, 95—114, 1978.

87. Genzel, R., Downes, D., Moran, J. M., Johnston, K. J., Spencer, J. H., Walker, R. C., Haschick, A. D., Matveyenko, L. I., Kogan, L. R., Kostenko, V. I., Ronnang, B., Rydbeck, O. E. H., and Moiseev, I. G., Structure and kinematics of H_2O sources in clusters of newly formed OB stars, *Astron. Astrophys.*, 66, 13—29, 1978.

88. Genzel, R., Downes, D., Moran, J. M., Johnston, K. J., Spencer, J. H., Matveyenko, L. I., Kogan, L. R., Kostenko, V. I., Ronnang, B., Haschick, A. D., Reid, M. J., Walker, R. C., Giuffrida, T. S., Burke, B. F., and Moiseev, I. G., H_2O in W51 main: an expanding bubble around a young massive star?, *Astron. Astrophys.*, 78, 239—247, 1979.

89. Downes, D., Genzel, R., Moran, J. M., Johnston, K. J., Matveyenko, L. I., Kogan, L. R., Kostenko, V. I., and Ronnang, B., New VLBI maps of H_2O sources in different stages of evolution, *Astron. Astrophys.*, 79, 233—242, 1979.

90. Reid, M. J., Haschick, A. D., Burke, B. F., Moran, J. M., Johnston, K. J., and Swenson, G. W., The structure of interstellar hydroxyl masers: VLBI synthesis observations of W3(OH), *Astrophys. J.*, 239, 89—111, 1980.

91. Goldreich, P. and Kwan, J., Astrophysical masers. V. Pump mechanisms of H_2O masers, *Astrophys. J.*, 191, 93—100, 1974.

92. Litvak, M. M., Non-Equilibrium processes in interstellar molecules, in *Atoms and Molecules in Astrophysics*, Carson, T. R. and Roberts, M. J., Eds., Academic Press, London, 1972, 201—276.

93. Kegel, W. H., Radiative transport effects in OH maser sources, *Astron. Astrophys. Suppl.*, 38, 131—150, 1979.

94. Litvak, M. M. and Dickinson, D. F., OH infrared stars, *Astrophys. Lett.*, 12, 113—117, 1972.

95. Elitzur, M., Goldreich, P., and Scoville, N., OH-IR stars. II. A model for the 1612 MHz masers, *Astrophys. J.*, 205, 384—396, 1976.

96. Elitzur, M., OH main lines masers I: OH/IR stars, *Astron. Astrophys.*, 62, 305—309, 1978.

97. Litvak, M. M., Infrared pumping of interstellar OH, *Astrophys. J.*, 156, 471—492, 1969.

98. **Deguchi, S.**, Water masers and envelopes of infrared stars, *Pub. Astron. Soc. Jpn.*, 29, 669—681, 1977.
99. **Litvak, M. M.**, Hydroxyl and water masers in protostars, *Science*, 165, 855—861, 1969.
100. **de Jong, T.**, Water masers in a protostellar gas cloud, *Astron. Astrophys.*, 26, 297—313, 1973.
101. **Geballe, T. R. and Townes, C. H.**, Infrared pumping processes for SiO masers, *Astrophys. J.*, 191, L37—41, 1974.
102. **Kwan, J. and Scoville, N.**, Radiative trapping and population inversions of the SiO maser, *Astrophys. J.*, 194, L97—101, 1974.
103. **Forster, J. R., Welch, W. J., and Wright, M. C. H.**, Accurate H_2O source positions in W3, *Astrophys. J.*, 215, L121—125, 1977.
104. **Wilson, W. J. and Barrett, A. H.**, Characteristics of OH emission from infrared stars, *Astron. Astrophys.*, 17, 385—402, 1972.
105. **Bowers, P. F.**, A large scale OH survey at 1612 MHz Part 1. The observations, *Astron. Astrophys. Suppl. Ser.*, 31, 127—45, 1978.
106. **Baud, B., Habing, H. J., Matthews, H. E., and Winnberg, A.**, A systematic search at 1612 MHz for OH maser sources. I. Survey near the galactic center, *Astron. Astrophys. Suppl. Ser.*, 35, 179—192, 1979.
107. **Baud, B., Habing, H. J., Matthews, H. E., and Winnberg, A.**, A systematic search at 1612 MHz for OH maser sources. II. A large scale survey between $10° < \ell < 150°$ and $|b| < 4°.2$, *Astron. Astrophys. Suppl. Ser.*, 36, 193—211, 1979.
108. **Weaver, H. F., Dieter, N. H., and Williams, D. R. W.**, Observations of OH emission in W3, NGC6334, W49, W51, W75, and Ori A, *Astrophys. J. Suppl. Ser.*, 16, 219—274, 1968.
109. **Turner, B. E.**, A survey of OH near the galactic plane, *Astron. Astrophys. Suppl. Ser.*, 37, 1—332, 1979.
110. **Robinson, B. J., Goss, W. M., and Manchester, R. N.**, 18 cm observations of galactic OH emission from longitudes 350° to 50°, *Aust. J. Phys.*, 23, 363—404, 1970.
111. **Goss, W. M., Manchester, R. N., and Robinson, B. J.**, 18 cm observations of galactic OH emission from longitudes 305° to 334°, *Aust. J. Phys.*, 23, 559—573, 1970.
112. **Manchester, R. N., Robinson, B. J., and Goss, W. M.**, 18 cm observations of galactic OH from longitudes 128° to 300°, *Aust. J. Phys.*, 23, 751—775, 1970.
113. **Palmer, P. and Zuckerman, B.**, Observations of interstellar OH at 4660 MHz, *Astrophys. J.*, 161, L199—201, 1970.
114. **Zuckerman, B. and Palmer, P.**, Observations of the $^2\Pi_{1/2}$, J = ½, State of interstellar OH, *Astrophys. J.*, 159, L197—201, 1970.
115. **Zuckerman, B., Yen, J. L., Gottlieb, C. A., and Palmer, P.**, Observations of the $^2\Pi_{3/2}$, J = 5/2 State of Interstellar OH, *Astrophys. J.*, 177, 59—78, 1972.
116. **Knowles, S. H., Caswell, J. L., and Goss, W. M.**, Excited OH Radiation from the $^2\Pi_{3/2}$, J = 5/2 State in Southern HII Regions, *Mon. Not. R. Astron. Soc.*, 175, 537—555, 1976.
117. **Turner, B. E., Palmer, P., Zuckerman, B.**, Detection of the $^2\Pi_{3/2}$, J = 7/2 State of Interstellar OH at a Wavelength of 2.2 cm, *Astrophys. J.*, 160, L125—129, 1970.
118. **Kleinmann, S. G., Dickinson, D. F., and Sargent, D. G.**, Stellar H_2O masers, *Astron. J.*, 83, 1206—1213, 1978.
119. **Dickinson, D. F.**, Water emission from infrared stars, *Astrophys. J. Suppl. Ser.*, 30, 259—271, 1976.
120. **Cheung, A. C., Rank, D. M., Townes, C. H., Thornton, D. D., and Welch, W. J.**, Detection of water in interstellar regions by its microwave radiation, *Nature*, 221, 626—628, 1969.
121. **Genzel, R. and Downes, D.**, H_2O in the galaxy: sites of newly formed OB stars, *Astron. Astrophys. Suppl. Ser.*, 30, 145—168, 1977.
122. **Genzel, R. and Downes, D.**, H_2O in the galaxy. II. Duration of the maser phase and galactic distribution of H_2O sources, *Astron. Astrophys.*, 72, 234—240, 1979.
123. **Genzel, R. and Downes, D.**, H_2O in Orion: outflow of matter in the last stage of star formation, *Astron. Astrophys.*, 61, 117—126, 1977.
124. **Batchelor, R. A., Caswell, J. L., Goss, W. M., Haynes, R. F., Knowles, S. H., and Wellington, K. J.**, Galactic plane H_2O masers — a southern survey, *Aust. J. Phys.*, 33, 139—157, 1980.
125. **Kaufmann, P., Zisk, S., Scalise, E., Schaal, R. E., and Gammon, R. H.**, Survey of water vapor sources in the southern hemisphere, *Astron. J.*, 82, 577—586, 1977.
126. **Blitz, L. and Lada, C. J.**, H_2O masers near OB associations, *Astrophys. J.*, 227, 152—158, 1979.
127. **Rodriguez, L. F., Moran, J. M., Ho, P. T. P., and Gottlieb, E. W.**, Radio observations of water vapor, hydroxyl, silicon monoxide, ammonia, carbon monoxide, and compact HII regions in the vicinities of suspected Herbig-Haro objects, *Astrophys. J.*, 235, 845—865, 1980.
128. **Cesarsky, C. J., Cesarsky, D. A., Churchwell, E., and Lequeux, J. A.**, A survey of H_2O masers in dense interstellar clouds, *Astron. Astrophys.*, 68, 33—39, 1978.
129. **Lo, K. Y., Burke, B. F., and Haschick, A. D.**, H_2O sources in regions of star formation, *Astrophys. J.*, 202, 81—91, 1975.

130. Knapp, G. R. and Morris, M., H₂O maser emission associated with T Tauri and other regions of star formation, *Astrophys. J.*, 206, 713—717, 1976.

131. Scalise, E. and Lepine, J. R. D., Detection of a new transition of SiO in OH/IR stars, *Astron. Astrophys.*, 65, L7—8, 1978.

132. Buhl, D., Snyder, L. E., Lovas, F. J., and Johnson, D. R., Silicon monoxide: detection of maser emission from the second vibrationally excited state, *Astrophys. J.*, 192, L97—100, 1974.

133. Thaddeus, P., Mather, J., Davis, J. H., and Blair, G. N., Detection of the J = 1→0 rotational transition of vibrationally excited silicon monoxide, *Astrophys. J.*, 192, L33—36, 1974.

134. Snyder, L. E. and Buhl, D., Detection of new stellar sources of vibrationally excited silicone monoxide maser emission at 6.96 millimeters, *Astrophys. J.*, 197, 329—340, 1975.

135. Balister, M., Batchelor, R. A., Haynes, R. F., Knowles, S. H., McCulloch, M. G., Robinson, B. J., Wellington, K. J., and Yabsley, D. E., Observations of SiO masers at 43 GHz with the Parkes radio telescope, *Mon. Not. R. Astron. Soc.*, 180, 415—427, 1977.

136. Snyder, L. E., Dickinson, D. F., Brown, L. W., and Buhl, D., Detection of a weak maser emission pedistal associated with the SiO maser, *Astrophys. J.*, 224, 512—519, 1978.

137. Kaifu, N., Buhl, D., and Snyder, L. E., Vibrationally excited SiO: a new type of maser sources in the millimeter wavelength region, *Astrophys. J.*, 195, 359—366, 1975.

138. Davis, J. H., Blair, G. N., Van Till, H., and Thaddeus, P., Vibrationally excited silicon monoxide in the Orion nebula, *Astrophys. J.*, 190, L117—119, 1973.

139. Buxton, R. B., Barrett, A. H., Ho, P. T. P., and Schneps, M. H., Search for methanol masers, *Astron. J.*, 82, 985—988, 1977.

140. Barrett, A. H., Schwartz, P. R., and Waters, J. W., Detection of methyl alcohol in orion at a wavelength of 1 cm, *Astrophys. J.*, 168, L101—106, 1971.

141. Matsakis, D. N., Cheung, A. C., Wright, M. C. H., Askne, J. A., Townes, C. H., and Welch, W. J., An interferometric and multitransitional study of the Orion methanol masers, *Astrophys. J.*, 236, 481—491, 1980.

142. Cahn, J. H. and Elitzur, M., A correlation between SiO and stellar luminosities in long-period variables and the nature of the SiO maser pump mechanism, *Astrophys. J.*, 231, 124—127, 1979.

143. Schwartz, P. R., Waak, J. A., and Bologna, J. M., The relative intensity and velocity of SiO J = 1→0, v = 1 and 2 masers, *Astron. J.*, 84, 1349—1356, 1979.

144. Barrett, A. H., Ho, P. T. P., and Martin, R. N., Time variations and spectral structure of the methanol maser in orion A, *Astrophys. J.*, 198, L119—122, 1975.

145. Chui, M. F., Cheung, A. C., Matsakis, D., Townes, C. H., and Cardiasmenos, A. G., The methanol source in Orion at 1.2 centimeters, *Astrophys. J.*, 187, L19—21, 1974.

146. Moran, J. M., Burke, B. F., Barrett, A. H., Roger, A. E. E., Carter, J. C., Ball, J. A., and Cudaback, D. D., The structure of the OH source in W3, *Astrophys. J.*, 152, L97—101, 1968.

147. Harvey, P. J., Booth, R. S., Davies, R. D., Whittet, D. C. B., and McLaughlin, W., Interferometric observations of the structure of main-line OH sources, *Mon. Not. R. Astron. Soc.*, 169, 545—576, 1974.

148. Wynn-Williams, C. G., Werner, M. W., and Wilson, W. J., Accurate positions of OH sources, *Astrophys. J.*, 187, 41—44, 1974.

149. Moran, J. M., Papadopoulos, G. D., Burke, B. F., Lo, K. Y., Schwartz, P. R., Thacker, D. L., Johnston, K. J., Knowles, S. H., Reisz, A. C., and Shapiro, I. I., Very long baseline interferometric observations of the H₂O sources in W49N, W3(OH), Orion A, and VY Canis Majoris, *Astrophys. J.*, 185, 535—567, 1973.

150. Burke, B. F., Johnston, K. J., Efanov, V. A., Clark, B. G., Kogan, L. R., Kostenko, V. I., Lo, K. Y., Matveyenko, L. I., Moiseev, I. G., Moran, J. M., Knowles, S. H., Papa, D. C., Papadopoulos, G. D., Rogers, A. E. E., and Schwartz, P. R., Observations of maser radio sources with angular resolution of 0.0002 arcseconds, *Sov. Astron.-AJ*, 16, 379—382, 1972.

151. Haschick, A. D., Moran, J. M., Rodriguez, L. F., Burke, B. F., Greenfield, P., and Garcia-Barreto, J. A., Observations of a compact H II region and molecular sources in the vicinity of the Herbig-Haro objects 7-11, *Astrophys. J.*, 237, 26—37.

152. Elmegreen, B. G., Genzel, R., Moran, J. M., Reid, M. J., and Walker, R. C., VLBI observations of the H₂O masers in Sgr B2, *Astrophys. J.*, 241, 1007—1013, 1980.

153. Moran, J. M., Johnston, K. J., Spencer, J. H., and Schwartz, P. R., Observations of the SiO and H₂O masers in Orion A, *Astrophys. J.*, 217, 434—441, 1977.

154. Genzel, R., Moran, J. M., Lane, A. P., Predmore, C. R., Ho, P. T. P., Hansen, S. S., and Reid, M. J., VLBI observations of the SiO maser in Orion, *Astrophys. J.*, 231, L73—76, 1979.

155. Hills, R., Pankonin, V., and Landecker, T. L., Evidence for maser action in the 1.2 cm transitions of methanol in Orion, *Astron. Astrophys.*, 39, 149—153, 1975.

156. Barrett, A. H., Bologna, J. M., Cheung, A. C., Chui, M. F., Ho, P. T. P., Johnston, K. J., Martin, R. N., Matsakis, D., Moran, J. M., and Schwartz, P. R., A lower limit to the angular size of the methanol source in Orion, *Astrophys. Lett.*, 18, 13—14, 1976.

157. **Kukolich, S. G.,** Measurement of the molecular g values in H_2O and D_2O and hyperfine structure in H_2O, *J. Chem. Phys.,* 50, 3751—3755, 1969.

158. **Lees, R. M., Lovas, F. J., Kirchhoff, W. H., and Johnson, D. R.,** Microwave spectra of molecules of astrophysical interest. III. Methanol, *J. Phys. Chem. Ref. Data,* 2, 205—214, 1973.

159. **Wynn-Williams, C. G., Becklin, E. E., and Neugebaurer, G.,** Infrared sources in the H II Region W3, *Mon. Not. R. Astron. Soc.,* 160, 1—14, 1972.

160. **Turner, B. E., Zuckerman, B., Fourikis, N., Morris, M., and Palmer, P.,** Microwave detection of interstellar HDO, *Astrophys. J.,* 198, L125—128, 1975.

161. **Jewell, P. R., Elitzur, M., Webber, J. C., and Snyder, L. E.,** Monitoring of OH maser emission from late-type stars, *Astrophys. J. Suppl. Ser.,* 41, 191—207, 1979.

162. **Destombes, J. L., Marliere, C., Baudry, A., and Brillet, J.,** The exact hyperfine structure and Einstein A-coefficients of OH: consequences in simple astrophysical models, *Astron. Astrophys.,* 60, 55—60, 1977.

163. **Genzel, R., Reid, M. J., Moran, J. M., and Downes, D.,** Proper motions and distances of H_2O maser sources I: the outflow in Orion-KL, *Ap. J.,* 244, 844, 1980.

164. **Dupree, A. K. and Goldberg, L.,** Radio frequency recombination lines, *Annu. Rev. Astron. Astrophys.,* 8, 231—264, 1970.

165. **Mumma, M. J., Buhl, D., Chin, G., Drake, D., Espenak, F., and Kostiuk, T.,** Discovery of natural gain amplification in the 10-micrometer carbon dioxide laser bands on Mars: A natural laser, *Science,* 212, 45—49, 1981.

Section 6
Laser Safety

6. LASER SAFETY

David H. Sliney

The health and safety hazards associated with the use of lasers are often broken into three general categories: laser radiation hazards, electrical hazards, and hazards from associated contaminants. This section is therefore divided into three chapters which emphasize these three types of hazards.

The hazards from laser radiation are confined largely to the eye and, to a smaller extent, the skin. Few serious eye injuries due to lasers have been reported in the 18 years since the appearance of commercial devices. The accident rate is not that low because the ocular exposure limits are overly conservative; they are not. Instead, the possibility of accidental exposure of the eye to a collimated beam is extremely remote if a few rudimentary commonsense precautions are followed.

Electrical hazards so far have proven more serious. At least five laser workers have been electrocuted. Procedures for handling high voltages safely are given in the second chapter of this section.

Hazards from airborne contaminants, such as vaporized target materials, cryogenic fluids, noise, and explosive mixtures are also of concern in some specialized applications and in some research laboratories. These hazards are explained in the third chapter of this section.

6.1 OPTICAL RADIATION HAZARDS*

David H. Sliney

INJURY TO THE EYE AND SKIN

The eye is generally considered the structure of the human body most vulnerable to laser injury. In the visible and near-infrared portions of the spectrum this is the result of the focusing properties of the eye. In the ultraviolet and infrared portions of the spectrum, this susceptibility is more the result of the fact that the cornea and conjunctive (outer layers of the human eye) are virtually the only living tissue exposed directly to the environment. By contrast, the living tissue of the skin is protected by its outer, dead horny layer (the *stratum corneum*). Nevertheless, at sufficient exposure levels, even the skin may suffer laser-induced injury. Figure 6.1.1 illustrates schematically the spectral absorption properties of the eye for three types of exposure.

In addition to the simple absorption properties of the eye, one must also consider the relative sensitivity of different structures to injury, the potential for repair of the injury, and the impact of any injury on the functional state of the eye. Injury may be due to thermal processes, or to photochemical, or other damage mechanisms. The nature of the damage mechanism plays an important role in defining the degree of hazard and whether the effect is additive over a period of time and whether a delayed effect upon function may be possible. These considerations are too involved to discuss at length here. The interested reader is referred to the literature which provides the rationale for laser radiation protection standards.[1-5]

The potential biological effects of optical radiation are often categorized as a function of wavelength as shown in Figure 6.1.2. In the ultraviolet region, the adverse

---

*   The opinions or assertions contained herein are the private views of the author and are not to be construed as official or as reflecting the views of the Department of the Army or Department of Defense.

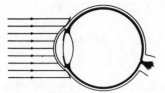

a. Far Infrared and Far Ultraviolet Radiation

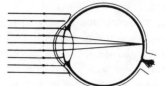

b. Light and Near Infrared Radiation

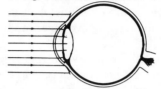

c. Near Ultraviolet Radiation

FIGURE 6.1.1.   Absorption of optical radiation by the human eye. The site of principal absorption of optical radiation varies significantly depending upon the spectral band of the incident radiation.

effects are initiated by a photochemical reaction. As is characteristic for any photochemical reaction, the reciprocity of irradiance (exposure rate) and exposure duration generally holds for durations ranging from microseconds to hours. There is a strong dependence of susceptibility with wavelength. Of those photons which penetrate to living tissue, the photons of shorter wavelength and greater energy are more effective in elliciting an adverse response. The inflammatory responses such as sunburn (erythema) or welder's flash (UV photokeratitis) are generally delayed in onset for a few hours. A photochemical retinal injury (one type of retinal ''burn'') is also possible from lengthy staring into a bright blue-light source.

At longer visible and near-infrared wavelengths retinal injury can only occur from a thermal damage mechanism. Likewise, retinal injury from pulsed light sources is also predominately the result of a rapid temperature rise in the retina. Retinal burns have generally occupied the center of attention of those concerned about laser safety in the laboratory. At wavelengths in the far infrared, only thermal or thermomechanical injury mechanisms are known. Thermal injury of the skin is possible at virtually all wavelengths from pulsed lasers.

To date, lasers that emit principally in the ultraviolet region of the spectrum have been relatively uncommon and have been used principally in the laboratory environment. Extensive experience with the potential hazards of such lasers therefore does not exist. Studies of the biological effects of UV laser radiation are also very limited. Fortunately, there is an extensive literature related to studies of the effects of incoherent UV radiation upon both the eye and the skin. The thresholds of injury for such effects have been used to establish limits of personnel exposure to UV laser radiation. In general, one can say that if injury occurs from a single-pulse exposure of less than 1-

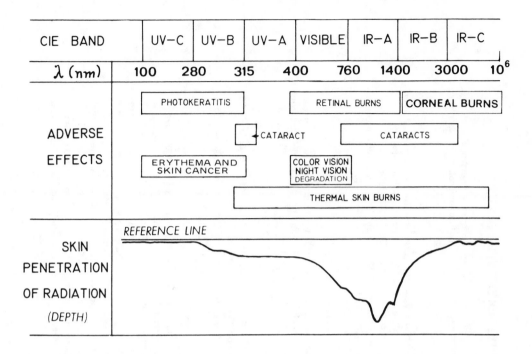

FIGURE 6.1.2.   Optical radiation hazards. The approximate spectral ranges of the adverse biological effects of laser radiation are shown. The spectral bands shown at the top are those of the CIE (the International Commission on Illumination).

μsec duration, the damage mechanism is likely to be thermally initiated. If injury occurs only after prolonged exposure, then the injury mechanism is almost certainly photochemical in nature. Table 6.1.1 summarizes the adverse biological effects of these types of lasers.

Lasers which emit in the visible and near-infrared spectral region (400 to 1400 nm) have created the greatest concern. The focusing properties of the human eye create a very serious hazard when the eye is located within the beam (i.e., intrabeam viewing). If the eye is relaxed — focused at infinity — the retinal irradiance is increased by a factor of approximately 100,000 times. For example, the retinal irradiance in a 10- to 20-μm diameter focal spot would be 10 kW/cm^2 for a corneal irradiance of 100 mW/cm^2 from a He-Ne or Ar laser operating in the visible region. Such a retinal irradiance would create a tiny thermal lesion in the retina. If this tiny lesion were in the central (macular) area of the retina a small blind spot — a scotoma — would be noticed by the individual who looked directly into the beam with his eye focused at infinity. Normally, ones eyes are not relaxed and focused at infinity indoors and this worst-case condition would therefore not always apply. Pupil size also influences the retinal irradiance, but not as dramatically as one might expect. For small pupillary sizes, characteristic of outdoor daylight conditions (i.e., 2 to 3 mm), the eye is almost diffraction limited. For large pupillary diameters the spherical aberrations of the corneal surface produce a larger retinal image, with the result that the retinal irradiance (W/cm^2) is not noticeably increased.

Short-wavelength visible lasers are the most dangerous to the retina. Less energy is absorbed in the retina at longer wavelengths with the result that thresholds of retinal injury are typically a factor of five to ten times greater for the Nd:YAG (1064-nm) wavelength than for visible wavelengths. The maximum permissible exposure (MPE) limits for ocular exposure are adjusted accordingly. For lengthy retinal (greater than 10-sec) exposures the only laser injury mechanism of concern is photochemical. Inas-

Table 6.1.1

SUMMARY OF BIOLOGICAL EFFECTS AND LIMITS FOR DIFFERENT SPECTRAL BANDS

Wavelength region	Representative lasers	Adverse effects	Symptoms	Representative maximal levels of exposure for a collimated laser beam
190 nm to 315 nm Actinic UV (UV-B&C)	ArF, KrF XeF, (repetitively pulsed)	(A)Photokeratoconjunctivitis, normally reversible within 48 hr. (B)Erythema (sunburn), normally reversible within 1 week. (C)Skin cancer. (D)Cataract of lens (300-315 nm).	(A)Uncontrolled blinking, painful sensation of "sand" in eyes, inflammation of the cornea and conjunctiva, delayed in onset 6-12 hr. (B)Reddening of skin (after delay). (C,D)Delayed in onset by many years.	3 mJ/cm² for 200 nm to 302 nm for single or multiple pulse exposure; 0.56 $t^{1/4}$ J/cm² for 1 nsec to 10 sec; 1 J/cm² for single pulse or pulse train at 315 nm
315 nm to 380-400 nm Near UV (UV-A)	He-Cd Argon	Same as above, except that threshold doses are orders of magnitude greater.	Same as above. Some accelerated tanning of skin. Very low risk of skin cancer.	0.56 $t^{1/4}$ J/cm² for 1 nsec to 10 sec; 1 J/cm² for 10 sec to 1000 sec (16 min); 1 mW/cm² for 1000-sec exposure or greater
400 nm to 550 nm Blue	He-Cd Argon	Retinal photochemical injury ("Blue-Light Injury"). Permanent scotoma in severe cases; some recovery after two to six weeks.	Delayed onset of 24-48 hr. Retinal lesion visible by ophthalmoscope; blind-spot (scotoma) also develops in 24-48 hr.	10 mJ/cm² for t = 10 sec to 10,000 sec; 1 μW/cm² for t = 10,000 sec (2.8 hr)
400 nm to 1400 nm IR-A and visible (VIS)	Argon, Ruby He-Ne, Nd:YAG, Ga-As	(A)Retinal thermal injury; permanent scotoma. (B)Thermal burns of the skin.	(A)Rapid appearance of ophthalmoscopically visible retinal lesion with associated blind spot that does not improve. (B)Rapid appearance of red spot (erythema) at site of exposure. Generally healing (sometimes with scar) within 4-6 weeks.	5 μJ/cm² at 1064 nm or 0.5 μJ/cm² for t = 1 nsec to 18 μsec at 400 nm to 700 nm; 1.8 $t^{3/4}$ C_A mJ/cm² for 100 μs to 10 sec; 320 C_A μW/cm² for 700 nm to 1400 nm at t = 1000 sec; 2 $C_A \times 10^{-2}$ J/cm² for 1 nsec to 100 nsec; 1.1 C_A $t^{1/4}$ J/cm² for 100 nsec to 10 sec; 200 C_A mW/cm² for t = 10 sec (small spots)

| 1400 nm to 1 mm | HF, DF, Er, Ho, CO, CO_2 | Corneal thermal burns, skin burns. | White flaky patch from pulsed laser near threshold; erythema from CW exposure to deep burn at much higher levels. Corneal vacuolization. | 10 mJ/cm² for 1 nsec to 100 nsec (exception: @ 1.54 µm, 1 J/cm², 1-100 nsec)
0.56 $t^{1/4}$ J/cm² for 100 nsec to 10 sec
100 mW/cm² for t > 10 sec |

Note: The expressions for exposure limits with time t raised to a fractional exponent of ¼ or ¾ require the use of seconds for t.

Ultrashort Pulses: No exposure limits for subnanosecond pulses have been set; however, a conservative approach recommended by the IEC would require the exposure in terms of irradiance (W/cm²) not to exceed the irradiance limit for a 1-nsec pulse. There is some limited experimental evidence to support that approach.

Extended Sources: The limits given above for wavelengths between 400 and 1400 nm apply strictly only to intrabeam (direct) exposure for the eye. More lenient limits not shown apply to laser radiation exposure of the eye from extended source, scattered radiation.

IR-B NS IR-C

much as short-wavelength (blue) light is responsible for this type of injury, and red and infrared wavelengths are almost totally ineffective in producing that form of injury, retinal injury from lengthy or repeated exposures to visible lasers is only of real concern for lasers such as He-Cd and Ar lasers — not from He-Ne or Ga-As lasers. This is reflected in the MPEs for lengthy (1 hr exposures) to be presented later in this chapter.

Still another type of potential retinal hazard has been reported in recent years.[6] Very prolonged staring at visible laser speckle patterns may have adverse effects upon visual function. Although this effect has only been demonstrated in monkeys and has not been detected in humans, the potential may still exist and caution should be exercised when viewing laser display patterns for long periods (hours) wherein the speckle is noticeable.

Near-infrared radiation (IR-A 760 nm to 1400 nm) reaches the retina, and as previously noted, can cause retinal thermal burns. Some of this radiation is absorbed in the anterior structures of the eye, such as the cornea, lens, aqueous humor, and vitreous humor. Although thermal injury to these structures — particularly to the lens — is theoretically possible and has been of concern, present evidence supports the argument that the retina is still the most vulnerable ocular structure. Thermal injury wben it occurs is almost completely evident within 5 min of the laser exposure.

Middle Infrared (IR-B: 1400 nm to 3000 nm) radiation does not reach the retina. It is absorbed principally in the cornea and aqueous. Far Infrared (IR-C: 3000 nm to 1 mm) is absorbed completely in the cornea. Corneal thermal injury can occur from pulsed lasers. It is unlikely that thermal injury would occur from prolonged exposure from CW lasers since the pain from heating would be severe, and the blink reflex would protect the cornea. Prolonged exposure to IR-B and IR-C radiation at levels below the pain threshold does tend to accelerate the drying of the cornea and can lead to irritation. See also Table 6.1.2.

THE RISK OF EXPOSURE

Although lasers are now used widely throughout industry, the potential for hazardous exposure to laser radiation or other ancillary hazards is still greatest for the laser research worker and laser servicing personnel. This is primarily the result of the need for flexibility in the arrangement of laser systems used in research and the requirements for unenclosed, high-power laser beams in some of these specialized operations.

In the research laboratory, administrative safety measures are relied upon instead of engineering control measures. Likewise, protective clothing and protective eyewear have found their greatest usage in the laboratory environment. By contrast, the use of medium-power and high-power lasers in most industrial production lines rely on the use of engineering control measures (e.g., enclosures) to prevent exposure of workers in the vicinity of the equipment. For maximum flexibility, few research scientists and engineers consider the installation of permanent beam enclosures, interlocks and similar fixtures, since optical beam-paths can change daily, if not hourly. Many specialized procedures or tests cannot be accomplished within enclosures. Under these laser operating conditions, eye protection is normally worn, despite the associated reduction in visual capabilities, and possible discomfort.

These generalities do not apply to many of the laser devices used in commercial instruments. For example, the lasers employed in specialized holographic instruments, laser particle sizing instruments, and Raman spectrometers are generally well enclosed and there is no risk to the user. These products, like other complete laser systems, normally include sufficient engineering controls to preclude any risk.

In only a few cases, primarily in the high-energy laser laboratory and in materials-

Table 6.1.2
INTERNATIONAL GUIDELINES OR STANDARDS RELATING TO LASERS

Source	Standard or guide
International Electrotechnical Commission (IEC), Geneva	"Radiation Safety of Laser Products and Equipment" Still in Committee TC-76 as of January 1981. Classification and performance standard similar to BRH regulation, and exposure limits similar to ANSI; applies to both manufacturer and user. Committee Chairman: G. Wilkening, U.S.A.
International Radiological Protection Association (IRPA)	"Overviews on Non-Ionizing Radiation," April 1977 published by U.S. Dept. of Health, Education, and Welfare, Rockville; a general information booklet explaining the biological effects of all non-ionizing radiations, including radio-frequency radiation. Chairman: H. Jammet, France.
International Standards Organization (ISO), Geneva	"Filters and Eye Protectors Against Laser Radiation," ISO standard expected approval by 1980; density limits, optical quality of lenses, standard markings. Chairman of ISO/TC 94/SC 6/WG 3: E. Sutter, West Germany.
World Health Organization (WHO) European Office, Copenhagen	"Optical Radiation, with Particular Reference to Lasers", a chapter in a manual for nonionizing radiation, Report ICP/CEP 803, 1977. Authors: Goldman, L., Rockwell, R. J., Michaelson, S. M., Sliney, D. H., Tengroth, B. M., and Wolbarsht, M. based upon a working group meeting held in Dublin in October 1977. Informative, general guidelines on laser bioeffects, hazard evaluation and controls; gives exposure limits that are the same as ANSI Z-136.1, 1976; Available free from WHO European Office.

(From Sliney, D. H. and Wolbarsht, M. L., *Safety with Lasers and Other Optical Source,* Plenum Press, New York, 1980. With permission.)

working, is there a large probability of accidental exposure. If a diffuse reflection is so bright that it is hazardous, a viewer is susceptible to injury if the source is close enough to be viewed as an extended source rather than a point source, since the probability of one's eye being positioned within a narrow beam of light at the right moment and of the eye being relaxed and oriented along the beam axis obviously is infinitesimal compared with the high probability of viewing a diffuse reflection. A second case of susceptible exposure occurs if the nature of the laser operation requires an individual to fix his eye upon a source of direct or reflected laser radiation. This is the case with many materials-working operations which require an operator to use his central vision constantly to view — with the fovea — the source of reflected laser radiation.

Ruby and neodymium laser irradiances capable of causing surface ablation in any material are almost always above safe exposure limits for viewing diffuse reflections. Furthermore, these irradiances are orders of magnitude above limits for viewing a specular, or mirrorlike, reflection. There is risk, therefore, in viewing any materials-processing operation with a laser emitting at 400 to 1,400 nm at a distance sufficiently close to observe a diffuse reflection as an extended source.

HAZARD EVALUATION AND LASER CLASSIFICATION

Three major aspects must be considered in evaluating any type of laser hazard. These relate to:

1.   The laser equipment itself, and how hazardous it may be
2.   The environment in which the laser system is used, i.e., indoors, outdoors, etc.
3.   The people who are potentially exposed to laser radiation and ancillary hazards, and the people who operate the laser

The first aspect is dealt with primarily by using a laser hazard classification scheme that is now almost universally accepted. The latter two aspects may have to be considered if the laser is not in a fool-proof enclosure. Such an evaluation requires some experience and judgment which must be exercised by the user or a Laser Safety Officer (LSO). Safety standards for lasers exist and aid considerably in hazard analysis and controls. The most highly regarded standard was first produced in 1973 and has since been revised: American National Standards Institute, "Safe Use of Lasers", standard Z-136.1 (1980).[7] A new edition of the ANSI standard was issued in 1980. Several other organizations have followed suit with safety standards which are compatible: the American Conference of Governmental Industrial Hygienists (in 1973 and 1976),[8] the World Health Organization,[9] and the International Electrotechnical Commission.[10] These are summarized in Table 6.1.1. More importantly, the U.S. Department of Health, and Human Services, Food and Drug Administration's Bureau of Radiological Health (BRH) now regulates the manufacture and sale of laser products in the U.S.A. By this regulation the manufacturer must classify the laser product and this greatly aids the user in determining what control measures, if any, are necessary.

The basic concepts of the classification schemes of ANSI, ACGIH, WHO, IEC, and BRH are as follows:

- Class I laser products are considered essentially safe, and are generally enclosed laser systems which do not emit hazardous levels.
- Class II laser products are limited to visible lasers which are safe for momentary viewing, but should not be stared into continuously unless the exposure is within the recommended ocular exposure limits (EL's); the dazzle of the brilliant visible light source would normally preclude staring into the source.
- Class III laser products are not safe even for momentary viewing and procedural controls and protective equipment are normally required in their use.
- Class IV laser products are normally considered much more hazardous than Class III products since they may represent a significant fire hazard or skin hazard and may also product hazardous diffuse reflections. If the laser operates between 400 nm and 1400 nm, hazardous diffuse reflections are particularly dangerous since the probability of a hazardous retinal exposure is far greater than if the exposure were only possible from specular reflections.

General concepts of laser safety are dealt with quite adequately in the aforementioned standards and elsewhere.[1] The BRH emission limits (upper limits) for each class are given in Tables 6.1.3 through 6.1.7. Table 6.1.8 summarizes the hazard classification of many common lasers. Table 6.1.9 summarizes the performance requirements of BRH and the ANSI standard as a function of hazard classification.

In contrast to laser classification limits, ELs are used far less frequently. This results from the fact that in most laser applications no personnel are intentionally exposed. Table 6.1.10 provides the ELs for some of the most common laser exposure conditions.

Table 6.1.3
CLASS I ACCESSIBLE EMISSION LIMITS FOR LASER RADIATION

Wavelength (nm)	Emission duration (sec)	Class I - Accessible emission limits
> 250 but ≤ 400	≤ 3.0×10⁴	$2.4×10^{-5}k_1k_2$ J[a]
	> 3.0×10⁴	$8.0×10^{-10}k_1k_2r$ J
	> 1.0×10⁻⁹ to 2.0×10⁻⁵	$2.0×10^{-7}k_1k_2$ J
	> 2.0×10⁻⁵ to 1.0×10¹	$7.0×10^{-4}k_1k_2r^{3/4}$ J
	> 1.0×10¹ to 1.0×10⁴	$3.9×10^{-3}k_1k_2$ J
> 400 but ≤ 1400	> 1.0×10⁴	$3.9×10^{-7}k_1k_2r$ J
	OR	
	> 1.0×10⁻⁹ to 1.0×10¹	$10k_1k_2r^{1/3}$ Jcm⁻²sr⁻¹
	> 1.0×10¹ to 1.0×10⁴	$20k_1k_2$ Jcm⁻²sr⁻¹
	> 1.0×10⁴	$2.0×10^{-3}k_1k_2r$ Jcm⁻²sr⁻¹
> 1400 but ≤ 13000	> 1.0×10⁻⁹ to 1.0×10⁻⁷	$7.9×10^{-5}k_1k_2$ J
	> 1.0×10⁻⁷ to 1.0×10¹	$4.4×10^{-3}k_1k_2^{1/4}$ J
	> 1.0×10¹	$7.9×10^{-4}k_1k_2r$ J

[a] Class I accessible emission limits for the wavelength range of greater than 250 nm but less than or equal to 400 nm shall not exceed the Class I accessible emission limits for the wavelength range of greater than 1400 nm but less than or equal to 13000 nm with a k_1 and k_2 of 1.0 for comparable sampling intervals.

Table 6.1.4
CLASS II ACCESSIBLE EMISSION LIMITS FOR LASER RADIATION

Wavelength (nm)	Emission duration (sec)	Class II — Accessible emission limits
> 400 but ≤ 710	> 2.5 × 10⁻¹	$1.0 × 10^{-3}k_1k_2r$ J

There is typically a factor of three to ten between the EL and the level which produces a clearly visible reaction in tissue. The limits were designed to be sufficiently low to preclude adverse functional or any delayed effects.[1-5] The limits cannot be considered as fine lines between safe and dangerous conditions.

SAFETY MEASURES

Enclosed Lasers and Enclosed Facilities

Clearly the soundest and safest laser applications are those where the laser and hazardous beam paths are completely enclosed. As previously mentioned, this approach may not always be possible in all research facilities. When a laser is used in the open and the laser is Class III or Class IV, the laser operation should be carried out in a completely controlled (closed-off) facility. This is not to say that the facility must be airtight and "light tight". As long as there is no direct beam path outside of the facility there should be no problem. The use of diffuse baffles, door interlocks, and window shades are called for in most situations. It is unnecessary to seal light leaks beneath doors or at vents well above eye level.

Table 6.1.5
CLASS III ACCESSIBLE EMISSION LIMITS FOR LASER RADIATION

Wavelength (nm)	Emission duration (sec)	Class III — Accessible emission limits
>250 but	$\leqslant 2.5 \times 10^{-1}$	$3.8 \times 10^{-4} k_1 k_2$ J
$\leqslant 400$	$>2.5 \times 10^{-1}$	$1.5 \times 10^{-3} k_1 k_2 r$ J
>400 but	$>1.0 \times 10^{-9}$ to 2.5×10^{-1}	$10 k_1 k_2 r^{1/3}$ J cm^{-2}
$\leqslant 1400$		to a maximum value of 10 J cm^{-2}
	$>2.5 \times 10^{-1}$	$5.0 \times 10^{-1} r$ J
>1400 but	$>1.0 \times 10^{-9}$ to 1.0×10^{1}	10 J cm^{-2}
$\leqslant 13000$	$>1.0 \times 10^{1}$	$5.0 \times 10^{-1} r$ J

Table 6.1.6
VALUES OF WAVELENGTH-DEPENDENT CORRECTION FACTORS k_1 AND k_2

Wavelength (nm)	k_1			k_2		
250 to 302 4	10			10		
>302 4 to 315	$_{10}[(\lambda - 302\ 4)/5]$			10		
>315 to 400	3300			10		
>400 to 700	10			10		
>700 to 800	$_{10}[(\lambda-700)/515]$	if t > [10100/λ-699] then $k_2 = 10$		if 10100/(λ−699) < r $\leqslant 10^4$ then $k_2 = + (\lambda-699)/10100$	if r > 10^4 then $k_2 = (\lambda-699)/1.01$	
>800 to 1060	$_{10}[(\lambda-700)/515]$	if r $\leqslant$ 100 then $k_2 = 10$ 10		if 100 < r $\leqslant 10^4$ then $k_2 = t/100$	if r > 10^4 then $k_2 = 100$	
>1050 to 1400	5.0					
>1400 to 1535	10			10		
>1535 to 1545	r $\leqslant 10^{-7}$ $k_1 = 100.0$ r > 10^{-7} $k_1 = 10$			10		
>1545 to 13000	10			10		

Note: The variables in the expressions are the magnitudes of the sampling interval (r), in units of seconds, and the wavelength (λ), in units of nanometers.

Open Beam Paths for Laboratory Experiments

Several general safety procedures are particularly useful when working with open beams. Open beams are necessary where optical elements along the beam path must be continuously adjusted. The beam paths should be enclosed where feasible. A variable output device (polarizer or variable beam splitter) can be installed as close to the beam port as possible to limit power output to the minimum necessary for the task. In essence, the laser safety filter is placed over the beam instead of over the eyes. If the beam remains static during the experiment it may not be necessary to wear eye protection, particularly if it is not necessary to view the beam path once the experimental setup has been aligned.

A common complaint of research workers who have used argon lasers is that the commercially available eye protection filters have such enormous optical densities as to make the beam not only safe, but also, invisible. In many of these instances, the

Table 6.1.7
SELECTED NUMERICAL SOLUTIONS FOR k_1 AND k_2

Wavelength (nm)	k_1	r≤100	r = 300	r = 1000	r = 3000	r≥10,000
250	1.0					
300	1.0					
302	1.0					
303	1.32					
304	2.09					
305	3.31					
306	5.25					
307	8.32					
308	13.2					
309	20.9					
310	33.1			1.0		
311	52.5					
312	83.2					
313	132.0					
314	209.0					
315	330.0					
400	330.0					
401	1.0					
500	1.0					
600	1.0					
700	1.0					
710	1.05	1	1	1.1	3.3	11.0
720	1.09	1	1	2.1	5.3	21.0
730	1.14	1	1	3.1	9.3	31.0
740	1.20	1	1.2	4.1	12.0	41.0
750	1.25	1	1.5	5.0	15.0	50.0
760	1.31	1	1.8	6.0	18.0	60.0
770	1.37	1	2.1	7.0	21.0	70.0
780	1.43	1	2.4	8.0	24.0	80.0
790	1.50	1	2.7	9.0	27.0	90.0
800	1.56	1	3.0	10.0	30.0	100.0
850	1.95	1	3.0	10.0	30.0	100.0
900	2.44	1	3.0	10.0	30.0	100.0
950	3.05	1	3.0	10.0	30.0	100.0
1000	3.82	1	3.0	10.0	30.0	100.0
1050	4.78	1	3.0	10.0	30.0	100.0
1060	5.00	1	3.0	10.0	30.0	100.0
1100	5.00	1	3.0	10.0	30.0	100.0
1400	5.00	1	3.0	10.0	30.0	100.0
1500	1.0					
1540	100.0[a]			1.0		
1600	1.0					
13000	1.0					

Note: The variable (r) is the magnitude of the sampling interval in units of seconds.

[a] The factor $k_1 = 100.0$ when $r \le 10^{-7}$ and $k_1 = 1.0$ when $r > 10^{-7}$.

installation of a fixed or variable attenuator to reduce the laser beam power to 1 mW would reduce the risk of eye injury from momentary viewing, while still permitting visibility of the beam. An alternative method for visualizing the position of the beam through eye protectors makes use of a fluorescent card. The longer-wavelength fluorescence of the card can be seen through the protective filter. Image converter monoculars or goggles may also permit safe viewing of the beam's location and may even render normally invisible ultraviolet or near-infrared beams visible. The disadvantages

Table 6.1.8

CLASSIFICATION FOR NONENCLOSED, NONSCANNING LASERS[a]

Type of laser device	Wavelength	Temporal output mode	BRH Standard: 21 CFR 1040			ANSI standard Z 136.1 and ACGIH guide
			k_1	k_2	Classification	
Argon ion (Ar³⁺)	488 nm and 514.5 nm	CW	1.0	1.0	III if $\Phi \leq 0.5$ W IV if $\Phi > 0.5$ W	Same as BRH
Carbon-dioxide (CO_2)	10.6 μm	CW	1.0	1.0	III if $\Phi \leq 0.5$ W IV if $\Phi > 0.5$ W	Same as BRH
Carbon-monoxide (CO)	5 μm	CW	1.0	1.0	III if $\Phi \leq 0.5$ W IV if $\Phi > 0.5$ W	Same as BRH
Dye laser, argon-pumped	400 nm to 780 nm	CW	1.0	1.0	III if $\Phi \leq 0.5$ W IV if $\Phi > 0.5$ W	Same as BRH
Dye laser, xenon-flashlamp pumped	450 nm to 780 nm	1 μsec	1.0	1.0	III if $H_o \leq 0.1$ J/cm² IV if $H_o > 0.1$ J/cm²	III if $H_o \leq 0.31$ J/cm² IV if $H_o > 0.31$ J/cm²
Erbium (ER³⁺)	1540 nm or 1640 nm	20 nsec	100	100	III if $H_o \leq 10$ J/cm² IV if $H_o > 10$ J/cm²	Same as BRH
Gallium-aluminum-arsenide (GaAlAs)	850 nm	100 nsec @ 10 Hz to 100 kHz	1.96	100ᶜ	I if $\Phi_a \leq 0.076$ mW and $Q_p \leq 0.4$ μJ III if $\Phi_a > 0.076$ mW or $Q_p > 0.4$ μJ	I if $\Phi_a \leq 0.63$ mW and $Q_p \leq 0.4 \, C_p$ μJ III if $\Phi_a > 0.63$ mW or $Q_p > 0.4 \, C_p$ μJ
Gallium-arsenide (GaAs)[b]	905 nm	100 nsec @ 10 Hz to 100 kHz	2.5	100ᶜ	I if $\Phi_a \leq 0.0975$ mW and $Q_p \leq 0.5$ μJ III if $\Phi_a > 0.0975$ mW or $Q_p > 0.5$ μJ	I if $\Phi_a \leq 0.8$ mW and $Q_p \leq 0.5 \, C_p$ μJ III if $\Phi_a > 0.8$ mW or $Q_p > 0.5 \, C_p$ μJ
Hydrogen-cyanide (HCN)	115 μm	CW	N/A	N/A	Not defined by BRH	III if $\Phi \leq 0.5$ W IV if $\Phi > 0.5$ W
Helium-cadmium (HeCd)	325 nm	CW	330	1.0	III if $\Phi \leq 0.5$ W and $\Phi > 0.26$ μW	III if $\Phi \leq 0.5$ W and $\Phi > 8.0$ μW
Helium-cadmium (HeCd)	441.6 nm	CW	1.0	1.0	II if $\Phi \leq 1.0$ mW III if $\Phi > 1.0$ mW	Same as BRH
Helium-neon (HeNe)	632.8 nm	CW	1.0	1.0	II if $\Phi \leq 1.0$ mW III if $\Phi > 1.0$ mW	Same as BRH
Holmium (Ho³⁺)	850 nm [or 2.06 μm]	100 nsec	1.0	1.0	III if $H_o \leq 0.91$ J/cm² [10 J/cm²] IV if $H_o > 0.91$ J/cm² [10 J/cm²]	III if $H_o \leq 93$ mJ/cm² [10 J/cm²] IV if $H_o > 93$ mJ/cm² [10 J/cm²]

			k1	k2	BRH	ANSI
Krypton ion (Kr⁺⁺)	568 nm and 647 nm	CW	1.0		III if $\Phi \leq 0.5$ W IV if $\Phi > 0.5$ W	Same as BRH
Neodymium long pulse (Nd³⁺)	1060 nm to 1064 nm	1 msec	1.0		III if $H_o \leq 5.0$ J/cm² I IV if $H_o > 5.0$ J/cm²	III if $H_o \leq 10$ J/cm² IV if $H_o > 10$ J/cm²
Neodymium, Q-switched (Nd³⁺)	1060 nm to 1064 nm	20 nsec	1.0		III if $H_o \leq 0.136$ J/cm² IV if $H_o > 0.136$ J/cm²	III if $H_o \leq 0.43$ J/cm² IV if $H_o > 0.43$ J/cm²
Neodymium:YAG (Nd:YAG)	1064 nm	CW	1.0	100ᶜ	III if $\Phi \leq 0.5$ W IV if $\Phi > 0.5$ W	Same as BRH
Ruby, long pulse (Al₂O₃:Cr³⁺)	694.3 nm	1 msec	1.0		III if $H_o \leq 1.0$ J/cm² IV if $H_o > 1.0$ J/cm²	III if $H_o \leq 3.1$ J/cm² IV if $H_o > 3.1$ J/cm²
Ruby, Q-switched (Al₂O₃:Cr³⁺)	694.3 nm	20 nsec	1.0		III if $H_o \leq 0.027$ J/cm² IV if $H_o > 0.027$ J/cm²	III if $H_o \leq 0.084$ J/cm² IV if $H_o > 0.084$ J/cm²
TEA Laser, CO₂ [or CO]	10.6 μm [or 5 μm]	100 nsec	1.0		III if $H_o \leq 10$ J/cm² IV if $H_o > 10$ J/cm²	Same as BRH
Water vapor (H₂O) [or HCN]	118 μm [or 337 μm]	CW	N/A	N/A	N/A	I if $\Phi \leq 80$ mW; III if $\Phi \leq 0.5$ W

a   The assumption is made that the laser's output is typical — within the expected laser power output for present commercial lasers and would normally not fall into Class 1.

b   Assumes a point source which is generally true for a single diode.

c $k_2 = 100$ only for sampling intervals greater than 10,000s (i.e., 2.8 hr).

(From Sliney, D. H. and Wolbarsht, M. L., *Safety with Lasers and Other Optical Sources*, Plenum Press, New York, 1980. With permission.)

Table 6.1.9
LASER PRODUCT SAFETY REQUIREMENTS IN THE ANSI AND BRII STANDARDS

Performance requirement or recommendation	ANSI Class I	BRII Class I	ANSI Class 2a	BRII Class IIa	ANSI Class 2	BRII Class II	ANSI Class 3a	BRII Class IIIa	ANSI Class 3b	BRII Class IIIb	ANSI Class 4	BRII Class IV
Classification label or warning label.	No	Yes	Yes	Yes	Yes	Yes	Yes	Yes	Yes	Yes	Yes	Yes
Protective housing to limit accessible radiation to lowest achievable class required for application.	a	Yes	a	Yes	a	Yes	Yes	Yes	Yes	Yes	Yes	Yes
Safety interlocks for protective housing to assure retention of hazard classification if cover(s) are removed.	a	Yes	a	Yes	a	Yes	a	Yes	Yes	Yes	Yes	Yes
Scanning safeguard for scanning lasers to maintain class limit in event of scan failure.	b	Yes	b	Yes	b	Yes	b	yes	b	Yes	b	No
Remote-control connector to permit use of door or ancillary safety interlocks.	No	No	No	No	No	No	No	Yes	No	Yes	c	Yes
Key actuated master control so that laser is inoperable when key is removed.	No	No	No	No	No	Yes	No	Yes	c	Yes	Yes	Yes
Laser-radiation emission indicator with no delay.	No	No	No	No	No	No	No	Yes	c	Yes	c	Yes
Laser-radiation emission indicator with a delay to warn.	No	No	No	No	No	No	No	Yes	c	Yes	c	Yes
Permanently attached attenuator to reduce to Class I	No	No	No	No	No	Yes	c	Yes	c	Yes	c	Yes
Controls located to reduce the chance of operator exposure.	No	No	No	Yes	No	Yes	No	Yes	No	Yes	No	Yes
Protective viewing optics so that exposure is less than Class I.	a	Yes	a	Yes	a	Yes	Yes	Yes	Yes	Yes	Yes	Yes

Safety information must be furnished with the laser system.   No   Yes   Yes   Yes   Yes   Yes   Yes   Yes   Yes   Yes   Yes   Yes

a Applicable if higher class is enclosed.

b Applicable only to entertainment applications, i.e., laser light shows.

c Advisory (i.e., a "should" is used by ANSI for advisory recommendations; a "shall" is used to indicate a requirement of the standard).

(From Sliney, D. H. and Wolbarsht, M. L., *Safety with Lasers and Other Optical Source*, Plenum Press, New York, 1980. With permission.)

Table 6.1.10A
INTRABEAM ELs WHICH ARE APPLICABLE TO MANY COMMON CW LASERS FOR EYE AND SKIN EXPOSURE TO LASER RADIATION

Laser type	Primary wavelength(s) (nm)	Exposure limit valve	
		Eye	Skin
Helium-Cadmium	441.6	(a) 2.5 m W·cm⁻² for 0.25 sec	0.2 W·cm⁻²
Helium-Neon	632.8	(b) 10 C_B mJ·cm⁻² for 10–10,000 sec	
Argon	488.514.5 }		
Krypton	647.1		
Frq. Doubled ND:YAG	532	(c) $C_B\mu$W·cm⁻² for > 10,000 sec	0.2 W·cm⁻²
Neodymium:YAG	1064	1.6 mW·cm⁻² for t > 100 sec	1.0 W·cm⁻²
Gallium-Arsenide at room temp.	905	0.8 mW·cm⁻² for t > 100 sec	0.5 W·cm⁻²
Helium-Cadmium	325 }	(a) 1 J·cm⁻² for 10 to 1000 sec	(a) 1 J·cm⁻² for 10 to 1000 sec
Nitrogen	337.1 }	(b) 1 mW·cm⁻² for t > 1000 sec	(b) 1 mW·cm⁻² for t > 1000 sec
Carbon-dioxide (and other lasers 1.4 μm to 1000 μm)	10.6 μm (10,600 nm)	0.1 W·cm⁻² for t > 10 sec	0.1 W·cm⁻² for t > 10 sec

Table 6.1.10B
INTRABEAM ELs WHICH ARE APPLICABLE TO MANY PULSED LASERS FOR EYE AND SKIN EXPOSURE TO LASER RADIATION

Laser type	Primary wavelength(s) (nm)	Pulse duration	Exposure limit value	
			Eye	Skin
Normal-pulsed ruby	694.3	~1 msec	10⁻⁵ J·cm⁻²	0.2 J·cm⁻²
Q-switched ruby	694.3	5–100 nsec	5×10⁻⁷ J·cm	0.02 J·cm⁻²
Rhodamine 6G dye laser	~500–700	0.5–20 μsec	5×10⁻⁷ J·cm⁻²	0.03–0.07 J·cm⁻²
Normal pulsed neodymium	1064	~1 msec	5×10⁻⁸ J·cm⁻²	1.0 J·cm⁻²
Q-switched neodymium	1064	5–100 nsec	5×10⁻⁶ J·cm⁻²	0.1 J·cm⁻²

Note: Indicates (a), (b), and (c) apply to each laser type in the grouping.

of image converters — or for that matter, of closed circuit television systems — are: some loss of detail, a limited field-of-view, and a monochrome presentation. Vidicons and image converter tubes may also be damaged by direct laser exposure at levels comparable to retinal damage levels, and this factor must be considered prior to use where direct intrabeam illumination of the tube is expected. Such television and converter systems are becoming more common where tunable dye lasers are in use. On the other hand, the use of a single dye does not always require such an approach since each dye has a limited spectral range.

Controlled Entry

For open-beam Class IV laser operations (and even some higher power Class III operations) it is often customary to limit entry to only those persons actually engaged in laser research during the laser firing. When this approach is followed the laboratory is termed a "controlled area".[7] This approach is advisable where eye protection is not the principal control measure and where laboratory personnel require instruction in basic laser safety procedures, and are well familiarized with the experimental arrangement and potentially dangerous conditions. At one time this practice was more common when such a restriction was necessary to limit those requiring eye examinations. In any case, a laboratory director is often receptive to such a policy if for no other reason than to keep away the curious.

Entry restrictions are administrative. They are enforced through the use of signs, special door latches, and interlocks. Class IV laser installations should have door interlocks which are connected to the remote control connector required by the BRH Laser Product Performance Standard. Such a system precludes accidental injury of unprotected personnel entering the laser facility during laser operation; the laser beam is either blocked or shut off when the door is opened. This may be undesirable in specialized laser experiments. Entry control and barriers between the laser and the entrance may be alternative control methods.

REFERENCES

1. Sliney, D. H. and Wolbarsht, M. L., *Safety with Lasers and Other Optical Sources,* Plenum Press, New York, 1980.
2. Sliney, D. H., The development of laser safety criteria — biological considerations, in *Laser Applications in Medicine and Biology,* Vol. I, Wolbarsht, M. L., Ed., Plenum Press, New York, 1971, 163.
3. Sliney, D. H. and Freasier, B. C., The evaluation of optical radiation hazards, *Appl. Opt.,* 12, 1, 1973.
4. Wolbarsht, M. L. and Sliney, D. H., The formulation of protection standards for lasers, in *Laser Applications in Medicine and Biology,* Vol. II, Wolbarsht, M. L., Ed., Plenum Press, New York, 1974, 125.
5. Clarke, A. M., Ocular hazards from lasers and other optical sources, *Crit. Rev. Environ. Control,* 1, 307, 1970.
6. Zwick, H. and Jenkins, D., Effects of coherent light on retinal receptor processes of pseudemys, *Invest. Ophthal. Vis. Sci.,* 17(Suppl.), 172, 1978.
7. American National Standards Institute (ANSI), "The Safe Use of Lasers," ANSI Standard Z-136.1 (1980), ANSI, New York, 1980.
8. American Conference of Governmental Industrial Hygienists (ACGIH), "Threshold Limit Values for Chemical Substances and Physical Agents in the Workroom Environment with Intended Changes for 1979," and "A Guide for Control of Laser Hazards (1976)," ACGIH, Cincinnati, 1979.

9. Goldman, L., Rockwell, R. J., Michaelson, S. M., Sliney, D. H., Tengroth, B. M., and Wolbarsht, M. L., "Optical Radiation, with Particular Reverence to Lasers," Rep. ICP/CEP 803, World Health Organization Regional Office for Europe, Copenhagen, 1977.
10. U.S. Department of Health, Education and Welfare, *Code of Federal Regulations,* Title 21, Chapter 1, subchapter J, part 1040, Performance Standards for Light-Emitting Products, Laser Products, Washington, D.C., 1979.

6.2 ELECTRICAL HAZARDS FROM LASER POWER SUPPLIES*

James K. Franks

INTRODUCTION

The high power or high energy electrical sources used to excite lasers are not uniquely hazardous. The same potential for lethal injury exists with any similar type of power supply. The laser, in general, is not a very efficient device so that modest amounts of optical power require much larger expenditures of excitation energy. The majority of laser power supplies, under appropriate shock circuit parameters, are capable of lethal injury.

Since 1970 four persons have been electrocuted in the United States while working with laser power supplies[1] and at least one person in Europe. These deaths in the U.S. account for much less than 0.25% of the total number of electrical fatalities occurring annually. Approximately 25% of electrical fatalities each year occur from natural lightning.

Commercial laser devices, unless defective, do not present an electrical hazard unless interlocks are defeated and covers are removed. The four deaths from laser power supplies in the U.S. occurred in a research laboratory environment. Here is where laser power supplies are built and "tinkered" with and where new lasers are "bread-boarded".

PHYSIOLOGICAL EFFECTS

General

Most of the studies on the physiological effect of electric current were performed during the 1930s and 1940s at the University of California at Berkeley, Columbia University, and Johns Hopkins University.[2-4]

The studies used to determine a "let-go" threshold were performed on human volunteers. Those studies on thresholds for ventricular fibrillation or uncoordinated heart action were performed on other mammals. In these studies it was determined that several factors establish the severity of injury associated with electric shock. These factors are

1. Current magnitude
2. Current path
3. Duration of shock or discharge
4. Body weight/general physical development
5. Frequency if alternating current

Current Magnitude Effects

Current magnitude, for a given applied voltage, is determined by the shock circuit resistance. Shock circuit resistance is the sum of two contact resistances and internal body resistance. Internal body resistance varies between 200 and 1000 ohms depending on the current path and is usually small compared to the total resistance in a shock circuit.

---

* The opinions or assertions contained herein are the private views of the author and are not to be construed as official or as reflecting the views of the Department of the Army or Department of Defense.

Table 6.2.1
TYPICAL RESISTANCE VALUES FOR
SKIN CONTACT SITUATIONS

| | Resistance (Ω) | |
Situations	Dry	Wet
Finger touch	0.13 M-1.1 M	20k
Hand holding wire	50k	8k
Finger-thumb grip	20k	4k
Hand holding pliers	8k	1k
Palm touch	5k	700
Hand gripping 1.5-in. pipe	1k	300
Hand immersed	—	200
Foot immersed		100

Table 6.2.2
RESISTANCE VALUES OF 130 CM²
AREA OF MATERIAL OR BULK
MISCELLANEOUS MATERIALS

Material	Resistance (Ω)
Rubber gloves, soles, heels	20 - 50 M
Dry wood	10 - 20M
Dry concrete — above ground	1 — 2 M
Dry concrete on the ground	300 500k
Dry leather sole foot	100k
Wet leather sole foot	10k
Wet concrete on ground	1 - 2k

Table 6.2.1 gives values for typical skin contact situations and Table 6.2.2 gives values of resistance of a 130 cm² area of various materials. These tables are adapted from Reference 5.

It is useful to distinguish quantitatively between at least four levels of effects due to continuous currents.

1.   Nonperceptible electrical currents
2.   Perceptible (perhaps painful) currents below the ''let-go-threshold''
3.   Currents above the ''let-go-threshold''
4.   Currents that cause ventricular fibrillation (discoordinated heart action)

The threshold of human perception to 60 Hz alternating current has been well established[6] at approximately 1 mA^{RMS}. The perception threshold for women is about ⅔ of that value or 0.7 mA^{RMS} at 60 Hz. Higher levels of current result in muscular contractions, pain, and heat sensation. Finally a value of current is reached where voluntary release of the conductor is no longer possible. This level is important because a person can tolerate repeated exposure to levels below this ''let-go-threshold'' with no serious aftereffects. Figure 6.2.1, adapted from Dalziel[7] shows the results of a ''let-go-threshold'' test[a] on 134 men and 28 women. The results may be considered artificially high because of the competitive spirit among the men tested and certainly ''let-go-thresholds'' depend on physical development of hands and forearms. The women's ''let-go-threshold'' was lower and this may be attributed to lack of physical development and

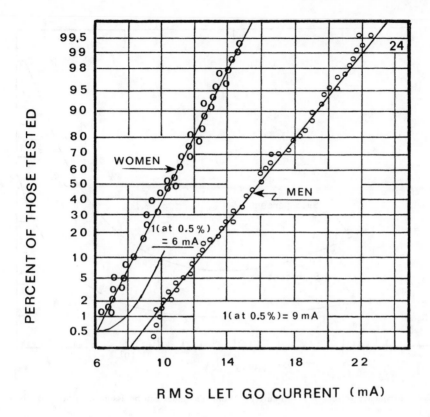

FIGURE 6.2.1.   Let-go current threshold for 60 Hz A.C. (From Dalziel, C. F., *IRE Trans. Med. Electr.*, PGM-5, 1956. With permission.)

a reported lack of competition among women subjects. A reasonable value for the "let-go-threshold" that would protect the total population would be the 0.5% level for women, which is 6 mA. Current levels just above this "let-go-threshold" are dangerous because long exposure to these levels can produce physical exhaustion and possible death.

At current levels of five times the "let-go-threshold" respiration will cease if the current path is through the chest. At still higher levels we approach a dangerous threshold — ventricular fibrillation or discoordinated heart action.

During this discoordinated heart beat normal blood flow is absent and death will result if a defibrillating machine is not applied. Fibrillation thresholds are sensitive to frequency and for short shocks, energy content and timing of the shock with respect to cardiac cycle. The fibrillation threshold for 60 cycles A.C. for men and women is 100 mA. Experiments to determine fibrillation thresholds were performed in 1936 by Ferris et al. at Columbia University[3] and Bell Telephone Laboratories in New York, in the 1950s by Kouwenhoven et al.[4] at Johns Hopkins University and Kiselev of the U.S.S.R. Academy of Sciences in 1963. Dalziel and Lee[6] summarized these experiments. For short shocks from 8.3 msec to 5 sec in length for 60 Hz, they found that the fibrillation threshold followed $I = 0.116/\sqrt{T}$ A (rms).

Kiselev[8] suggests that exposure to longer than 5 sec shocks at 60 Hz results in an approximately constant threshold value. For shock durations less than 100 msec, the shock must also be at least partially coincident with the T-wave of the heart cycle. This portion of the cycle occupies approximately one quarter of the total cycle duration. The cycle interval in man is about 0.75 sec.

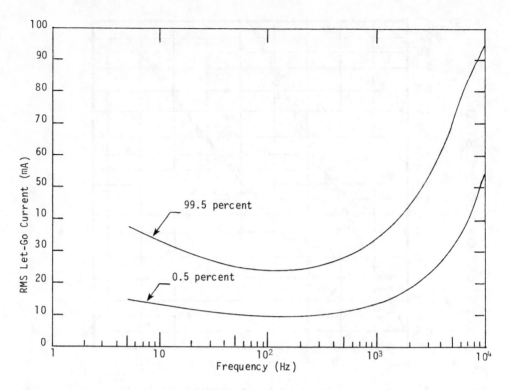

FIGURE 6.2.2.   Frequency effects of let-go current. Women's values can be assumed to be ⅔ of these values.

Frequency Effects

Capacitive Discharge

Capacitive discharge power supplies are commonly used with high energy laser power supplies. Most capacitors used in this application are charged to voltages in excess of 600 V. At voltage levels greater than 600 V skin contact resistance becomes negligible since a hole is actually punched through the skin. The RC time constant for capacitors charged above 600 V would then be determined by the capacitance alone. The pulse width for a 1 μF capacitor for five time constants discharging into 1000 Ω would be 5 msec. A pulse this short, even with an equal amount of energy ($\frac{1}{2}CV^2$) would not be as dangerous as a pulse lasting 330 msec, which would make the entire heart cycle susceptible.

One of the safety problems associated with the use of capacitive discharge power supplies is charge buildup in unshorted capacitors. This charge buildup is due to the release of charge trapped by absorption in the dielectric. It has been found by G.P. Boicourt at Los Alamos Scientific Laboratory that unshorted capacitors may charge to 6 to 10% of operating voltage.

D.C. Threshold Values

The D.C. let-go threshold for men is 76 mA. This is about five times the 60 Hz threshold. Experimental data for D.C. let-go thresholds for women has not been determined and a value of 51 mA or ⅔ the male value has been assumed. For 3 sec shocks the D.C. threshold level for fibrillation is 500 mA or again about five times the 60 Hz value.

A.C. Threshold Values

Figure 6.2.2 adapted from Reference 7 summarizes the effect of frequency on A.C.

Table 6.2.3
QUANTITATIVE EFFECTS OF ELECTRIC CURRENT ON MAN[a]

Physiologic effect	Direct current (mA) Men	Women	Alternating current (mA) 60 Hz Men	Women	10,000 Hz Men	Women
Slight sensation on hand	1	0.6	0.4	0.3	7	5
Perception threshold, median	5.2	3.5	1.1	0.7	12	8
Shock — not painful and muscular control not lost	9	6	1.8	1.2	17	11
Painful shock — muscular control lost by ½ %	62	41	9	6	55	37
Painful shock — let-go threshold, median	76	51	16	10.5	75	50
Painful and severe shock — breathing difficult, muscular control lost by 99 ½%	90	60	23	15	94	63
Possible ventricular fibrillation						
Three-second shocks	500	500	100	100		
Short shocks (T in seconds)			$116/\sqrt{T}$	$116/\sqrt{T}$		
Capacitor discharges	50*	50*				

[a] Energy in W-sec. Adapted from Dalziel, Reference 6.

let-go-thresholds. Perception thresholds and fibrillation thresholds are also effected by frequency and follows similar curves. Table 6.2.3 adapted from Reference 6 summarizes current magnitude effects.

Current Path

To be lethal, the current path through the body must involve a critical organ. The organ most susceptible to electric shock is the heart so that current paths including the heart can be dangerous at lower current magnitude. Relatively high currents flowing in the region of the heart may produce cardiac arrest. Currents flowing through nerve centers controlling breathing may produce respiratory inhibition. Higher currents may produce fatal damage to the central nervous system. Currents sufficient to raise the body temperature substantially produce immediate death.

RECOMMENDATIONS

- Consider live parts of circuits and components with peak open circuit potential over 42.5 V as hazardous unless limited to less than 0.5 mA. Such circuits require positive protection against contact.
- Install interlock switches (and capacitor bleeder resistors if applicable), or their equivalent for equipment intended for general use, to remove the voltage from accessible live parts to permit servicing operation. Bleeder resistors should be of such size and rating to carry the capacitor discharge current without burnout or mechanical damage.
- Cover or enclose circuits and components with peak open circuit potentials of 2.5 kV or more, if an appreciable capacitance is associated with the circuits.
- Use a solid metal grounding rod to assure discharge of high voltage capacitors if servicing of equipment requires entrance into an interlocked enclosure. The grounding rod should be firmly attached to ground prior to contact with the potentially live point. A resistor grounding rod (e.g., a large wattage ceramic resistor) may be used prior to application of the aforementioned solid conductor

grounding rod to protect circuit components from overly rapid discharge, but not as a replacement for the solid conductor rod.

- Ground the frames, enclosures, and other accessible metal noncurrent-carrying metallic parts of laser equipment. Grounding should be accomplished by providing a reliable, continuous, metallic connection between the parts to be grounded and the grounding conductor of the power wiring system.

- Provide information on electrical hazard prevention to all field maintenance personnel working with lasers and all laser laboratory personnel.

- Encourage the use of shock-prevention shields, power supply enclosures, and shielded leads in laboratory experimental arrangements, despite the temporary nature of some high voltage circuits.

- Supply safety devices such as safety glasses, rubber gloves and insulating mats.

- Provide metering, control, and auxiliary circuits that are suitably protected from possible high potentials, even during fault conditions.

- Perform routine inspection for deformed or leaky capacitor containers.

- Where feasible, wait 24 hr before attempting any work on circuits involving high energy capacitors.

FIRST AID FOR SHOCK VICTIMS

If a person has stopped breathing or his heart has stopped beating, heart-lung resuscitation should be started at once. If the person is not breathing, do the following:

1. Clear the throat. Wipe out any foreign matter in his mouth with your fingers or cloth wrapped around your fingers.
2. Place victim on his back. Place on a firm surface such as the floor or ground, not on a bed or sofa.
3. Tilt his head straight back. Extend the neck up as far as possible (this will automatically keep the tongue out of the airway).
4. Open your mouth wide and place it tightly over the victim's mouth. At the same time pinch the victim's nostrils shut or close the nostrils with your cheek. Or close the victim's mouth and place your mouth over his nose. This latter method is preferable with babies and small children. Blow into the victim's mouth or nose with a smooth steady action until the victim's chest is seen to rise.
5. Remove mouth. Listen for the return of air that indicates air exchange.
6. Repeat. Continue with relatively shallow breaths, appropriate for victim's size, at the rate of one breath each five seconds.

NOTE: If you are not getting air exchange, quickly recheck position of head, turn victim on his side and give several sharp blows between the shoulder blades to jar foreign matter free. Sweep fingers through mouth to remove foreign matter.

After four or five breaths, stop and determine if heart is beating by checking the pulse. If the heart is beating, return to the mouth-to-mouth resuscitation and continue until breathing starts or until a physician tells you to stop. If the heart has stopped, begin heart massage.*

1. Place the heel of one hand on the lower third of the breastbone, the other hand on top of the first.
2. Thrust downward from your shoulders with enough force to depress the breastbone about 1½ to 2 in.

* Adapted from the Los Alamos Scientific Laboratory safety manual.

3. Relax at the end of each stroke to permit natural expansion of the chest.
4. Repeat at the rate of about 1/sec.

If you are alone with the victim, you must alternate mouth-to-mouth breathing with heart massage at the ratio of about 2 to 15 (two breaths, then 15 heart compressions).

If you have help, the ratio is 1 to 5. After five heart compressions, pause slightly to allow your partner to breathe once into the lungs of the victim.

Call for help. Continue one or both of the above while the victim is being transported to the hospital, or until he revives or until told to stop by a physician.

REFERENCES

1. Franks, J. K. and Sliney, D. H., *Electro-Opt. Syst. Des.*, 7 (12), 1975.
2. Dalziel, C. F., Lagen, J. B., and Thurston, J. L., Electric shock, *AIEE Trans. (Electr. Eng.)*, 60, 1073—1079, 1941.
3. Ferris, L. P., King, B. G., Spence, P. W., and Williams, H. B., Effect of electric shock on the heart, *Electr. Eng.*, 55, 498—515, 1936.
4. Kouwenhoven, W. B., Chesnut, R. W., Knickerbocker, G. G., Milnor, W. R., and Sass, D. J., A-C shocks of varying parameters affecting the heart, *AIEE Trans. Commun. Electr.*, 78, 42, 163—69, 1959.
5. Lee, Ralph H., Electrical safety in industrial plants, *IEEE Trans. Ind. Gen. Appl.*, Vol. IGA-7, No. 1, 1971.
6. Dalziel, C. F. and Lee, W. R., Lethal electric currents, *IEEE Spectrum*, 6—2, 44—50, 1969.
7. Dalziel, C. F., The effects of electric shock on man , *IRE Trans. Med. Electr.*, PGM-5, 1956.
8. Kiselev, A. P., Threshold values of safe current at mains frequency in Problems of Electrical Equipment, *Electr. Suppl. Electr. Meas.*, (in Russian), SbMIIT, 171, 47—58, 1963.

6.3 HAZARDS FROM ASSOCIATED AGENTS*

Robin K. DeVore

INTRODUCTION

The potential health hazards associated with research and noncommercial laser systems are not limited to electric shock and optical radiation. Development and operation of laser systems include certain industrial hygiene hazards: toxicity or carcinogenicity of gases, vapors, fumes; and physical hazards of noise. These hazards may originate from the gases, solvents, or dyes used in the laser or generated by the laser; fumes given off by target materials; and noise generated from high-power lasers such as the gas-dynamic lasers, chemical lasers, or TEA lasers.

AIRBORNE CONTAMINANTS

Many materials used in laser systems and operations require or produce products that are potentially hazardous and require adequate controls to prevent personnel exposure to toxic or carcinogenetic materials. Materials of concern are solvents and dyes used in dye lasers, gases in gaseous media lasers, gases exhausting from gas lasers, and fumes generated by laser interaction with targets. Exposure to chemical agents in the work place is regulated by health limits established by the American Conference of Governmental Industrial Hygienists (ACGIH), the Occupational Safety and Health Administration (OSHA), and American National Standards Institute (ANSI), and also various state occupational health agencies.

CONTAMINANT CONCENTRATIONS

Some allowable time-weighted exposure concentrations for probable, specific components of laser fuels, target byproducts, and the products of beam generation are listed in Table 6.3.1. The allowable time-weighted exposures are from the OSHA standards, Code of Federal Regulations subpart G, Section 1910.1000 (U.S. Department of Labor, 1977). It is important to recognize that most of these values are for an 8-hr work day as part of a 40-hr work week, and that for some contaminants, occasional excursion above these values are permitted. General guidelines have been established to limit the magnitude of these excursions, and formulas for adding the permissible exposure limits (TLVs) of a combination of these airborne contaminants are also available, but definitive information is available only for a limited number. The formula most often used for calculating the threshold limit value (TLV) for a mixture of airborne contaminants that have similar toxicological effects is

$$\frac{c_1}{T_1} + \frac{c_2}{T_2} + \frac{c_3}{T_3} + \ldots = 1 \tag{1}$$

where c is the concentration in parts per million (ppm) and T is the TLV.

We do not wish to leave the reader with the deceptive notion that the measurements and calculations associated with the analysis of hazards from airborne contaminants are simple. These are not. Formula 1 is given only as an example, and Table 6.3.1 is

---

* The opinions or assertions contained herein are the private views of the author and are not to be construed as official or as reflecting the views of the Los Alamos Scientific Laboratory or the Department of Energy.

Table 6.3.1
REPRESENTATIVE CONTAMINANTS ASSOCIATED WITH LASER OPERATION

Contaminants	Probable source	OSHA allowable time weighted exposure[b]	OSHA ceiling value[b]
Asbestos	Target backstop	5 fibers/cc[a]	
Benzene			
Beryllium	Firebrick target	0.002 mg/m^3	
Cadmium oxide fume	Metal target	0.1 (0.04) mg/m^3	3 (0.2) mg/m^3
Carbon monoxide	Laser gas	50 (35) ppm	(200 ppm)
Carbon dioxide	Active laser medium	5,000 (10,000) ppm	50,000 ppm
Chromium	Metal targets	0.5 (0.025) mg/m^3	(0.05 mg/m^3)
Cobalt, metal fume, and dust	Metal targets	0.1 mg/m^3	
Copper fume	Metal targets	0.1 mg/m^3	
DMF-DCE			
Fluorine	HF chemical laser	0.1 ppm	
Hydrogen fluoride	Active medium of laser	3 (2.5) ppm	(5 ppm)
Iron oxide fume	Metal targets	10 (5[c]) mg/m^3	
Manganese	Metal targets		5 mg/m^3
Nickel	Metal targets	1 (0.015) mg/m^3	
Nitrogen dioxide	GDL discharge	5 ppm	(1 ppm)
Ozone	Target and Marx generators	0.1 ppm	
Sulfur dioxide	Laser exhaust	5 (0.5) ppm	
Sulfur hexafluoride	Saturable absorber	1,000 ppm	
Uranium (soluble/insoluble)	Target	0.05/0.25 mg/m^3	
Vanadium fume (dust/fume)	Target		0.5/0.1 mg/m^3
Zinc oxide fume	Target	5 mg/m^3	

[a] > 5 μm in length.
[b] Values in parentheses denote changes recommended by NIOSH or ACGIH.
[c] Denotes change recommended by ACGIH.

provided to give the reader a "feel" for TLVs. It lists OSHA permissible exposure limits. For those contaminants for which no excursion above a certain value is permitted, the values in Table 6.3.1 are termed "ceiling concentrations." This means that time-weighting is not permitted and that all exposure levels should fluctuate below the designated value. Those ceiling concentrations are analogous to exposure limits (MPEs) for pulsed laser exposures where no excursions above those limits are permitted. Values in parentheses are NIOSH or ACGIH recommendations for changes in the OSHA limits.

The use of asbestos in laboratory operations is increasingly becoming of concern to the health professionals because of the lowering of the permissible exposure limit (PEL) and evidence that asbestos not only causes fibrogenic lung diseases but also can produce lung cancer. All asbestos products should be suspected of being hazardous. This may include even working with asbestos gloves and curtains. In view of the recent availability of suitable substitute materials that are nonasbestos, it is better to avoid using asbestos altogether.

Several liquid laser dyes and their common solvents are toxic but may not be properly labeled to indicate this. Also, for many dyes and solvents there is inadequate toxicity data. The manufacturer should normally be consulted on this question. The tests of one research group showed that the cyanine and carbocyanine (or polymethine) compounds were the most toxic of the 150 dyes used most extensively in the red and infrared portions of the spectrum. The cyanines are generally dissolved in dimethylsulfoxide (DMSO), a solvent that is by itself quite dangerous because of its ability to facilitate the transfer of molecules (which could be any airborne or splash contaminant) through biological membranes such as skin. The wisest approach in using any dye in a laser system is to wear gloves, and in those rare instances where large quantities are used, accompany them with additional precautionary measures, such as the use of face masks, rubber aprons, and rubber sleeves. Many of the dye solvents are also highly flammable and fire protective controls should be included in the laboratory design.

In addition to the hazards of the dye, the solvent itself may be toxic. Common solvents that are toxic are ethanol, methanol, and 1,2-dichlorethane.

Two types of materials used in electrical equipment have produced concern in recent years: polychlorinated biphenyls (PCB) are high viscosity liquid insulators used in electrical transformers because of their low vapor pressure. Related compounds, polybromenated biphenyls are used as fire retardants. Current control measures should be consulted if these substances are encountered. All sources of PCBs require labeling so that proper cleanup procedures are followed for spills.

CONTROL OF CHEMICAL HAZARDS

Personnel exposure to hazardous materials must be kept below the established or regulatory health limits. This can be accomplished by engineering controls, administrative controls, or personnel protective devices. Engineering or administrative controls are preferred whenever feasible. Personnel protection should not be required unless as a last resort; however, the nature of research operations often is such that it is impractical to implement costly engineering controls for a temporary operation.

Dilution ventilation is one of the most effective and frequently used control methods. For example, in an outdoor environment, over-exposure to airborne contaminants from vaporized target materials would not be expected because of ample dilution ventilation. For indoor operations local exhaust ventilation is probably the most effective control method. Airborne contaminants should be captured near the point of evolution even before they have a chance to be diluted in room air. In general, mechanical dilution (fan ventilation) is not an acceptable method for indoor control of the more toxic materials or where contaminants are generated from a point source. The more toxic materials in this regard are those with exposure limits below 100 ppm. For indoor operations, local exhaust (hood) ventilation is usually a more economical method of control than dilution ventilation since the required air volume flow rate is substantially less. Local exhaust systems should be designed to provide, at the point of contaminant evolution, a capture velocity in the direction of the exhaust inlet of 100 to 150 linear-ft-per-min. The exhaust inlet should be designed to minimize contaminant escape. For this purpose, total enclosure is optimal but not always obtainable. Thus, efforts should be made to enclose as much of the contaminant source as is practical. The location of the exhaust inlet should be selected to take advantage of the natural movement of the contaminant which, however, should not be allowed to pass through an individual's breathing zone en route. Disposal of chemical laser exhausts from large HF and DF lasers sometimes requires the use of selfcontained laboratory scrubbing systems to remove SO_2 and HF vapors. Information particularly useful in the design of exhaust and dilution ventilation is extensively developed in the ACGIH industrial ventilation manual (1974) and the ANSI standard entitled "Fundamentals Governing the Design Operation of Local Exhaust Systems."

RESPIRATORY PROTECTIVE DEVICES

In addition to detection equipment, respiratory protective devices for emergency use must also be provided in certain installations. Such safety equipment should be approved by the Bureau of Mines or the National Institute for Occupational Safety and Health (NIOSH) for use with many contaminant(s) likely to be encountered. Many of the contaminants associated with high power laser operations are described as simple asphyxiants; their principal action is the displacement of oxygen (e.g., nitrogen). Other materials are explosive (e.g., hydrogen) or toxic (e.g., ozone) or both. Some of the larger chemical and gas-dynamic laser systems include a large number of tanks or bulk storage facilities. Here a large leak or vessel rupture can result in the release of enormous quantities of hazardous material. Available emergency equipment should include approved, selfcontained or air-supplied breathing devices, preferably of the pressure-demand type. All respiratory devices should be located in an area where they can be donned in a contaminant-free atmosphere and should be properly maintained and inspected as required by the OSHA standard in Section 1910.134 (U.S. Department of Labor, 1977).

In addition to the target site and the gas effluent from a flowing-gas laser, airborne contaminants are sometimes produced in unexpected locations. A common source of such airborne contaminants is beam interaction with optical elements in the laser system. For instance, infrared-transmitting optical materials (such as zinc-selenide, cadmium-telluride or selenium fluoride) occasionally degrade or vaporize under very high irradiances and decompose into toxic contaminants. Metal and plastic target materials may also produce airborne contaminants.

CONTAMINANT DETECTION

Provisions should be made for contaminant detection around the larger laser systems, especially those where gas or liquid bulk storage facilities are maintained. It is important not only to maintain contaminant levels below allowable exposure limits, but also to provide early detection of system leaks. Since the presence of many of these contaminants, even in trace amounts, most often denotes leakage, detectors are generally located near the potential leakage sites and adjusted to respond to levels only slightly above ambient. These contaminant detection devices should be selected with great care; the sensitivity, accuracy, drift, maintenance requirements, and specificity should be considered as well as the operating environment (e.g., outdoors exposed to weather).

The fluoride lines that are used to feed HF and DF laser systems sometimes leak at welded joints. This is particularly common if one attempts to maintain a pressure of fluorine in the line above one atmosphere. Rupture of a fluorine line can result in a dangerous fire. In heavily occupied laboratory buildings, double-containment lines have been used. Leakage of fluorine into the interstitial space can be monitored by filling the jacket (interstitial space) with nitrogen and monitoring any change in nitrogen pressure.

Gloves and other protective garments are frequently required for work with dye solvents such as DMSO and gas systems such as HF and fluorine. Selection of gloves or garments for protection must be made carefully since the protection afforded varies considerably. For example, a glove suitable for ethanol will not provide any protection against DMSO or vice versa.

Table 6.3.2
THRESHOLD LIMIT
VALUES FOR STEADY-
STATE NOISE

Duration per day (hr)	Sound level dBA
16	80
8	85
4	90
2	95
1	100
½	105
¼	110
1/8	115[a]

[a] No exposure to continuous or intermittent in excess of 115 dBA.

NOISE HAZARDS

Hearing conservation measures, including the use of protective devices, should be practiced when steady-state noise levels exceed 90 dB(A) for 8 hr per day (by Federal regulation). However, an exposure of 115 dB(A) is permissible for a period of less than 15 min. For a pulsed laser impact that produces impulsive noise (i.e., less than 0.5), the maximum recommended exposure is 140 dB(A). The complete requirements for hearing conservation of the Occupational Health and Safety Administration (OSHA) are covered in sub part G of Section 1910.95 (U.S. Department of Labor, 1977). Table 6.3.2 lists the ACGIH (1979) threshold limit values (TLVs) for noise.

As TEA lasers normally produce large impact noises, acoustical control measures are generally required. Often hazardous noise levels exist only in the immediate vicinity of the laser or target where other safety recommendations exclude personnel, thus obviating the need for control measures.

The sound level shall be determined by a sound level meter, conforming as a minimum to the requirements of the American National Standard Specification for Sound Level Meters, S1.4 (1971) Type S2A, and set to use the A-weighted network with slow meter response. Duration of exposure shall not exceed that shown in Table 6.3.2.

When the daily noise exposure is composed of two or more periods of noise exposure of different levels, their combined effect should be considered, rather than the individual effect of each. If the sum of the following fractions:

$$\frac{C_1}{T_1} + \frac{C_2}{T_2} + \ldots \frac{C_n}{T_n}$$

exceeds unity, then, the mixed exposure should be considered to exceed the threshold limit value, C_1 indicates the total duration of exposure at a specific noise level, and T_1 indicates the total duration of exposure permitted at that level. All on-the-job noise exposures of 80 dBA or greater shall be used in the above calculations.

It is recommended that exposure to impulsive or impact noise shall not exceed the limits listed in Table 6.3.3. No exposures in excess of 140 decibels peak sound pressure levels are permitted. Impulsive or impact noise is considered to be those variations in noise levels that involve maxima at intervals of greater than 1/sec. Where the intervals are less than 1 sec, it should be considered continuous.

It should be recognized that the application of the TLV for noise will not protect

Table 6.3.3
THRESHOLD LIMIT VALUES FOR
IMPULSIVE OR IMPACT NOISE

Peak sound pressure level dB	Permitted number of impulses or impacts per day
140	100
130	1,000
120	10,000

all workers from the adverse effects of noise exposure. A hearing conservation program with audiometric testing is necessary when workers are exposed to noise at or above the TLV levels.

FIRE HAZARDS

As previously noted, solvents used in dye lasers are extremely flammable and, according to a report from Stanford Research Institute, have led to at least eight fires within a two-year period in U.S. laboratories. Most of these fires were started by a high-voltage pulse through an alcohol solvent, although in one case a hot xenon arc lamp tube ignited the solvent. Probably the best control measure for this fire hazard is to restrict the operation of high-energy coaxial flash lamps to the average power levels recommended by the manufacturers and to use only nonvolatile solvents. Another approach to the problem is, if possible, to replace the alcohol solvent with a water-alcohol mixture of more than 50% water. When alcohol must be used, the laser housing should enclose any exploding liquid (Grant and Hawley, 1975). If a fire occurs, dye pumps and laser pumps must be shut down quickly, which may be accomplished by the installation of a fluid-pressure or microphone-actuated switch.

It is common practice to increase the flashlamp voltage above the recommended operating level to compensate for the decreased output of aging lamps. This aging process results in an increase in the breakdown voltage of xenon-oxygen gas mixtures and is caused by the decomposition of the silicon dioxide in the glass or quartz envelope from repeated firing. The oxygen created by this decomposition increases the lamp threshold voltage. Therefore to prevent explosions, the threshold voltage can be monitored by placing a resistor in parallel with the lamp. One can choose the shunt resistance value at a power rating that will assure the burn-up of the resistor once the lamp breakdown voltage passes the maximal level.

The explosion hazard of dye laser solvents is shared by high-pressure xenon flash lamps. One model of a solid-state pulsed laser manufactured between 1965 and 1970 was notorious for explosions, apparently because the flashlamp manufacturer's specifications were exceeded in the design of the high-voltage power supply. The enclosure around the flashlamp should always be sufficiently rugged to contain any explosion, although small pressure relief areas or holes should be provided.

High-power CW infrared lasers can produce substantial fire hazards from the direct beam. Unexpected specular reflections from metal surfaces can ignite flammable materials in the area of the laser operation. Fire-resistant sheeting and other materials should be used to the maximum extent possible in the immediate vicinity of very high power laser operations. The fire fighting equipment required for laser operations can vary. Dry chemical fire extinguishers are often found in laboratories. However, laboratory personnel should be aware of the fact that such dry chemicals coat and etch optical surfaces, and may damage electronic components. Therefore, dry chemical extinguishers are less preferred than gas extinguishers, such as carbon dioxide or

Freon® (Dupont). Other fire-fighting gas materials such as Halon 1301® (also a product of Dupont) are quite effective in this regard. Water shares the disadvantages of dry chemical fire extinguishers in damaging some optical and electronic components and, in addition, may exacerbate an electrical fire, as well as presenting a shock hazard.

CRYOGENIC HAZARDS

Liquid nitrogen and other cryogenic fluids are used to cool some lasers and many high sensitivity photodetectors. Examples of cryogenically cooled lasers are CW solid state ruby lasers with continuous pumping by a high intensity lamp, or high-power gallium-arsenide laser arrays. Cryogenic cooling is most often encountered in infrared detectors such as the mercury-cadmium-telluride type. The signal-to-noise ratio of these detectors is greatly improved by the lower temperatures achieved with cryogenic cooling.

When cryogenic fluids evaporate they displace breatheable oxygen and thus should be used only in areas of good ventilation. Another safety hazard associated with the use of cryogenic fluids is the possibility of explosion from ice collecting in a valve or a connector. This explosive hazard is most common when the plumbing is not specifically designed to operate with cryogenic materials. The more hazardous cryogenic fluids such as liquid oxygen and liquid hydrogen are not typically used in laser research laboratories. Nevertheless, liquid oxygen, which is both a serious fire and explosive hazard, is often produced from room air. As the condensation temperature of oxygen is approximately 13° higher than the boiling point of liquid nitrogen, it may collect in open Dewar flasks of liquid nitrogen.

Both protective clothing and face shields should be used when handling large quantities of liquid nitrogen. Often protective gloves are not used in handling extremely small quantities of liquid nitrogen; in fact, the normal moisture and oil on the surface of the skin is sufficient to protect the hand from cold injury from a few small drops of liquid nitrogen. The small quantities of liquid nitrogen used to cool infrared sensitive detectors do not ordinarily present a serious health hazard.

There are a number of safety procedures required in the use of gas cannisters and cryogenic Dewar flasks to prevent serious accidents. It is now common practice to chain all gas cannisters to the wall or instrument system. This precludes a dangerous explosion should the cannister fall, and break the very vulnerable valve, with a resultant sudden release of the gas.

RADIATION HAZARDS

The high voltage power supplies used with some lasers often produce a limited quantity of low energy X radiation. This generally originates from the high-voltage vacuum tubes, such as rectifiers, thyratrons, and crowbars used in laser power supplies. With the increased use of solid state electronics the X-ray potential is reduced or eliminated. Power supplies producing potentials greater than 15 kV with vacuum tubes may produce a sufficient quantity of X-rays to present a significant health hazard. Normally such tubes have sufficient shielding to prevent the escape of hazardous levels. To measure such low energy X radiation it is important not to use a common geiger counter or similar instrument designed for very high energy ionizing radiation. A thin-window instrument, well shielded from the radio-frequency radiation often present in power supplies, should be used for this purpose.

In addition to the power supply itself, X-rays are produced in large quantities from electron beam (E-beam) laser systems. However, the X-ray hazard from electron-beam machines is always anticipated in the design of such commercial equipment, and one should expect that adequate shielding has been placed around the system to preclude hazardous leakage levels. Safety personnel, however, should be on the lookout for modification, or poor maintenance practices, in an electron-beam system that could affect the efficiency of the shielding. If the operator of an E-beam laser is not familiar with the X radiation hazard, he might unknowingly remove a portion of the shielding, not realizing its value or purpose. An industrial X-ray facility that produces in excess of 100 mrad in any 1 hr is considered an "open protective" installation: some of the hazard control measures used for such X-ray installations can also be applied directly to an electron-beam laser system. Voltages of up to 500 kV may be generated by Marks-bank techniques and the X-rays produced from such systems can present more serious control design problems since much heavier shielding materials must be used.

Index

INDEX

WITHDRAWAL